D0145713

THE RIEMANN ZETA–FUNCTION

THEORY AND APPLICATIONS

Aleksandar Ivić

University of Belgrade
Yugoslavia

DOVER PUBLICATIONS, INC.
Mineola, New York

Bibliographical Note

This Dover edition, first published in 2003, is an unabridged republication of
the work first published in 1985 as *The Riemann Zeta–Function: The Theory of
the Riemann Zeta–Function with Applications* by John Wiley & Sons, Inc., New
York. A list of errata has been included in this edition.

Library of Congress Cataloging-in-Publication Data

Ivic, A., 1949–
    The Riemann zeta–function : theory and applications / Aleksander Ivic.
      p. cm.
    Originally published: New York : Wiley, c1985.
    Includes bibliographical references and index.
    ISBN 0-486-42813-3 (pbk.)
     1. Functions, Zeta. I. Title.

QA351.I85 2003
515'.56—dc21

2003043460

Manufactured in the United States of America
Dover Publications, Inc., 31 East 2nd Street, Mineola, N.Y. 11501

PREFACE

The aim of this book is to present the theory of the Riemann zeta-function $\zeta(s)$ and some of its applications. The basis for this text are the author's research lecture notes (1983b), although this work contains several completely new chapters, and some of the chapters of the 1983b work have been revised.

The Riemann zeta-function occupies a central place in analytic number theory. Many problems from multiplicative number theory directly depend on properties of the zeta-function, such as power moments, the zero-free region, zero-density estimates, and so on. Therefore a better understanding of the theory of $\zeta(s)$ helps number theorists to obtain various arithmetical results. Another important aspect of the theory of the Riemann zeta-function is that $\zeta(s)$ is the simplest of a large class of the Dirichlet series known as zeta-functions. General zeta-functions occur in many branches of mathematics, including algebraic and analytic number theory, and results from the theory of $\zeta(s)$ in many instances generalize to other zeta-functions.

In the literature there exist two well-known texts devoted solely to $\zeta(s)$. The fine work of H. M. Edwards (1974) develops the subject along historical lines, while the great classic work of E. C. Titchmarsh (1951) contains an extensive treatment of the theory of $\zeta(s)$ up to 1951. In recent years many important contributions to the subject have been made, which are not to be found in the aforementioned works. Therefore the time seemed appropriate for a comprehensive new text on $\zeta(s)$. This book is organized to provide first an ample account of classical material, and then to lead the reader to the wealth of new results, many of which appear in book form for the first time. I hope that the book will be of interest both to the readers wishing to acquire elements of zeta-function theory, and to experts seeking a unified account of the latest research results.

Special care has been taken to provide some of the most important applications of zeta-function theory. Chapter 12 is devoted to the distribution of primes, and although a full treatment of this topic certainly requires a separate volume, the reader will nevertheless find many applications of $\zeta(s)$ to prime number theory. Chapter 13 deals with the general divisor problem, while other divisor problems of multiplicative number theory are discussed in Chapter 14.

A glance through the contents will tell more about the material presented in the book, so only some general remarks and a brief description of the contents is given here. The first chapter contains the basic theory of the zeta-function and fulfills the needs of those who have no intention to pursue the subject any further. A moderate knowledge of elementary number theory and complex analysis is required of the reader, although efforts have been made to make the text as self-contained as possible. Most of the topics discussed in Chapter 1 are treated again in later chapters, and in many cases the sharpest known results are proved. In Section 1.8 of Chapter 1 some common zeta-functions besides $\zeta(s)$ are mentioned, although a more detailed account of this subject (especially of L-functions) does not fall within the scope of this book. Chapter 1 is intended to serve as the basis for a short course on zeta-function theory, and such a course may be expanded by supplementing selected topics from later chapters.

Zeta-function theory abounds in unproved conjectures. The most famous ones are Lindelöf's $[\zeta(\frac{1}{2} + it) \ll |t|^{\varepsilon}]$ and Riemann's [all complex zeros of $\zeta(s)$ have real parts equal to $\frac{1}{2}$]. These conjectures are discussed in Section 1.9 of Chapter 1, and at the present time neither conjecture is proved nor disproved. As new results from zeta-function theory constantly continue to accrue, it becomes evident that the methods needed for settling the Lindelöf and Riemann hypotheses must be much deeper than was assumed in Riemann's times and even at the beginning of this century. However, these conjectures have been the driving force behind much of the research on $\zeta(s)$. As a rule the results proved in this text are unconditional, that is, they do not depend on any hitherto unproved hypothesis.

Chapters 2 and 3 are independent of other chapters and contain material on exponential sums and the Voronoi summation formula, respectively. This digression is necessary to prepare the groundwork for a more advanced account of zeta-function theory, which begins with Chapter 4. This chapter extensively treats approximate functional equations, an important tool of zeta-function theory.

Mean values and power moments play an important role both in the theory and application of $\zeta(s)$. A large part of the text is devoted to this subject, starting with the fourth power moment in Chapter 5. The sharpest known results concerning mean values and higher power moments are proved in Chapters 7 and 8, while Chapter 15 contains the proof and applications of Atkinson's formula for $\int_0^T |\zeta(\frac{1}{2} + it)|^2 dt$. Atkinson's formula was left for the end of the text because of its difficulty and connections with the ordinary Dirichlet divisor problem, discussed in Chapter 13. Some omega results for $\zeta(s)$ and lower bounds for power moments are discussed in Chapter 9.

The location and distribution of zeros of $\zeta(s)$ is also one of the most remarkable parts of zeta-function theory. The sharpest known zero-free region for $\zeta(s)$ is obtained in Chapter 6 by the Vinogradov–Korobov method for the estimation of exponential sums. The distribution of zeros on the critical line $\sigma = \frac{1}{2}$ is investigated in Chapter 10, while zero-density estimates are extensively discussed in Chapter 11.

One must be selective in the choice of topics in a work such as this one. While many important aspects of zeta-function theory could not be properly discussed, each chapter is followed by Notes, where many results not treated in the body of the text are mentioned with appropriate references. These Notes also contain historical discussion, elucidation of certain details in the proofs, and other material. Whenever possible, standard notation (explained at the beginning of the text) is used. References and the Indexes are at the back of the book together with the Appendix, which contains miscellaneous results used constantly in the proofs.

In concluding I wish to thank all number theorists who have read (parts of) the manuscript and made valuable remarks. In particular, I wish to thank B. C. Berndt, H. G. Diamond, S. W. Graham, D. R. Heath-Brown, M. N. Huxley, M. Jutila, J. Pintz, and H.-E. Richert. My thanks also go to the staff of John Wiley & Sons, for their unfailing courtesy and help.

ALEKSANDAR IVIĆ

Belgrade Yugoslavia
January 1985

CONTENTS

NOTATION   *xiii*

ERRATA   *xvii*

1. ELEMENTARY THEORY   *1*

1.1. Definition of $\zeta(s)$ and Elementary Properties, 1
1.2. The Functional Equation, 8
1.3. The Hadamard Product Formula, 12
1.4. The Riemann–von Mangoldt Formula, 17
1.5. An Approximate Functional Equation, 21
1.6. Mean Value Theorems, 26
1.7. Various Dirichlet Series Connected with $\zeta(s)$, 30
1.8. Other Zeta-Functions, 38
1.9. Unproved Hypotheses, 44

2. EXPONENTIAL INTEGRALS AND EXPONENTIAL SUMS   *55*

2.1. Exponential Integrals, 55
2.2. Exponential Sums, 67
2.3. The Theory of Exponent Pairs, 72
2.4. Two-Dimensional Exponent Pairs, 81

3. THE VORONOI SUMMATION FORMULA   *83*

3.1. Introduction, 83
3.2. The Truncated Voronoi Formula, 86
3.3. The Weighted Voronoi Formula, 88
3.4. Other Formulas of the Voronoi Type, 93

4. THE APPROXIMATE FUNCTIONAL EQUATIONS 97

4.1. The Approximate Functional Equation for $\zeta(s)$, 97
4.2. The Approximate Functional Equation for $\zeta^2(s)$, 104
4.3. The Approximate Functional Equation for
     Higher Powers, 110
4.4. The Reflection Principle, 122

5. THE FOURTH POWER MOMENT 129

5.1. Introduction, 129
5.2. The Mean Value Theorem for Dirichlet Polynomials, 130
5.3. Proof of the Fourth Power Moment Estimate, 135

6. THE ZERO-FREE REGION 143

6.1. A Survey of Results, 143
6.2. The Method of Vinogradov–Korobov, 145
6.3. Estimation of the Zeta Sum, 152
6.4. The Order Estimate of $\zeta(s)$ Near $\sigma = 1$, 160
6.5. The Deduction of the Zero-Free Region, 162

7. MEAN VALUE ESTIMATES OVER SHORT INTERVALS 171

7.1. Introduction, 171
7.2. An Auxiliary Estimate, 172
7.3. The Mean Square When σ Is in the Critical Strip, 174
7.4. The Mean Square When $\sigma = \frac{1}{2}$, 178
7.5. The Order of $\zeta(s)$ in the Critical Strip, 188
7.6. Third and Fourth Power Moments in Short Intervals, 194

8. HIGHER POWER MOMENTS 199

8.1. Introduction, 199
8.2. Some Convexity Estimates, 202
8.3. Power Moments for $\sigma = \frac{1}{2}$, 205
8.4. Power Moments for $\frac{1}{2} < \sigma < 1$, 213
8.5. Asymptotic Formulas for Power Moments When
     $\frac{1}{2} < \sigma < 1$, 221

9. OMEGA RESULTS 231

9.1. Introduction, 231
9.2. Omega Results When $\sigma \geqslant 1$, 232

9.3. Lemmas on Certain Order Results, 238
9.4. Omega Results for $\frac{1}{2} \leqslant \sigma \leqslant 1$, 241
9.5. Lower Bounds for Power Moments When $\sigma = \frac{1}{2}$, 244

10. ZEROS ON THE CRITICAL LINE 251

10.1. Levinson's Method, 251
10.2. Zeros on the Critical Line in Short Intervals, 254
10.3. Consecutive Zeros on the Critical Line, 260

11. ZERO-DENSITY ESTIMATES 269

11.1. Introduction, 269
11.2. The Zero-Detection Method, 270
11.3. The Ingham–Huxley Estimates, 273
11.4. Estimates for σ Near Unity, 276
11.5. Reflection Principle Estimates, 281
11.6. Double Zeta Sums, 283
11.7. Zero-Density Estimates for $\frac{3}{4} < \sigma < 1$, 288
11.8. Zero-Density Estimates for σ Close to $\frac{3}{4}$, 290

12. THE DISTRIBUTION OF PRIMES 297

12.1. General Remarks, 297
12.2. The Explicit Formula for $\psi(x)$, 300
12.3. The Prime Number Theorem, 304
12.4. The Generalized von Mangoldt Function and the Möbius Function, 309
12.5. Von Mangoldt's Function in Short Intervals, 315
12.6. The Difference between Consecutive Primes, 321
12.7. Almost Primes in Short Intervals, 330
12.8. Sums of Differences between Consecutive Primes, 333

13. THE DIRICHLET DIVISOR PROBLEM 351

13.1. Introduction, 351
13.2. Estimates for $\Delta_2(x)$ and $\Delta_3(x)$, 353
13.3. Estimates of $\Delta_k(x)$ by Power Moments of the Zeta-Function, 354
13.4. Estimates of $\Delta_k(x)$ When k Is Very Large, 356
13.5. Estimates of β_k, 357
13.6. Mean-Square Estimates of $\Delta_k(x)$, 361
13.7. Large Values and Power Moments of $\Delta_k(x)$, 366
13.8. The Circle Problem, 372

14. VARIOUS OTHER DIVISOR PROBLEMS   385

    14.1. Summatory Functions of Arithmetical Convolutions, 385
    14.2. Some Applications of the Convolution Method, 393
    14.3. Three-Dimensional Divisor Problems, 397
    14.4. Powerful Numbers, 407
    14.5. Nonisomorphic Abelian Groups of a Given Order, 413
    14.6. The General Divisor Function $d_z(n)$, 420
    14.7. Small Additive Functions, 429

15. ATKINSON'S FORMULA FOR THE MEAN SQUARE   441

    15.1. Introduction, 441
    15.2. Proof of Atkinson's Formula, 443
    15.3. Modified Atkinson's Formula, 460
    15.4. The Mean Square of $E(t)$, 465
    15.5. The Connection Between $E(T)$ and $\Delta(x)$, 471
    15.6. Large Values and Power Moments of $E(T)$, 480

APPENDIX   483

REFERENCES   497

AUTHOR INDEX   511

SUBJECT INDEX   515

NOTATION

Owing to the nature of this text, absolute consistency in notation could not be attained, although whenever possible standard notation is used. Notation commonly used throughout the text is explained here, while specific notation introduced in the proof of a theorem or lemma is given at the proper place in the body of the text.

k, l, m, n	Natural numbers (positive integers).
p	A prime number.
$A, B, C, C_1, \ldots$	Absolute positive constants (not necessarily the same at each occurrence in a proof).
s, z, w	Complex variables ($\operatorname{Re} s$ and $\operatorname{Im} s$ denote the real and imaginary part of s, respectively; common notation is $\sigma = \operatorname{Re} s$ and $t = \operatorname{Im} s$).
t, x, y	Real variables.
$\operatorname*{Res}_{s=s_0} F(s)$	The residue of $F(s)$ at the point $s = s_0$.
$\zeta(s)$	Riemann's zeta-function defined by $\zeta(s) = \sum_{n=1}^{\infty} n^{-s}$ for $\operatorname{Re} s > 1$ and otherwise by analytic continuation.
$\Gamma(z)$	The gamma-function defined by $\Gamma(z) = \int_0^{\infty} t^{z-1} e^{-t}\, dt$ for $\operatorname{Re} z > 0$, otherwise by analytic continuation.
γ	Euler's constant, defined by $\gamma = -\int_0^{\infty} e^{-x} \log x\, dx = 0.5772157\ldots$.
$\chi(s)$	The function defined by $\zeta(s) = \chi(s)\zeta(1-s)$, so that by the functional equation for the zeta-function $\chi(s) = (2\pi)^s/(2\Gamma(s)\cos(\pi s/2))$.

$\rho = \beta + i\gamma$	A complex zero of $\zeta(s)$; $\beta = \mathrm{Re}\,\rho$, $\gamma = \mathrm{Im}\,\rho$.
$N(T)$	The number of zeros ρ of $\zeta(s)$ for which $0 < \beta < 1$, $0 < \gamma \leqslant T$.
$N_0(T)$	The number of zeros $\rho = \frac{1}{2} + i\gamma$ of $\zeta(s)$ for which $0 < \gamma \leqslant T$.
$N(\sigma, T)$	The number of zeros ρ of $\zeta(s)$ for which $\beta \geqslant \sigma$, $|\gamma| \leqslant T$.
$\mu(\sigma)$	For σ real is defined by $\mu(\sigma)$ $$= \limsup_{t \to \infty} \frac{\log |\zeta(\sigma + it)|}{\log t}.$$
$\exp z$	$= e^z$.
$e(z)$	$= e^{2\pi iz}$.
$\log x$	$= \mathrm{Log}_e x\ (= \ln x)$.
$[x]$	The greatest integer not exceeding the real number x.
$\displaystyle\sum_{n \leqslant x} f(n)$	A sum taken over all natural numbers n not exceeding x; the empty sum is defined to be zero.
$\displaystyle{\sum_{n \leqslant x}}' f(n)$	The same as above, only $'$ denotes that when x is an integer one should take the last term in the sum as $\frac{1}{2}f(x)$ and not as $f(x)$.
$\displaystyle\prod_{j}$	A product taken over all possible values of the index j; the empty product is defined to be unity.
$\psi(x)$	Equals $x - [x] - \frac{1}{2}$, but in Chapter 3 $\psi(z) = \Gamma'(z)/\Gamma(z)$, while in Chapter 12 $\psi(x) = \displaystyle\sum_{p^m \leqslant x} \log p$.
$\Lambda(n)$	The von Mangoldt function defined by $\Lambda(n) = \log p$ if $n = p^m$ and zero otherwise.
$\mu(n)$	The Möbius function, defined as $\mu(n) = (-1)^k$ if $n = p_1 \cdots p_k$ (the p_j's being different primes) and zero otherwise, and $\mu(1) = 1$.
$\displaystyle\sum_{d \mid n}$	A sum taken over all positive divisors of n.
$\Lambda_k(n)$	The generalized von Mangoldt function defined by $\Lambda_k(n) = \displaystyle\sum_{d \mid n} \mu(d)(\log n/d)^k$; $\Lambda_1(n) = \Lambda(n)$, the ordinary von Mangoldt function.

$\psi_k(x)$ $\qquad = \sum\limits_{n \leqslant x} \Lambda_k(n).$

$\pi(x)$ $\qquad = \sum\limits_{p \leqslant x} 1$, the number of primes not exceeding x.

$\theta(x)$ $\qquad = \sum\limits_{p \leqslant x} \log p.$

$M(x)$ $\qquad = \sum\limits_{n \leqslant x} \mu(n).$

li x $\qquad = \int_0^x \dfrac{dt}{\log t} = \lim\limits_{\varepsilon \to 0} \left(\int_0^{1-\varepsilon} \dfrac{dt}{\log t} + \int_{1+\varepsilon}^x \dfrac{dt}{\log t} \right).$

$d_k(n)$ $\qquad$ The number of ways n can be written as a product of $k \geqslant 2$ fixed factors; $d_2(n) = d(n)$ is the number of divisors of n.

$r(n)$ $\qquad$ The number of ways n can be written as a sum of two integer squares.

$\Delta_k(x)$ $\qquad = \sum\limits_{n \leqslant x} d_k(n) - \operatorname*{Res}\limits_{s=1} x^s \zeta^k(s)s^{-1}$ and $\Delta_2(x) = \Delta(x)$ [but see Chapter 3 for a slightly modified definition of $\Delta(x)$].

$P(x)$ $\qquad = \sum\limits_{n \leqslant x}{}' r(n) - \pi x - 1$ (but see Chapter 13 for a slightly modified definition).

$\varphi(n)$ $\qquad$ Euler's function defined as $\varphi(n) = n \prod\limits_{p \mid n} \left(1 - \dfrac{1}{p} \right)$, where the product is over all prime divisors of n.

$\omega(n)$ $\qquad$ The number of distinct prime divisors of n.

$\Omega(n)$ $\qquad$ The number of all prime divisors of n.

$J_p(z), K_p(z), Y_p(z)$ $\qquad$ Bessel functions of index p, defined in Chapter 3.

$E(T)$ $\qquad \int_0^T \left| \zeta(\tfrac{1}{2} + it) \right|^2 dt - T \log(T/2\pi)$ $- T(2\gamma - 1).$

ar sinh z $\qquad = \log\left(z + \sqrt{z^2 + 1} \right).$

$(\varkappa, \lambda)$ $\qquad$ An exponent pair (a certain pair of real numbers for which $0 \leqslant \varkappa \leqslant \tfrac{1}{2} \leqslant \lambda \leqslant 1$; precise definition and properties are given in Section 2.3 of Chapter 2).

$f(x) \sim g(x)$ as $x \to x_0$ $\qquad$ Means $\lim\limits_{x \to x_0} \dfrac{f(x)}{g(x)} = 1$, with x_0 possibly infinite.

$f(x) = O(g(x))$	Means $\|f(x)\| \leqslant \dot{C}g(x)$ for $x \geqslant x_0$ and some absolute constant $C > 0$. Here $f(x)$ is a complex function of the real variable x and $g(x)$ is a positive function of x for $x \geqslant x_0$.
$f(x) \ll g(x)$	Means the same as $f(x) = O(g(x))$.
$f(x) \gg g(x)$	Means the same as $g(x) = O(f(x))$.
$f(x) \asymp g(x)$	Means that both $f(x) \ll g(x)$ and $g(x) \gg f(x)$ hold.
(a, b)	Means the interval $a < x < b$.
$[a, b]$	Means the interval $a \leqslant x \leqslant b$.
δ, ε	An arbitrarily small number, not necessarily the same at each occurrence in the proof of a theorem or lemma.
$C^r[a, b]$	The class of functions having a continuous rth derivative in $[a, b]$.
$f(x) = o(g(x))$ as $x \to x_0$	Means $\lim\limits_{x \to x_0} \dfrac{f(x)}{g(x)} = 0$, with x_0 possibly infinite.
$f(x) = \Omega_+(g(x))$	Means that there exists a suitable constant $C > 0$ such that $f(x) > Cg(x)$ holds for a sequence $x = x_n$ such that $\lim\limits_{n \to \infty} x_n = \infty$.
$f(x) = \Omega_-(g(x))$	Means that there exists a suitable constant $C > 0$ such that $f(x) < -Cg(x)$ holds for a sequence $x = x_n$ such that $\lim\limits_{n \to \infty} x_n = \infty$.
$f(x) = \Omega_\pm(g(x))$	Means that both $f(x) = \Omega_+(g(x))$ and $f(x) = \Omega_-(g(x))$ holds.
$f(x) = \Omega(g(x))$	Means that $\|f(x)\| = \Omega_+(g(x))$.

ERRATA

p. V l. 9 should have a point after "results".

p. 13 l. 1 should be $\log |z|$ and $f(0) = 1$ l. -14 "is" should be "if"

p. 17 l. 1, absolute value signs are not needed in $|r_n|^{-1}$.

p. 17 l. -2 better: Then, if T is not an ordinate of a zero,

p. 18 l. 11 should be (A.33), not (A.35).

p. 20 l. 1 should be (1.38), not (1.48).

p. 27 better put: (1.71) (for $n \geq n_0(\varepsilon)$) follows

p. 36 l. 7 should be (brackets missing): $\frac{1}{2}(j+2)(j+1)$

p. 48 l. 6 should be $\sigma > \frac{1}{2}$ (not $\geq$).

p. 62 and 63: in (2.19) and (2.20) should have $\int_a^{x_0}$ and not $\int_a^{x_a}$

p. 69 in (2.40) should have $|S|^2$

p. 71 at the end of (2.44) bracket is missing: should be ... $+ 2))$.

p. 74 in (2.56) should be (absolute value signs missing) $|g^{(r)}(x)|$

p. 77 l. -12 should be $r \geq 1$, l. -8 delete comma between $(\frac{1}{6}, \frac{2}{3})$ and $(\frac{4}{18}, \frac{11}{18})$.

p. 80 l. 1 should have $u > 1$ (not $<$).

p. 87 l. 5 should be x^{-a} and not x^{-2a}

p. 88 l. 17 should be: obtains, for $N \ll x$,

p. 89 l. 15 should be "convergent" and not "covergent"

p. 91 in (3.31) replace $+B_q e^{-4\pi\sqrt{nx}}$ and the next line by

$$+B_q \cos(4\pi\sqrt{nx} - \pi/4 - \pi q/2)\Big\}$$

$$+C_q e^{-4\pi\sqrt{nx}} + O\left((nx)^{-q/2-5/4}\right) + O\left((nx)^{-2}\right),$$

where A_q, B_q ans C_q are uniformly bounded.

p. 91 l. -8 replace $a, b > 0$, by $0 < a < b$,

p. 98 l. 1 should be $\zeta(\bar{s}) =$ (delete 1)

p. 106 l. 2 should be $D(x) = \sum_{n \leq x} d(n)$ (replace $+$ by $=$)

p. 111 l. 5 replace s by $\frac{s}{2}$, should have $\frac{\Gamma'}{\Gamma}\left(\frac{s}{2}\right)$

p. 121 l. -8 replace "than" by "then"

p. 130 in (5.4) $\frac{1}{i}$ is missing, should have $\frac{1}{i}a_m \bar{a}_n$

p. 134 one absolute value sign missing in (5.17), should be $|F'(t)|$.

p. 140 l. -11 should be "which at" and not "which as".

p. 172 $\Gamma(w)$ missing in (7.3), should be $\Gamma(w)dw =$

p. 175 l. 1 should be $N < N' \leq 2N$

p. 200 l. 12 should not be $t_m =$ but $|\zeta(\frac{1}{2} + it_m)| =$ and two lines below delete $T^{1/6}+$ and close-up, namely should have $\ll \sum\limits_{r}$

p. 226 l. 12 should be 14.6 not 13.6

p. 238 l. -2 should be $\frac{1}{2} < \sigma > 1$ not $\frac{1}{2} \leq \sigma \leq 2$

p. 239 l. -1 ...$p^{-\sigma})^2$ should be ...$p^{-r\sigma})^2$

p. 257 l. -9 should have $du \ll U \exp$

p. 263 in (10.40) should have $(2x)^4/$, four lines below should be $M \gg T^{1/4}$

p. 265 l. 5 should have $\xi^{(m)}$ instead of $\zeta^{(m)}$

p. 267 l. -6 should have $T^{1/(2k+6)}$ replaced by $T^{1/(6k+6)}$

p. 270 (11.7) should have

$$(2\pi i)^{-1} \int_{\mathrm{Re}=1/2-\beta} = o(1) + (2\pi)^{-1} \int$$

and then all as before.

p. 294 l. 7 should have: If we now use $R_2 \ll$

p. 295 l. -1 and l. -2, S should be S_1

p. 306 in (12.35) we should have $\int_1^\infty$ not $\int_i^\infty$

p. 315 l. -3 should end with $\left(\sum\limits_{x<n^2\leq x+h} 1 + 1\right)$

p. 334 l. -2 should be: holds, then we obtain

p. 334 l.-1 should be: $\Delta(y) = -yU^{-1}$

p. 336 l. 3 two bars are missing in the denominator, it should be $\rho_2\bar{\rho}_3\bar{\rho}_4($

p. 350 l. 9 should be: $A < 1$ not $A > 1$

p. 352 in (13.9) delete twice $+\varepsilon$ from the exponents, should be x^{a_k} and x^{1+2b_k}

p. 363 l. 1 should be "the first sum", l. 10 (13.27) should be (13.26)

p. 383 l. -13 should have : Hafner (1981a) and not only Hafner (1981)

p. 386 l. -11 replace $\gamma_0 < 1$ by $-1 < \gamma_0 < 1$

p. 387, 389, 391 in heading on top of the page "Summary" should be "Summatory"

p. 419 l. 1 in the middle expression it should be $e^{it(a(n)-k)}$ not $e^{it(a(n)k)}$

p. 425 l. 3 should be $du = \log^{A-1} x \int_{r \log x}^{\eta \log x} v^{-A} e^{-v} dv$

p. 432 l. -10 should be: $g(n,z)n^{-1}$

p. 437 l. -18 "Pintz (in press)" should be "Pintz (1985)"

p. 438 l. -13 (1982) should be (1972)

p. 440 l. -4 "Renyi's problem" should be "Rényi's problem"

p. 445 l.-12 $|s|^{-v}$ should be $|s^{-v}|$

p. 448 l. -5 should be $e(y)dy$

p. 449 l. 10 should be $-\log\Gamma(1-u) + \log\Gamma(u)$, l. -6 (A.34) should be (A.33)

p. 452 l. 3 $\frac{1}{2} + T/2\pi x^2 + \frac{1}{4}$ should be $\frac{1}{2} + (T/2\pi x^2 + \frac{1}{4})$

p. 470 l. -8 should have $+2\int_{-GL}^{GL}$ (factor 2 is missing)

p. 487 l. -7 $\Gamma(a+b+1)$ should be $\Gamma(a+b)$

p. 490 last formula in (A.24) should have $\frac{P_{2n+1}(t)}{(2n+1)!}$

p. 492 (A.35) should have $-1/(2s)$

p. 497 ref. R.C. Baker replace (in press) by: **46**(1985), 73-79.

p. 498 ref B.C. Berndt and R.J. Evans (1983) (in press) should be **92**(1983), 67-96

p. 499 ref. P. Deligne replace **53** by **43**, (1975) by (1974)

p. 500 ref. S.W. Graham replace (in press) by: Austin, 1985, pp. 96-126.

p. 502 l. 7 replace "on von" by "of von"

p. 503 first ref. Karacuba replace **119** by **112** last ref. Kolesnik replace (in press) by: **45**(1985), 115-143.

p. 508 delete the ref. D. Suryanarayana which is twice printed, namely (1973b)

The Riemann
Zeta–Function

/ CHAPTER ONE /
ELEMENTARY THEORY

1.1 DEFINITION OF $\zeta(s)$ AND ELEMENTARY PROPERTIES

The Riemann zeta-function $\zeta(s)$ is defined as

$$(1.1) \qquad \zeta(s) = \sum_{n=1}^{\infty} n^{-s} \qquad (\operatorname{Re} s > 1),$$

and the series in (1.1) converges absolutely and uniformly in the half-plane $\operatorname{Re} s \geqslant 1 + \varepsilon$, since $|n^{-s}| \leqslant n^{-1-\varepsilon}$ and $\sum_{n=1}^{\infty} n^{-1-\varepsilon}$ converges. Here, n^{-s} for complex s is defined as $e^{-s \log n}$, where $\log n$ is the natural logarithm of n. The zeta-function seems to have been studied first by L. Euler (1707–1783), who considered only real values of s. The present notation and the notion of $\zeta(s)$ as a function of the complex variable s are due to B. Riemann (1826–1866), who made a number of startling discoveries about $\zeta(s)$. Riemann wrote $s = \sigma + it$ (σ, t real) for the complex variable s, and this tradition still persists, although some authors prefer the more logical notation $s = \sigma + i\tau$. Uniform convergence for $\sigma = \operatorname{Re} s \geqslant 1 + \varepsilon$ of the series in (1.1) implies by a well-known theorem of Weierstrass that $\zeta(s)$ is regular for $\sigma > 1$, and its derivatives in this region may be obtained by termwise differentiation of the series in (1.1).

Historically, the zeta-function arose from the need for an analytic tool capable of dealing with problems involving prime numbers. It was observed already by Euler that

$$(1.2) \qquad \sum_{n=1}^{\infty} n^{-s} = \prod_{p} (1 - p^{-s})^{-1}$$

for real $s > 1$, where the product is taken over all primes. The identity (1.2) plays a fundamental role in the analytic theory of primes and it in fact holds for $\operatorname{Re} s > 1$, being a corollary of the following:

THEOREM 1.1. Let $f(n)$ be a real or complex-valued multiplicative function [i.e., $f(mn) = f(m)f(n)$ if $(m, n) = 1$] such that $\sum_{n=1}^{\infty}|f(n)| < \infty$. Then

$$(1.3) \qquad \sum_{n=1}^{\infty} f(n) = \prod_p \left(1 + f(p) + f(p^2) + f(p^3) + \cdots \right).$$

Proof of Theorem 1.1. The product in (1.3) is called an Euler product, and each of its factors is an absolutely convergent series. Hence we may multiply a finite number of these series to obtain

$$\prod_{p \leqslant x} \left(1 + f(p) + f(p^2) + \cdots \right) = \sum_n {}' f(n),$$

where $\sum_n'$ denotes the summation over all n having no prime factor larger than x. Since

$$\sum_n {}' f(n) = \sum_{n \leqslant x} f(n) + R(x),$$

where

$$|R(x)| \leqslant \sum_{n > x} |f(n)|,$$

we obtain

$$\lim_{x \to \infty} R(x) = 0,$$

because $\sum_{n=1}^{\infty}|f(n)|$ converges. Taking $f(n) = n^{-s}$ in (1.3) we obtain (1.2) for $\sigma > 1$.

From (1.2) we have, for $\sigma > 1$,

$$|\zeta(s)| > \prod_p (1 + p^{-\sigma})^{-1} = \exp\left(-\sum_p \log(1 + p^{-\sigma})\right)$$

$$= \exp\left(\sum_p \left(-p^{-\sigma} + \tfrac{1}{2}p^{-2\sigma} - \tfrac{1}{3}p^{-3\sigma} + \cdots \right)\right) > \exp\left(-\sum_p p^{-\sigma}\right) > 0,$$

since $\sum_p p^{-\sigma}$ converges for $\sigma > 1$. This shows that $\zeta(s) \neq 0$ for $\sigma > 1$. The regions in the complex plane where $\zeta(s) \neq 0$ are known as "zero-free regions" for $\zeta(s)$. A little later in Theorem 1.5 we shall see that $\sigma \geqslant 1$ is a zero-free region for $\zeta(s)$, and it is of great importance to extend the zero-free region to the left of $\sigma = 1$ as much as possible. This problem will be reconsidered in Chapter 6, where we shall obtain the best known zero-free region for $\zeta(s)$.

Euler's identity (1.2) can be recast into another, often more convenient form —that is, take the logarithm of both sides in (1.2) and differentiate to obtain

$$(1.4) \qquad -\frac{\zeta'(s)}{\zeta(s)} = \sum_{p} \frac{\log p}{p^s - 1} = \sum_{p} \sum_{n=1}^{\infty} (\log p) p^{-ns}$$

for $s = \sigma > 1$ real. This procedure is justified, since the first series in (1.4) is majorized by $\sum_p (\log p)/(p^{1+\varepsilon} - 1)$ for $\sigma \geqslant 1 + \varepsilon$ and hence it converges uniformly for $\sigma \geqslant 1 + \varepsilon$. Changing the order of summation in (1.4) we obtain, for $s = \sigma > 1$,

$$(1.5) \qquad -\frac{\zeta'(s)}{\zeta(s)} = \sum_{n=1}^{\infty} \Lambda(n) n^{-s},$$

where $\Lambda(n)$ is known as the von Mangoldt function, and may be defined as

$$(1.6) \qquad \Lambda(n) = \begin{cases} \log p, & n = p^k \quad (p \text{ prime}, k \geqslant 1), \\ 0, & \text{otherwise.} \end{cases}$$

However, the series in (1.5) converges uniformly and absolutely for $\operatorname{Re} s = \sigma \geqslant 1 + \varepsilon$; thus by analytic continuation (1.5) holds for $\operatorname{Re} s = \sigma > 1$.

The importance of the zeta-function comes from the fact that the variable s in (1.1) is a complex number, so that the function represented by the series $\sum_{n=1}^{\infty} n^{-s}$ may possess an analytic continuation outside the region $\sigma > 1$, where the series converges. In fact in most applications of zeta-function theory, it is precisely the information about $\zeta(s)$ when $\sigma \leqslant 1$ which is of crucial importance. The simplest facts about the analytic character of $\zeta(s)$ are contained in

THEOREM 1.2. The function $\zeta(s)$, defined by (1.1) for $\sigma > 1$, admits of analytic continuation over the whole complex plane having as its only singularity a simple pole with residue 1 at $s = 1$.

Proof of Theorem 1.2. For $x > 1$, we have

$$\sum_{n \leqslant x} n^{-s} = \int_{1-0}^{x} t^{-s} d[t] = [x] x^{-s} + s \int_{1}^{x} [t] t^{-s-1} dt$$

$$= O(x^{1-\sigma}) + s \int_{1}^{x} ([t] - t) t^{-s-1} dt + \frac{s}{s-1} - \frac{sx^{1-s}}{s-1}.$$

If $\sigma > 1$ and $x \to \infty$, it follows that

$$(1.7) \qquad \zeta(s) = \frac{s}{s-1} + s \int_{1}^{\infty} ([t] - t) t^{-s-1} dt.$$

Since $|[t] - t| \leqslant 1$ it is seen that the integral in (1.7) is uniformly convergent for $\sigma \geqslant \delta$ and any fixed $\delta > 0$. Thus this integral represents an analytic function for $\sigma > 0$, and (1.7) provides the analytic continuation of $\zeta(s)$ over the half-plane $\sigma > 0$, its only pole for $\sigma > 0$ being $s = 1$ with residue 1. More generally, the Euler–Maclaurin summation formula [see (A.24)] gives, for $\sigma > 1$ and $n \geqslant 1$ fixed,

$$(1.8) \qquad \zeta(s) = \frac{1}{s-1} + \frac{1}{2} + \sum_{k=1}^{n} \frac{B_{2k}\Gamma(s+2k-1)}{(2k)!\Gamma(s)}$$

$$- \frac{\Gamma(s+2n+1)}{\Gamma(s)} \int_{1}^{\infty} P_{2n+1}(t) t^{-s-2n-1} \, dt.$$

Using the fact that $P_{2n+1}(t) = O(1)$ we obtain by analytic continuation that (1.8) holds for $\sigma + 2n > 0$. Since n is arbitrary, the conclusion of the theorem follows.

Another useful representation of $\zeta(s)$ may be obtained if one observes that, for $\sigma > 1$,

(1.9)

$$1 - 2^{-s} + 3^{-s} - 4^{-s} + 5^{-s} - \cdots = \zeta(s) - 2\sum_{n=1}^{\infty}(2n)^{-s} = (1 - 2^{1-s})\zeta(s),$$

so that

$$(1.10) \qquad \zeta(s) = (1 - 2^{1-s})^{-1} \sum_{n=1}^{\infty} (-1)^{n-1} n^{-s} \qquad (\sigma > 0),$$

since the alternating series in (1.9) converges for $\sigma > 0$. Thus (1.10) also provides the analytic continuation of $\zeta(s)$ for $\sigma > 0$, and shows in particular that $\zeta(\sigma) < 0$ for $0 < \sigma < 1$.

Theorem 1.2 shows that the Laurent expansion of $\zeta(s)$ in the neighborhood of its pole $s = 1$ is

$$(1.11) \qquad \zeta(s) = \frac{1}{s-1} + \gamma_0 + \gamma_1(s-1) + \gamma_2(s-1)^2 + \cdots .$$

It is possible to evaluate explicitly the coefficients γ_k in (1.11), as shown by

THEOREM 1.3. If $\gamma_0, \gamma_1, \gamma_2, \ldots$ are defined by (1.11), then

$$(1.12) \qquad \gamma_k = \frac{(-1)^k}{k!} \lim_{N \to \infty} \left(\sum_{m \leqslant N} \frac{1}{m} \log^k m - \frac{\log^{k+1} N}{k+1} \right),$$

and, in particular,

(1.13) $\quad \gamma = \gamma_0 = \lim_{N \to \infty} \left(1 + \frac{1}{2} + \cdots + \frac{1}{N} - \log N \right) = 0.5772157\ldots$

is Euler's constant.

Proof of Theorem 1.3. Let $k, r \geqslant 0$ be integers and let

$$c_r = - \int_{1-0}^{\infty} t^{-1} \log^r t \, d\psi(t), \qquad \psi(t) = t - [t] - \tfrac{1}{2}.$$

By the Stieltjes integral representation we have

$$c_r = - \lim_{N \to \infty} \int_{1-0}^{N} t^{-1} \log^r t \, d\psi(t)$$

$$= \lim_{N \to \infty} \left(\int_{1-0}^{N} t^{-1} \log^r t \, d[t] - \int_{1-0}^{N} t^{-1} \log^r t \, dt \right)$$

$$= \lim_{N \to \infty} \left(\sum_{n \leqslant N} n^{-1} \log^r n - (r+1)^{-1} \log^{r+1} N \right).$$

Further let $S_r(x) = \sum_{n \leqslant x} n^{-1} (\log(x/n))^r$. Then

$$S_r(x) = \int_{1-0}^{x} t^{-1} \log^r(x/t) \, d[t]$$

$$= \int_{1}^{x} t^{-1} \log^r(x/t) \, dt - \int_{1-0}^{x} t^{-1} \log^r(x/t) \, d\psi(t)$$

$$= (r+1)^{-1} \log^{r+1} x - \int_{1-0}^{\infty} t^{-1} (\log x - \log t)^r \, d\psi(t)$$

$$+ \int_{x}^{\infty} t^{-1} \log^r(x/t) \, d\psi(t)$$

$$= (r+1)^{-1} \log^{r+1} x + \sum_{k=0}^{r} (-1)^k \binom{r}{k} c_k \log^{r-k} x + R_r(x)$$

say. We want to prove that $\gamma_k = (-1)^k c_k / k!$ for $0 \leqslant k \leqslant r$. Observe that

$$\int_{1}^{\infty} t^{-s} \, dS_r(t) = r \int_{1}^{\infty} t^{-s-1} S_{r-1}(t) \, dt = -rs^{-1} \int_{1}^{\infty} S_{r-1}(t) \, dt^{-s}$$

$$= rs^{-1} \int_{1}^{\infty} t^{-s} \, dS_{r-1}(t) = \cdots = r! s^{-r} \int_{1}^{\infty} t^{-s} \, dS_0(t)$$

$$= r! s^{-r} \sum_{n=1}^{\infty} n^{-s-1} = r! s^{-r-1} + \sum_{k=0}^{\infty} \gamma_k r! s^{k-r},$$

since the γ_k's are defined by (1.11). On the other hand, using the expression for $S_r(x)$ it is seen that

$$\int_1^\infty t^{-s} dS_r(t) = r!s^{-r-1} + \sum_{k=0}^\infty a_k s^{k-r},$$

where for $0 \leqslant k \leqslant r$ we have

$$s^{k-r}a_k = \int_1^\infty (-1)^k \binom{r}{k} c_k t^{-s} d(\log^{r-k} t)$$

$$= (-1)^k \binom{r}{k} c_k (r-k) \int_1^\infty t^{-s-1} \log^{r-k-1} t \, dt$$

$$= s^{k-r}(-1)^k \binom{r}{k} c_k (r-k) \int_0^\infty e^{-v} v^{r-k-1} \, dv = s^{k-r}(-1)^k r! c_k/k!.$$

Comparing the two series expansions for $\int_1^\infty t^{-s} dS_r(t)$, we obtain $\gamma_k = (-1)^k c_k/k!$ for $0 \leqslant k \leqslant r$, and since r may be arbitrary (1.12) follows for $k = 0, 1, 2, \ldots$.

A formula such as (1.8) may be used for numerical evaluation of $\zeta(s)$, but for $s = 2k$ already Euler established a simple, explicit formula which we state here as

THEOREM 1.4. If $k = 1, 2, \ldots$ and B_j denotes the jth Bernoulli number, then

(1.14) $$\zeta(2k) = \frac{(-1)^{k+1}(2\pi)^{2k} B_{2k}}{2(2k)!}.$$

Proof of Theorem 1.4. Substituting $x = -\frac{1}{2}iu$ into the well-known identity

$$\sin x = x \prod_{k=1}^\infty (1 - x^2/\pi^2 k^2)$$

one obtains by logarithmic differentiation

(1.15) $$\frac{1}{e^u - 1} = \frac{1}{u} - \frac{1}{2} + \sum_{k=1}^\infty \frac{2u}{4k^2\pi^2 + u^2} \qquad (u \neq \pm 2m\pi i).$$

Recall that the Bernoulli numbers B_k are defined by

$$\frac{z}{e^z - 1} = \sum_{k=0}^\infty B_k z^k/k! \qquad (|z| < 2\pi),$$

so that $B_0 = 1$, $B_1 = -\frac{1}{2}$, and $B_{2k+1} = 0$ for $k \geqslant 1$. Hence, for $|u| < 2\pi$, (1.15) gives

$$\sum_{k=1}^{\infty} \frac{B_{2k}u^{2k}}{(2k)!} = \frac{u}{e^u - 1} - 1 + \frac{u}{2} = 2\sum_{r=1}^{\infty} \frac{u^2}{4r^2\pi^2 + u^2}$$

$$= 2\sum_{r=1}^{\infty} \left(\frac{u}{2r\pi}\right)^2 \sum_{k=0}^{\infty} (-1)^k \left(\frac{u}{2r\pi}\right)^{2k}$$

$$= 2\sum_{k=1}^{\infty} (-1)^{k+1} \left(\sum_{r=1}^{\infty} (2r\pi)^{-2k}\right) u^{2k}$$

$$= \sum_{k=1}^{\infty} \left\{2(2\pi)^{-2k}\zeta(2k)(-1)^{k+1}\right\} u^{2k},$$

and equating coefficients of u^{2k} we obtain (1.14). In particular,

(1.16) $\qquad \zeta(2) = \sum_{n=1}^{\infty} n^{-2} = \frac{\pi^2}{6}, \qquad \zeta(4) = \sum_{n=1}^{\infty} n^{-4} = \frac{\pi^4}{90}.$

No one has yet succeeded in obtaining a formula as simple as (1.14) for $\zeta(2k + 1)$. In fact, besides the result of R. Apéry that $\zeta(3)$ is irrational, almost nothing is known about the arithmetical structure of $\zeta(2k + 1)$ for $k > 1$.

We shall now use Theorem 1.2 to prove the classical result of J. Hadamard and C. J. de la Vallée-Poussin that $\zeta(s)$ does not vanish on the line $\sigma = 1$. This is

THEOREM 1.5. For t real, $\zeta(1 + it) \neq 0$.

Proof of Theorem 1.5. The proof is by contradiction. Since $s = 1$ is a pole of $\zeta(s)$, assume that $1 + it_1$ $(t_1 \neq 0)$ is a zero of order m. Then $1 + it_1$ is a first-order pole with residue $m \geqslant 1$ for the function $\zeta'(s)/\zeta(s)$. Thus for $\sigma > 1$, but sufficiently close to 1, we have

(1.17) $\qquad \dfrac{\zeta'(\sigma + it_1)}{\zeta(\sigma + it_1)} = (1 + o(1))\dfrac{m}{\sigma - 1} \qquad (\sigma \to 1 + 0).$

Using Theorem 1.2 we have

(1.18) $\qquad \dfrac{\zeta'(\sigma)}{\zeta(\sigma)} = (1 + o(1))\dfrac{-1}{\sigma - 1} \qquad (\sigma \to 1 + 0),$

and also

(1.19) $\qquad \dfrac{\zeta'(\sigma + 2it_1)}{\zeta(\sigma + 2it_1)} = (1 + o(1))\dfrac{k}{\sigma - 1} \qquad (\sigma \to 1 + 0).$

In (1.19) the left-hand side is bounded if $\zeta(1 + 2it_1) \neq 0$, hence in that case we may set $k = 0$, while if $\zeta(1 + 2it_1) = 0$ we have $k \geq 1$. In view of $\Lambda(n) \geq 0$, (1.5) gives

$$(1.20) \quad \mathrm{Re}\left(\frac{3\zeta'(\sigma)}{\zeta(\sigma)} + \frac{4\zeta'(\sigma + it_1)}{\zeta(\sigma + it_1)} + \frac{\zeta'(\sigma + 2it_1)}{\zeta(\sigma + 2it_1)} \right)$$

$$= - \sum_{n=1}^{\infty} \Lambda(n)n^{-\sigma}\{3 + 4\cos(t_1\log n) + \cos(2t_1\log n)\}$$

$$= - \sum_{n=1}^{\infty} 2\Lambda(n)n^{-\sigma}\{1 + \cos(t_1\log n)\}^2 \leq 0.$$

But from (1.17)–(1.19) we see that

$$(1.21) \qquad \mathrm{Re}\left(\frac{3\zeta'(\sigma)}{\zeta(\sigma)} + \frac{4\zeta'(\sigma + it_1)}{\zeta(\sigma + it_1)} + \frac{\zeta'(\sigma + 2it_1)}{\zeta(\sigma + 2it_1)} \right)$$

$$= (1 + o(1))\frac{(-3 + 4m + k)}{\sigma - 1} > 0$$

if σ is sufficiently close to 1, since $4m \geq 4$ and $k \geq 0$. Thus (1.21) contradicts (1.20) and proves the theorem.

1.2   THE FUNCTIONAL EQUATION

The functional equation for the Riemann zeta-function is given by

THEOREM 1.6.   For all complex s

$$(1.22) \qquad \pi^{-s/2}\Gamma(s/2)\zeta(s) = \pi^{-(1-s)/2}\Gamma((1 - s)/2)\zeta(1 - s).$$

This is the symmetric form of the functional equation, which represents one of the fundamental results of zeta-function theory. It was discovered and proved by B. Riemann, and represents one of his most remarkable achievements. The functional equation shows that $\zeta(-2n) = 0$ for $n = 1, 2, \ldots$, since the gamma-function has poles at nonpositive integers and for $s = 0$ the pole of $\zeta(1 - s)$ cancels the pole of $\Gamma(s/2)$, and in fact dividing (1.22) by $\Gamma(s/2)$ and letting $s \to 0$ we obtain $\zeta(0) = -\frac{1}{2}$. The zeros $s = -2n$ are often called the "trivial zeros" of $\zeta(s)$, since it will be shown a little later in Theorem 1.7 that $\zeta(s)$ has an infinity of complex zeros, whose precise location remains a mystery even today. The distribution of zeros represents one of the major problems in zeta-function theory. Later, in Chapter 10, it will be seen that there are infinitely many zeros on the "critical line" $\sigma = \frac{1}{2}$, and some results concerning their distribution will be proved.

There are several equivalent ways in which the functional equation (1.22) may be written. Using standard properties of the gamma-function (see Appendix) one may write (1.22) as

(1.23) $\zeta(s) = \chi(s)\zeta(1-s)$, $\chi(s) = (2\pi)^s / (2\Gamma(s)\cos(\pi s/2))$,

or equivalently as

(1.24) $\zeta(s) = 2^s \pi^{s-1} \sin(\pi s/2)\Gamma(1-s)\zeta(1-s)$.

From Stirling's formula (A.34) it follows that

(1.25) $\chi(s) = (2\pi/t)^{\sigma + it - 1/2} e^{i(t + \pi/4)} (1 + O(t^{-1}))$ $(t \geqslant t_0 > 0)$,

which is for most purposes a sufficiently sharp approximation. The condition $t > 0$ in (1.25) is sufficient, since from (1.1) $\overline{\zeta(\bar{s})} = \zeta(s)$ for $\sigma > 1$, and from (1.7) and (1.22) this property holds for other values of s too.

Proof of Theorem 1.6. There exist in the literature many proofs of the functional equation, and here we shall present two of them. The first is classical, while the second one is partially new and almost elementary. The first proof of (1.22) starts from (1.7) for $0 < \sigma < 1$, which may be rewritten as

(1.26)

$$\zeta(s) = -s\int_1^\infty \psi(x)x^{-s-1}\,dx + (s-1)^{-1} + \tfrac{1}{2},\qquad \psi(x) = x - [x] - \tfrac{1}{2}.$$

Actually, it turns out that (1.26) provides the analytic continuation of $\zeta(s)$ for $\sigma > -1$ [and incidentally yields at once $\zeta(0) = -\frac{1}{2}$]. To see this let $P(x) = \int_1^x \psi(y)\,dy$, and observe that $P(x)$ is bounded because $\int_n^{n+1} \psi(x)\,dx = 0$. Hence

$$\int_A^B \psi(x)x^{-s-1}\,dx = P(x)x^{-s-1}\Big|_A^B + (s+1)\int_A^B P(x)x^{-s-2}\,dx.$$

If $\sigma > -1$, then this expression tends to zero as $A, B \to \infty$, proving that the integral in (1.26) is bounded; hence (1.26) holds for $\sigma > -1$. Further, for $\sigma < 0$, we have

$$-s\int_0^1 \psi'(x)x^{-s-1}\,dx = s\int_0^1 (-x + \tfrac{1}{2})x^{-s-1}\,dx = (s-1)^{-1} + \tfrac{1}{2};$$

hence

(1.27) $\zeta(s) = -s\int_0^\infty \psi(x)x^{-s-1}\,dx$ $(-1 < \sigma < 0)$.

Recalling the Fourier expansion $\psi(x) = -\sum_{n=1}^{\infty}(n\pi)^{-1}\sin(2n\pi x)$ (x not an integer), we have, by termwise integration of (1.27),

$$(1.28) \qquad \zeta(s) = \pi^{-1}s \sum_{n=1}^{\infty} n^{-1} \int_0^{\infty} \sin(2n\pi x) x^{-s-1}\, dx$$

$$= \pi^{-1}s \sum_{n=1}^{\infty} (2n\pi)^s n^{-1} \int_0^{\infty} (\sin y) y^{-s-1}\, dy$$

$$= \pi^{-1}s(2\pi)^s \{-\Gamma(-s)\}\sin(\pi s/2)\zeta(1-s).$$

This holds for $-1 < \sigma < 0$ and reduces easily to (1.24), and by analytic continuation it holds for other values of s too. To justify term-by-term integration observe that the series for $\psi(x)$ is boundedly convergent; hence term-by-term integration in any finite interval is permissible. Thus it will be sufficient to show

$$(1.29) \qquad \lim_{y \to \infty} \sum_{n=1}^{\infty} n^{-1} \int_y^{\infty} \sin(2n\pi x) x^{-s-1}\, dx = 0$$

for $-1 < \sigma < 0$. An integration by parts gives

$$\int_y^{\infty} \sin(2n\pi x) x^{-s-1}\, dx = -\cos(2n\pi x)(2n\pi x^{s+1})^{-1} \Big|_y^{\infty}$$

$$-(s+1)(2n\pi)^{-1}\int_y^{\infty} \cos(2n\pi x) x^{-s-2}\, dx$$

$$\ll n^{-1}y^{-\sigma-1} + n^{-1}\int_y^{\infty} x^{-\sigma-2}\, dx \ll n^{-1}y^{-\sigma-1},$$

whence (1.29) easily follows.

Our second proof of the functional equation starts from the elementary identity

$$(1 - e^{-x2^{-n}}) \prod_{k=1}^{n} (1 + e^{-x2^{-k}}) = 1 - e^{-x}.$$

Logarithmic differentiation gives

$$\sum_{k=1}^{n} \frac{-2^{-k}e^{-x2^{-k}}}{1 + e^{-x2^{-k}}} = \frac{e^{-x}}{1 - e^{-x}} - \frac{2^{-n}e^{-x2^{-n}}}{1 - e^{-x2^{-n}}} \qquad (x > 0)$$

or

$$\sum_{k=1}^{n} \frac{2^{-k}}{e^{x2^{-k}} + 1} = \frac{1}{x}\frac{x2^{-n}}{e^{x2^{-n}} - 1} - \frac{1}{e^x - 1} \qquad (x > 0).$$

Letting $n \to \infty$ we obtain the identity

(1.30) $\qquad \displaystyle\sum_{k=1}^{\infty} \frac{2^{-k}}{e^{x2^{-k}} + 1} = \frac{1}{x} - \frac{1}{e^x - 1} \qquad (x > 0),$

which is the starting point for the proof of the functional equation. Consider now, for $0 < \sigma < 1$,

$$\int_0^\infty \left(\frac{1}{e^x - 1} - \frac{1}{x} \right) x^{s-1} dx = \sum_{k=1}^{\infty} 2^{-k} \sum_{n=1}^{\infty} (-1)^n \int_0^\infty e^{-n2^{-k}x} x^{s-1} dx$$

$$= \Gamma(s) \sum_{k=1}^{\infty} 2^{-k} 2^{ks} \sum_{n=1}^{\infty} (-1)^n n^{-s}$$

$$= \Gamma(s) \frac{2^{s-1}}{1 - 2^{s-1}} (2^{1-s} - 1) \zeta(s) = \Gamma(s) \zeta(s),$$

where we used (1.10) and the fact that the order of summation and integration may be inverted by absolute convergence. The change of variable $\sqrt{2\pi}\, y = x$ gives now, for $0 < \sigma < 1$,

(1.31) $\qquad F(s) = \zeta(s)\Gamma(s)(2\pi)^{-s/2} = \displaystyle\int_0^\infty f(x) x^{s-1} dx,$

where

(1.32) $\qquad f(x) = \left(e^{\sqrt{2\pi}\, x} - 1 \right)^{-1} - (\sqrt{2\pi}\, x)^{-1}.$

If we now use the fact that $f(x)$ is self-reciprocal with respect to sine transforms, that is,

(1.33) $\qquad f(t) = (2/\pi)^{1/2} \displaystyle\int_0^\infty f(x)\sin(tx)\, dx,$

then (1.22) easily follows from (1.31). Namely, we have

(1.34) $\quad F(s) = \displaystyle\int_0^\infty f(x) x^{s-1} dx = (2/\pi)^{1/2} \int_0^\infty f(y) \left(\int_0^\infty x^{s-1}\sin(xy)\, dx \right) dy$

$$= (2/\pi)^{1/2} \Gamma(s) \sin(\pi s/2) \int_0^\infty f(y) y^{-s} dy$$

$$= (2/\pi)^{1/2} \Gamma(s) F(1 - s)\sin(\pi s/2) = F(s),$$

and the last equality also holds by analytic continuation outside the strip

$0 < \sigma < 1$. Here the inversion of the order of integration is justified, since it is readily seen that

$$\lim_{\substack{\delta \to 0 \\ \Delta \to \infty}} \int_0^\infty f(y)\, dy \left(\int_0^\delta + \int_\Delta^\infty \right) x^{s-1} \sin(xy)\, dx = 0.$$

Using the first identity in (A.30), we obtain then from (1.34)

$$\zeta(s)\Gamma(s)(2\pi)^{-s/2} = (2/\pi)^{1/2} \sin(\pi s/2)\Gamma(s)\zeta(1-s)(2\pi)^{(s-1)/2}\Gamma(1-s)$$

$$= \frac{\zeta(1-s)(2\pi)^{s/2}}{2\cos(\pi s/2)},$$

which is exactly (1.23). Finally, to see that (1.33) holds we use (1.15) and (1.32) to obtain

$$\int_0^\infty f(x)\sin(xt)\, dx = \int_0^\infty \left\{ \left(e^{\sqrt{2\pi}\, x} - 1 \right)^{-1} - \left(\sqrt{2\pi}\, x \right)^{-1} \right\} \sin(xt)\, dx$$

$$= -(2\pi)^{-1/2} \int_0^\infty x^{-1} \sin(xt)\, dx + \sum_{k=1}^\infty \int_0^\infty e^{-kx\sqrt{2\pi}} \sin(xt)\, dx$$

$$\quad - \tfrac{1}{4}\sqrt{2\pi} + \sum_{k=1}^\infty \frac{t}{2\pi k^2 + t^2}$$

$$= -\tfrac{1}{4}\sqrt{2\pi} + \tfrac{1}{2}\sqrt{2\pi} \left\{ \left(e^{\sqrt{2\pi}\, t} - 1 \right)^{-1} - \left(\sqrt{2\pi}\, t \right)^{-1} + \tfrac{1}{2} \right\}$$

$$= (\pi/2)^{1/2} f(t).$$

1.3 THE HADAMARD PRODUCT FORMULA

From the discussion made so far it follows that the only zeros of $\zeta(s)$ lying outside the strip $0 < \sigma < 1$ are the so-called trivial zeros $s = -2n$ ($n = 1, 2, \ldots$). The strip $0 < \sigma < 1$ is called the "critical strip" and the line $\sigma = \tfrac{1}{2}$ the "critical line" in zeta-function theory. To show that $\zeta(s)$ has indeed many zeros in the critical strip we must first prepare the groundwork by developing a suitable product formula, due to J. Hadamard, from complex function theory. We begin with

LEMMA 1.1. (Jensen's formula) Let $f(s)$ be a function of the complex variable $s = re^{i\theta}$ (r, θ real) which is regular in $|s| \leq R$ with no zeros on $|s| = R$ and which satisfies $f(0) = 1$. Then

(1.35) $$(2\pi)^{-1} \int_0^{2\pi} \log|f(Re^{i\theta})|\, d\theta = \int_0^R r^{-1} n(r)\, dr,$$

where $n(r)$ is the number of zeros of $f(s)$ inside the circle $|s| = r$.

Proof of Lemma 1.1. Since $\operatorname{Re}\log z = \log|z|$, the left-hand side of (1.35) is

$$(2\pi)^{-1}\int_0^{2\pi}\left(\int_0^R \operatorname{Re}\!\left(\frac{df}{f}\right) dr\right) d\theta = (2\pi)^{-1}\operatorname{Re}\int_0^{2\pi}\left(\int_0^R \frac{f'(re^{i\theta})}{f(re^{i\theta})}e^{i\theta}\, dr\right) d\theta$$

$$= \operatorname{Re}\int_0^R r^{-1}(2\pi i)^{-1}\left(\int_{|s|=r}\frac{f'(s)}{f(s)}\, ds\right) dr$$

$$= \int_0^R r^{-1}n(r)\, dr.$$

If $s_1, s_2, \ldots, s_n$ are the zeros of $f(s)$ in $|s| < R$ and $|s_1| = r_1, \ldots, |s_n| = r_n$, $f(0) = 1$, then

$$\int_0^R r^{-1}n(r)\, dr = \log(r_2/r_1) + 2\log(r_3/r_2) + \cdots + n\log(R/r_n)$$

$$= \log\frac{R^n}{|s_1 \cdots s_n|},$$

which is an alternative way of writing Jensen's formula.

A function $f(s)$, regular over the whole complex plane is called an integral function of finite order is

$$\log|f(s)| \ll |s|^A$$

for some finite constant A as $s \to \infty$. The order of $f(s)$ is the lower bound of those A for which the above inequality holds. The study of integral functions of finite order was developed at the end of 19th century by J. Hadamard, who showed that these functions can be written as an infinite product containing factors of the form $s - s_0$ corresponding to the zero s_0 of the function in question. The integral function useful in zeta-function theory is

(1.36) $$\xi(s) = s(s-1)\pi^{-s/2}\Gamma(s/2)\zeta(s),$$

which is an integral function of order 1. To see this, note that from (1.7) we obtain

$$\log|(1-s)\zeta(s)| \ll \log|s| + 1$$

uniformly in $\sigma \geqslant \frac{1}{2}$, and so by Stirling's formula (A.34)

(1.37) $$\log|\xi(s)| \ll |s|(\log|s| + 1)$$

uniformly in $\sigma \geqslant \frac{1}{2}$. Using the functional equation (1.22) we have $\xi(1 - s) = \xi(s)$, so the bound in (1.37) holds for $\sigma \leqslant \frac{1}{2}$ too. Further, by Stirling's formula for real $s \to \infty$ we have $\log \xi(s) \sim \frac{1}{2}s \log s$, which shows that the order of $\xi(s)$ is exactly 1.

An immediate consequence of Jensen's formula (1.35) and (1.37) is the bound

$$(1.38) \qquad N(T) \ll T \log T,$$

where $N(T)$ is the number of zeros of $\zeta(s)$ in the region $0 < \sigma < 1, 0 < t \leqslant T$. From the definition of $\xi(s)$ in (1.36) it follows that the zeros of $\xi(s)$ are the nontrivial zeros of $\zeta(s)$, since in (1.36) the trivial zeros $s = -2n$ $(n = 1, 2, \ldots)$ of $\zeta(s)$ are canceled by the poles of $\Gamma(s/2)$, while $s\Gamma(s/2)$ has no zeros and the zero of $s - 1$ cancels the pole of $\zeta(s)$. Taking $R = 4T$, $T \geqslant T_0$, we have

$$N(T) \ll n(R/2) \int_{R/2}^{R} r^{-1} dr \leqslant \int_{R/2}^{R} r^{-1} n(r) \, dr \ll \int_0^{2\pi} R \log R \, d\theta$$

$$\ll R \log R \ll T \log T$$

on using (1.35) and (1.37). The bound $T \log T$ is actually the correct order of magnitude for $N(T)$, as will be shown by the Riemann–von Mangoldt asymptotic formula (1.44).

Let $f(s)$ from now on denote an integral function of order 1, where we have in mind the eventual application to $f(s) = \xi(s)$. Observe first that using (1.35) we have

$$(1.39) \qquad n(R)\log 2 = n(R) \int_R^{2R} r^{-1} dr \leqslant \int_R^{2R} r^{-1} n(r) \, dr \ll R^{1+\varepsilon/2}$$

for any $\varepsilon > 0$ if $f(s)$ is an integral function of order 1. If $r_1, r_2, \ldots$ are the moduli of zeros $\rho_1, \rho_2, \ldots$ of $f(s)$, then using (1.39) we have

$$\sum_{n=1}^{\infty} r_n^{-1-\varepsilon} = \int_0^{\infty} r^{-1-\varepsilon} \, dn(r) = (1 + \varepsilon) \int_0^{\infty} r^{-2-\varepsilon} n(r) \, dr$$

$$\ll 1 + \int_1^{\infty} r^{-1-\varepsilon/2} \, dr \ll 1$$

for any fixed $\varepsilon > 0$. Hence the product

$$P(s) = \prod_{n=1}^{\infty} (1 - s/\rho_n) e^{s/\rho_n}$$

is either finite or it converges absolutely for all s, and therefore represents an integral function with zeros (of the appropriate multiplicities) at $\rho_1, \rho_2, \ldots$. If we set

$$f(s) = P(s)F(s),$$

then $F(s)$ is an integral function without zeros, and we wish to show that $F(s) = e^{A+Bs}$ with some suitable constants A and B. To achieve this we shall consider $g(s) = \log F(s)$ and show that $g(s)$ is a first-degree polynomial in s. We shall need the fact that $F(s)$ is an integral function of order at most 1, so that for $|s| = R$,

$$\operatorname{Re} g(s) = \log|F(s)| \leqslant R^{1+\varepsilon},$$

and $g(s)$ can be defined to be single valued [since $F(s)$ has no zeros], and consequently it is also an integral function. If we write

$$g(s) = \sum_{n=0}^{\infty} (a_n + ib_n)s^n \qquad (a_n, b_n \text{ real}),$$

then without loss of generality we may assume $g(0) = 0$, so that if $s = Re^{i\theta}$, then

$$\operatorname{Re} g(s) = \sum_{n=0}^{\infty} a_n R^n \cos n\theta - \sum_{n=1}^{\infty} b_n R^n \sin n\theta,$$

which is a Fourier series in θ. Thus

$$\pi a_n R^n = \int_0^{2\pi} \cos(n\theta) \operatorname{Re} g(Re^{i\theta}) \, d\theta;$$

hence

$$|a_n| \ll R^{-n} \int_0^{2\pi} |\operatorname{Re} g(Re^{i\theta})| \, d\theta \ll R^{1+\varepsilon-n} \to 0$$

for $n \geqslant 2$ if R is large enough, and a similar argument shows that $b_n = 0$ for $n \geqslant 2$. Therefore $g(s) = A + Bs$, and it remains to show that $F(s)$ has order at most 1, which will follow from

(1.40) $$|P(s)| > \exp(-R^{1+3\varepsilon}) \qquad (|s| = R, R \to \infty).$$

To prove (1.40) observe that the total length of the intervals $(r_n - r_n^{-2}, r_n + r_n^{-2})$ is finite because $\sum r_n^{-2}$ converges, so that there exist arbitrarily large values of R for which

$$|R - r_n| > r_n^{-2}$$

for all n. Let

$$P(s) = \prod_{n=1}^{\infty} (1 - s/\rho_n)e^{s/\rho_n} = P_1(s)P_2(s)P_3(s)$$

say, where in P_1, $|\rho_n| < R/2$; in P_2, $R/2 \leqslant |\rho_n| \leqslant 2R$; and in P_3, $|\rho_n| > 2R$. For the factors of P_1 we have, on $|s| = R$,

$$\left|(1 - s/\rho_n)e^{s/\rho_n})\right| \geqslant (|s/\rho_n| - 1)e^{-|s/\rho_n|} > \exp(-R/r_n),$$

so that using (1.39) we obtain

$$|P_1(s)| > \exp\left(-R \sum_{|\rho_n| < R/2} r_n^{-1}\right) > \exp(-R^{1+2\varepsilon}).$$

For the factors of P_2,

$$\left|(1 - s/\rho_n)e^{s/\rho_n}\right| \geqslant e^{-2}|s - \rho_n|/2R > CR^{-3},$$

where C is a positive constant. Hence by (1.39)

$$|P_2(s)| > (CR^{-3})^{R^{1+\varepsilon}} > \exp(-R^{1+2\varepsilon}),$$

while finally for P_3,

$$\left|(1 - s/\rho_n)e^{s/\rho_n}\right| > \exp\left(-C(R/r_n)^2\right),$$

since $|s/\rho_n| < \frac{1}{2}$. Therefore

$$|P_3(s)| > \exp\left(-CR^2 \sum_{r_n > 2R} r_n^{-2}\right) > \exp\left(-CR^{1+\varepsilon} \sum_{n=1}^{\infty} r_n^{-1-\varepsilon}\right) > \exp(-R^{1+2\varepsilon}),$$

since $\Sigma r_n^{-1-\varepsilon}$ converges. Hence

$$|P(s)| = |P_1(s)||P_2(s)||P_3(s)| > \exp(-R^{1+3\varepsilon}),$$

and (1.40) follows. The net result is that

(1.41) $$f(s) = e^{A+Bs} \prod_{n=1}^{\infty} (1 - s/\rho_n)e^{s/\rho_n}$$

for suitable constants A, B if $f(s)$ is an integral function of order 1 with zeros $\rho_1, \rho_2, \ldots$.

We shall now apply the product formula (1.41) to the function $\xi(s)$. From the fact that (up to a numerical constant) the estimate (1.37) is the best possible, it follows on using Jensen's formula that the series $\Sigma |r_n|^{-1}$ diverges if $r_1, r_2 \cdots$ are the moduli of the zeros $\rho_1, \rho_2, \ldots$ of $\xi(s)$, proving incidentally that $\zeta(s)$ has an infinity of complex (i.e., nontrivial) zeros, which must then all lie in the critical strip $0 < \sigma < 1$. Logarithmic differentiation of (1.41) with $f = \xi$ gives

$$(1.42) \qquad \frac{\xi'(s)}{\xi(s)} = B + \sum_{\rho} \left(\frac{1}{s - \rho} + \frac{1}{\rho} \right),$$

which combined with (1.36) gives

(1.43)

$$\frac{\zeta'(s)}{\zeta(s)} = B - \frac{1}{s - 1} + \frac{1}{2}\log \pi - \frac{1}{2} \left(\frac{\Gamma'(s/2 + 1)}{\Gamma(s/2 + 1)} \right) + \sum_{\rho} \left(\frac{1}{s - \rho} + \frac{1}{\rho} \right),$$

where the summation is over all nontrivial zeros of $\zeta(s)$. Letting $s \to 0$ in (1.43) we obtain

$$\frac{\zeta'(0)}{\zeta(0)} = B + 1 + \frac{1}{2}\log \pi - \frac{\Gamma'(1)}{2\Gamma(1)}.$$

Now we have $\Gamma'(1) = -\gamma$, $\Gamma(1) = 1$, while using the functional equation for $\zeta(s)$ and logarithmic differentiation we find that $\zeta'(0)/\zeta(0) = \log 2\pi$, which gives

$$B = \log 2 + \tfrac{1}{2}\log \pi - 1 - \tfrac{1}{2}\gamma.$$

1.4 THE RIEMANN – VON MANGOLDT FORMULA

The formula in question was conjectured by B. Riemann in 1859 and proved by H. von Mangoldt more than 30 years later. It furnishes a precise expression for $N(T)$, the number of zeros of $\zeta(s)$ in the region $0 < \sigma < 1$, $0 < t \leqslant T$, and represents an important tool in zeta-function theory. We state the result here as

THEOREM 1.7.

$$(1.44) \qquad N(T) = \frac{T}{2\pi}\log \frac{T}{2\pi} - \frac{T}{2\pi} + O(\log T).$$

Proof of Theorem 1.7. Let $\mathcal{D}$ be the rectangle with vertices $2 \pm iT$, $-1 \pm iT$. Then

$$(1.45) \qquad N(T) = (4\pi)^{-1}\mathrm{Im}\left(\int_{\mathcal{D}} \frac{\zeta'(s)}{\zeta(s)} \, ds \right),$$

where $\xi(s)$ is defined by (1.36). With $\eta(s) = \pi^{-s/2}\Gamma(s/2)\zeta(s)$ we therefore may write

$$\frac{\xi'(s)}{\xi(s)} = \frac{1}{s} + \frac{1}{s-1} + \frac{\eta'(s)}{\eta(s)}.$$

Observe first that

$$\text{Im}\left[\int_{\mathscr{D}}\left(\frac{1}{s} + \frac{1}{s-1}\right)ds\right] = 4\pi,$$

while $\eta(s) = \eta(1-s)$ and $\eta(\sigma \pm it)$ are conjugates, so that

$$\text{Im}\left(\int_{\mathscr{D}}\frac{\eta'(s)}{\eta(s)}ds\right) = 4\,\text{Im}\left(\int_{\mathscr{L}}\frac{\eta'(s)}{\eta(s)}ds\right),$$

where $\mathscr{L}$ consists of the segments $[2, 2+iT]$ and $[2+iT, \frac{1}{2}+iT]$. Therefore

$$\text{Im}\left(\int_{\mathscr{L}}\frac{\eta'(s)}{\eta(s)}ds\right) = \text{Im}\left[\int_{\mathscr{L}}\left(-\frac{1}{2}\log\pi + \frac{1}{2}\frac{\Gamma'(s/2)}{\Gamma(s/2)} + \frac{\zeta'(s)}{\zeta(s)}\right)ds\right]$$

$$= -\frac{1}{2}(\log\pi)T + \text{Im}\left(\int_{\mathscr{L}}\frac{\Gamma'(s/2)}{2\Gamma(s/2)}ds + \int_{\mathscr{L}}\frac{\zeta'(s)}{\zeta(s)}ds\right).$$

Using Stirling's formula (A.35) and

$$\text{Im}\left(\int_{\mathscr{L}}\frac{\Gamma'(s/2)}{2\Gamma(s/2)}ds\right) = \text{Im}\log\Gamma\left(\frac{1}{4} + \frac{1}{2}iT\right),$$

we obtain

$$\text{Im}\left(\int_{\mathscr{L}}\frac{\Gamma'(s/2)}{2\Gamma(s/2)}ds\right) = \frac{1}{2}T\log\left(\frac{T}{2}\right) - \frac{T}{2} - \frac{\pi}{8} + O\left(\frac{1}{T}\right).$$

Thus from the above estimates we have

$$N(T) = \frac{T}{2\pi}\log\frac{T}{2\pi} - \frac{T}{2\pi} + \frac{7}{8} + \frac{1}{\pi}\text{Im}\left(\int_{\mathscr{L}}\frac{\zeta'(s)}{\zeta(s)}ds\right) + O\left(\frac{1}{T}\right),$$

and to prove the theorem it remains to show that

(1.46)
$$\text{Im}\left(\int_{1/2+iT}^{2+iT}\frac{\zeta'(s)}{\zeta(s)}ds\right) = O(\log T),$$

since the integral over the other segment of $\mathscr{L}$ is clearly bounded. To prove (1.46) we shall need some estimates involving ζ'/ζ. To begin we suppose that $t \geqslant 2$, $1 \leqslant \sigma \leqslant 2$, so that the gamma term in (1.43) is $O(\log t)$ and, consequently,

$$(1.47) \quad - \operatorname{Re} \frac{\zeta'(s)}{\zeta(s)} < C \log t - \sum_{\rho} \operatorname{Re}\left(\frac{1}{s - \rho} + \frac{1}{\rho} \right) \qquad (1 \leqslant \sigma \leqslant 2, t \geqslant 2).$$

In this formula we take now $s = 2 + iT$. Since $\dfrac{\zeta'}{\zeta}(2 + iT) \ll 1$, we obtain

$$(1.48) \quad \sum_{\rho} \operatorname{Re}\left(\frac{1}{s - \rho} + \frac{1}{\rho} \right) < C \log T.$$

If $\rho = \beta + i\gamma$ is a nontrivial zero of $\zeta(s)$, then $\operatorname{Re} \rho^{-1} = \beta(\beta^2 + \gamma^2)^{-1} > 0$ and

$$\operatorname{Re} \frac{1}{s - \rho} = \frac{2 - \beta}{(2 - \beta)^2 + (T - \gamma)^2} \geqslant \frac{1}{4 + (T - \gamma)^2};$$

hence (1.48) gives

$$(1.49) \quad \sum_{\rho} \frac{1}{1 + (T - \gamma)^2} = O(\log T),$$

where the summation is over all nontrivial zeros ρ of $\zeta(s)$. The last bound gives immediately

$$(1.50) \quad N(T + 1) - N(T) \ll \sum_{\rho} \frac{1}{1 + (T - \gamma)^2} \ll \log T,$$

that is, each strip $T < t \leqslant T + 1$ contains fewer than $C \log T$ zeros of $\zeta(s)$ for some absolute $C > 0$. Using again (1.43) with s and $2 + it$ (where $t > 2$ is not the ordinate of any ρ) and subtracting, we obtain

$$(1.51) \quad \frac{\zeta'(s)}{\zeta(s)} = O(\log t) + \sum_{\rho} \left(\frac{1}{s - \rho} - \frac{1}{2 + it - \rho} \right).$$

Here for terms with $|\gamma - t| \geqslant 1$ we have

$$\left| (s - \rho)^{-1} - (2 + it - \rho)^{-1} \right| = (2 - \sigma)|(s - \rho)(2 + it - \rho)|^{-1}$$

$$\leqslant 3|\gamma - t|^{-2}$$

if $-1 \leq \sigma \leq 2$. Hence by (1.48) the portion of the sum in (1.51) for which $|\gamma - t| \geq 1$ is $O(\log t)$, and for $|\gamma - t| < 1$ we have $|2 + it - \rho| \geq 1$, and the number of such ρ is $O(\log t)$ by (1.50). Thus we obtain

$$(1.52) \qquad \frac{\zeta'(s)}{\zeta(s)} = \sum_{\rho, |t-\gamma| < 1} \frac{1}{s - \rho} + O(\log t) \qquad (-1 \leq \sigma \leq 2).$$

Now the proof of (1.46) easily follows, since by (1.52)

$$\mathrm{Im}\left(\int_{1/2+iT}^{2+iT} \frac{\zeta'(s)}{\zeta(s)} \, ds \right) = O(\log T) + \mathrm{Im}\left(\int_{1/2+iT}^{2+iT} \sum_{\rho, |T-\gamma| < 1} \frac{ds}{s - \rho} \right)$$

$$= O(\log T) + \sum_{\rho, |T-\gamma| < 1} \Delta \arg(s - \rho) = O(\log T),$$

since $|\Delta \arg(s - \rho)| < \pi$ on $[\frac{1}{2} + iT, 2 + iT]$ and (1.50) holds.

This finishes the proof of the Riemann–von Mangoldt formula (1.44). Some immediate corollaries are

$$(1.53) \qquad \sum_{|\gamma| \leq T} \frac{1}{|\gamma|} \ll \log^2 T, \qquad \sum_{|\gamma| > T} \frac{1}{\gamma^2} \ll \frac{\log T}{T},$$

and

$$(1.54) \qquad \gamma_n \sim \frac{2\pi n}{\log n} \qquad (n \to \infty),$$

where $0 < \gamma_1 \leq \gamma_2 \leq \gamma_3 \leq \cdots$ are consecutive ordinates of nontrivial zeros $\rho = \beta + i\gamma$ of $\zeta(s)$. The bounds in (1.53) follow by partial summation from (1.44), while (1.54) follows from (1.44) and the obvious inequalities

$$N(\gamma_n - 1) < n \leq N(\gamma_n + 1).$$

We conclude this section by remarking that (1.47) can be used to prove that $\zeta(\sigma + it) \neq 0$ for

$$(1.55) \qquad \sigma \geq 1 - C/\log t, \qquad t \geq 2,$$

where $C > 0$ is an absolute constant. To see this, take $s = \sigma + it$ or $s = \sigma + 2it$, $\rho = \beta + i\gamma$, and $1 \leq \sigma < 2$. In each case the sum over ρ in (1.48) is clearly positive, hence (1.47) gives

$$-\mathrm{Re} \frac{\zeta'(\sigma + 2it)}{\zeta(\sigma + 2it)} < A \log t, \qquad -\mathrm{Re} \frac{\zeta'(\sigma + it)}{\zeta(\sigma + it)} < A \log t - \frac{1}{\sigma - \beta}$$

for some absolute $A > 0$. Moreover, we have

$$-\frac{\zeta'(\sigma)}{\zeta(\sigma)} < \frac{1}{\sigma - 1} + A,$$

since $s = 1$ is a simple pole of $\zeta(s)$. If we put the above inequalities in (1.20), then (1.55) follows after some simplification with $\sigma = 1 + \delta/\log t$, where δ is a suitable positive constant. The zero-free region (1.55) was the one that was independently obtained by J. Hadamard and C. J. de la Vallée-Poussin around 1900 in their search for a proof of the prime number theorem. (See Chapter 12 for the strongest known version of the prime number theorem.) The estimate (1.55) is primarily of historical interest now, since better zero-free regions are known at present (see Chapter 6), and for this reason (1.55) was not formulated as a theorem.

1.5 AN APPROXIMATE FUNCTIONAL EQUATION

Although very important, the functional equation (1.22) for $\zeta(s)$ has the shortcoming of not expressing $\zeta(s)$ explicitly, but only in terms of $\zeta(1 - s)$ and the gamma-function. Many formulas exist in the literature which express $\zeta(s)$ [or $\zeta^k(s)$] as a number of finite sums involving the function n^{-s}. These formulas have been nicknamed "approximate functional equations," and in many instances they lead to very precise results about $\zeta(s)$. Approximate functional equations for $\zeta(s)$ will be discussed extensively in Chapter 4. Our aim here is to prove only one of the simplest approximate functional equations for $\zeta(s)$ which we present as

THEOREM 1.8. For $0 < \sigma_0 \leqslant \sigma \leqslant 2$, $x \geqslant |t|/\pi$, $s = \sigma + it$,

$$(1.56) \qquad \zeta(s) = \sum_{n \leqslant x} n^{-s} + \frac{x^{1-s}}{s - 1} + O(x^{-\sigma}),$$

where the O-constant depends only on σ_0.

For the proof of this theorem we shall need a simple result concerning exponential sums, which is the subject of Chapter 2. This is

LEMMA 1.2. Let $f(x)$ be a real-valued function on the interval $[a, b]$, and let $f'(x)$ be continuous and monotonic on $[a, b]$ and $|f'(x)| \leqslant \delta < 1$. Then

$$(1.57) \qquad \sum_{a < n \leqslant b} e(f(n)) = \int_a^b e(f(x)) \, dx + O((1 - \delta)^{-1}),$$

where the O-constant is absolute.

Proof of Lemma 1.2. We may suppose that $f'(x) \geqslant 0$ on $[a, b]$ by taking the conjugate of the sum in (1.57) if $f'(x) < 0$. Let $\varphi_n(x) = e(f(n + x))$ for

$0 < x < 1$ and let us extend the definition of $\varphi_n(x)$ over the whole real line by making it periodic with period 1 and

$$\varphi_n(0) = \varphi_n(1) = \tfrac{1}{2}\{e(f(n)) + e(f(n + 1))\}.$$

Then $\varphi_n(x)$ may be expanded into a Fourier series of the form

(1.58) $$\varphi_n(x) = \sum_{k=-\infty}^{\infty} u_k e(kx),$$

where $u_0 = \int_0^1 e(f(n + x))\, dx$ and for $k \neq 0$ an integration by parts yields

$$u_k = u_k(n) = \int_0^1 \varphi_n(x)e(-kx)\, dx = \frac{e(f(n + x))e(-kx)}{-2k\pi i}\bigg|_0^1$$

$$+ k^{-1}\int_0^1 f'(n + x)e(f(n + x) - kx)\, dx.$$

Therefore setting $x = 1$ in (1.58) we obtain

$$\tfrac{1}{2}\{e(f(n)) + e(f(n + 1))\} = \sum_{k=-\infty}^{\infty} u_k(n)$$

$$= \int_0^1 e(f(n + x))\, dx + \sum_{k=-\infty,\, k\neq 0}^{\infty} u_k(n)$$

$$= \int_n^{n+1} e(f(x))\, dx + \sum_{k=-\infty,\, k\neq 0}^{\infty} k^{-1}\int_n^{n+1} f'(x)e(f(x) - kx)\, dx.$$

Summation over n gives then

$$\sum_{a < n \leqslant b} e(f(n)) = \int_a^b e(f(x))\, dx + \sum_{k=-\infty,\, k\neq 0}^{\infty} k^{-1}R_k + O(1),$$

where

$$R_k = \int_{[a]+1}^{[b]} f'(x)e(f(x) - kx)\, dx$$

$$= (2\pi i)^{-1}\int_{[a]+1}^{[b]} \frac{f'(x)}{f'(x) - k}\, d\{e(f(x) - kx)\}.$$

Recall that the second mean value for real integrals asserts that there is a ξ in the interval $[a, b]$ such that

(1.59) $$\int_a^b u(x)v(x)\, dx = u(a)\int_a^\xi v(x)\, dx + u(b)\int_\xi^b v(x)\, dx,$$

providing that u, v are Riemann integrable on $[a, b]$ and u is monotonic. Thus we apply (1.59) to the real and imaginary part of the integral for R_k to obtain

$$R_k \ll (|k| - \delta)^{-1}.$$

Hence

$$\sum_{k \neq 0} k^{-1} R_k \ll (1 - \delta)^{-1} + \sum_{k=2}^{\infty} (k(k - \delta))^{-1} \ll (1 - \delta)^{-1},$$

and the lemma follows.

Proof of Theorem 1.8. We have, for $\operatorname{Re} s > 1$ and $N \geqslant 2$,

$$\sum_{n > N} n^{-s} = \int_{N}^{\infty} \tau^{-s} d[\tau] = -N^{1-s} + s \int_{N}^{\infty} [\tau] \tau^{-s-1} d\tau$$

$$= \frac{N^{1-s}}{s - 1} - \frac{1}{2} N^{-s} - s \int_{N}^{\infty} \psi(\tau) \tau^{-s-1} d\tau.$$

Therefore

$$(1.60) \quad \zeta(s) = \sum_{n \leqslant N} n^{-s} + \sum_{n > N} n^{-s}$$

$$= \sum_{n \leqslant N} n^{-s} + \frac{N^{1-s}}{s - 1} - \frac{1}{2} N^{-s} - s \int_{N}^{\infty} \psi(\tau) \tau^{-s-1} d\tau,$$

and by analytic continuation this is valid for $\sigma > 0$, the last summand being $\ll (1 + |t|) N^{-\sigma}$. If $u \geqslant x$ we set

$$A(u) = \sum_{x < n \leqslant u} n^{-it}$$

and apply Lemma 1.2 with $f(x) = (2\pi)^{-1} |t| \log x$, $\delta = \frac{1}{2}$, provided that $x \geqslant |t| / \pi$. From (1.57) we have

$$A(u) = \int_{x}^{u} y^{-it} dy + O(1) = \frac{u^{1-it} - x^{1-it}}{1 - it} + O(1).$$

For $x \leqslant N$, partial summation gives

$$\sum_{x < n \leqslant N} n^{-s} = \sigma \int_{x}^{N} u^{-\sigma-1} A(u) \, du + A(N) N^{-\sigma}$$

$$= \sigma \int_{x}^{N} \frac{u^{-s} - u^{-\sigma-1} x^{1-it}}{1 - it} \, du + O(x^{-\sigma}) + O(xN^{-\sigma}) + \frac{N^{1-\sigma-it}}{1 - it}$$

$$= \frac{N^{1-s}}{1 - s} - \frac{x^{1-s}}{1 - s} + O(xN^{-\sigma}) + O(x^{-\sigma}).$$

Substituting this estimate in (1.60) we finally have, for $\sigma_0 \leqslant \sigma \leqslant 2$,

$$\zeta(s) = \sum_{n \leqslant x} n^{-s} + \frac{x^{1-s}}{s-1} + O(x^{-\sigma}) + O(xN^{-\sigma}) + O(|t|N^{-\sigma}),$$

and letting $N \to \infty$ (1.56) follows.

We are now in a position to say something about the order of $|\zeta(s)|$, which is one of the most important and difficult problems in zeta-function theory. From (1.56) we have, for $1 \leqslant \sigma \leqslant 2$, uniformly in σ,

$$(1.61) \qquad \zeta(\sigma + it) \ll \sum_{n \leqslant t} n^{-1} + 1 \ll \log t \qquad (1 \leqslant \sigma \leqslant 2, t \geqslant 2).$$

There is no loss of generality in supposing $t > 0$, since $\zeta(\sigma \pm it)$ are conjugates. Furthermore, for $\sigma > 2$ we have $|\zeta(s)| \leqslant \zeta(2) = \frac{1}{6}\pi^2$, while using (1.23) and (1.25) we have

$$(1.62) \qquad \zeta(it) \ll t^{1/2}\log t, \qquad \zeta(\sigma + it) \ll t^{1/2-\sigma}\log t \qquad (\sigma < 0, t > 0).$$

A useful way to compare the orders of $|\zeta(s)|$ at two different values of σ is given by the following:

LEMMA 1.3. For $0 \leqslant \sigma_1 \leqslant \sigma_0 \leqslant 3/2, t \geqslant t_0$, we have uniformly in σ_0

$$(1.63) \qquad \zeta(\sigma_0 + it) \ll 1 + \max_{|v| \leqslant \log^2 t} |\zeta(\sigma_1 + it + iv)|.$$

Proof of Lemma 1.3. Let $\mathscr{D}$ be the rectangle with vertices $\sigma_1 + it \pm i \log^2 t$, $3 + it \pm i \log^2 t$. Then $f(s) = \zeta(s)\Gamma(s - s_0 + 2)$ ($s_0 = \sigma_0 + it$) is regular in the domain bounded by $\mathscr{D}$, and therefore by the maximum modulus principle

$$|\zeta(s_0)| = |f(s_0)| \leqslant \max_{s \in \mathscr{D}} |\zeta(s)\Gamma(s - s_0 + 2)|.$$

On the horizontal sides of $\mathscr{D}$ we have $|\text{Im}(s - s_0 + 2| = \log^2 t$, and trivially [e.g., from (1.56)] $\zeta(s) \ll t^2$; hence the maximum of $|f(s)|$ over these sides is $o(1)$ by Stirling's formula (A.34), while the maximum on the side with $\sigma = 3$ is clearly $O(1)$. Also by Stirling's formula we have

$$\max_{|v| \leqslant \log^2 t} |\zeta(\sigma_1 + it + iv)\Gamma(\sigma_1 - \sigma_0 + 2 + iv)|$$

$$\ll \max_{|v| \leqslant t_0} |\zeta(\sigma_1 + it + iv)|$$

$$+ \max_{t_0 \leqslant |v| \leqslant \log^2 t} |\zeta(\sigma_1 + it + iv)||v|^{3/2 + \sigma_1 - \sigma_0} e^{-\pi|v|/2}$$

$$\ll 1 + \max_{|v| \leqslant \log^2 t} |\zeta(\sigma_1 + it + iv)|e^{-|v|}.$$

To obtain an order result for $0 \leqslant \sigma \leqslant 1$ we use $\zeta(1 + it) \ll \log t$, $\zeta(it) \ll t^{1/2}\log t$ and apply the Phragmén–Lindelöf principle (see Section A.8) to the function

$$f(s) = \frac{\zeta(s)(s - 1)}{(s + 2)^{(3-\sigma)/2}\log(s + 2)}$$

in the strip $0 \leqslant \sigma \leqslant 1$. This function is regular in the strip and bounded for $\sigma = 0$ and $\sigma = 1$ by (1.61) and (1.62). Thus we obtain

$$(1.64) \qquad \zeta(\sigma + it) \ll t^{(1-\sigma)/2}\log t \qquad (0 \leqslant \sigma \leqslant 1, \; t \geqslant t_0 > 0).$$

Therefore it makes sense to define the function $\mu(\sigma)$ for each real σ as the infimum of number $c \geqslant 0$ such that $\zeta(\sigma + it) \ll t^c$, or alternatively as

$$(1.65) \qquad \mu(\sigma) = \limsup_{t \to \infty} \frac{\log|\zeta(\sigma + it)|}{\log t}.$$

From (1.61) and (1.62) we have $\mu(\sigma) = 0$ for $\sigma \geqslant 1$ and $\mu(\sigma) = \frac{1}{2} - \sigma$ for $\sigma \leqslant 0$, but the exact value of $\mu(\sigma)$ for any $0 < \sigma < 1$ remains to this day unknown [by the functional equation (1.22) it is sufficient to consider only the values $\sigma \geqslant \frac{1}{2}$]. From Lemma 1.3 we see immediately that $\mu(\sigma)$ is nonincreasing, and again by the Phragmén–Lindelöf principle (see also Section 8.2 Chapter 8 for a more direct proof) it is seen to be convex downwards, in the sense that no arc of the curve $y = \mu(\sigma)$ has any point above its chord. From convexity, or from general properties of the Dirichlet series (series of the form $\sum_{n=1}^{\infty} a_n n^{-s}$), it follows that $\mu(\sigma)$ is continuous. Therefore we may formulate

THEOREM 1.9. For $t \geqslant t_0 > 0$ uniformly in σ,

$$(1.66) \qquad \zeta(\sigma + it) \ll \begin{cases} 1 & \text{for } \sigma \geqslant 2, \\ \log t & \text{for } 1 \leqslant \sigma \leqslant 2, \\ t^{(1-\sigma)/2}\log t & \text{for } 0 \leqslant \sigma \leqslant 1, \\ t^{1/2-\sigma}\log t & \text{for } \sigma \leqslant 0, \end{cases}$$

and if $\mu(\sigma)$ is defined by (1.65), then $\mu(\sigma)$ is continuous, nonincreasing and for $\sigma_1 \leqslant \sigma \leqslant \sigma_2$,

$$(1.67) \qquad \mu(\sigma) \leqslant \mu(\sigma_1)\frac{\sigma_2 - \sigma}{\sigma_2 - \sigma_1} + \mu(\sigma_2)\frac{\sigma - \sigma_1}{\sigma_2 - \sigma_1}.$$

More advanced methods will be employed in Chapters 6 and 7 to improve on (1.66), while omega results for the order of $\zeta(s)$ are discussed in Chapter 9.

1.6 MEAN VALUE THEOREMS

Mean value theorems will be the subject in later chapters, so here we intend to give only some simple results concerning mean values. As was pointed out in the last section, the problem of the order of $\zeta(s)$ in the critical strip is yet unsolved. The easier problem of the average order, or mean value, in its simplest form has been solved completely. This concerns the behavior of

$$T^{-1}\int_1^T |\zeta(\sigma + it)|^2 \, dt \qquad (T \to \infty),$$

for any given value of σ. If $\sigma > 1$, then one can consider the mean value of $\zeta^k(s)$ for any fixed integer $k \geq 1$. Observe that

(1.68) $\qquad \zeta^k(s) = \left(\sum_{n=1}^{\infty} n^{-s} \right)^k = \sum_{n=1}^{\infty} d_k(n) n^{-s} \qquad (\sigma > 1),$

where

$$d_k(n) = \sum_{n_1 n_1 \cdots n_k = n} 1$$

is the number of ways n can be written as a product of k fixed factors, so that $d_k(n)$ is clearly a multiplicative function of n. Now we can formulate our first result involving mean values, which we state as

THEOREM 1.10. For $\sigma > 1$ fixed and $k \geq 1$ a fixed integer we have

(1.69) $\qquad \int_0^T |\zeta(\sigma + it)|^{2k} \, dt = \left(\sum_{n=1}^{\infty} d_k^2(n) n^{-2\sigma} \right) T + O(T^{2-\sigma+\varepsilon}) + O(1).$

Proof of Theorem 1.10. First, we shall prove that the series in (1.69) converges by showing that

(1.70) $\qquad d_k(n) < \exp(C(k)\log n/\log\log n), \qquad (n \geq 3),$

where $C(k) > 0$ is a computable constant depending only on k. It will be enough to show

(1.71) $\qquad d(n) = d_2(n) < \exp(C\log n/\log\log n), \qquad (n \geq 3),$

and then to use induction and the relation

$$d_k(n) = \sum_{\delta | n} d_{k-1}(\delta),$$

which follows by equating coefficients in the identity

$$\zeta^k(s) = \zeta(s)\zeta^{k-1}(s) \qquad (\sigma > 1),$$

or by simple combinatorial arguments.

To prove (1.71) let $n = p_1^{\alpha_1} \cdots p_r^{\alpha_r}$ be the canonical decomposition of n. Since p^α has exactly $\alpha + 1$ divisors we have

$$d(n)n^{-\delta} = \prod_{j=1}^{r} (\alpha_j + 1)p_j^{-\alpha_j\delta},$$

where $\delta > 0$ is a number which will be chosen in a moment. Now $(\alpha + 1)p^{-\alpha\delta} \leqslant 1$ for $p \geqslant 2^{1/\delta}$ and

$$(\alpha + 1)p^{-\alpha\delta} \leqslant 1 + (\delta \log 2)^{-1}$$

for all primes p and integers $\alpha \geqslant 1$, hence

$$d(n)n^{-\delta} \leqslant \left(1 + \frac{1}{\delta \log 2}\right)^{2^{1/\delta}}.$$

Therefore (1.71) follows with $C = (1 + \varepsilon)\log 2$ if we choose

$$\delta = ((1 + \varepsilon/2)\log 2)/\log\log n.$$

Using (1.70) it follows that for $\sigma > 1$, $0 \leqslant t \leqslant T$ and any $\varepsilon > 0$ we may write

$$\zeta^k(\sigma + it) = \sum_{n \leqslant T} d_k(n)n^{-\sigma-it} + O(T^{1-\sigma+\varepsilon}) = S + R$$

say, and by the Cauchy–Schwarz inequality we may write

(1.72) $\displaystyle \int_0^T |\zeta(\sigma + it)|^{2k}\, dt = \int_0^T |S|^2\, dt + \int_0^T |R|^2\, dt$

$$+ O\left(\left(\int_0^T |S|^2\, dt\right)^{1/2}\left(\int_0^T |R|^2\, dt\right)^{1/2}\right).$$

But trivially

(1.73) $\displaystyle \int_0^T |R|^2\, dt \ll \int_0^T T^{2-2\sigma+2\varepsilon}\, dt = T^{3-2\sigma+2\varepsilon} \leqslant T^{2-\sigma+2\varepsilon},$

while

$$\int_0^T |S|^2 \, dt = \sum_{n \leqslant T} d_k^2(n) n^{-2\sigma} T$$

$$+ \sum_{1 \leqslant n \neq m \leqslant T} d_k(m) d_k(n) (mn)^{-\sigma} \frac{(m/n)^{iT} - 1}{\log m/n}$$

$$= \left(\sum_{n=1}^{\infty} d_k^2(n) n^{-2\sigma} \right) T + O(T^{2-2\sigma+\varepsilon}) + O\left(\sum_{1 \leqslant n < m \leqslant T} (mn)^{\varepsilon-\sigma} (\log m/n)^{-1} \right).$$

Setting $m = n + r$ and observing that $(\log m/n)^{-1} \ll 1$ for $m \geqslant 2n$, we have

$$\sum_{1 \leqslant n < m \leqslant T} (mn)^{\varepsilon-\sigma} (\log m/n)^{-1} \ll \sum_{1 \leqslant n < m \leqslant T} (mn)^{\varepsilon-\sigma}$$

$$+ \sum_{n \leqslant T, 1 \leqslant r \leqslant n} nr^{-1} n^{2\varepsilon-2\sigma} \ll 1$$

if $\varepsilon > 0$ is small enough. Hence

(1.74)
$$\int_0^T |S|^2 \, dt = \left(\sum_{n=1}^{\infty} d_k^2(n) n^{-2\sigma} \right) T + O(1),$$

and

(1.75)
$$\int_0^T |S|^2 \, dt \int_0^T |R|^2 \, dt \ll T^{4-2\sigma+2\varepsilon},$$

so that Theorem 1.10 follows from (1.72)–(1.75).

We proceed now to the strip $\frac{1}{2} \leqslant \sigma < 1$ and prove the following:

THEOREM 1.11. For $\frac{1}{2} < \sigma < 1$ fixed

(1.76)
$$\int_1^T |\zeta(\sigma + it)|^2 \, dt = \zeta(2\sigma) T + O(T^{2-2\sigma} \log T).$$

Proof of Theorem 1.11. For $\frac{1}{2} T \leqslant t \leqslant T$ we may choose $x = T$ in Theorem 1.8 to obtain

$$\zeta(\sigma + it) = \sum_{n \leqslant T} n^{-\sigma-it} + O(T^{-\sigma}) = S + R$$

say. Analogously as in (1.72), we may write

(1.77)

$$\int_{T/2}^{T} |\zeta(\sigma + it)|^2 \, dt = \int_{T/2}^{T} |S|^2 \, dt + \int_{T/2}^{T} |R|^2 \, dt$$

$$+ O\left(\left(\int_{T/2}^{T} |S|^2 \, dt\right)^{1/2} \left(\int_{T/2}^{T} |R|^2 \, dt\right)^{1/2}\right).$$

Here trivially

$$\int_{T/2}^{T} |R|^2 \, dt \ll T^{1-2\sigma},$$

while

$$\int_{T/2}^{T} |S|^2 \, dt = \frac{1}{2} T \sum_{n \leqslant T} n^{-2\sigma} + \sum_{T/2 < n \neq m \leqslant T} (mn)^{-\sigma} \frac{(m/n)^{iT} - (m/n)^{iT/2}}{\log m/n}$$

$$= \frac{1}{2} \zeta(2\sigma) T + O(T^{2-2\sigma}) + O\left(\sum_{T/2 \leqslant n < m \leqslant T} (mn)^{-\sigma} (\log m/n)^{-1}\right).$$

In the last sum we consider again the ranges $m > 2n$ and $m \leqslant 2n$, when we set $m = n + r, \, 1 \leqslant r \leqslant n$. This gives

$$\sum_{T/2 \leqslant n < m \leqslant T} (mn)^{-\sigma} (\log m/n)^{-1} \ll \left(\sum_{n \leqslant T} n^{-\sigma}\right)^2 + \sum_{n \leqslant T} \sum_{r \leqslant n} n^{1-2\sigma} r^{-1}$$

$$\ll T^{2-2\sigma} \log T.$$

Hence Theorem 1.11 follows from (1.77) on replacing T by $T/2, \, T/2^2, \ldots,$ and adding all the results.

We could have also considered the boundary cases $\sigma = \frac{1}{2}$ and $\sigma = 1$ in the foregoing proof to obtain

(1.78)
$$\int_{1}^{T} |\zeta(1 + it)|^2 \, dt = \zeta(2) T + O(\log^2 T)$$

and

(1.79)
$$\int_{1}^{T} |\zeta(\tfrac{1}{2} + it)|^2 \, dt \ll T \log T.$$

This shows that our method breaks down for $\sigma = \frac{1}{2}$ in the sense that it does not produce an asymptotic formula for the integral in (1.79), but just an upper bound. However, this upper bound is of the right order of magnitude, since following more closely the proof given above we actually obtain

$$(1.80) \qquad \int_1^T \left| \zeta\left(\tfrac{1}{2} + it\right) \right|^2 dt = T \log T + O\left(T \log^{1/2} T\right)$$

by noting that the crucial sum

$$\sum_{T/2 \leqslant n \neq m \leqslant T} (mn)^{-1/2} \frac{(m/n)^{iT} - (m/n)^{iT/2}}{\log m/n}$$

is $O(T)$ by appealing to Theorem 5.2. Also with the aid of Theorem 5.2 one may remove $\log T$ from the error term in (1.76). The asymptotic formula for the integral in (1.80) is of great importance in zeta-function theory, and in Chapter 15 advanced methods will be used to furnish a result which greatly supersedes (1.80).

1.7 VARIOUS DIRICHLET SERIES CONNECTED WITH $\zeta(s)$

By a Dirichlet series generated by the arithmetical function $f(n)$ we mean the series

$$(1.81) \qquad F(s) = \sum_{n=1}^{\infty} f(n) n^{-s},$$

provided that such a series converges for some $s = s_0$. If this is so, then $F(s)$ converges uniformly for $\sigma \geqslant \sigma_0 + \varepsilon$ ($\sigma_0 = \operatorname{Re} s_0$), representing consequently a regular function in the half-plane $\sigma \geqslant \sigma_0 + \varepsilon$. To see this let $M < N$ and use partial summation to obtain

$$\sum_{M \leqslant n \leqslant N} f(n) n^{-s} = \sum_{M \leqslant n \leqslant N} f(n) n^{-s_0} n^{s_0 - s}$$

$$= N^{s_0 - s} \sum_{M \leqslant n \leqslant N} f(n) n^{-s_0} - \int_M^N \left(\sum_{M \leqslant n \leqslant t} f(n) n^{-s_0} \right) (s_0 - s) t^{s_0 - s - 1} dt$$

$$\ll N^{\sigma_0 - \sigma} + \frac{|s - s_0|}{|\sigma - \sigma_0|} M^{\sigma_0 - \sigma} \to 0$$

uniformly as $M \to \infty$, proving uniform convergence of $F(s)$ for $\sigma \geqslant \sigma_0 + \varepsilon$. Thus with every Dirichlet series $F(s)$ we may associate a number σ_c, defined as the greatest lower bound of numbers σ_0 such that $F(s)$ converges for $\sigma > \sigma_0$, and one calls σ_c the abscissa of convergence of $F(s)$. The abscissa of absolute convergence σ_a of $F(s)$ is defined to be the abscissa of convergence of $\sum_{n=1}^{\infty} |f(n)| n^{-s}$, and obviously $\sigma_c \leqslant \sigma_a$. Uniform convergence enables us to conclude that the Dirichlet series possess a uniqueness property in the following sense: If

$$F(s) = \sum_{n=1}^{\infty} f(n)n^{-s}, \qquad G(s) = \sum_{n=1}^{\infty} g(n)n^{-s}$$

for $\sigma > \sigma_0$ and $F(\sigma) = G(\sigma)$, then $f(n) = g(n)$ for all n. To see this note that

$$\sum_{n=1}^{\infty} f(n)n^{-\sigma} = \sum_{n=1}^{\infty} g(n)n^{-\sigma}$$

and since both series converge uniformly for $\sigma \geq \sigma_0 + \varepsilon$, we obtain $f(1) = g(1)$ by letting $\sigma \to \infty$. Supposing that $f(n) = g(n)$ for $n = 1, 2, \ldots, N - 1$ we have

$$f(N) = \lim_{\sigma \to \infty}\left(N^{\sigma}\sum_{n=N}^{\infty} f(n)n^{-\sigma}\right) = \lim_{\sigma \to \infty}\left(N^{\sigma}\sum_{n=N}^{\infty} g(n)n^{-\sigma}\right) = g(N);$$

hence $f(n) = g(n)$ for all n.

The uniqueness property of the Dirichlet series is a very useful tool for establishing identities between arithmetical functions. Suppose that $F(s)$ and $G(s)$ are two Dirichlet series which both converge absolutely for $\sigma > \sigma_1$. Then we can write

$$(1.82) \quad F(s)G(s) = \sum_{k=1}^{\infty} f(k)k^{-s}\sum_{l=1}^{\infty} g(l)l^{-s} = \sum_{n=1}^{\infty} h(n)n^{-s} = H(s),$$

where $H(s)$ also converges absolutely for $\sigma > \sigma_1$. The arithmetical function $h(n)$ may be expressed through $f(n)$ and $g(n)$ as

$$(1.83) \quad h(n) = \sum_{kl=n} f(k)g(l) = \sum_{d|n} f(d)g(n/d) = \sum_{d|n} g(d)f(n/d);$$

because of absolute convergence the series in (1.82) may be multiplied out and the terms arranged arbitrarily. By grouping together terms with $kl = n$ we obtain (1.83), and by the uniqueness property (1.83) may be viewed as an equivalent of (1.82). The function $h(n)$ in (1.83) is called the convolution (or Dirichlet convolution, to distinguish it from other arithmetical convolutions) of $f(n)$ and $g(n)$, and it is easily seen to be multiplicative if both $f(n)$ and $g(n)$ are multiplicative.

Many common arithmetical functions generate a Dirichlet series which can be in some way written as a finite or infinite product of zeta-functions. One such example in Section 1.6 was the divisor function $d_k(n)$, which generates $\zeta^k(s)$, and in this section we shall consider some other arithmetical functions of interest and their Dirichlet series.

The Möbius Functions

For $\mathrm{Re}\, s > 1$ we have

$$(1.84) \qquad 1/\zeta(s) = \prod_p (1 - p^{-s}) = \sum_{n=1}^{\infty} \mu(n)n^{-s},$$

where $\mu(n)$ is called the Möbius function. From the product representation we deduce that

$$\mu(n) = \begin{cases} 1 & \text{if } n = 1, \\ (-1)^k & \text{if } n \text{ is a product of } k \text{ different primes,} \\ 0 & \text{if } p^2|n \text{ for some prime } p, \end{cases}$$

while the relation $1 = \zeta(s)(1/\zeta(s))$ has the arithmetical interpretation

$$(1.85) \qquad \sum_{d|n} \mu(d) = \begin{cases} 1 & \text{if } n = 1, \\ 0 & \text{otherwise.} \end{cases}$$

The function $\mu(n)$ is multiplicative, and further the relations

$$(1.86) \qquad f(n) = \sum_{d|n} g(d)$$

and

$$(1.87) \qquad g(n) = \sum_{d|n} \mu(d)f(n/d)$$

are equivalent, which follows on using (1.85).

k-Free Numbers

A natural number n is square-free if it is a product of different primes, and more generally n is called k-free ($k \geq 2$ fixed) if in the canonical decomposition $n = p_1^{\alpha_1} \cdots p_r^{\alpha_r}$ we have $\alpha_1 \leq k - 1, \ldots, \alpha_r \leq k - 1$. Therefore if $f_k(n) = 1$ when n is k-free and zero otherwise, then $f_k(n)$ is the characteristic function of the set of k-free numbers, and for $\sigma > 1$

$$(1.88) \qquad \sum_{n=1}^{\infty} f_k(n)n^{-s} = \prod_p (1 + p^{-s} + \cdots + p^{-(k-1)s})$$

$$= \prod_p \frac{1 - p^{-ks}}{1 - p^{-s}} = \frac{\zeta(s)}{\zeta(ks)}.$$

Note that $f_2(n) = \mu^2(n)$ and

(1.89)
$$\mu^2(n) = \sum_{d^2|n} \mu(d).$$

Powerful Numbers

Let $k \geqslant 2$ be a fixed integer. A positive integer is said to be powerful if it contains as its factors primes of at least kth power. More precisely, let $G(k)$ denote the set of all positive integers with the property that if a prime p divides an element of $G(k)$, then p^k divides it also. In other words the set of powerful (or k-full) numbers $G(k)$ contains 1 and the numbers whose canonical representation is

(1.90) $n = p_1^{\alpha_1} p_2^{\alpha_2} \cdots p_r^{\alpha_r}$ $(\alpha_1 \geqslant k, \alpha_2 \geqslant k, \ldots, \alpha_r \geqslant k)$.

If we set

(1.91) $f_k(n) = \begin{cases} 1 & n \in G(k) \\ 0 & n \notin G(k) \end{cases}$, $F_k(s) = \sum_{n=1}^{\infty} f_k(n) n^{-s}$,

then $f_k(n)$ is the characteristic function of $G(k)$, and for $\operatorname{Re} s > 1/k$ we have

(1.92) $F_k(s) = \prod_p (1 + p^{-ks} + p^{-(k+1)s} + \cdots) = \prod_p \left(1 + \frac{p^{-ks}}{1 - p^{-s}} \right).$

For further factoring of $F_k(s)$ we note that for $k \geqslant 2$ and $K = \frac{1}{2}(3k^2 + k - 2)$ there exist constants $a_{r,k}$ $(2k + 2 < r \leqslant K)$ such that

(1.93) $\left(1 + \frac{v^k}{1 - v} \right)(1 - v^k)(1 - v^{k+1}) \cdots (1 - v^{2k-1})$

$$= 1 - v^{2k+2} + \sum_{r=2k+3}^{K} a_{r,k} v^r,$$

which follows from the fact that the product of the first two factors on the left-hand side of (1.93) equals

$$(1 - v + v^k)(1 + v + \cdots + v^{k-1}) = 1 + v^{k+1} + v^{k+2} + \cdots + v^{2k-1},$$

and multiplying out the remaining factors we obtain (1.93). Taking $v = p^{-s}$ in (1.93) and using (1.92) it follows that

(1.94)
$$F_2(s) = \frac{\zeta(2s)\zeta(3s)}{\zeta(6s)},$$

and for $k > 2$,

$$(1.95) \quad F_k(s) = \zeta(ks)\zeta((k+1)s) \cdots$$

$$\times \zeta((2k-1)s)\prod_p\left(1 - p^{-(2k+2)s} + \sum_{r=2k+3}^{K} a_{r,k}p^{-rs}\right)$$

$$= \zeta(ks)\zeta((k+1)s) \cdots \zeta((2k-1)s)\zeta^{-1}((2k+2)s)\Phi_k(s),$$

where the abscissa of absolute convergence of $\Phi_k(s)$ is equal to $1/(2k+3)$. Therefore we may write

$$(1.96) \qquad\qquad F_k(s) = G_k(s)H_k(s),$$

where

$$(1.97) \quad H_k(s) = \sum_{n=1}^{\infty} h_k(n)n^{-s}$$

$$= \zeta(ks)\zeta((k+1)s) \cdots \zeta((2k-1)s) \qquad (\mathrm{Re}\, s > 1/k),$$

and

$$(1.98) \qquad G_k(s) = \sum_{n=1}^{\infty} g_k(n)n^{-s} = \Phi_k(s)/\zeta((2k+2)s)$$

is a Dirichlet series which converges absolutely for $\mathrm{Re}\, s > 1/(2k+2)$. One can write $G_k(s)$ as an infinite product of zeta-functions, for example,

$$G_3(s) = \frac{\zeta(13s)\zeta(14s)\zeta(21s)\zeta^2(22s)\zeta^2(23s)\zeta(24s) \cdots}{\zeta(8s)\zeta(9s)\zeta(10s)\zeta(17s)\zeta(18s)\zeta(19s)\zeta(25s)\zeta^3(26s) \cdots},$$

and further deduce that the line $\sigma = 0$ is the natural boundary for $G_k(s)$.

The Generalized von Mangoldt Function

This function is defined by

$$(1.99) \qquad\qquad \Lambda_k(n) = \sum_{d|n}\mu(d)(\log n/d)^k.$$

For $\sigma > 1$ we have

$$(1.100) \qquad\qquad \zeta^{(k)}(s) = (-1)^k \sum_{n=1}^{\infty} (\log n)^k n^{-s},$$

where $k \geqslant 1$ if a fixed integer. Hence from (1.82)–(1.84) and (1.100) we have

$$(1.101) \qquad \sum_{n=1}^{\infty} \Lambda_k(n)n^{-s} = (-1)^k \zeta^{(k)}(s)/\zeta(s) \qquad (\sigma > 1).$$

Taking $k = 1$ in (1.101) and comparing with (1.5) we find that $\Lambda_1(n) = \Lambda(n)$, the von Mangoldt function, so that $\Lambda_k(n)$ indeed generalizes the von Mangoldt function. From the identity

$$\left(\frac{\zeta^{(k)}(s)}{\zeta(s)}\right)' = \frac{\zeta^{(k+1)}(s)}{\zeta(s)} - \frac{\zeta^{(k)}(s)}{\zeta(s)} \frac{\zeta'(s)}{\zeta(s)}$$

and (1.101) it follows on comparing the coefficients of n^{-s} that

$$(1.102) \qquad \Lambda_{k+1}(n) = \Lambda_k(n)\log n + \sum_{d\mid n} \Lambda_k(d)\Lambda_1(n/d),$$

and then by induction we infer that $\Lambda_k(n) = 0$ whenever n has more than k distinct prime factors. Another useful identity is

$$(1.103) \qquad \log^k n = \sum_{d\mid n} \Lambda_k(d),$$

which follows from (1.99) and the Möbius inversion formulas (1.86) and (1.87).

The Number of Divisors Functions

It was mentioned in Section 1.6 that

$$\zeta^k(s) = \sum_{n=1}^{\infty} d_k(n)n^{-s} \qquad (\sigma > 1)$$

for integers $k \geqslant 1$, so that in particular

$$(1.104) \qquad \zeta^2(s) = \sum_{n=1}^{\infty} d(n)n^{-s},$$

where $d(n) = d_2(n) = \sum_{d\mid n} 1$ is the number of divisors of n. Here we record two further identities involving $d(n)$, namely,

$$(1.105) \qquad \frac{\zeta^3(s)}{\zeta(2s)} = \sum_{n=1}^{\infty} d(n^2)n^{-s} \qquad (\sigma > 1)$$

and

$$(1.106) \qquad \frac{\zeta^4(s)}{\zeta(2s)} = \sum_{n=1}^{\infty} d^2(n)n^{-s} \qquad (\sigma > 1).$$

The proofs of these identities are straightforward. For example, for $\sigma > 1$ we obtain, using (1.2),

$$\frac{\zeta^4(s)}{\zeta(2s)} = \prod_p \frac{1 - p^{-2s}}{(1 - p^{-s})^4} = \prod_p \frac{1 + p^{-s}}{(1 - p^{-s})^3}$$

$$= \prod_p \left((1 + p^{-s}) \sum_{j=0}^{\infty} (-1)^j \binom{-3}{j} p^{-js} \right)$$

$$= \prod_p (1 + p^{-s}) \left(1 + \sum_{j=1}^{\infty} \tfrac{1}{2} j + 2(j + 1) p^{-js} \right)$$

$$= \prod_p \left(1 + \sum_{j=1}^{\infty} (j + 1)^2 p^{-js} \right) = \sum_{n=1}^{\infty} d^2(n)n^{-s},$$

since $d^2(n)$ is multiplicative and $d(p^j) = j + 1$. The identity (1.105) is established similarly, and the identity

$$(1.107) \qquad \frac{\zeta^2(s)}{\zeta(2s)} = \sum_{n=1}^{\infty} 2^{\omega(n)}n^{-s} \qquad (\sigma > 1),$$

where $\omega(n) = \sum_{p|n} 1$ is the number of distinct prime factors of n, is even easier.

The Euler Function

From the elementary relation

$$n = \sum_{d|n} \varphi(d)$$

and the Möbius inversion formula (1.87) we obtain

$$(1.108) \qquad \varphi(n) = \sum_{d|n} \frac{n}{d} \mu(d) = n \prod_{p|n} (1 - p^{-1}).$$

Therefore $\varphi(n)$ is the convolution of n with $\mu(n)$ and consequently

$$(1.109) \qquad \sum_{n=1}^{\infty} \varphi(n)n^{-s} = \frac{\zeta(s - 1)}{\zeta(s)} \qquad (\sigma > 2).$$

Sum of Divisors Function

For any complex a we define

$$\sigma_a(n) = \sum_{d \mid n} d^a,$$

and in particular $\sigma(n) = \sigma_1(n) = \sum_{d \mid n} d$ is the sum of divisors of n. Thus $\sigma_a(n)$ is the convolution of 1 with n^a, and consequently

(1.110) $\displaystyle\sum_{n=1}^{\infty} \sigma_a(n) n^{-s} = \zeta(s)\zeta(s - a)$ $\qquad (\sigma > 1, \sigma > \mathrm{Re}\, a + 1).$

A generalization of this formula is the identity

$$\sum_{n=1}^{\infty} \sigma_a(n)\sigma_b(n) n^{-s} = \frac{\zeta(s)\zeta(s - a)\zeta(s - b)\zeta(s - a - b)}{\zeta(2s - a - b)}$$

$$\big[\sigma > \max(1, \mathrm{Re}\, a + 1, \mathrm{Re}\, b + 1, \mathrm{Re}(a + b) + 1)\big],$$

which may be proved by the same technique used for the proof of (1.106), since $f(n) = \sigma_a(n)\sigma_b(n)$ is a multiplicative function.

Enumerating Functions of Algebraic Structures

It turns out that certain common algebraic structures have enumerating functions whose Dirichlet series have a simple product representation involving the zeta-function. This fact establishes an analytic approach towards the study of these functions, and provides another application of zeta-function theory. As an illustrative example we shall consider $a(n)$, the number of nonisomorphic abelian groups with n elements. It is well known from group theory that $a(n)$ is a multiplicative function such that $a(p^k) = P(k)$ for every prime p and $k \geqslant 1$, where $P(k)$ is the number of (unrestricted) partitions of k. By the classical Hardy–Ramanujan formula

(1.111) $\quad P(k) = (1 + o(1))(4\sqrt{3}\, k^{-1}) \exp\!\big(\pi (2k/3)^{1/2}\big)$ $\qquad (k \to \infty),$

and an argument which is only slightly more complicated than the proof of (1.71) gives also

(1.112) $\qquad a(n) < \exp\!\left(\left(\frac{\log 5}{4} + \varepsilon\right)\frac{\log n}{\log \log n}\right)$ $\qquad (n \geqslant n_0(\varepsilon)).$

Since for $|x| < 1$ we have

$$1 + \sum_{k=1}^{\infty} P(k)x^k = (1 - x)^{-1}(1 - x^2)^{-1}(1 - x^3)^{-1}\ldots ,$$

it follows at once that

$$(1.113) \qquad \sum_{n=1}^{\infty} a(n)n^{-s} = \zeta(s)\zeta(2s)\zeta(3s)\zeta(4s) \cdots \qquad (\sigma > 1).$$

As another example we mention the function $S(n)$, the number of nonisomorphic semisimple rings with n elements. Here one obtains

$$\sum_{n=1}^{\infty} S(n)n^{-s} = \prod_{r\geqslant 1,\, m\geqslant 1}\prod_{p}\left(1 - p^{-rm^2s}\right)^{-1} = \prod_{r\geqslant 1,\, m\geqslant 1}\zeta(rm^2s) \qquad (\sigma > 1),$$

and an upper bound analogous to (1.112) also holds.

1.8 OTHER ZETA-FUNCTIONS

Here we shall give a brief survey of some of the most important generalizations of $\zeta(s)$, which are also known as "zeta-functions." These functions arise naturally in many branches of analytic and algebraic number theory, and their study has many important applications. Many of these functions admit an analogue of the functional equation (1.22) for $\zeta(s)$, have an infinity of complex zeros, and bear other similarities to $\zeta(s)$. A vast literature concerning these functions exists [e.g., L-series have been investigated for many years as extensively as $\zeta(s)$], but our survey aims only at providing an insight into some of the various possibilities for constructing zeta-functions, so many important examples had to be omitted.

Dirichlet L-Series

These are probably the most common zeta-functions besides $\zeta(s)$ and are defined by

$$(1.114) \qquad L(s,\chi) = \sum_{n=1}^{\infty} \chi(n)n^{-s} \qquad (\sigma > 1),$$

where for a fixed modulus q ($\geqslant 1$), $\chi(n)$ is the arithmetical function known as a character modulo q [for $q = 1$, $L(s,\chi) = \zeta(s)$]. Each character $\chi(n)$ is a totally multiplicative function [i.e., $\chi(mn) = \chi(m)\chi(n)$ for all m and n] which is complex valued and satisfies $|\chi(n)| \leqslant 1$. It also has the following periodic property: $\chi(a) = \chi(b)$ if $a \equiv b \pmod{q}$, while $\chi(a) = 0$ if $(a,q) > 1$ and $\chi(a) \neq 0$ if $(a,q) = 1$. There exist $\varphi(q)$ distinct characters mod q, and they form an abelian group (under pointwise multiplication) which is isomorphic to the multiplicative group of the reduced system of residues mod q. A special character is the principal character, denoted by $\chi_1(n)$ [or $\chi_0(n)$], defined by $\chi_1(a) = 1$ if $(a, q) = 1$ and zero otherwise. Characters mod q satisfy the orthogonality relations

(1.115)
$$\sum_{n(\mathrm{mod}\ q)} \chi(n) = \begin{cases} \varphi(q) & \chi = \chi_1, \\ 0 & \chi \neq \chi_1, \end{cases}$$

and

(1.116)
$$\sum_{\chi(\mathrm{mod}\ q)} \chi(n) = \begin{cases} \varphi(q) & n \equiv 1 \ (\mathrm{mod}\ q), \\ 0 & n \not\equiv 1 \ (\mathrm{mod}\ q), \end{cases}$$

which are dual to one another. An application of Theorem 1.1 gives

(1.117) $$L(s, \chi) = \sum_{n=1}^{\infty} \chi(n) n^{-s} = \prod_{p} (1 - \chi(p) p^{-s})^{-1} \qquad (\sigma > 1),$$

while for $\chi = \chi_1$,

(1.118)
$$L(s, \chi_1) = \zeta(s) \prod_{p|q} (1 - p^{-s}) \qquad (\sigma > 1).$$

Thus $L(s, \chi_1)$ has a first-order pole at $s = 1$ just like $\zeta(s)$ and it behaves similarly to $\zeta(s)$ in many other ways, while $L(s, \chi)$ for $\chi \neq \chi_1$ [as defined by (1.114)] is regular for $\sigma > 0$. Moreover, $L(s, \chi)$ had an analytic continuation over the whole complex plane and $L(s, \chi) \neq 0$ for $\sigma \geq 1$. If χ is a character mod k and $k|q$, the χ induces a character Ψ mod q which may be defined by setting $\Psi(n) = 0$ if $(n, q) > 1$ and $\Psi(n) = \chi(n)$ if $(n, q) = 1$. A character mod q is said to be primitive if it is not induced by any character mod k with $k < q$. Dirichlet series associated with primitive characters satisfy a functional equation similar to (1.22), although this one is naturally more complicated. If χ is a nonprincipal character mod q, let $\alpha = 1$ if $\chi(-1) = -1$ and $\alpha = 0$ if $\chi(-1) = 1$. Further let

$$G(\chi, \eta) = \sum_{r(\mathrm{mod}\ q)} \chi(r) \eta^r, \quad \eta = e(1/q) = e^{2\pi i q^{-1}},$$

$$E(\chi) = \begin{cases} G(\chi, \eta) q^{-1/2} & \text{if } \alpha = 0, \\ -iG(\chi, \eta) q^{-1/2} & \text{if } \alpha = 1. \end{cases}$$

Then the functional equation for $L(s, \chi)$ takes the form

(1.119)

$$\xi(s, \chi) = E(\chi) \xi(1 - s, \bar{\chi}), \xi(s, \chi) = (\pi/q)^{-(s+\alpha)/2} \Gamma(s/2 + \alpha/2) L(s, \chi),$$

which may be used in deriving various properties of $L(s, \chi)$. If χ is a primitive, nonprincipal character $\bmod q$ and $\chi(-1) = 1$, then $L(s, \chi)$ has simple "trivial" zeros at $s = 0, -2, -4, \ldots$, while if $\chi(-1) = -1$, then $L(s, \chi)$ has simple "trivial" zeros at $s = -1, -3, -5, \cdots$. For any function $L(s, \chi)$ one may, in analogy with $\zeta(s)$, define $N(T)$ as the number of zeros $\rho = \beta + i\gamma$ of $L(s, \chi)$ for which $0 < \beta < 1, 0 < \gamma \leqslant T$. Then the analogue of the Riemann–von Mangoldt formula (1.44) takes the form

$$N(T) = \frac{1}{2\pi} T \log T + A(q)T + O(\log qT),$$

where $A(q)$ is a constant depending on the modulus q of χ such that $A(q) \ll \log q$. Thus $L(s, \chi)$ has an infinity of complex zeros, and moreover it may be shown that an infinity of them lie on the critical line $\sigma = \frac{1}{2}$.

Just as the study of $\zeta(s)$ leads to results in prime number theory (see Chapter 12), the study of $L(s, \chi)$ furnishes results about primes in arithmetic progressions. If $(l, q) = 1$,

$$\pi(x; q, l) = \sum_{p \leqslant x,\, p \equiv l (\bmod q)} 1,$$

then an important result, known as the Siegel–Walfisz theorem, may be obtained for $\pi(x; q, l)$ with the aid of the theory of L-functions. Their result states that

$$(1.120) \quad \pi(x; q, l) = \frac{\mathrm{li}\, x}{\varphi(q)} + O\big(x \exp\big(-C\sqrt{\log x}\big)\big) \qquad (C > 0),$$

uniformly for $3 \leqslant q \leqslant \log^4 x$, $l \leqslant q$, where $A > 0$ is any fixed constant and $\mathrm{li}\, x = \int_0^x (dt/\log t)$. A principal ingredient in the proof of (1.120) is Siegel's theorem, which states that for any $\varepsilon > 0$ there is an $A = A(\varepsilon) > 0$ such that if $q > A$ and χ is a real-valued, nonprincipal character $\bmod q$, then $L(s, \chi) \neq 0$ for $\sigma \geqslant 1 - q^{-\varepsilon}$.

The Hurwitz Zeta-Function

If χ is a character $\bmod q$, then for $\sigma > 1$,

$$L(s, \chi) = \sum_{n=1}^{\infty} \chi(n) n^{-s} = \sum_{a=1}^{q-1} \chi(a) \sum_{n=0}^{\infty} (qn + a)^{-s}$$

$$= q^{-s} \sum_{a=1}^{q-1} \chi(a) \sum_{n=0}^{\infty} (n + a/q)^{-s} = q^{-s} \sum_{a=1}^{q-1} \chi(a)\zeta(s, a/q),$$

where

$$(1.121) \qquad \zeta(s, x) = \sum_{n=0}^{\infty} (n + x)^{-s} \qquad (\sigma > 1, 0 < x \leqslant 1)$$

is the Hurwitz zeta-function. This is a generalization of the Riemann zeta-function, since obviously $\zeta(s) = \zeta(s, 1)$ $(\sigma > 1)$. The Hurwitz zeta-function has also an analytic continuation over the whole complex plane, with a simple pole at $s = 1$ with residue 1. Its Laurent series in the neighborhood of $s = 1$ is

$$(1.122) \quad \zeta(s, x) = (s - 1)^{-1} + \sum_{n=0}^{\infty} \gamma_n(x)(s - 1)^n \qquad (0 < x \leqslant 1),$$

where the coefficients $\gamma_n(x)$ may be evaluated explicitly as

$$(1.123) \quad \gamma_n(x) = \frac{(-1)^n}{n!} \lim_{N \to \infty} \left(\sum_{m=0}^{N} \frac{\log^n(m + x)}{m + x} - \frac{\log^{n+1}(N + x)}{n + 1} \right).$$

The functional equation (1.23) for $\zeta(s)$ may be regarded as a special case of a functional equation for $\zeta(s, x)$. Namely, if $1 \leqslant h \leqslant k$ are integers, then for all s

$$(1.124) \qquad \zeta\left(1 - s, \frac{h}{k}\right) = \frac{2\Gamma(s)}{(2\pi k)^s} \sum_{r=1}^{k} \cos\left(\frac{\pi s}{2} - \frac{2\pi rh}{k}\right) \zeta\left(s, \frac{r}{k}\right),$$

which clearly reduces to (1.23) if $h = k = 1$. Except for $\zeta(s, 1) = \zeta(s)$ and $\zeta(s, \frac{1}{2}) = (2^s - 1)\zeta(s)$, the Hurwitz zeta-function $\zeta(s, x)$ does not have an Euler product similar to (1.2) for $x \neq 1, \frac{1}{2}$, and its behavior is in many aspects different from that of $\zeta(s)$.

The Lerch Zeta-Function

This is defined as

$$(1.125) \qquad\qquad \varphi(x, a, s) = \sum_{n=0}^{\infty} e(nx)(n + a)^{-s},$$

where x is real, $0 < a \leqslant 1$, and $\sigma > 1$ if x is an integer and $\sigma > 0$ otherwise. Note that $\varphi(0, a, s) = \zeta(s, a)$ is the Hurwitz zeta-function and that $\varphi(0, 1, s) = \zeta(s)$ is the Riemann zeta-function. For $0 < x < 1$, the Lerch zeta-function $\varphi(x, a, s)$ has an analytic continuation to the whole complex plane. A noteworthy property of $\varphi(x, a, s)$, valid for all s, is the functional equation

$$\varphi(x, a, 1 - s) = (2\pi)^{-s}\Gamma(s)\{e(\tfrac{1}{4}s - ax)\varphi(-a, x, s)$$
$$+ e(-\tfrac{1}{4}s + a(1 - x))\varphi(a, 1 - x, s)\},$$

which shows that one can go relatively far in generalizing $\zeta(s)$ but still retaining an analogue of the functional equation (1.22).

The Epstein Zeta-Function

Let $Q(x, y)$ be a positive definite quadratic form with discriminant D. Then for $\sigma > 1$ the Epstein zeta-function $\zeta(s, Q)$ is defined by

$$(1.126) \qquad \zeta(s, Q) = \sideset{}{'}\sum_{m, n = -\infty}^{\infty} (Q(m, n))^{-s},$$

where the prime $'$ denotes that $m = n = 0$ is excluded from summation. This function has an analytic continuation to the whole complex plane that is regular, except for a simple pole at $s = 1$ with residue $\pi D^{-1/2}$. The Epstein zeta-function also satisfies a functional equation, which may be written as

$$\pi^{-s}\Gamma(s)\zeta(s, Q) = D^{-1/2}\pi^{s-1}\Gamma(1 - s)\zeta(1 - s, Q^{-1}),$$

where Q^{-1} is the inverse of Q.

The Dedekind Zeta-Function

Let $\mathbf{K}$ denote an algebraic number field of degree $n \geqslant 3$ over the rationals. Then the Dedekind zeta-function $\zeta_{\mathbf{K}}(s)$ is

$$(1.127) \qquad \zeta_{\mathbf{K}}(s) = \sum_{m=1}^{\infty} a_{\mathbf{K}}(m)m^{-s} \qquad (\sigma > 1),$$

where $a_{\mathbf{K}}(m)$ is the number of nonzero integral ideals in $\mathbf{K}$ with norm m. The Dedekind zeta-function has an analytic continuation to the whole complex plane, and it has only a simple pole at $s = 1$. It satisfies the functional equation

$$\Gamma^{r_1}(s/2)\Gamma^{r_2}(s)\zeta_{\mathbf{K}}(s) = C^{2s-1}\Gamma^{r_1}((1 - s)/2)\Gamma^{r_2}(1 - s)\zeta_{\mathbf{K}}(1 - s),$$

where C is a constant depending on the field $\mathbf{K}$, r_1 is the number of real conjugates in $\mathbf{K}$, and $2r_2$ is the number of imaginary conjugates in $\mathbf{K}$, so that

$r_1 + 2r_2 = n$. There exist numerous generalizations of $\zeta_K(s)$ in other (number) fields.

Zeta-Functions Associated with Cusp Forms

These are the functions of the form

$$(1.128) \qquad \varphi(s) = \sum_{n=1}^{\infty} c(n)n^{-s},$$

where the $c(n)$'s are the Fourier coefficients of a cusp form f of weight k for the full modular group. Thus for Im $z > 0$

$$f(z) = \sum_{n=1}^{\infty} c(n)e(nz),$$

f is regular in the half-plane Im $z > 0$, and for $k \geqslant 12$ an even integer

$$f\left(\frac{az + b}{cz + d}\right) = (cz + d)^k f(z),$$

whenever $\begin{pmatrix} a & b \\ c & d \end{pmatrix}$ is an integral matrix of unit determinant. The series in (1.128) is absolutely convergent for $\sigma > \frac{1}{2}(k + 1)$ and can be continued analytically to the whole complex plane. It satisfies the functional equation

$$(1.129) \quad (2\pi)^{-s}\Gamma(s)\varphi(s) = (-1)^{k/2}(2\pi)^{s-k}\Gamma(k - s)\varphi(k - s),$$

which is the analogue of the functional equation (1.22) for $\zeta(s)$. As an illustrative example, let us consider the Ramanujan zeta-function

$$F(s) = \sum_{n=1}^{\infty} \tau(n)n^{-s} \qquad (\sigma > \tfrac{13}{2}),$$

where $\tau(n)$ is the well-known multiplicative function of Ramanujan, defined by

$$x\left[(1 - x)(1 - x^2)(1 - x^3)\cdots\right]^{24} = \sum_{n=1}^{\infty} \tau(n)x^n \qquad (|x| < 1).$$

In this case $k = 12$, and (1.129) becomes

$$(2\pi)^{-s}\Gamma(s)F(s) = (2\pi)^{s-12}\Gamma(12 - s)F(12 - s).$$

The Ramanujan zeta-function $F(s)$ has the Euler product

$$F(s) = \prod_p \left(1 - \tau(p)p^{-s} + p^{11-2s}\right)^{-1} \qquad (\sigma > \tfrac{13}{2});$$

it has the "trivial" zeros $s = 0, -1, -2, -3, \ldots$, but no other zeros for $\sigma \leqslant \tfrac{11}{2}$ and $\sigma \geqslant \tfrac{13}{2}$. The "critical strip," corresponding to the strip $0 < \sigma < 1$ for $\zeta(s)$, is the strip $\tfrac{11}{2} < \sigma < \tfrac{13}{2}$ in the case of Ramanujan's zeta-function, and there are problems concerning the behavior of $F(s)$ in this strip corresponding to all familiar problems about $\zeta(s)$.

The order of magnitude of $\tau(n)$ is given by the inequality

$$|\tau(n)| \leqslant n^{11/2}d(n),$$

which was conjectured by S. Ramanujan in 1916 and proved almost 60 years later by P. Deligne. This is very sharp, since it is known that

$$\tau(n) = \Omega\left\{ n^{11/2}\exp\left(C\log^{2/3}n(\log\log n)^{-5/3}\right)\right\} \qquad (C > 0).$$

1.9 UNPROVED HYPOTHESES

We shall briefly discuss here some of the most important conjectures of zeta-function theory. As remarked already in the Preface, almost all the proofs in this book are unconditional, that is, they do not depend on any hitherto unproved hypothesis. However, for the sake of completeness and historical perspective it seems appropriate to present here some of the most important conjectures, with a short account of their mutual dependence. The truth (or falsity) of some of these conjectures would have very important consequences in many applications of zeta-function theory.

The most famous conjecture in zeta-function theory is doubtlessly the Riemann hypothesis. Stated in modern terminology it asserts that all nontrivial (i.e., complex) zeros of $\zeta(s)$ lie on the critical line $\sigma = \tfrac{1}{2}$. All attempts to prove the Riemann hypothesis have failed so far, and it is not even known whether there is a σ_0 such that $\tfrac{1}{2} \leqslant \sigma_0 < 1$ and $\zeta(s) \neq 0$ for $\mathrm{Re}\,s > \sigma_0$. Incidentally, it is not known whether all nontrivial zeros of $\zeta(s)$ are simple or not, although it seems reasonable that they are all simple. The general progress in numerical calculation in recent times made it possible to establish that the first 300,000,001 complex zeros of $\zeta(s)$ are simple and on the critical line. No doubt the numerical data will continue to accrue, but number theory is unfortunately one of the branches of mathematics where numerical evidence does not count for much. A classical example is the inequality $\pi(x) < \mathrm{li}\,x$, conjectured originally by C. F. Gauss. This conjecture is supported by all available numerical evidence, but nevertheless J. E. Littlewood proved that it is false for an infinity of x's (which tend to ∞), and S. Skewes found that the first sign change occurs for some x not exceeding $10_4(3)$, where $10_1(x) = 10^x$, $10_2(x) = 10^{10_1(x)}$, and so on.

The second well-known conjecture is the Lindelöf hypothesis, which says that $\zeta(\frac{1}{2} + it) \ll |t|^\varepsilon$ for any $\varepsilon > 0$. In terms of the function $\mu(\sigma)$ defined in Section 1.5 this is equivalent to $\mu(\frac{1}{2}) = 0$, or to the statement that $\mu(\sigma) = \frac{1}{2} - \sigma$ for $\sigma \leqslant \frac{1}{2}$ and zero otherwise. An immediate consequence of the Lindelöf hypothesis is the bound

$$(1.130) \qquad \int_1^T \left|\zeta(\tfrac{1}{2} + it)\right|^{2k} dt \ll T^{1+\varepsilon}$$

for any fixed integer $k \geqslant 1$. However (1.130) is in fact equivalent to the Lindelöf hypothesis, since by Lemma 7.1 we have

$$|\zeta(\tfrac{1}{2} + iT)|^{2k} \ll \log T\left(1 + \int_{T-\log^2 T}^{T+\log^2 T} \left|\zeta(\tfrac{1}{2} + it)\right|^{2k} dt\right) \ll T^{1+2\varepsilon}$$

if (1.130) holds, and this clearly implies the Lindelöf hypothesis if k is sufficiently large. Thus anyone wishing to disprove the Lindelöf hypothesis might reasonably try disproving (1.130); however, the present methods which give lower bounds for the integral in (1.130) (see Chapter 9) seem incapable of achieving this. Other equivalents of the Lindelöf hypothesis are

$$(1.131)$$

$$\int_1^T \left|\zeta(\sigma + it)\right|^{2k} dt = (1 + o(1)) \sum_{n=1}^{\infty} d_k^2(n) n^{-2\sigma} T \qquad (\sigma > \tfrac{1}{2}, \, k = 1, 2, \ldots)$$

and

$$(1.132) \qquad \zeta^k(s) = \sum_{n \leqslant t^\delta} d_k(n) n^{-s} + O(t^{-\lambda})$$

$$[\lambda = \lambda(k, \delta, \sigma) > 0, \, \sigma > \tfrac{1}{2}, \, k = 1, 2, \ldots],$$

where $0 < \delta < 1$ is any fixed number. A further equivalent of the Lindelöf hypothesis is

$$(1.133) \qquad \beta_k = (k-1)/2k \qquad (k = 2, 3, 4, \ldots),$$

where β_k is the constant defined in Chapter 13, which appears in the general Dirichlet divisor problem. The Lindelöf hypothesis is weaker than the Riemann hypothesis, which implies

$$(1.134) \qquad \zeta(s) \ll t^\varepsilon \quad \text{and} \quad 1/\zeta(s) \ll t^\varepsilon \qquad (t \geqslant t_0)$$

for every $\sigma > \frac{1}{2}$. This follows from a stronger statement, namely,

THEOREM 1.12. If the Riemann hypothesis is true, then uniformly for $\frac{1}{2} < \sigma_0 \leqslant \sigma \leqslant 1$ and $t \geqslant 2$

$$(1.135) \qquad\qquad \log \zeta(s) \ll (\log t)^{2-2\sigma+\varepsilon}.$$

Proof of Theorem 1.12. If the Riemann hypothesis is true, then $\log \zeta(s)$ [and not only $\zeta(s)$!] is regular for $\sigma > \frac{1}{2}$, except at $s = 1$. We apply (6.55) to the function $f(z) = \log \zeta(z)$, $z_0 = 2 + it$, $R = \frac{3}{2} - \frac{1}{2}\delta$, and $r = \frac{3}{2} - \delta$ $(0 < \delta < \frac{1}{2})$. On the larger circle,

$$\operatorname{Re} \log \zeta(z) = \log|\zeta(z)| \ll \log t;$$

hence (6.55) yields for z on the smaller circle,

$$(1.136) \qquad\qquad |\log \zeta(z)| \ll \delta^{-1} \log t.$$

Now we apply Hadamard's three-circles theorem (Section A.9) to the circles C_1, C_2, and C_3 with center $\sigma_1 + it$ $(1 < \sigma_1 < \log t)$ passing through the points $1 + \eta + it$, $\sigma + it$, and $\frac{1}{2} + \delta + it$ $(0 < \eta < \frac{1}{2})$, respectively. Thus the radii of the circles are, respectively, $r_1 = \sigma_1 - 1 - \eta$, $r_2 = \sigma_1 - \sigma$, and $r_3 = \sigma_1 - \frac{1}{2} - \delta$. If M_1, M_2, and M_3 are the maxima of $|\log \zeta(z)|$ on C_1, C_2, and C_3, then

$$M_2 \leqslant M_1^{1-a} M_3^a,$$

where

$$a = \frac{\log r_2/r_1}{\log r_3/r_1} = \log\left(1 + \frac{1 + \eta - \sigma}{\sigma_1 - 1 - \eta}\right) \Big/ \log\left(1 + \frac{\frac{1}{2} + \eta - \delta}{\sigma_1 - 1 - \eta}\right)$$

$$= \frac{1 + \eta - \sigma}{\frac{1}{2} + \eta - \delta} + O\left(\frac{1}{\sigma_1}\right) = 2 - 2\sigma + O(\delta) + O(\eta) + O\left(\frac{1}{\sigma_1}\right).$$

By (1.136), $M_3 \ll \delta^{-1} \log t$, and moreover

$$M_1 \leqslant \max_{x \geqslant 1+\eta} \left|\log \zeta(x + iy)\right| = \max_{x \geqslant 1+\eta} \left|\sum_{n=2}^{\infty} h(n) n^{-x-iy}\right| \ll \eta^{-1},$$

since the function $h(n)$ obviously satisfies $|h(n)| \leqslant 1$. Therefore

$$|\log \zeta(\sigma + it)| < (C\eta^{-1})^{1-a} (\delta^{-1} \log t)^a$$

$$\ll \eta^{a-1} \delta^{-a} (\log t)^{2-2\sigma+O(\delta)+O(\eta)+O(1/\sigma_1)}.$$

Choosing $\sigma_1 = \delta^{-1} = \eta^{-1} = \log \log t$ we obtain

$$\log \zeta(s) \ll \log \log t (\log t)^{2-2\sigma} \qquad \left[\tfrac{1}{2} + (\log \log t)^{-1} \leqslant \sigma \leqslant 1\right];$$

hence (1.135) follows. Since the exponent of $\log t$ in (1.135) is less than unity if ε is small enough, it follows that

$$-\varepsilon \log t < \log|\zeta(s)| < \varepsilon \log t \qquad [t \geqslant t_0(\varepsilon)],$$

which implies (1.134). It is, however, yet unknown whether the Lindelöf hypothesis implies the Riemann hypothesis.

Another important conjecture is the density hypothesis, which asserts that

$$(1.137) \qquad\qquad N(\sigma, T) \ll T^{2-2\sigma+\varepsilon} \qquad (\tfrac{1}{2} \leqslant \sigma \leqslant 1),$$

where $N(\sigma, T)$ is the number of zeros $\rho = \beta + i\gamma$ of $\zeta(s)$ for which $\beta \geqslant \sigma$, $|\gamma| \leqslant T$. (Sometimes in the formulation of the density hypothesis the factor T^ε is even replaced by a log power.) Estimates of $N(\sigma, T)$ play an important role in many applications of zeta-function theory, and for some applications concerning primes the reader is referred to Chapter 12. Bounds for $N(\sigma, T)$ will be extensively discussed in Chapter 11. Using (1.130) as an equivalent of the Lindelöf hypothesis, it is easily seen from (11.16), (11.20) with $M(A) = 1 + \varepsilon$, $A = r$, and the bound

$$R_1 \ll T^\varepsilon (T + M) M^{1-2\sigma}$$

that one obtains

$$N(\sigma, T) \ll T^\varepsilon \left(Y^{2r(1-\sigma)} + TY^{r^2(1-2\sigma)/(r+1)} \right) \qquad (1/2 \leqslant \sigma \leqslant 1)$$

for $r \geqslant 2$. Therefore taking $Y = T^{1/r}$ and r sufficiently large (1.137) follows, which means that the Lindelöf hypothesis implies the density hypothesis. Actually, the Lindelöf hypothesis appears to be much stronger than the density hypothesis, since it implies even

$$(1.138) \qquad\qquad N(\sigma, T) \ll T^\varepsilon \qquad (\sigma \geqslant \tfrac{3}{4} + \delta),$$

where $\varepsilon = \varepsilon(\delta)$ may be made arbitrarily small for any $\delta > 0$.
    In 1897 F. Mertens conjectured

$$(1.139) \qquad |M(x)| \leqslant x^{1/2} \qquad \left(M(x) = \sum_{n \leqslant x} \mu(n), x > 1 \right)$$

and in 1983 H. J. J. te Riele and A. Odlyzko finally disproved the Mertens hypothesis employing a technique involving the use of computers. The Mertens hypothesis is much stronger than the Riemann hypothesis, which is equivalent to $M(x) \ll x^{1/2+\varepsilon}$. To see this equivalence note that, for $\sigma > 1$, we have

$$1/\zeta(s) = \sum_{n=1}^{\infty} \mu(n)n^{-s} = \int_{1/2}^{\infty} t^{-s} dM(t) = s \int_{1}^{\infty} t^{-s-1} M(t)\, dt.$$

But if $M(x) \ll x^{1/2+\varepsilon}$, then the second integral above is absolutely convergent for $\sigma > \frac{1}{2} + \varepsilon$ and represents a regular function in this region. Consequently, $1/\zeta(s)$ can have no poles for $\sigma > \frac{1}{2} + \varepsilon$, which means that all zeros of $\zeta(s)$ lie in the half-plane $\sigma \leqslant \frac{1}{2}$, and this is the Riemann hypothesis.

Conversely, if the Riemann hypothesis holds, then by (1.134) we have $1/\zeta(s) \ll t^{\varepsilon}$ for $\sigma \geqslant \frac{1}{2}$, $t \geqslant t_0$. Using then the inversion formula for Dirichlet series [see (A.10)] we obtain, for $1 \leqslant T \leqslant x$ and $c = 1 + (\log x)^{-1}$,

$$M(x) = (2\pi i)^{-1} \int_{c-iT}^{c+iT} \frac{x^s \, ds}{s\zeta(s)} + O(xT^{-1}\log x)$$

$$= (2\pi i)^{-1} \left(\int_{c-iT}^{1/2+\delta-iT} + \int_{1/2+\delta-iT}^{1/2+\delta+iT} + \int_{1/2+\delta+iT}^{c+iT} \right) \frac{x^s \, ds}{s\zeta(s)}$$

$$+ O(xT^{-1}\log x) \ll \int_{-T}^{T} (1 + |v|)^{-1+\varepsilon} x^{1/2+\delta} \, dv + T^{\varepsilon-1}x$$

$$\ll T^{\varepsilon}x^{1/2+\delta} + xT^{\varepsilon-1} \ll x^{1/2+\varepsilon}$$

if $T = x^{1/2}$, $\delta = \varepsilon/2$. The best known O-result for $M(x)$ under the truth of the Riemann hypothesis is

$$(1.140) \qquad M(x) = O\left(x^{1/2}\exp\left(\frac{A\log x}{\log\log x} \right) \right) \qquad (A > 0),$$

while in the other direction it is known unconditionally that

$$(1.141) \qquad \limsup_{x \to \infty} |M(x)| x^{-1/2} > 0.557.$$

Finally, from numerous other conjectures in zeta-function theory we mention the pair-correlation hypothesis of H. L. Montgomery. Montgomery assumes the Riemann hypothesis and proves

$$(1.142) \qquad \sum_{0 < \gamma \leqslant T, \, \rho \text{ simple}} 1 \geqslant \left(\frac{2}{3} + o(1) \right) \frac{T}{2\pi} \log T \qquad (T \to \infty).$$

Moreover, he conjectures, for fixed $\alpha < \beta$,

$$(1.143) \qquad \sum_{0 < \gamma, \, \gamma' \leqslant T, \, \alpha/L \leqslant \gamma - \gamma' \leqslant \beta/L} 1$$

$$= (1 + o(1)) \left\{ \int_{\alpha}^{\beta} \left(1 - \left(\frac{\sin \pi u}{\pi u} \right)^2 \right) du + \delta(\alpha, \beta) \right\} TL,$$

where $L = \log T/(2\pi)$, $T \to \infty$, $\delta(\alpha, \beta) = 1$ if $0 \in [\alpha, \beta]$ and zero otherwise, and γ, γ' denote imaginary parts of zeros on the critical line. If the pair-correlation hypothesis (1.143) is true, then almost all zeros of $\zeta(s)$ are simple, that is, (1.142) holds with $\frac{2}{3}$ replaced by 1.

Notes

Besides (1.2) L. Euler discovered several other identities from zeta-function theory. His paper (1768) contains the assertion

$$\frac{1 - 2^{n-1} + 3^{n-1} - 4^{n-1} + 5^{n-1} - \cdots}{1 - 2^{-n} + 3^{-n} - 4^{-n} + 5^{-n} - \cdots} = \frac{-1 \times 2 \times 3 \cdots \times (n-1)(2^n - 1)}{(2^{n-1} - 1)\pi^n} \cos\frac{n\pi}{2},$$

which he verified for $n = 1$ and $n = 2k$. E. Landau (1906) wrote Euler's identity as

$$\frac{\lim\limits_{x \to 1} \sum\limits_{n=1}^{\infty} (-1)^{n+1} n^{s-1} x^{n-1}}{\lim\limits_{x \to 1} \sum\limits_{n=1}^{\infty} (-1)^{n+1} n^{-s} x^{n-1}} = \frac{-\Gamma(s)(2^s - 1)}{(2^{s-1} - 1)\pi^s} \cos\frac{\pi s}{2},$$

proved its validity, and showed its equivalence with the functional equation (1.22).

B. Riemann wrote only one paper (1859) on number theory, but that one was truly epoch making and justifies $\zeta(s)$ being called the "Riemann zeta-function." Riemann proved Theorem 1.2 and Theorem 1.6, but he went much further beyond this. Besides conjecturing Theorem 1.7 and the famous Riemann hypothesis (discussed in Section 1.9), he also conjectured the product formula (1.41) for $\xi(s)$ and an explicit formula for $\psi(x) = \Sigma_{n \leq x} \Lambda(n)$ (see Chapter 12 for more details about this). It took more than 30 years for these last two conjectures to be proved by J. Hadamard and H. von Mangoldt, respectively. It seems unknown how Riemann was led to most of his conjectures, but it is apparent that he knew much more about $\zeta(s)$ than he cared to publish. This is best witnessed by the Riemann–Siegel formula (4.3)–(4.4). The result was proved by C. L. Siegel (1932), who was inspired by Riemann's notes on $\zeta(s)$ which are preserved in the library of Göttingen University.

A detailed discussion of the constants γ_k in (1.12) is given by M. I. Israilov (1979, 1981). The first three values are

$$\gamma_1 = 0.07281\ldots, \qquad \gamma_2 = -0.00485\ldots, \qquad \gamma_3 = -0.00034\ldots.$$

The constants γ_k are often called the Stieltjes constants, because T. J. Stieltjes (1905) was the first to prove (1.2). Curiously enough, this relatively simple formula was "rediscovered" several times during the last 100 years. For a historical survey of some of these proofs and a discussion of some interesting formulas involving series with $\zeta(s)$ (due to S. Ramanujan), see B. C. Berndt and R. J. Evans (1983). γ_k also will denote the ordinate of the kth zero of $\zeta(s)$ of the form $\frac{1}{2} + it$, $t > 0$ (see Chapter 10), but no confusion will arise with the γ_k's defined by (1.12). Both notations are standard.

Theorem 1.4 is a classical formula, proved first by L. Euler. For an elementary evaluation of $\zeta(2k)$, see B. C. Berndt (1975b). A simple way (due to Euler) to show that $\zeta(2) = \pi^2/6$ is to divide the product formula for $\sin x$ [the equation preceding (1.15)] by x and to equate coefficients of x^2 of both sides.

In connection with (1.14) it may be mentioned that S. Ramanujan (1957) (Vol. II, p. 117, no. 21) stated a remarkable formula for $\zeta(2n + 1)$. If $\alpha, \beta > 0$, $\alpha\beta = \pi^2$, and $n \geq 1$ is an integer, then Ramanujan's formula is

$$\alpha^{-n} \left(\frac{1}{2}\zeta(2n + 1) + \sum_{k=1}^{\infty} \frac{k^{-2n-1}}{e^{2\alpha k} - 1} \right) = (-\beta)^{-n} \left(\frac{1}{2}\zeta(2n + 1) + \sum_{k=1}^{\infty} \frac{k^{-2n-1}}{e^{2\beta k} - 1} \right)$$

$$-2^{2n} \sum_{k=0}^{n+1} (-1)^k \frac{B_{2k} B_{2n+2-2k}}{(2k)!(2n + 2 - 2k)!} \alpha^{n+1-k} \beta^k.$$

Ramanujan never produced a proof, but the result has been established by several mathematicians, including B. C. Berndt (1977).

R. Apéry (1981) proved that $\zeta(3)$ is irrational. This is a remarkable achievement, although his proof does not show whether $\zeta(3)$ is algebraic or transcendental, nor does it seem possible to generalize the method to prove that $\zeta(2k + 1)$ ($k \geqslant 2$) is irrational.

The seemingly mysterious inequality $3 + 4\cos\varphi + \cos 2\varphi \geqslant 0$ was used in (1.20) because in (1.21) we need $4m > 3$ to obtain the desired contradiction. Other simple trigonometrical inequalities, such as

$$5 + 8\cos\varphi + 4\cos 2\varphi + \cos 3\varphi = (1 + \cos\varphi)(1 + 2\cos\varphi)^2 \geqslant 0,$$

would also suffice for the proof.

Our first proof of the functional equation (1.22) for $\zeta(s)$ is one of the seven proofs given by E. C. Titchmarsh (1951) in Chapter 2. The second proof is in essence the proof given also by Titchmarsh in Section 2.7 of the same work, but the crucial relation (1.30) is proved in an elementary way following J. van de Lune (1979).

The functional equation for $\zeta(s)$ in a certain sense characterizes it completely. This was established by H. Hamburger (1921, 1922a, 1922b) [see also Chapter 2 of K. Chandrasekharan (1970) for a detailed proof], who proved the following result: Let G be an integral function of finite order, P a polynomial, and let

$$f(s) = G(s)/P(s) = \sum_{n=1}^{\infty} a_n n^{-s}$$

converge absolutely for $\sigma > 1$. If

$$\pi^{-s/2}\Gamma(s/2)f(s) = \pi^{-(1-s)/2}\Gamma\left(\tfrac{1}{2} - s/2\right)g(s),$$

where $g(1 - s) = \sum_{n=1}^{\infty} b_n n^{s-1}$ converges absolutely for $\sigma < -\alpha < 0$, then

$$f(s) = a_1\zeta(s) = g(s).$$

Differentiating (1.26) and estimating the resulting expression it is found that $\zeta(\sigma)$ is decreasing for $0 < \sigma < 1$. The details are given, for example, by J. Koekoek (1983), who also evaluates elementarily $\zeta'(0) = -\tfrac{1}{2}\log(2\pi)$, a fact readily obtained by logarithmic differentiation of the functional equation and $\Gamma'(1) = -\gamma$.

The contents of Section 1.3 are standard results from complex function theory, and may be found, for example, in H. Davenport (1980), K. Chandrasekharan (1970), K. Prachar (1967), and A. A. Karacuba (1975). For our purposes it was sufficient to consider only integral functions of order one, although it would not be much more difficult to treat the general case too.

The inequality (1.71) may be improved to

$$d(n) \leqslant \exp\left\{ \frac{\log 2 \log n}{\log\log n} + O\left(\frac{\log n}{(\log\log n)^2} \right) \right\},$$

as shown first by S. Ramanujan (1962, p. 80), and the equality holds asymptotically if n is the product of the first k primes.

For arithmetical convolutions different from the Dirichlet convolution (1.83), see J. Knopfmacher (1975) and M. V. Subbarao (1972).

Formula (1.84) was stated to hold for $\sigma > 1$, while in fact it also holds at all points of the line $\sigma = 1$, namely,

$$1/\zeta(s) = \sum_{n=1}^{\infty} \mu(n)n^{-s} \qquad (\sigma \geqslant 1)$$

and in particular $\sum_{n=1}^{\infty} \mu(n)n^{-1} = 0$, proved first by H. von Mangoldt (1897). This follows from an application of Perron's formula, and the use of the zero-free region (1.55) which yields for some $C_1, C_2 > 0$, $1/\zeta(s) \ll \log^{C_1} t$ for $\sigma \geqslant 1 - C_2/\log t$, $t \geqslant t_0$. The details of proof are given in Theorem 3.13 of E. C. Titchmarsh (1951).

Powerful numbers have been extensively investigated by many authors. The present notation is from the paper by the author and P. Shiu (1982), where many references to earlier works may be found.

For some additional properties and identities involving the generalized von Mangoldt function $\Lambda_k(n)$, see author's papers (1973b, 1975, 1977a). The identity (1.102) for $k = 1$ plays an important part in elementary proofs of the prime number theorem [see Chapter 1 of K. Chandrasekharan (1970) or the fine survey article of H. Diamond (1982)]. The functions $\Lambda_k(n)$ also play an important role in E. Bombieri's elementary proof (1962) of the prime number theorem with the error term $O(x(\log x)^{-A})$ for any fixed $A > 0$.

For may enumerating functions of algebraic structures besides $a(n)$ and $S(n)$ see J. Knopf-macher (1975). The bound (1.112) was proved by E. Krätzel (1970), and improved by W. Schwarz and E. Wirsing (1973). The bound in (1.112) is asymptotically attained for $n = (p_1 \cdots p_k)^4$, where p_j is the jth prime.

The Hardy–Ramanujan formula (1.111) [see S. Ramanujan (1962)] for the partition function $P(k)$ is superseded by a formula of H. Rademacher [see Chapter 7 of K. Chandrasekharan, (1970)]. Rademacher's formula is one of the few instances in analytic number theory where the problem has been completely solved, since his expression for $P(k)$ is an exact formula, which may be asymptotically evaluated to any degree of accuracy.

For an extensive account of the theory of L-functions (up to 1957) see K. Prachar (1957). For the Dedekind zeta-function and related ideal theory see E. Landau (1949). P. Epstein's article (1907) gives an account of what is now customarily called the Epstein zeta-function. For modular functions and cusp forms the reader is referred to the book of T. M. Apostol (1976b), while for an account of Ramanujan's τ-function see Chapter 10 of G. H. Hardy (1962).

There exist several proofs of (1.123). One is given by B. C. Berndt (1972a), who also proved the bound

$$\left| \gamma_n(x) - \frac{(-1)^n \log^n x}{n!x} \right| \leqslant \begin{cases} 4/(n\pi^n) & \text{if } n \text{ is even,} \\ 2/(n\pi^n) & \text{if } n \text{ is odd,} \end{cases}$$

where $n \geqslant 1$ and $0 < x \leqslant 1$.

The functional equation for the Lerch zeta-function was proved by M. Lerch (1887), while two simpler proofs have been found by B. C. Berndt (1972b).

Ramanujan's inequality $|\tau(n)| \leqslant n^{11/2}d(n)$ is a special case of the more general Ramanujan–Petersson conjecture, proved by P. Deligne (1975). The omega result for $\tau(n)$ is due to M. Ram Murty (1982). The results of S. Ramanujan, found in his notebooks (1957) after his untimely death, concern many branches of analytic number theory, including $\zeta(s)$. They are systematically investigated by B. C. Berndt (1977, 1978, 1980, 1983), B. C. Berndt and R. J. Evans (1983), B. C. Berndt and B. M. Wilson (1981a, 1981b), and B. C. Berndt et al. (1981).

For an excellent account of the Riemann, Lindelöf, and Mertens hypothesis, see E. C. Titchmarsh (1951), where many consequences and statements equivalent to these hypotheses are extensively discussed; H. L. Montgomery's work on (1.142) and (1.143) is to be found in his article (1973). Variants of Montgomery's pair-correlation hypothesis may be used to yield conditional results concerning problems involving $p_{n+1} - p_n$, the difference between consecutive primes. For this type of application see J. Mueller (1981), D. R. Heath-Brown (1982), and D. A. Goldston (1981).

Concerning the falsity of $\pi(x) < \text{li } x$, it should be mentioned [see K. Prachar (1967)] that J. E. Littlewood (1914) proved

$$\pi(x) = \text{li } x + \Omega_{\pm}\left(\frac{x^{1/2}}{\log x} \log\log\log x \right),$$

while S. Skewes (1955) found effectively that the first sign change occurs below $10_4(3)$ and R. Sherman Lehman (1966a) improved this value to 1.65×10^{1165}, which is still a very large number.

The theoretical layout concerning calculations relating to the complex zeros of $\zeta(s)$ may be found in H. M. Edwards (1974) or E. C. Titchmarsh (1951). The ordinates of the first six zeros are approximately 14.13, 21.02, 25.01, 30.42, 32.93, and 37.58. At the time of writing of this text, the best published result is that of J. van de Lune and H. J. J. te Riele (1983), whose very elaborate computations show that $\zeta(s)$ has 300,000,001 nontrivial zeros in the region $0 < t < 119,590,809.282$. All these zeros are simple and lie on the critical line, and their number is very accurately given by the main terms in the Riemann–von Mangoldt asymptotic formula (1.44). Numerical investigations related to the Riemann hypothesis were initiated by Riemann himself and have subsequently been of great interest. The number of the first N zeros that are simple and satisfy $t > 0$, $\sigma = \frac{1}{2}$ was historically established as follows:

J. Gram (1903)	$N = \quad 15$
R. Backlund (1914)	$N = \quad 79$
J. I. Hutchinson (1925)	$N = \quad 138$
E. C. Titchmarsh (1935b, 1936)	$N = \quad 1,041$
D. H. Lehmer (1956a, 1956b)	$N = \quad 25,000$
N. A. Meller (1958)	$N = \quad 35,337$
R. Sherman Lehman (1966b)	$N = 250,000$
J. B. Rosser et al. (1969)	$N = \quad 3,500,000$
R. P. Brent (1979)	$N = \quad 81,000,001$
J. van de Lune et al. (1981)	$N = 200,000,001$
J. van de Lune et al. (1983)	$N = 300,000,001$

Despite these impressive numerical data one may assess the difficulties of *proving* the Riemann hypothesis by observing that it took about 60 years to reduce the bound $\mu(\frac{1}{2}) \leqslant \frac{1}{6}$ to $\mu(\frac{1}{2}) \leqslant \frac{1}{6} - \frac{1}{216}$ (see Chapter 7), and the proofs of the recent bounds of $\mu(\frac{1}{2})$ seem to be among the most difficult proofs in analytic number theory in general. The Riemann hypothesis, via its consequence, the Lindelöf hypothesis, implies $\mu(\frac{1}{2}) = 0$ [in fact $\mu(\frac{1}{2}) = 0$ is the Lindelöf hypothesis]. It seems appropriate here to use the quote at the end of S. Chowla's review (MR 28 #1179) of the very long and difficult article of W. Haneke (1962–1963), who proves $\mu(\frac{1}{2}) \leqslant \frac{6}{37}$: "One may ask: If the proof for the exponent $\frac{6}{37}$ needs 74 pages, how many pages will be required to achieve the exponent ε of Lindelöf?".

Theorem 1.12 was proved by J. E. Littlewood (1912), and the proof given in the text follows the one in Chapter 14 of E. C. Titchmarsh (1951).

The density hypothesis (1.137) holds for $\sigma = \frac{1}{2}$ by the Riemann–von Mangoldt formula (1.44), and it will be shown in Chapter 11 that it holds also for $\sigma \geqslant \frac{11}{14}$, while for values of σ closer to one even sharper results are known to be true. The conditional bound (1.138) is due to G. Halász and P. Turán (1969).

P. Turán obtained a number of interesting conditional results involving zero-free regions of $\zeta(s)$ and the Lindelöf hypothesis [see the review articles of G. Halász (1980) and P. Turán (1971)]. A typical result of his is the following one: Let $a, b, c > 0$; $t \geqslant t_0$, and

$$\left| \sum_{N < n \leqslant N'} \Lambda(n) n^{it} \right| \leqslant N t^{-b} \log^c N$$

whenever $t^a \leqslant N < N' \leqslant 2N$. Then $\zeta(\sigma + it) \neq 0$ for $\sigma > 1 - c'b^3 a^{-2}$ and some absolute constant $c' > 0$.

J. Pintz (1982a) proved for $M(x) = \sum_{n \leqslant x} \mu(n)$,

$$\max_{Y/(100 \log Y) \leqslant x \leqslant Y} |M(x)| \geqslant CY^{\beta_0}|\rho_0|^{-3} \qquad (Y > e^{|\gamma_0|+4})$$

where $\rho_0 = \beta_0 + i\gamma_0$ is any complex zero of $\zeta(s)$ and $C > 0$ is an absolute constant. For other results of J. Pintz concerning oscillation results for $M(x)$, see his papers (1980e, 1981a).

F. Mertens (1897, p. 779) conjectured (1.139), and his hypothesis remained one of the classical problems of zeta-function theory until the recent disproof of H. J. J. te Riele and A. Odlyzko. The disproof of the Mertens hypothesis is announced in the short note of te Riele (1983), who kindly informed me of this remarkable achievement while the manuscript of this text was nearing completion. The method of disproof does not actually give a counterexample for (1.139). Instead, it is shown by elaborate computation involving high-speed computers that $\sigma_n(t) = 1.061 \cdots$ for some explicit $t = t_0$ (consisting of 65 digits!); this disproves (1.139) by the work of W. Jurkat and A. Peyerimhoff (1976), where $\sigma_n(t)$ is precisely defined.

The bound in (1.141) is due to R. J. Anderson and H. M. Stark (1981). This was the sharpest result of its kind before the aforementioned investigations of te Riele and Odlyzko, who will very likely improve on (1.141).

/ *CHAPTER TWO* /
EXPONENTIAL INTEGRALS AND EXPONENTIAL SUMS

2.1 EXPONENTIAL INTEGRALS

This chapter is about one of the most important and difficult parts of analytic number theory. Exponential integrals and exponential sums occur in a large number of problems whose solutions ultimately depend on asymptotic formulas or good O-bounds for these integrals or sums. Van der Corput found a deep method for dealing with exponential integrals and sums in the 1920s. This is the so-called saddle-point method or the method of the stationary phase, which has greatly advanced analytic number theory and brought on remarkable improvements in many classical problems, such as divisor problems, circle problem, and the order of the zeta-function in the critical strip. This method is systematized here in Section 2.3 in a simplified form as the theory of (one-dimensional) exponent pairs, and applications of this method are encountered frequently in later chapters.

Albeit the theory of exponent pairs is, in general, superseded by two-dimensional and multidimensional methods for the estimation of exponential sums, this theory is nevertheless fairly simple to use in practice. Furthermore, the best existing theories of multidimensional methods are very difficult and have not yet produced dramatic improvements over results obtainable by the classical (one-dimensional) theory of exponent pairs. Therefore we restrict ourselves primarily to the classical theory of exponent pairs, devoting this section to the estimation of certain exponential integrals and Section 2.2 to general results on exponential sums. It is only in Section 2.4 that we give a brief survey of B. R. Srinivasan's theory of two-dimensional exponent pairs. An important part of the theory of exponential sums is embodied in Voronoi's summation formula, which is discussed extensively in Chapter 3.

We begin our treatment of exponential integrals with the simple

LEMMA 2.1. Let $F(x)$ be a real differentiable function such that $F'(x)$ is monotonic and $F'(x) \geqslant m > 0$ or $F'(x) \leqslant -m < 0$ for $a \leqslant x \leqslant b$. Then

(2.1)
$$\left| \int_a^b e^{iF(x)} \, dx \right| \leqslant 4m^{-1}.$$

Proof of Lemma 2.1. Since the conjugate of $\int_a^b e^{iF(x)} \, dx$ is $\int_a^b e^{-iF(x)} \, dx$, this means that in most problems involving exponential integrals and sums we may suppose that $F'(x) > 0$. Now $e^{iF(x)} = \cos F(x) + i \sin F(x)$, and by the second mean value theorem for real integrals

(2.2)
$$\int_a^b f(x) g(x) \, dx = \begin{cases} f(b) \int_c^b g(x) \, dx, & \text{if } f(x) \geqslant 0, \ f'(x) \geqslant 0 \text{ in } [a, b], \\ f(a) \int_a^c g(x) \, dx, & \text{if } f(x) \geqslant 0, \ f'(x) \leqslant 0 \text{ in } [a, b], \end{cases}$$

where c is some number satisfying $a < c < b$. Writing

$$\int_a^b \cos F(x) \, dx = \int_a^b (F'(x))^{-1} d(\sin F(x))$$

and using (2.2) it is seen that

$$\left| \int_a^b \cos F(x) \, dx \right| \leqslant 2m^{-1},$$

because $(F'(x))^{-1}$ [as the reciprocal of $F'(x)$] is monotonic in $[a, b]$. The same bound also holds for the integral with $\sin F(x)$, hence (2.1) follows. Using (2.2) and the same argument it is also seen that

(2.3)
$$\left| \int_a^b G(x) e^{iF(x)} \, dx \right| \leqslant 4Gm^{-1},$$

where F is as in Lemma 2.1, and $G(x)$ is a positive, monotonic function for $a \leqslant x \leqslant b$ such that $|G(x)| \leqslant G$.

LEMMA 2.2. Let $F(x)$ be a real, twice-differentiable function in $[a, b]$ such that $F''(x) \geqslant m > 0$ or $F''(x) \leqslant -m < 0$. Then

(2.4)
$$\left| \int_a^b e^{iF(x)} \, dx \right| \leqslant 8m^{-1/2}.$$

Proof of Lemma 2.2. Assume that $F''(x) > 0$, so that $F'(x)$ is monotonically increasing and has at most one zero c, that is, $F'(c) = 0$ with $a < c < b$.

We write

$$\int_a^b e^{iF(x)}\,dx = \int_a^{c-u} + \int_{c-u}^{c+u} + \int_{c+u}^b = I_1 + I_2 + I_3,$$

say, where $u > 0$ will be suitably determined. Trivially, $|I_2| \leqslant 2u$, and for $u < c - a$ and $a \leqslant x \leqslant c - u$ we have $|F'(x)| = |\int_x^c F''(t)\,dt| \geqslant um$, so that Lemma 2.1 gives $|I_1| \leqslant 4(um)^{-1}$. A similar estimate holds for I_3 if $u < b - c$. If $u \geqslant c - a$ or $u \geqslant b - c$, or if F' has no zero in $[a, b]$, the analysis is similar, and in all cases leads to

$$\left| \int_a^b e^{iF(x)}\,dx \right| \leqslant 8(um)^{-1} + 2u = 8m^{-1/2}$$

if we take $u = 2m^{-1/2}$. Analogously to (2.3) one has

$$(2.5) \qquad \left| \int_a^b G(x) e^{iF(x)}\,dx \right| \leqslant 8Gm^{-1/2}$$

if F satisfies the hypothesis of Lemma 2.2, and $G(x)$ is a positive, monotonic function for $a \leqslant x \leqslant b$ such that $|G(x)| \leqslant G$.

Lemmas 2.1 and 2.2 are very general, but they have the feature that the estimates given for exponential integrals are only upper bounds which do not explicitly depend on the length of the interval of integration. We present now a "saddle-point" theorem, which shows that the main contribution to the exponential integral comes from its saddle point [i.e., the point where the first derivative $F'(x)$ in $e^{iF(x)}$ vanishes], provided that certain conditions are satisfied. This is

THEOREM 2.1. Suppose that $f(x)$ is a real-valued function such that $f(x) \in C^4[a, b]$, $f''(x) < 0$ for $x \in [a, b]$, and

$$m_2 \asymp |f''(x)|, \qquad |f^{(3)}(x)| \ll m_3, \qquad |f^{(4)}(x)| \ll m_4,$$

where $m_3^2 = m_2 m_4$. If $f'(c) = 0$ for some $a \leqslant c \leqslant b$, then

$$(2.6)$$

$$\int_a^b e(f(x))\,dx = e\left(f(c) - \tfrac{1}{8}\right)|f''(c)|^{-1/2} + O\left(m_2^{-1} m_3^{1/3}\right)$$

$$+ O\left(\min\left(m_2^{-1/2}, |f'(a)|^{-1}\right)\right) + O\left(\min\left(m_2^{-1/2}, |f'(b)|^{-1}\right)\right).$$

If $f''(x) > 0$ in $[a, b]$ and the other hypotheses hold, then the same result is obtained with $e(f(c) + \tfrac{1}{8})$ in place of $e(f(c) - \tfrac{1}{8})$.

Proof of Theorem 2.1. The cases $f'' < 0$ and $f'' > 0$ are analogous, so only the former is considered. We write

(2.7)

$$\int_a^b e(f(x))\,dx = \int_a^{c-u} e(f(x))\,dx + \int_{c-u}^{c+u} e(f(x))\,dx + \int_{c+u}^b e(f(x))\,dx$$

$$= I_1 + I_2 + I_3,$$

say, where we suppose that u satisfies $u \leqslant \min(c - a, b - c)$ and will be suitably determined later. By Lemma 2.1

$$I_1 \ll 1/|f'(c-u)| = 1 \Big/ \Big| \int_c^{c-u} f''(t)\,dt \Big| \ll (um_2)^{-1},$$

and similarly the same bound holds also for I_3.

Since $f'(c) = 0$ we use Taylor's formula to obtain with some θ for which $|\theta| \leqslant 1$,

$$(2.8) \quad I_2 = \int_{-u}^u e(f(x+c))\,dx$$

$$= \int_{-u}^u e\left(f(c) + \frac{x^2}{2!}f''(c) + \frac{x^3}{3!}f^{(3)}(c) + \frac{x^4}{4!}f^{(4)}(c+\theta x) \right) dx$$

$$= e(f(c)) \int_{-u}^u e\left(\tfrac{1}{2}x^2 f''(c) + \tfrac{1}{6}x^3 f^{(3)}(c) \right)\left(1 + O(|x|^4 m_4) \right) dx$$

$$= e(f(c)) \int_{-u}^u e\left(\tfrac{1}{2}x^2 f''(c) + \tfrac{1}{6}x^3 f^{(3)}(c) \right) dx + O(u^5 m_4).$$

Abbreviating $F = \tfrac{1}{3}\pi i f^{(3)}(c)$, the last integral in (2.8) becomes

(2.9)

$$\int_{-u}^u e\left(\tfrac{1}{2}x^2 f''(c) \right)\exp(Fx^3)\,dx = 2\int_0^u \exp(\pi i f''(c)x^2)\left(1 + \sum_{r=1}^\infty \frac{(Fx^3)^{2r}}{(2r)!} \right) dx,$$

since the integrals involving odd powers of r vanish identically. Making the change of variable $\pi|f''(c)|x^2 = y$ in the integrals appearing on the right-hand side of (2.9) we obtain

$$(2.10) \quad \left(\pi|f''(c)| \right)^{-1/2} \int_0^{\pi|f''(c)|u^2} e^{-iy} y^{-1/2}\,dy$$

$$+ \sum_{r=1}^\infty \frac{F^{2r}}{(2r)!} \int_0^{\pi|f''(c)|u^2} e^{-iy} y^{3r-1/2} \left(\pi|f''(c)| \right)^{-3r-1/2} dy.$$

Applying Cauchy's integral theorem to the function $\exp(-iz^2)$ and the sector of the circle of radius $x^{1/2}$, center at the origin, and endpoints $z_1 = x^{1/2}$, $z_2 = e(-1/8)x^{1/2}$, we obtain

$$\int_0^x e^{-iy}y^{-1/2}\,dy = 2\int_0^{x^{1/2}} e^{-iz^2}\,dz = \pi^{1/2}e^{-\pi i/4} + O(x^{-1/2}),$$

if we use $\int_0^\infty \exp(-z^2)\,dz = \tfrac{1}{2}\pi^{1/2}$. Therefore the first term in (2.10) is

$$e\left(-\tfrac{1}{8}\right)|f''(c)|^{-1/2} + O\left((um_2)^{-1}\right),$$

and the remaining terms are by (2.2)

$$\ll \sum_{r=1}^\infty \frac{F^{2r}}{(2r)!}m_2^{-3r-1/2}\left(m_2u^2\right)^{3r-1/2}$$

$$= (um_2)^{-1}\sum_{r=1}^\infty \frac{(Fu^3)^{2r}}{(2r)!} \ll (um_2)^{-1}\exp\left(Cu^3m_3\right)$$

for some absolute $C > 0$. Therefore we obtain

(2.11)
$$\int_a^b e(f(x))\,dx = e(f(c)) - \tfrac{1}{8})|f''(c)|^{-1/2} + O\left(u^{-1}m_2^{-1}\right) + O\left(u^5m_4\right)$$

$$+ O\left(u^{-1}m_2^{-1}\exp\left(Cu^3m_3\right)\right),$$

and choosing $u = (m_2m_4)^{-1/6} = (m_3)^{-1/3}$ the error terms above are of the same order of magnitude. This proves (2.6) when $u \leqslant \min(c - a, b - c)$. If this condition is not satisfied suppose first that $b - u < c \leqslant b$. Proceeding as above it is seen that there is an extra error term

(2.12) $\quad I_4 = e(f(c))\displaystyle\int_{b-c}^u e\left(\tfrac{1}{2}x^2f''(c) + \tfrac{1}{6}x^3f^{(3)}(c)\right)dx$

$$= e(f(c))\int_{b-c}^u e\left(\tfrac{1}{2}x^2f''(c)\right)dx + \sum_{r=1}^\infty \frac{F^r}{r!}\int_{b-c}^u e\left(\tfrac{1}{2}x^2f''(c)\right)x^{3r}\,dx$$

to be dealt with, where as before $F = \tfrac{1}{3}\pi i f^{(3)}(c)$. By (2.3) we have uniformly in r

(2.13) $\quad \displaystyle\int_{b-c}^u e\left(\tfrac{1}{2}x^2f''(c)\right)x^{3r}\,dx \ll u^{3r-1}m_2^{-1},$

while Lemma 2.1 and Lemma 2.2 yield

(2.14) $$\int_{b-c}^{u} e\left(\tfrac{1}{2}x^2 f''(c)\right) dx \ll \min\left(\left|f'(b)\right|^{-1}, m_2^{-1/2}\right),$$

since for $F(x) = \tfrac{1}{2}f''(c)x^2$ and $b - c \leqslant x \leqslant u$,

$$\left|F'(x)\right| = x\left|f''(c)\right| \gg (b - c)m_2 \gg f'(b),$$

because by the mean value theorem for some $c < \xi < b$ we have

$$f'(b) = f'(b) - f'(c) = (b - c)f''(\xi) \ll (b - c)m_2.$$

Therefore (2.13) and (2.14) give for I_4 in (2.12)

$$I_4 \ll \min\left(\left|f'(b)\right|^{-1}, m_2^{-1/2}\right) + \sum_{r=1}^{\infty} (um_2)^{-1} \frac{(Fu^3)^r}{r!}$$

$$\ll \min\left(\left|f'(b)\right|^{-1}, m_2^{-1/2}\right) + (um_2)^{-1}\exp\left(Cu^3 m_3\right),$$

and (2.6) follows again with $u = m_3^{-1/3}$. Similarly, $O(\min(|f'(a)|^{-1}, m_2^{-1/2}))$ appears in (2.6) if $c - u < a$, and (2.6) follows in all cases.

Theorem 2.1 is sharp when m_2 (i.e., the order of f'') is sufficiently large, and enough additional information about $f^{(3)}$ and $f^{(4)}$ is known. On the weaker assumption that $|f''(x)| \asymp m_2$, $f^{(3)}(x) \ll m_3$ for $a \leqslant x \leqslant b$, we can obtain by the method of proof of Theorem 2.1,

(2.15)

$$\int_a^b e(f(x)) dx = e\left(f(c) - \tfrac{1}{8}\right)\left|f''(c)\right|^{-1/2} + O\left(m_2^{-4/5}m_3^{1/5}\right)$$

$$+ O\left(\min\left(\left|f'(a)\right|^{-1}, m_2^{-1/2}\right)\right) + O\left(\min\left(\left|f'(b)\right|^{-1}, m_2^{-1/2}\right)\right),$$

where $f'(c) = 0$ and $f'' < 0$, and if $f'' > 0$ then $e(f(c) - \tfrac{1}{8})$ is to be replaced by $e(f(c) + \tfrac{1}{8})$. The proof of (2.15) (where no information about $f^{(4)}$ is needed) is easier than the proof of (2.6), since for (2.15) only the first three terms in Taylor's formula for I_2 are taken, while for (2.6) we needed the first four terms in Taylor's formula.

Next, we shall formulate and prove another result which is similar to Theorem 2.1. The main difference will be that instead of $f(x)$ we consider $f(z)$, where z is a complex variable lying in a suitable domain, and suppose that $f(z)$ is real when z is real and lies in $[a, b]$. The main term will turn out to be essentially the same one as in (2.6), but the error terms will be different and in certain applications sharper than those in (2.6). The result is

THEOREM 2.2. Let $f(z), \varphi(z)$ be two functions of the complex variable z and $[a, b]$ a real interval such that:

1. For $a \leqslant x \leqslant b$ the function $f(x)$ is real and $f''(x) > 0$.
2. For a certain positive differentiable function $\mu(x)$, defined on $a \leqslant x \leqslant b$, $f(z)$ and $\varphi(z)$ are analytic for $a \leqslant x \leqslant b, |z - x| \leqslant \mu(x)$.
3. There exist positive functions $F(x), \Phi(x)$ defined on $[a, b]$ such that for $a \leqslant x \leqslant b, |z - x| \leqslant \mu(x)$ we have

$$\varphi(z) \ll \Phi(x), \qquad f'(z) \ll F(x)\mu^{-1}(x),$$

$$|f''(z)|^{-1} \ll \mu^2(x) F^{-1}(x),$$

and the $\ll$ constants are absolute.

Let k be any real number, and if $f'(x) + k$ has a zero in $[a, b]$ denote it by x_0. Let the values of $f(x)$, $\varphi(x)$, and so on, at a, x_0, and b be characterized by the suffixes a, 0, and b respectively. Then

(2.16) $\quad \int_a^b \varphi(x) e(f(x) + kx)\, dx = \varphi_0\left(f_0''\right)^{-1/2} e\left(f_0 + kx_0 + \tfrac{1}{8}\right)$

$$+ O\left(\int_a^b \Phi(x) \exp\left[-C|k|\mu(x) - CF(x)\right]\left(dx + |d\mu(x)|\right)\right)$$

$$+ O\left(\Phi_0 \mu_0 F_0^{-3/2}\right) + O\left(\Phi_a\left(|f_a' + k| + f_a''^{1/2}\right)^{-1}\right)$$

$$+ O\left(\Phi_b\left(|f_b' + k| + f_b''^{1/2}\right)^{-1}\right).$$

If $f'(x) + k$ has no zero for $a \leqslant x \leqslant b$, then the terms involving x_0 are to be omitted.

Proof of Theorem 2.2. We shall consider only the more difficult case when $f'(x) + k$ has a zero in $[a, b]$, and we shall denote $\lambda(x) = \alpha\mu(x)$ for a suitable $0 < \alpha < \tfrac{1}{2}$ to be determined later. As in the proof of Theorem 2.1 we shall split the integral on the left-hand side of (2.16) into several integrals. By Cauchy's integral theorem we can replace the path of integration by the contour joining the points $a, a - \lambda_a(1 + i), x_0 - \lambda_0(1 + i), x_0 + \lambda_0(1 + i)$,

$b + \lambda_b(1 + i)$, b. Denoting the corresponding integrals by $I_1, \ldots, I_5$, respectively, we take I_1, I_3, and I_5 along straight lines, and I_2 and I_4 are to be taken along the loci of the points $x \pm \lambda(x)(1 + i)$, respectively.

Therefore for $z = x + (1 + i)y$, $-\lambda(x) \leqslant y \leqslant \lambda(x)$, $a \leqslant x \leqslant b$ we have

$$f(z) + kz = f(x) + kx + (1 + i)y(f'(x) + k) + iy^2f''(x) + \theta(y),$$

where by Taylor's formula

$$\theta(x) \ll F(x) \sum_{n=3}^{\infty} |z - x|^n |f^{(n)}(x)| F^{-1}(x)/n! \ll F(x)|y|^3\mu^{-3}(x),$$

since by (3) and Cauchy's formula for derivatives of analytic functions

$$f^{(n)}(x) \ll F(x)\mu^{-n}(x), \ n = 2, 3, \ldots .$$

Hence by taking α sufficiently small, we obtain $|\theta(y)| < \frac{1}{2}y^2f''(x)$, which gives

$$(2.17) \quad \mathrm{Re}\{2\pi i(f(z) + kz)\} < -2\pi y(f'(x) + k) - \pi y^2 f''(x),$$

and

(2.18)

$$I_1 \ll \int_0^{\lambda_a} \Phi_a \exp\left(-2\pi y|f_a' + k| - \pi y^2 f_a''\right) dy \ll \Phi_a\left(|f_a' + k| + f_a''^{1/2}\right)^{-1},$$

and a similar bound holds for I_5 with the suffix a replaced by b. By the same argument

(2.19)

$$I_2 \ll \int_a^{x_a} \Phi(x)\exp\left(-2\pi\lambda(x)|f'(x) + k| - \pi\lambda^2(x)f''(x)\right)\left(dx + |d\lambda(x)|\right).$$

Now if $|k| \leqslant 2|f'(x)|$, then by (3)

$$\lambda(x)|f'(x) + k| \ll \lambda(x)|f'(x)| \ll F(x),$$

$$k\mu(x) \ll f'(x)\mu(x) \ll F(x),$$

while for $|k| > 2|f'(x)|$

$$\lambda(x)|f'(x) + k| \geqslant \lambda(x)(|k| - |f'(x)|) \gg |k|\mu(x);$$

and since $\lambda^2(x)f''(x) \gg F(x)$ by (3), it is seen that in any event

$$-2\pi\lambda(x)|f'(x) + k| - \pi\lambda^2(x)f''(x) < -C|k|\mu(x) - CF(x).$$

This gives

$$(2.20) \quad I_2 \ll \int_a^{x_a} \Phi(x) \exp\left(-C|k|\mu(x) - CF(x)\right)\left(dx + |d\mu(x)|\right),$$

and a corresponding bound for I_4.

From (2.18) and (2.20) it is seen that it remains yet to show

$$(2.21) \qquad I_3 = \varphi_0 f_0''^{-1/2} e\left(f_0 + kx_0 + \tfrac{1}{8}\right) + O\left(\Phi_0 \mu_0 F_0^{-3/2}\right),$$

and then the proof will be finished. To accomplish this we write

$$I_3 = \int_{-\lambda_0(1+i)}^{\lambda_0(1+i)} \varphi(x_0 + y) e\left(f(x_0 + y) + kx_0 + ky\right) dy$$

$$= \int_{-\lambda_0(1+i)}^{-v(1+i)} + \int_{-v(1+i)}^{v(1+i)} + \int_{v(1+i)}^{\lambda_0(1+i)} = I_{31} + I_{32} + I_{33},$$

say, where we choose

$$(2.22) \qquad\qquad\qquad v = \lambda_0\left(1 + F_0^{1/3}\right)^{-1}.$$

The integrals I_{31} and I_{33} are estimated analogously and yield the error term in (2.21), while the main term in (2.21) will come from I_{32}. By (2.17) and the change of variable $\pi y^2 f_0'' = x$ we have

$$(2.23) \quad I_{33} \ll \Phi_0 \int_v^\infty \exp\left(-\pi y^2 f_0''\right) dy \ll \Phi_0 v^{-1} f_0''^{-1} \int_{\pi v^2 f_0''}^\infty e^{-x} dx$$

$$= \Phi_0\left(v f_0''\right)^{-1} \exp\left(-\pi v^2 f_0''\right).$$

Hence

$$v^2 f_0'' = \lambda_0^2 f_0''\left(1 + F_0^{1/3}\right)^{-2} \gg F_0\left(1 + F_0^{1/3}\right)^{-2}$$

by (3), and also

$$\left(v f_0''\right)^{-1} = v\left(v^2 f_0''\right)^{-1} \ll \mu_0 F_0^{-1}\left(1 + F_0^{1/3}\right).$$

Therefore for $F_0 \geqslant 1$

$$I_{33} \ll \Phi_0 \mu_0 F_0^{-2/3} \exp\left(-CF_0^{1/3}\right),$$

while for $F_0 < 1$,

$$I_{33} \ll \Phi_0 \mu_0 F_0^{-1}$$

so that in any case

$$(2.24) \qquad\qquad\qquad I_{31} + I_{33} \ll \Phi_0 \mu_0 F_0^{-3/2}.$$

The estimation of I_{32} bears resemblance to the estimation of I_2 in (2.8). In both cases Taylor's formula is used and the fact that the first derivative vanishes at a certain point ("saddle point"). In I_{32} we write

(2.25)
$$e(f(x_0 + y) + k(x_0 + y))$$
$$= e\left(f_0 + f_0'y + \tfrac{1}{2}f_0''y^2 + \tfrac{1}{6}f_0^{(3)}y^3 + \sum_{r=4}^{\infty} \frac{f_0^{(r)}y^r}{r!} + kx_0 + ky \right).$$

Now by hypothesis $f'(x_0) + k = f_0' + k = 0$, and using $f^{(r)}(x) \ll \mu^{-r}(x)F(x)$ and $e^{iu} = 1 + iu + O(u^2)$ (u real), we see that the left-hand side of (2.25) is equal to

(2.26)
$$e\left(f_0 + kx_0 + \tfrac{1}{2}f_0''y^2 \right)\left\{ 1 + \tfrac{1}{3}\pi i f_0^{(3)}y^3 + O\left(y^6 F_0^2 \mu_0^{-6} \right) + O\left(y^4 F_0 \mu_0^{-4} \right) \right\}.$$

Next, by a change of variable $y = (1 + i)Y$ we have

$$\int_{-v(1+i)}^{v(1+i)} y^{2k}\exp(i\pi y^2 f'')\,dy = \int_{-v}^{v} Y^{2k}\exp(-2\pi Y^2 f_0'')(1 + i)^{2k+1}\,dy$$
$$\ll \left(f_0'' \right)^{-k-1/2} \ll F_0^{-k-1/2}\mu_0^{2k+1}$$

for $k > 0$ a fixed integer, so that the contribution of the error terms in (2.26) to I_{32} will be (since the integrals with odd powers vanish)

(2.27)
$$\ll \Phi_0 \mu_0 F_0^{-3/2}.$$

As regards the terms that remain in (2.26) we have (in view of $\varphi_0' \ll \Phi_0 \mu_0^{-1}$)

$$\varphi(x_0 + y)\left(1 + \tfrac{1}{6}y^3 f_0^{(3)}\right) = \varphi_0 + y\varphi_0' + \tfrac{1}{6}y^3 \varphi_0 f_0^{(3)} + O\left(y^2 \Phi_0 \mu_0^{-2} \right)$$
$$+ O\left(y^4 \Phi_0 F_0 \mu_0^{-4} \right).$$

Arguing as before it is seen that we are left with

(2.28)
$$\int_{-v(1+i)}^{v(1+i)} \varphi(x_0 + y)\left(1 + \tfrac{1}{6}y^3 f_0^{(3)}\right)\exp(\pi i y^2 f_0'')\,dy$$
$$= \varphi_0 \int_{-v(1+i)}^{v(1+i)} \exp(\pi i y^2 f_0'')\,dy + O\left(\Phi_0 \mu_0 F_0^{-3/2} \right) + O\left(\Phi_0 F_0 \mu_0^{-4} f_0''^{-5/2} \right)$$
$$= \varphi_0 \int_{-\infty(1+i)}^{\infty(1+i)} \exp(\pi i y^2 f_0'')\,dy + O\left(\Phi_0 \left| \int_{v(1+i)}^{\infty(1+i)} \exp(\pi i y^2 f_0'')\,dy \right| \right)$$
$$+ O\left(\Phi_0 \mu_0 F_0^{-3/2} \right) = \varphi_0 \left(f_0'' \right)^{-1/2} e^{\pi i/4} + O\left(\Phi_0 \mu_0 F_0^{-3/2} \right),$$

since by (A.38) and $y = (1 + i)Y$,

$$\int_{-\infty(1+i)}^{\infty(1+i)} \exp(\pi i y^2 f_0'') \, dy = (1 + i) \int_{-\infty}^{\infty} \exp(-2\pi Y^2 f_0'') \, dY$$

$$= (1 + i)(\pi/2\pi f_0'')^{1/2} = (f_0'')^{-1/2} e(\tfrac{1}{8}),$$

and where as in (2.23),

$$\int_{v(1+i)}^{\infty(1+i)} \exp(\pi i y^2 f_0'') \, dy \ll \mu_0 F_0^{-3/2}.$$

This completes the proof of (2.16) in case $f'(x) + k$ vanishes in $[a, b]$. In the other case we take the contour of integration as $a, a \pm \lambda_a(1 + i), b \pm \lambda_b(1 + i)$, b depending on whether $f'(x) + k \geqslant 0$ or $\leqslant 0$ in $[a, b]$, and then there is no term corresponding to I_3. Also, similarly as in Theorem 2.1, if all the hypotheses of Theorem 2.2 hold but $f''(x) < 0$ in $[a, b]$, then the main term in (2.16) is

$$\varphi_0 |f_0''|^{-1/2} e(f_0 + kx_0 - \tfrac{1}{8}).$$

Theorem 2.1 is essentially the same as Theorem 2.2 with $f(x)$ instead of $f(x) + k$, $\varphi(x) \equiv 1$ and different hypotheses on f which lead to different error terms.

We finish this section by presenting a lemma which involves double exponential integrals with no saddle point. This is

LEMMA 2.3. Let $f(z)$ and $g(z)$ be two functions of the complex variable z such that:

1. $f(x)$ is real for $a \leqslant x \leqslant b$.
2. $f(z)$ and $g(z)$ are regular for $|z - x| \leqslant \mu$ for some $\mu > 0$ and all $x \in [a, b]$.
3. $g(z) \ll G$, $|f'(z)| \asymp M$ for $|z - x| \leqslant \mu$.

Let $0 < U < \tfrac{1}{2}(b - a)$. Then for some absolute $A > 0$

(2.29)

$$U^{-1} \int_0^U \left(\int_{a+u}^{b-u} g(x) e(f(x)) \, dx \right) du \ll G e^{-A\mu M}(b - a + \mu) + GM^{-2}U^{-1}.$$

Proof of Lemma 2.3. By (3) and continuity, $f'(x)$ is of the same sign, say, positive, in $[a, b]$. Let $C(u)$ denote the contour of segments joining the points $a + u, a + u + i\alpha\mu, b - u + i\alpha\mu, b - u$, where $0 < \alpha < \tfrac{1}{2}$ is a number which will be specified in a moment. By (3) and Cauchy's formula for derivatives of analytic functions, we have $f^{(n)}(z) \ll M\mu^{1-n}$ for $n \geqslant 2$, and hence by Taylor's formula for $z = x + iy \in C(u)$,

$$f(z) = f(x) + iyf'(x) + \theta(x, y),$$

$$\theta(x, y) \ll \sum_{n=2}^{\infty} |z - x|^n |f^{(n)}(x)|/n! \ll My^2\mu^{-1},$$

and so

(2.30) $\qquad \qquad \operatorname{Im} f(x + iy) \gg My \qquad [x + iy \in C(u)]$

if α is chosen sufficiently small, since $|y| \leqslant \alpha\mu$ for $z \in C(u)$.
  By Cauchy's integral theorem

(2.31)

$$U^{-1} \int_0^U \int_{a+u}^{b-u} g(x) e(f(x))\, dx\, du = U^{-1} \int_0^U \left(\int_{C(u)} g(z) e(f(z))\, dz \right) du.$$

In view of (2.30) the integral over the horizontal side of $C(u)$ is

$$\int_{a+u+i\alpha\mu}^{b-u+i\alpha\mu} g(z) e(f(z))\, dz \ll (b - a) G e^{-A\mu M}$$

uniformly in u, with some absolute $A > 0$. For the vertical side joining $a + u$ and $a + u + i\alpha\mu$ we have

(2.32) $\qquad \left| U^{-1} \int_0^U \left(\int_{a+u}^{a+u+i\alpha\mu} g(z) e(f(z))\, dz \right) du \right|$

$$= \left| U^{-1} \int_0^{\alpha\mu} \left(\int_{a+iy}^{a+U+iy} g(z) e(f(z))\, dz \right) dy \right|,$$

if we write $z = x + iy = a + u + iy$, $dz = idy$, and $0 \leqslant y \leqslant \alpha\mu$ and invert the order of integration. An application of Cauchy's integral theorem to the rectangle with vertices $a + iy, a + i\alpha\mu, a + U + i\alpha\mu$, and $a + U + iy$ gives in view of (3) and (2.30) that

$$\int_{a+iy}^{a+U+iy} g(z) e(f(z))\, dz \ll G\left(\int_y^{\infty} e^{-AMv}\, dv + U e^{-A\mu M} \right)$$

$$\ll G(M^{-1}e^{-AMy} + U e^{-A\mu M}),$$

and therefore the left-hand side of (2.32) is

$$\ll GU^{-1} \int_0^{\alpha\mu} (M^{-1}e^{-AMy} + U e^{-A\mu M})\, dy \ll GM^{-2}U^{-1} + G\mu e^{-A\mu M}.$$

  A similar estimate can be obtained for the vertical side joining $b - u + i\alpha\mu$ and $b - u$, and the case $f'(x) < 0$ is dealt with analogously by taking the contour in the lower half-plane. This proves (2.29).

2.2 EXPONENTIAL SUMS

By exponential sums we shall mean here sums of the type $\Sigma_{a<n\leqslant b}e(f(n))$, where $f(x)$ is real for $a \leqslant x \leqslant b$, $b - a \geqslant 1$, and f is sufficiently many times differentiable. We begin with a result which transforms an exponential sum into a sum of exponential integrals, which are easier to estimate in view of the results of the preceding section. This is

LEMMA 2.4. Let $f(x)$ be a real function for $a \leqslant x \leqslant b$ such that $f(x) \in C^2[a, b]$ and $f''(x) < 0$ in $[a, b]$, and let $f'(b) = \alpha$, $f'(a) = \beta$. Then for any η with $0 < \eta < 1$ we have

(2.33)

$$\sum_{a<n\leqslant b} e(f(n)) = \sum_{\alpha-\eta<m<\beta+\eta} \int_a^b e(f(x) - mx)\, dx + O(\log(\beta - \alpha + 2)).$$

Proof of Lemma 2.4. By the Euler–Maclaurin summation formula (A.23)

(2.34)

$$\sum_{a<n\leqslant b} e(f(n)) = \int_a^b e(f(x))\, dx + 2\pi i \int_a^b \psi(x)f'(x)e(f(x))\, dx + O(1),$$

where $\psi(x) = x - [x] - \frac{1}{2}$. Without loss of generality we may suppose $\eta - 1 < \alpha \leqslant \eta$ (so that $m \geqslant 0$), for if k is an integer such that $\eta - 1 < \alpha - k \leqslant \eta$, then (2.33) becomes with $h(x) = f(x) - kx$,

(2.35) $$\sum_{a<n\leqslant b} e(f(n)) = \sum_{a<n\leqslant b} e(h(n))$$

$$= \sum_{\alpha'-\eta<m-k<\beta'+\eta} \int_a^b e(h(x) - (m - k)x)\, dx$$

$$+ O(\log(\beta' - \alpha' + 2)),$$

where $\alpha' = \alpha - k$ and $\beta' = \beta - k$. Thus (2.35) implies (2.33), and $m - k \geqslant 0$ by the choice of k. Using the Fourier expansion (A.27) for $\psi(x)$ it can be seen that the second integral in (2.34) is equal to

$$-2i \sum_{m=1}^{\infty} \int_a^b m^{-1}\sin(2\pi mx)e(f(x))f'(x)\, dx$$

$$= \sum_{m=1}^{\infty} m^{-1} \int_a^b (e(-mx) - e(mx))e(f(x))f'(x)\, dx$$

$$= \sum_{m=1}^{\infty} (2\pi im)^{-1} \int_a^b \frac{f'(x)}{f'(x) - m}\, d(e(f(x) - mx))$$

$$- \sum_{m=1}^{\infty} (2\pi im)^{-1} \int_a^b \frac{f'(x)}{f'(x) + m}\, d(e(f(x) + mx)).$$

By hypothesis $f'(x)$ is monotonically decreasing for $a \leqslant x \leqslant b$, and so is then also $f'(x)/(f'(x) + m)$. An application of (2.2) to the second integral above shows that it is $\ll \beta/(\beta + m)$ uniformly in m, so that the whole sum is

$$\ll \sum_{m=1}^{\infty} \beta m^{-1} (\beta + m)^{-1} \ll \sum_{m \leqslant \beta} m^{-1} + \sum_{m > \beta} \beta m^{-2} \ll 1 + \log(\beta + 2).$$

Similarly, it is seen that

$$\sum_{m \geqslant \beta + \eta} m^{-1} \int_a^b \frac{f'(x)}{f'(x) - m} d(e(f(x) - mx)) \ll \sum_{m \geqslant \beta + \eta} m^{-1}(m - \beta)^{-1} \beta$$

$$\ll \sum_{\beta + \eta \leqslant m \leqslant 2\beta} (m - \beta)^{-1} + \sum_{m \geqslant 2\beta} m^{-2} \beta \ll 1 + \log(\beta + 2).$$

It remains yet to estimate

$$\sum_{1 \leqslant m \leqslant \beta + \eta} (2\pi i m)^{-1} \int_a^b \frac{f'(x)}{f'(x) - m} d(e(f(x) - mx))$$

$$= \sum_{1 \leqslant m < \beta + \eta} m^{-1} \int_a^b f'(x) e(f(x) - mx) \, dx$$

$$= (2\pi i)^{-1} \sum_{1 \leqslant m < \beta + \eta} m^{-1} \int_a^b d(e(f(x) - mx))$$

$$+ \sum_{1 \leqslant m < \beta + \eta} \int_a^b e(f(x) - mx) \, dx$$

$$= O(\log(\beta + \eta)) + \sum_{1 \leqslant m < \beta + \eta} \int_a^b e(f(x) - mx) \, dx.$$

Taking into account the first integral on the right-hand side of (2.34) we finally obtain

$$\sum_{a < n \leqslant b} e(f(n)) = \sum_{0 \leqslant m < \beta + \eta} \int_a^b e(f(x) - mx) \, dx + O(\log(\beta + 2)),$$

which by the discussion made concerning (2.35) proves (2.33).

LEMMA 2.5. Let $f(x)$ be a real function for $a \leqslant x \leqslant b$ and let $H > 0$. Then

(2.36)

$$\sum_{a < n \leqslant b} e(f(n)) \ll (b - a)H^{-1/2} + H$$

$$+ \left((b - a)H^{-1} \sum_{h=1}^{H-1} \left| \sum_{a < n \leqslant b - h} e(f(n+h) - f(n)) \right| \right)^{1/2}.$$

Proof of Lemma 2.5. We may suppose that a and b are integers and $2 \leqslant H < b - a$, since trivially $|\sum_{a < n \leqslant b} e(f(n))| \leqslant b - a + 1$ and $b - a \ll (b - a)H^{-1/2}$ for $H < 2$, while for $H \geqslant b - a$ the left-hand side of (2.36) is trivially $O(H)$. Also we may suppose that H is an integer, since the right-hand side of (2.36) remains unchanged in magnitude if H is replaced by the integer nearest to it. Therefore the proof reduces to showing that

(2.37)

$$\sum_{a < n \leqslant b} e(f(n)) \ll (b - a)H^{-1/2}$$

$$+ \left((b - a)H^{-1} \sum_{h=1}^{H-1} \left| \sum_{a < n \leqslant b - h} e(f(n+h) - f(n)) \right| \right)^{1/2},$$

where H is an integer such that $2 \leqslant H \leqslant b - a$, and $a < b$ are integers. Observe that

(2.38) $$H \sum_{a < n \leqslant b} e(f(n)) = \sum_{m=1}^{H} \sum_{n=a-m+1}^{b-m} e(f(m+n)),$$

and define $f(k) = 0$ if k is an integer such that $k \leqslant a$ or $k > b$. Then writing $S = \sum_{a < n \leqslant b} e(f(n))$ and inverting the order of summation in (2.38) we obtain

(2.39) $$HS = \sum_{n=a+1-H}^{b-1} \sum_{m=1}^{H} e(f(m+n)),$$

so that n takes at most $b - a + H \leqslant 2(b - a)$ values. Applying the Cauchy–Schwarz inequality we obtain from (2.39),

(2.40) $$H^2 S^2 \leqslant 2(b - a) \sum_{n=a+1-H}^{b-1} \left| \sum_{m=1}^{H} e(f(m+n)) \right|^2.$$

Squaring out the modulus in (2.40) we obtain

$$(2.41) \qquad \sum_{n=a+1-H}^{b-1} \left| \sum_{m=1}^{H} e(f(m+n)) \right|^2$$

$$\leqslant 2(b-a)H + 2 \left| \sum_{n=a+1-H}^{b-1} \sum_{1 \leqslant r < s \leqslant H} e(f(n+s) - f(n+r)) \right|.$$

In the last sum in (2.41) for a fixed k, h such that $1 \leqslant h \leqslant H - 1$, $a < k \leqslant b - h$, we have $f(n+s) - f(n+r) = f(k+h) - f(k)$ exactly $H - h$ times, that is, for $r = 1, 2, \ldots, H - h$, $s = r + h$, $n = k - r$. Hence the modulus of the double sum in (2.41) does not exceed

$$(2.42) \qquad \left| \sum_{h=1}^{H-1} (H-h) \sum_{a < k \leqslant b-h} e(f(k+h) - f(k)) \right|$$

$$\leqslant H \sum_{h=1}^{H-1} \left| \sum_{a < n \leqslant b-h} e(f(n+h) - f(n)) \right|,$$

and thus (2.37) follows from (2.39) to (2.42).

LEMMA 2.6. Let $k \geqslant 2$ be a fixed integer and let $f(x) \in C^k[a, b]$. If $b \geqslant a + 1$, $K = 2^{k-1}$, and for $a \leqslant x \leqslant b$

$$0 < \lambda_k \leqslant f^{(k)}(x) \leqslant A\lambda_k \qquad \left[\text{or } \lambda_k \leqslant -f^{(k)}(x) \leqslant A\lambda_k, A > 1 \right],$$

then

$$(2.43) \qquad \sum_{a < n \leqslant b} e(f(n)) \ll A^{2/K}(b-a)\lambda_k^{1/(2K-2)} + (b-a)^{1-2/K}\lambda_k^{-1/(2K-2)}.$$

Proof of Lemma 2.6. We may suppose that $\lambda_k < 1$, for otherwise the result is trivial. If $k = 2$, then Lemma 2.2 and Lemma 2.4 give

$$\sum_{a < n \leqslant b} e(f(n)) \ll (\beta - \alpha + 1)\lambda_2^{-1/2} + \log(\beta - \alpha + 2)$$

$$\ll (\beta - \alpha + 1)\lambda_2^{-1/2} \ll |f'(a) - f'(b)|\lambda_2^{-1/2} + \lambda_2^{-1/2}$$

$$\ll \lambda_2 A(b-a)\lambda_2^{-1/2} + \lambda_2^{-1/2},$$

and this is (2.43) for $k = 2$.

If $k \geqslant 2$ we use induction, setting for h fixed $g(x) = f(x+h) - f(x)$. Then by the mean value theorem,

$$h\lambda_k \leqslant g^{(k-1)}(x) \leqslant Ah\lambda_k;$$

hence (2.43) with $k - 1$ for k gives

$$\sum_{a < n \leqslant b-h} e(g(n)) \ll A^{4/K}(b-a)(h\lambda_k)^{1/(K-2)} + (b-a)^{1-4/K}(h\lambda_k)^{-1/(K-2)}.$$

Summing this estimate for $h = 1, \ldots, H$, choosing

$$H = \left[\lambda_k^{-1/(K-1)}\right] + 1,$$

and using (2.37), we obtain after some simplifying (2.43). Instead of using induction one may apply (2.37) $k - 2$ times and then use (2.43) with $k = 2$. Choosing then the parameters in a suitable way one obtains that the sum in (2.43) is, for $k \geqslant 3$,

$$\ll (b-a)\lambda_k^{1/(2K-2)} + (b-a)^{1-1/K} + (b-a)^{1-4/K+4/K^2}\lambda_k^{-1/K}.$$

Here the $\ll$ constant depends on A only, and this bound is often sharper than (2.43).

Finally, we need a lemma which transforms an exponential sum into another exponential sum (plus error terms), and this new exponential sum is in many cases easier to estimate than the original sum. This is

LEMMA 2.7. Suppose that $f(x) \in C^4[a, b]$, $f'(x)$ is monotonically decreasing in $[a, b]$, $f'(b) = \alpha$, $f'(a) = \beta$. If x_ν is defined by $f'(x_\nu) = \nu$ ($\alpha < \nu \leqslant \beta$ and ν is an integer) and

$$|f''(x)| \asymp m_2, f^{(3)}(x) \ll m_3, f^{(4)}(x) \ll m_4 \qquad (m_3^2 = m_2 m_4),$$

then

(2.44)

$$\sum_{a < n \leqslant b} e(f(n)) = e(-\tfrac{1}{8}) \sum_{\alpha < \nu \leqslant \beta} |f''(x_\nu)|^{-1/2} e(f(x_\nu) - \nu x_\nu)$$

$$+ O(m_2^{-1/2}) + O((b-a)m_3^{1/3}) + O(\log((b-a)m_2 + 2)).$$

Proof of Lemma 2.7. We use Lemma 2.4, noting that by the mean value theorem

(2.45) $$\beta - \alpha \ll (b-a)m_2.$$

By Lemma 2.2 the limits of summation coming from Lemma 2.4 may be replaced by $\alpha + 1$ and $\beta - 1$ with an error $\ll m_2^{-1/2}$. An application of

Theorem 2.1 then gives

$$\sum_{\alpha+1<\nu<\beta-1} \int_a^b e(f(x) - \nu x)\, dx$$

$$= e(-\tfrac{1}{8}) \sum_{\alpha+1<\nu<\beta-1} |f''(x_\nu)|^{-1/2} e(f(x_\nu) - \nu x_\nu)$$

$$+ O\left(\sum_{\alpha+1<\nu<\beta-1} m_2^{-1} m_3^{1/3} \right)$$

$$+ O\left(\sum_{\alpha+1<\nu<\beta-1} \left((\nu - \alpha)^{-1} + (\beta - \nu)^{-1} \right) \right).$$

In view of (2.45) the first O-term above is $O((b - a)m_3^{1/3})$, and the second is $O(\log(\beta - \alpha + 2)) = O(\log((b - a)m_2 + 2))$, which ends the proof of (2.44), since again by Lemma 2.2 the limits of summation $(\alpha + 1, \beta - 1)$ may be changed to (α, β) with an error which is $\ll m_2^{-1/2}$. We should note that if we use (2.15) instead of (2.6) (with the appropriate hypotheses on f, of course), then we obtain (2.44) with the error term $O((b - a)m_3^{1/3})$ replaced by $O((b - a)m_2^{1/5}m_3^{1/5})$. Also, if f' is monotonically increasing in $[a, b]$ and the other hypotheses of Lemma 2.1 are the same but $f'(a) = \alpha$, $f'(b) = \beta$, then (2.44) remains true with $e(-\tfrac{1}{8})$ replaced by $e(\tfrac{1}{8})$.

2.3 THE THEORY OF EXPONENT PAIRS

We have now at our disposal two results, namely Lemma 2.5 and Lemma 2.7, which enable us to transform a given exponential sum into another exponential sum plus some (usually manageable) error terms. Lemma 2.5 requires practically no conditions on f, while Lemma 2.7 is much more restrictive and contains error terms. However, the conditions imposed on the derivatives of f in Lemma 2.7 allow us, for a large class of functions f (which occur in many important applications), to combine Lemmas 2.5 and 2.7 successfully several times and to obtain good upper bounds for the modulus of the exponential sum

$$(2.46) \qquad S = \sum_{B<n\leqslant B+h} e(f(n)) \qquad (B \geqslant 1, 1 < h \leqslant B).$$

The results of Section 2.2 suggest that the estimation of the sum S in (2.46) certainly depends on the number of summands, which is $\leqslant B$, and on the order of the first derivative of f. Therefore we shall suppose that

$$(2.47) \qquad A \ll |f'(x)| \ll A \qquad (A > \tfrac{1}{2}).$$

when $B \leqslant x \leqslant 2B$, and we seek an upper bound for $|S|$ of the form

(2.48)
$$S \ll A^{\varkappa}B^{\lambda}.$$

The pair of nonnegative real numbers $(\varkappa, \lambda)$ will be called an *exponent pair* if (2.47) and (2.48) hold and

(2.49)
$$0 \leqslant \varkappa \leqslant \tfrac{1}{2} \leqslant \lambda \leqslant 1.$$

Two remarks may be made here: (1) $(\varkappa, \lambda) = (0, 1)$ is trivially an exponent pair; and (2) exponent pairs obviously form a convex set. This is to be understood in the following sense: If $(\varkappa_1, \lambda_1)$ and $(\varkappa_2, \lambda_2)$ are arbitrary exponent pairs and $0 \leqslant t \leqslant 1$ is arbitrary, then

(2.50)
$$S = S'S^{1-t} \ll A^{\varkappa_1 t + (1-t)\varkappa_2} B^{\lambda_1 t + (1-t)\lambda_2},$$

which implies that

(2.51)
$$\left(\varkappa_1 t + (1-t)\varkappa_2, \lambda_1 t + (1-t)\lambda_2\right) \qquad (0 \leqslant t \leqslant 1)$$

is also an exponent pair. The above definition of exponent pairs is too general, and to obtain exponent pairs of practical value (via Lemma 2.7) we shall suppose that besides (2.47) $f(x) \in C^r[B, 2B]$ for some $r \geqslant 5$, and moreover that the derivatives of $f(x)$ for $B \leqslant x \leqslant 2B$ satisfy

(2.52)
$$AB^{1-r} \ll |f^{(r)}(x)| \ll AB^{1-r} \qquad (r = 1, 2, \ldots),$$

where the $\ll$ constants in (2.52) depend on r alone.

We may consider only the case $f'(x) > 0$ for $B \leqslant x \leqslant 2B$, since otherwise we may consider $\bar{S}$ instead of S with the effect that f is replaced by $-f$ and the sign of f' is thus changed. To obtain the first nontrivial exponent pair we apply Lemma 2.4 to S, estimating each integral as $\ll m_2^{-1/2} \ll (A/B)^{-1/2}$ by Lemma 2.2. This gives

(2.53) $\quad S \ll (\beta - \alpha)A^{-1/2}B^{1/2} + A^{-1/2}B^{1/2} + \log(A + 2) \ll (AB)^{1/2},$

since $\beta - \alpha = f'(a) - f'(b) \ll A$ and $A \gg 1$. Therefore it follows that $(\varkappa, \lambda) = (\tfrac{1}{2}, \tfrac{1}{2})$ is an exponent pair. Here (2.52) was used with $r = 1$ and $r = 2$ only. Thus we have so far $(0, 1)$ and $(\tfrac{1}{2}, \tfrac{1}{2})$ as exponent pairs, plus exponent pairs which may be formed from these two by convexity [in the sense of (2.51)]. We denote this set of exponent pairs by E_1. Further exponent pairs may be obtained by the following

LEMMA 2.8. If $(\varkappa, \lambda)$ is an exponent pair, then

$$(k, l) = \left(\frac{\varkappa}{2\varkappa + 2}, \frac{1}{2} + \frac{\lambda}{2\varkappa + 2} \right)$$

is also an exponent pair.

Proof of Lemma 2.8. First note that $0 \leqslant k \leqslant \frac{1}{2} \leqslant l \leqslant 1$, since $0 \leqslant \varkappa \leqslant \frac{1}{2} \leqslant \lambda \leqslant 1$ by hypothesis. An application of Lemma 2.5 gives

$$(2.54) \quad S^2 \ll B^2 H^{-1} + H^2 + BH^{-1} \sum_{j=1}^{H-1} \left| \sum_{B < n \leqslant B+h-j} e(f(n+j) - f(n)) \right|,$$

where $H > 0$ will be suitably chosen. For a fixed j we write

$$(2.55) \quad g(n) = f(n+j) - f(n),$$

and note that

$$g^{(r)}(x) = f^{(r)}(x+j) - f^{(r)}(x) = jf^{(r+1)}(x + \theta j) \qquad (r \geqslant 1, |\theta| \leqslant 1),$$

so that by (2.52), for $B \leqslant x \leqslant B + h - j$,

$$(2.56) \quad jAB^{-r} \ll g^{(r)}(x) \ll jAB^{-r} \qquad (r \geqslant 1).$$

Now we may suppose that $A > B^{1/2}$, since for $A \leqslant B^{1/2}$ we use the fact that $\lambda \geqslant \frac{1}{2}$ and $(\frac{1}{2}, \frac{1}{2})$ is an exponent pair to obtain

$$S \ll A^{1/2} B^{1/2} = A^{1/2} B^{1/2 + \lambda/(2\varkappa+2)} B^{-\lambda/(2\varkappa+2)} \ll A^{(1+\varkappa-2\lambda)/(2\varkappa+2)} B^l \ll A^k B^l.$$

The condition $A > B^{1/2}$ is needed in the case when for some suitable $c > 0$ we have $j < cBA^{-1}$, so that by (2.56) $|g'| < 1/2$, $g'' \asymp jAB^{-2} \ll 1$. Then by Lemma 1.2 and Lemma 2.2,

$$BH^{-1} \sum_{j < cB/A} \left| \sum_n g(n) \right| \ll BH^{-1} A^{-1/2} B \sum_{j < cB/A} j^{-1/2} \ll B^2 H^{-1},$$

since $A > B^{1/2}$. For the remaining j's in (2.54) we use the already existing exponent pair $(\varkappa, \lambda)$ and (2.56) to obtain

$$\sum_{B < n \leqslant B+h-j} e(f(n+j) - f(n)) \ll (jAB^{-1})^\varkappa B^\lambda.$$

Hence by (2.54)

$$(2.57) \quad S^2 \ll B^2 H^{-1} + H^2 + A^\varkappa B^{1-\varkappa+\lambda} H^\varkappa,$$

and the choice

$$H = B^{(1+\varkappa-\lambda)/(\varkappa+1)} A^{-\varkappa/(\varkappa+1)}$$

reduces (2.57) to

(2.58) $\qquad S^2 \ll A^{\varkappa/(1+\varkappa)} B^{(1+\varkappa+\lambda)/(1+\varkappa)} + B^{2(1+\varkappa-\lambda)/(1+\varkappa)} A^{-2\varkappa/(1+\varkappa)}.$

Since $0 \leqslant \varkappa \leqslant 1/2 \leqslant \lambda \leqslant 1$ we have

$$\frac{1+\varkappa+\lambda}{1+\varkappa} = 1 + \frac{\lambda}{1+\varkappa} \geqslant \frac{4}{3}$$

$$\geqslant 2\left(1 - \frac{\lambda}{1+\varkappa}\right) = \frac{2(1+\varkappa-\lambda)}{1+\varkappa},$$

so that in view of $A > \frac{1}{2}$ the second term in (2.58) does not exceed the first and Lemma 2.8 follows.

Now we denote by E_2 the set of exponent pairs obtainable from E_1, convexity, and repeated application of Lemma 2.8, which always produces a new exponent pair from a given pair $(\varkappa, \lambda)$. The proof of Lemma 2.8 shows that for the construction of E_2 we needed (2.52) with $r \leqslant 3$. The set E_2 does not exhaust our possibilities for constructing exponent pairs, and for what follows it will be useful to note that for $(\varkappa, \lambda) \in E_2$ we have

(2.59) $\qquad\qquad\qquad \varkappa + 2\lambda \geqslant \frac{3}{2}.$

This is trivial if $(\varkappa, \lambda) \in E_1$, and moreover convexity obviously preserves (2.59). With $k = \varkappa/(2\varkappa + 2)$ and $l = \frac{1}{2} + \lambda/(2\varkappa + 2)$ it is, however, readily checked that $k + 2l \geqslant \frac{3}{2}$ since $\lambda \geqslant \frac{1}{2}$.

Finally, the last possibility for constructing exponent pairs is furnished by

LEMMA 2.9. If $(\varkappa, \lambda)$ is an exponent pair for which (2.59) holds, then

$$(k, l) = \left(\lambda - \tfrac{1}{2}, \varkappa + \tfrac{1}{2}\right)$$

is also an exponent pair.

Proof of Lemma 2.9. The condition $0 \leqslant k \leqslant \frac{1}{2} \leqslant l \leqslant 1$ is trivial in view of (2.49). We shall apply Lemma 2.7 with $a = B$, $b = B + h$, $m_2 = AB^{-1}$, $m_3 = AB^{-2}$, and $m_4 = AB^{-3}$ so that $m_3^2 = m_2 m_4$ holds, and we may suppose that $f''(x) < 0$. The case $f''(x) > 0$ is discussed at the end of proof of Lemma 2.7 and will lead to the same final estimate. We have then

(2.60) $\quad S = e\left(-\tfrac{1}{8}\right) \sum_{\alpha < \nu \leqslant \beta} |f''(x_\nu)|^{-1/2} e\left(f(x_\nu) - \nu x_\nu\right) + O\left(A^{-1/2} B^{1/2}\right)$

$$+ O(\log(A + 2)) + O\left((AB)^{1/3}\right),$$

and the main task is to estimate

$$(2.61) \quad S_1 = \sum_{\alpha < \nu \leqslant \beta} |f''(x_\nu)|^{-1/2} e(g(\nu)), \qquad g(\nu) = f(x_\nu) - \nu x_\nu.$$

If we set $f'(x) = y$ and denote its inverse function by $x = h(y)$, then $g(y) = f(h(y)) - yh(y)$, which gives

$$g'(y) = f'(h(y))h'(y) - h(y) - yh'(y)$$

$$= f'(x)h'(y) - h(y) - yh'(y) = -h(y)$$

and

$$g''(y) = -h'(y) = -1/f''(x) = -1/f''(h(y))$$

if one uses $f'(h(y)) = y$, and likewise

$$g^{(3)}(y) = f^{(3)}(h(y))(f''(h(y)))^{-3}.$$

In general, $g^{(r)}(y)$ is found from $g''(y)f''(h(y)) = -1$ by applying Leibniz's rule for the rth derivative of a product. We have $h(y) \asymp B$ and therefore

$$(2.62) \qquad BA^{1-r} \ll |g^{(r)}(y)| \ll BA^{1-r} \qquad (r \leqslant 3),$$

and by induction it may be seen that the upper bound in (2.62) holds also for $r \geqslant 4$. Removing f'' by partial summation it further follows that S_1 in (2.61) is of the same type as S, only A and B are interchanged. Hence

$$(2.63) \qquad\qquad S_1 \ll A^{-1/2}B^{1/2}B^{\varkappa}A^{\lambda}$$

and

$$(2.64) \qquad S \ll A^kB^l + (AB)^{1/3} \qquad (k = \lambda - \tfrac{1}{2}, l = \varkappa + \tfrac{1}{2}),$$

and the proof will be finished if it can be shown that $(AB)^{1/3} \leqslant A^kB^l$. Since $l \geqslant \tfrac{1}{2}$, this is obvious if $k \geqslant \tfrac{1}{3}$. If $k < \tfrac{1}{3}$, then for $B \geqslant A^{(1-3k)/(3l-1)}$ we have

$$(AB)^{1/3} = A^kB^{1/3}A^{1/3-k} \leqslant A^kB^{1/3}B^{(3l-1)/3} = A^kB^l.$$

For $B < A^{(1-3k)/(3l-1)}$ we use $2k + l \geqslant 1$, which follows from $\varkappa + 2\lambda \geqslant \tfrac{3}{2}$. This gives

$$S \leqslant B = B^lB^{1-l} < B^lA^{(1-3k)(1-l)/(3l-1)} \leqslant A^kB^l,$$

ending the proof of Lemma 2.9, where (2.52) was needed for $r \leqslant 3$ and the upper bound of (2.52) for $r = 4$.

In view of the preceding discussion we formalize the concept of exponent pairs now even more by introducing E, the set of exponent pairs, as the set of all pairs $(\varkappa, \lambda)$ obtainable from E_2, convexity, and Lemma 2.9 applied a finite number of times. Nearly 60 years of research have not been able to produce any other exponent pairs, that is, any besides those of E. Though in the formulation of Lemmas 2.8 and 2.9 it was tacitly assumed that $(\varkappa, \lambda)$ belongs to E_1 and E_2, respectively, this is not necessarily true, as the proofs of these lemmas clearly show. It seems appropriate now to introduce three processes which will be denoted by A, B, and $C(t)$ [the letters A and B have no relation to (2.48) in this context], and which correspond to Lemma 2.8, Lemma 2.9, and convexity, respectively. Therefore if $(\varkappa, \lambda)$ and $(\varkappa_1, \lambda_1)$ are exponent pairs, let

$$(2.65) \qquad A(\varkappa, \lambda) = \left(\varkappa/(2\varkappa + 2), \tfrac{1}{2} + \lambda/(2\varkappa + 2)\right),$$

$$(2.66) \qquad B(\varkappa, \lambda) = (\lambda - \tfrac{1}{2}, \varkappa + \tfrac{1}{2}).$$

$$(2.67) \qquad C(t)(\varkappa, \lambda)(\varkappa_1, \lambda_1) = \left(\varkappa t + \varkappa_1(1 - t), \lambda t + \lambda_1(1 - t)\right)$$

$$(0 \leqslant t \leqslant 1).$$

Now we can restate the foregoing discussion and present formally the set of exponent pairs in

THEOREM 2.3. Let E denote the set of pairs of real numbers $(\varkappa, \lambda)$ such that $0 \leqslant \varkappa \leqslant \tfrac{1}{2} \leqslant \lambda \leqslant 1$ and $(\varkappa, \lambda)$ is obtained by a finite number of applications of the processes A, B, and $C(t)$ [defined by (2.65), (2.66), and (2.67)] to $(0, 1)$, which is to be considered as an element of E. Then E is the set of exponent pairs in the sense that (2.48) holds, provided that (2.52) holds with $r \geqslant 4$.

We end this section by listing several commonly used exponent pairs. These are: $(\tfrac{1}{2}, \tfrac{1}{2}) = B(0, 1)$; $(\tfrac{1}{6}, \tfrac{2}{3}) = A(\tfrac{1}{2}, \tfrac{1}{2})$; $(\tfrac{2}{7}, \tfrac{4}{7}) = BA(\tfrac{1}{6}, \tfrac{2}{3})$; $(\tfrac{4}{18}, \tfrac{11}{18}) = BA(\tfrac{2}{7}, \tfrac{4}{7})$; $(\tfrac{11}{30}, \tfrac{16}{30}) = BA^2(\tfrac{1}{6}, \tfrac{2}{3})$; $(\tfrac{13}{40}, \tfrac{22}{40}) = BA^2(\tfrac{2}{7}, \tfrac{4}{7})$; $(\tfrac{97}{251}, \tfrac{132}{251}) = BA^3(\tfrac{13}{40}, \tfrac{22}{40})$; $(\tfrac{13}{31}, \tfrac{16}{31}) = BAB(\tfrac{11}{30}, \tfrac{16}{30})$; $(\tfrac{5}{24}, \tfrac{15}{24}) = C(\tfrac{1}{4})(\tfrac{1}{6}, \tfrac{2}{3}), (\tfrac{4}{18}, \tfrac{11}{18})$; $(\tfrac{4}{11}, \tfrac{6}{11}) = C(\tfrac{12}{33})(\tfrac{1}{2}, \tfrac{1}{2})(\tfrac{2}{7}, \tfrac{4}{7})$.

It may be remarked that in actual problems where the theory of exponent pairs is applied it often seems unclear how to choose $(\varkappa, \lambda)$ in an optimal way, that is, to minimize a certain function $F(\varkappa, \lambda)$. In the case of the general F this problem is difficult and hitherto unsolved, but for $F(x, y) = x + y$ it has been solved by R. A. Rankin (1955), who showed that if $\alpha = 0.3290213568\ldots$, then $(\varkappa, \lambda) = (\tfrac{1}{2}\alpha + \varepsilon, \tfrac{1}{2} + \tfrac{1}{2}\alpha + \varepsilon)$ is an exponent pair for which (up to ε) $\varkappa + \lambda$ is minimal for all $(\varkappa, \lambda) \in E$. Graphically, the exponent pairs just discussed may be arranged in a table as follows:

x	0	$\frac{1}{2}$	$\frac{1}{6}$	$\frac{2}{7}$	$\frac{4}{18}$	$\frac{11}{30}$	$\frac{13}{40}$	$\frac{13}{31}$	$\frac{4}{11}$	$\frac{5}{24}$	$\frac{97}{251}$	$\alpha/2 + \varepsilon$
λ	1	$\frac{1}{2}$	$\frac{2}{3}$	$\frac{4}{7}$	$\frac{11}{18}$	$\frac{16}{30}$	$\frac{22}{40}$	$\frac{16}{31}$	$\frac{6}{11}$	$\frac{15}{24}$	$\frac{132}{251}$	$\frac{1}{2} + \alpha/2 + \varepsilon$

2.4 TWO-DIMENSIONAL EXPONENT PAIRS

We conclude this chapter by presenting an outline (without proofs) of B. R. Srinivasan's theory of two-dimensional exponent pairs for the estimation of exponential sums of the form

$$(2.68) \qquad S = \sum_{(m,n) \in \mathscr{D}} e(f(m,n)),$$

where $\mathscr{D}$ is a suitable region in the (x, y) plane. This theory is not the sharpest known, and the recent methods of G. Kolesnik supersede it. Nevertheless, it is in general stronger than the one-dimensional theory of exponent pairs, and besides it is relatively simple to apply.

The real function $g(x, y)$ is said to be an approximation of degree r to the real function $f(x, y)$ in a region $\mathscr{D}$ if f and g possess partial derivatives up to r orders in $\mathscr{D}$ and

$$\left| f_{x^p y^q} - g_{x^p y^q} \right| < c g_{x^p y^q}$$

for $(x, y) \in \mathscr{D}$ and $1 \leqslant p + q \leqslant r$, where c is a constant such that $0 < c < \frac{1}{2}$. In this case we write $f \overset{r}{\underset{\mathscr{D}}{\sim}} g$.

The pair of nonnegative reals (l_0, l_1) is a two-dimensional exponent pair if $0 \leqslant l_0, l_1 - l_0 \leqslant \frac{1}{6}$ and to every set of real numbers s, t such that $st \neq 0$,

$$(\mu + \mu_1)s + (\mu + \mu_2)t + \mu + \mu' + 1 \neq 0,$$

where $\mu, \mu' \geqslant 0$ are integers and μ_1, μ_2 is either 0 or 1, there exists an integer $r(\geqslant 6)$ depending on s, t such that the sum S in (2.68) is $O((zw)^{l_0}(ab)^{1-l_1})$ with respect to s, t whenever the following conditions hold. $\mathscr{D}$ is a region contained in the rectangle $a < m \leqslant ua$, $b < n \leqslant ub$ ($u > 1$),

$$z = |vs|a^{-s-1}b^{-t} \gg 1, \ w = |vt|a^{-s}b^{-t-1} \gg 1, \ (a, b \gg 1)$$

$$f \overset{r}{\underset{\mathscr{D}}{\sim}} vx^{-s}y^{-t}.$$

Trivially, $(0,0)$ is a two-dimensional exponent pair, while if (λ_0, λ_1) is a two-dimensional exponent pair, (l_0, l_1) is also, where for $K = 2^k$, $k \geqslant 1$,

$$l_0 = \frac{\frac{1}{2} - \lambda_1}{K + 4(K-1)(\frac{1}{2} - \lambda_1)}, \qquad l_1 = \frac{\frac{1}{2} - \lambda_0 + k(\frac{1}{2} - \lambda_1)}{K + 4(K-1)(\frac{1}{2} - \lambda_1)}.$$

Analogously to one-dimensional exponent pairs the set of two-dimensional exponent pairs also forms a convex set: If $0 \leqslant t \leqslant 1$ and (l_0', l_1'), (l_0'', l_1'') are two two-dimensional exponent pairs, then

$$(l_0, l_1) = \left(tl_0' + (1-t)l_0'', tl_1' + (1-t)l_1''\right)$$

is also a two-dimensional exponent pair.

Some commonly used two-dimensional exponent pairs are:

$$(l_0, l_1) = \left(\tfrac{1}{20}, \tfrac{3}{20}\right); \left(\tfrac{1}{44}, \tfrac{4}{44}\right); \left(\tfrac{1}{8}, \tfrac{1}{4}\right); \left(\tfrac{2}{24}, \tfrac{5}{24}\right); \left(\tfrac{2}{56}, \tfrac{7}{56}\right); \left(\tfrac{23}{250}, \tfrac{56}{250}\right).$$

Notes

Our first result on exponential sums and integrals is in fact Lemma 1.2, which was a prerequisite for the proof of Theorem 1.8.

The results presented in this chapter have their counterparts in Chapters 4 and 5 of E. C. Titchmarsh (1951), but the material given here is more extensive and the results sharper. In particular, Titchmarsh does not present the theory of exponent pairs, but stops at what is essentially Lemma 2.8 applied several times to the exponent pair $(\frac{1}{2}, \frac{1}{2})$; this is our Lemma 2.6 or Titchmarsh's Theorem 5.13.

The theory of exponent pairs and exponential integrals and sums was founded by J. G. van der Corput (1921, 1922) in the 1920s as one of the deepest theories of analytic theory ever made. Van der Corput (1922) contains the estimate

$$\Delta(x) \ll x^{33/100 + \varepsilon},$$

where $\Delta(x)$ is the error term in the classical divisor problem, for which the reader is referred to Chapters 3 and 13. The exponent $\frac{33}{100}$ improved on the previous value $\frac{1}{3}$, due to G. F. Voronoi. The exponent $\frac{1}{3}$ appeared also in the circle problem (i.e., determining θ such that $\sum_{m^2 + n^2 \leqslant x} 1 - \pi x \ll x^{\theta + \varepsilon}$; see Chapter 13 for a more extensive discussion of the circle problem). Until van der Corput's results appeared many competent mathematicians believed that $\frac{1}{3}$ was the natural limit of the exponent in the divisor and circle problem. Van der Corput's research opened a path in analytic number theory which leads to good bounds in many important problems, and forms the basis for more advanced methods.

In presenting the theory of (one-dimensional) exponent pairs we have followed primarily the work of E. Phillips (1933), since the original form of van der Corput's theory was rather complicated. Van der Corput's definition of exponent pairs involved a condition comparable with (2.52):$(\varkappa, \lambda)$ is an exponent pair if (2.49) holds and if, corresponding to every $s > 0$, there exist two numbers r and c depending only on s ($r \geqslant 4$ is an integer and $0 < c < \frac{1}{2}$) such that

$$(2.69) \qquad \sum_{a < n \leqslant b} e(f(n)) \ll z^\varkappa a^\lambda$$

holds with the $\ll$ constants depending only on s and u, where $u < 1, 1 \leqslant a < b < au, y > 0$, $z = ya^{-s} > 1$, $f(n)$ is any real function with the first r derivatives in the interval $a \leqslant n \leqslant b$ (a,b integers), and for $a \leqslant n \leqslant b, 0 \leqslant j \leqslant r - 1$,

(2.70)

$$\left| f^{(j+1)}(n) - (-1)^j ys(s+1) \cdots (s+j-1) n^{-s-j} \right| < cys(s+1) \cdots (s+j-1) n^{-s-j}.$$

It may be remarked that z is effectively $f'(a)$, so that (2.69) is in fact the same as (2.48), and the only difference is between (2.52) and (2.70), which both express the same type of inequalities for the derivatives of f. It is, however, (2.52) which is simpler to verify, and thus the definition of the exponent pair made in the text seems to be more practical than van der Corput's definition, although in most common applications [e.g., the divisor problem, the order of $\zeta(s)$ in the critical strip, etc.] the function f is easily seen to satisfy both definitions of exponent pairs.

In (2.47) one can suppose $A > \frac{1}{2}$, for otherwise S in (2.46) may be estimated by Lemmas 1.2 and 2.1. The condition $\frac{1}{2} \leqslant \lambda \leqslant 1$ in (2.49) must be satisfied for general exponent pairs. This follows by considering mean value estimates and Ω-results (see Chapters 5, 7, and 9).

Lemmas 2.1–2.6 (except Lemma 2.3) are to be found in Chapters 4 and 5 of E. C. Titchmarsh (1951), where also a proof of (2.15) may be found. Lemma 2.3, concerning certain integrals with no saddle points, is due to M. Jutila (1984a). This result will serve as a useful device for the truncation of certain series in Chapter 7, after application of the Voronoi summation formula to certain sums. The bound stated at the end of the proof of Lemma 2.6 was pointed out to the author by S. W. Graham.

Great simplifications in van der Corput's theory were introduced by E. Phillips (1933), whose proofs of Lemmas 2.7–2.9 are essentially given here, and the theory of exponent pairs is brought to a readily applicable form in Theorem 2.3 at the end of Section 2.3. Theorem 2.1 is also due to E. Phillips (1933), while Theorem 2.2 is due to F. V. Atkinson (1949) and will be used in Chapters 4 and 7 for transformations of certain Dirichlet polynomials. Theorem 2.2 also plays an important role in the proof of Atkinson's formula, discussed in Chapter 15.

The main step in the proof of Lemma 2.5 is the proof of the inequality (2.37), which is originally due to H. Weyl. This inequality is of a general nature and rests on a judicious use of the Cauchy–Schwarz inequality. In fact the sum appearing on the left-hand side of (2.42) is a double exponential sum (since $H - h$ can be removed by partial summation), so that Lemma 2.5 in fact transforms an (ordinary) exponential sum into a double exponential sum with the flexibility that H may be chosen suitably to minimize the estimates. Thus in (2.42) one may see the genesis of two- and multidimensional methods for the estimation of exponential sums.

The method of two-dimensional sums was developed in the 1930s by E. C. Titchmarsh (1931a, 1931b, 1932, 1934a, 1934b, 1934c, 1935a, 1942), where he obtained several improvements of exponents in classical problems such as the order of $\zeta(\frac{1}{2} + it)$ and the circle problem. One of his early results, which may be regarded as the two-dimensional analogue of Lemma 2.1 is as follows: Let $\mathscr{D}$ be a finite region bounded by $O(1)$ continuous monotonic arcs, which is included in the square $|x| \leqslant R, |y| \leqslant R$ for some $R \geqslant 2$. Let further $f_{xx}(x, y) > 0, f_{yy}(x, y) < 0$ (or $f_{xx} < 0$, $f_{yy} > 0$) and $f_{xy}(x, y) \geqslant b > 0$ throughout $\mathscr{D}$. Then

$$\iint_{\mathscr{D}} e(f(x, y)) \, dx \, dy \ll b^{-1}(\log R + |\log b|).$$

Later refinements of two-dimensional methods were effected by many mathematicians, including S. H. Min (1949), H.-E. Richert (1953), and W. Haneke (1962–1963), and the best methods at present are those of G. Kolesnik (1967, 1969, 1973, 1979, 1981a, 1981b, 1982, and in press). These advanced techniques, which do not seem to have appeared in book form yet, are very complicated and are in general based on the works of previous authors, so that anyone who wants to get acquainted with these techniques must already know some of the theory. An attempt to create the theory of n-dimensional exponent pairs has been made by B. R. Srinivasan (1963, 1965, 1973). As in the case of his theory of two-dimensional exponent pairs, outlined in Section 2.4, Srinivasan's theory is readily applicable, but the method does not seem to have been developed to its full potential. Namely, he fails to take full advantage of the Cauchy–Schwarz inequality (i.e., an appropriate multidimensional analogue of Lemma 2.5), and therefore his results are not as sharp as one would like. The methods of G. Kolesnik seem to be more powerful and in each specific problem lead to a sharper result then Srinivasan's methods do: Witness Kolesnik's recent improvement (1981b) of the exponent $\frac{105}{407} = 0.257985\ldots$ in the problem of finite nonisomorphic abelian groups of B. R. Srinivasan (1973) (see Chapter 14 for a discussion of this problem) to $\frac{97}{381} = 0.254593\ldots$. Kolesnik's methods may be perhaps still bettered, and his paper (in press) contains a presentation of a general theory of n-dimensional exponent pairs.

A discussion of results obtainable by the method of (one-dimensional) exponent pairs was made by R. A. Rankin (1955), where he showed that the best exponent (up to ε) that this method (at present) allows in the classical divisor problem is $0.3290213568\ldots$, while the best published result [due to G. Kolesnik (1982); see also Chapter 13] is $\frac{35}{108} = 0.32407407407\ldots$. All the existing two- and multidimensional methods do not improve very much on results obtainable by the classical method of exponent pairs, and this was one of the main reasons why our discussion in this chapter was confined mostly to the latter. It is probably the very generality of the existing methods for the estimation of exponential sums which prevents one from obtaining the results of the desired degree of sharpness in each specific problem (e.g., attaining the exponent $\frac{1}{4} + \varepsilon$ in the divisor problem), and very likely significant progress in specific problems in the future will depend on the ability to put to full advantage the particular structure of the exponential sum in question.

/ CHAPTER THREE /
THE VORONOI SUMMATION FORMULA

3.1 INTRODUCTION

At the beginning of this century G. F. Voronoi proved two remarkable formulas concerning the error term in the divisor problem. Besides giving an explicit expression for the error term $\Delta(x)$, Voronoi derived a general summation formula for sums involving the divisor function $d(n)$. The formulas of Voronoi express finite arithmetical sums by infinite series containing the Bessel functions, and they are

$$(3.1) \quad \Delta(x) = \sum_{n \leqslant x}{}' d(n) - x(\log x + 2\gamma - 1) - \tfrac{1}{4}$$

$$= -2\pi^{-1}x^{1/2} \sum_{n=1}^{\infty} d(n)n^{-1/2}\left(K_1(4\pi\sqrt{nx}) + \tfrac{1}{2}\pi Y_1(4\pi\sqrt{nx})\right),$$

and

$$(3.2)$$

$$\sum_{a \leqslant n \leqslant b}{}' d(n)f(n) = \int_a^b (\log x + 2\gamma)f(x)\,dx + \sum_{n=1}^{\infty} d(n)\int_a^b f(x)\alpha(nx)\,dx,$$

where

$$(3.3) \qquad \alpha(x) = 4K_0(4\pi x^{1/2}) - 2\pi Y_0(4\pi x^{1/2}).$$

Here $0 < a < b < \infty$, $f(x) \in C^2[a, b]$, and in $\sum'$ in (3.1) the last term is to be halved if x is an integer. Similarly, in (3.2) $\sum'$ means that if a or b is an integer, then $\tfrac{1}{2}d(a)f(a)$ or $\tfrac{1}{2}d(b)f(b)$ is to be counted instead of $d(a)f(a)$

and $d(b)f(b)$, respectively. The series in (3.1) and (3.2) are boundedly convergent when x or a and b lie in a fixed closed subinterval of $(0, \infty)$. The functions K_0, Y_0, K_1, and Y_1 are the familiar Bessel functions with power-series expansions

$$(3.4) \quad K_0(z) = -\left(\log\left(\frac{z}{2}\right) + \gamma\right)I_0(z) + \sum_{m=1}^{\infty} \frac{(z/2)^{2m}}{(m!)^2}\left(1 + \frac{1}{2} + \cdots + \frac{1}{m}\right),$$

$$(3.5) \quad Y_0(z) = \frac{2}{\pi}\left(\log\left(\frac{z}{2}\right) + \gamma\right)J_0(z)$$

$$+ \frac{2}{\pi}\sum_{m=1}^{\infty} \frac{(-1)^{m-1}(z/2)^{2m}}{(m!)^2}\left(1 + \frac{1}{2} + \cdots + \frac{1}{m}\right),$$

(3.6)

$$K_1(z) = \sum_{m=0}^{\infty} \frac{(z/2)^{2m+1}}{m!\,(m+1)!}\left\{\log\left(\frac{z}{2}\right) - \frac{1}{2}\psi(m+1) - \frac{1}{2}\psi(m+2)\right\} + z^{-1},$$

(3.7)

$$Y_1(z) = \pi^{-1}\sum_{m=0}^{\infty} (-1)^m \frac{(z/2)^{2m+1}}{m!\,(m+1)!}\left\{2\log\left(\frac{z}{2}\right) - \psi(m+1) - \psi(m+2)\right\}$$

$$- \frac{2}{\pi z}.$$

Here J_0 and I_0 are special cases of the Bessel functions J_p and I_p, which are defined for any fixed real p as

$$(3.8) \qquad J_p(z) = \sum_{k=0}^{\infty} \frac{(-1)^k(z/2)^{p+2k}}{k!\,\Gamma(p+k+1)},$$

$$I_p(z) = e\left(-\frac{p}{4}\right)J_p(iz) = \sum_{k=0}^{\infty} \frac{(z/2)^{p+2k}}{k!\,\Gamma(p+k+1)}.$$

Moreover, in this chapter only, ψ is defined as

$$(3.9) \qquad\qquad \psi(z) = \Gamma'(z)/\Gamma(z),$$

since this is standard notation in the theory of Bessel functions. From $\Gamma(1+z) = z\Gamma(z)$ and $\psi(1) = -\gamma$ [see (A.32)] it follows that for $n \geqslant 2$

$$\psi(n) = 1 + \tfrac{1}{2} + \cdots + 1/(n-1) - \gamma.$$

The general Bessel functions K_n and Y_n for any integer n are defined as

$$(3.10) \qquad K_n(z) = \frac{(-1)^n}{2} \left(\frac{\partial I_{-p}(z)}{\partial p} - \frac{\partial I_p(z)}{\partial p} \right)_{p=n},$$

$$Y_n(z) = \frac{1}{\pi} \left(\frac{\partial J_p(z)}{\partial p} - (-1)^n \frac{\partial J_{-p}(z)}{\partial p} \right)_{p=n},$$

and they satisfy many useful relations. In particular K_0, K_1 and Y_0, Y_1 are connected by

$$(3.11) \qquad \frac{d}{dx}(xK_1(x)) = -xK_0(x), \qquad \frac{d}{dx}(xY_1(x)) = xY_0(x).$$

The practical value of the formulas (3.1) and (3.2) lies in the fact that the Bessel functions appearing in them admit sharp asymptotic approximations involving elementary functions, which are valid for $|z|$ large and $|\arg z| < \pi$. Defining for $p \geqslant 0$ real and $m \geqslant 0$ an integer

$$(p, m) = \frac{\Gamma(p + m + \frac{1}{2})}{m!\,\Gamma(p - m + \frac{1}{2})}$$

$$= \frac{(4p^2 - 1^2)(4p^2 - 3^2) \cdots (4p^2 - (2m-1)^2)}{2^{2m}m!},$$

we have for, $|z|$ large and $|\arg z| < \pi$,

$$(3.12) \qquad K_p(z) \sim \left(\frac{\pi}{2z} \right)^{1/2} e^{-z} \sum_{m=0}^{\infty} (p, m)(2z)^{-m},$$

$$(3.13) \quad Y_p(z) \sim \left(\frac{2}{\pi z} \right)^{1/2} \left\{ \sin\left(z - \frac{\pi p}{2} - \frac{\pi}{4} \right) \sum_{m=0}^{\infty} (-1)^m (p, 2m)(2z)^{-2m} \right.$$

$$\left. + \cos\left(z - \frac{\pi p}{2} - \frac{\pi}{4} \right) \sum_{m=0}^{\infty} (-1)^m (p, 2m + 1)(2z)^{-2m-1} \right\},$$

$$(3.14) \quad J_p(z) \sim \left(\frac{2}{\pi z} \right)^{1/2} \left\{ \cos\left(z - \frac{\pi p}{2} - \frac{\pi}{4} \right) \sum_{m=0}^{\infty} (-1)^m (p, 2m)(2z)^{-2m} \right.$$

$$\left. - \sin\left(z - \frac{\pi p}{2} - \frac{\pi}{4} \right) \sum_{m=0}^{\infty} (-1)^m (p, 2m + 1)(2z)^{-2m-1} \right\}.$$

The symbol $\sim$ means here that in (3.12)–(3.14) equality holds if in the sums over m we stop at any finite term and multiply it by $1 + O(|z|^{-1})$. With the above formulas we obtain from (3.3)

$$(3.15) \quad \alpha(nx) = -2^{1/2}x^{-1/4}n^{-1/4}\left\{\sin(4\pi\sqrt{nx} - \pi/4)\right.$$

$$\left. - (32\pi)^{-1}(nx)^{-1/2}\cos(4\pi\sqrt{nx} - \pi/4)\right\} + O(n^{-5/4}x^{-5/4}),$$

which is sufficiently sharp for most applications of the summation formula (3.2).

3.2 THE TRUNCATED VORONOI FORMULA

There are no simple proofs of (3.1) and (3.2), where delicate questions of convergence are involved. The most difficult case in (3.1) is when x is an integer, but in most applications the distinction as to whether x is an integer or not is unimportant, since $d(n) \ll n^\varepsilon$ for any $\varepsilon > 0$. In practice it is useful to have a truncated form of (3.1), and it may be shown that

$$(3.16) \quad \Delta(x) = -2\pi^{-1}x^{1/2} \sum_{n \leqslant N} d(n)n^{-1/2}\left(K_1(4\pi\sqrt{nx}) + \frac{\pi}{2}Y_1(4\pi\sqrt{nx}) \right)$$

$$+ O(x^\varepsilon) + O(x^{1/2+\varepsilon}N^{-1/2}),$$

which in view of (3.12) and (3.13) may be replaced by the simpler expression

$$(3.17) \quad \Delta(x) = (\pi\sqrt{2})^{-1}x^{1/4} \sum_{n \leqslant N} d(n)n^{-3/4}\cos(4\pi\sqrt{nx} - \pi/4)$$

$$+ O(x^\varepsilon) + O(x^{1/2+\varepsilon}N^{-1/2}).$$

Here N is a (sufficiently large) parameter which may be suitably chosen. The choice $N = x^{1/3}$ implies immediately $\Delta(x) \ll x^{1/3+\varepsilon}$, while letting $N \to \infty$ in (3.16) we obtain (3.1) in a weaker form with the error term $O(x^\varepsilon)$ present. The proof of (3.16) starts from the inversion formula (A.10). We have

$$\sum_{n \leqslant x}' d(n) = (2\pi i)^{-1} \int_{c-iT}^{c+iT} \zeta^2(s)x^s s^{-1} ds + O\left(x^c T(c-1)^{-2}\right) + O(x^{1+\varepsilon}T^{-1}),$$

where $c = 1 + 1/\log x$, $T^2/(4\pi^2 x) = N + \frac{1}{2}$, and N is an integer. Here N is the same parameter which appears in (3.16) and (3.17), since in those formulas it is irrelevant whether N is an integer or not if $N \ll x^A$ for some fixed $A > 0$, which we henceforth assume. The contour in the above integral is replaced by the contour joining the points $c \pm iT$, $-a \pm iT$ ($a > 0$). Allowing for the poles at $s = 0$ and $s = 1$, we obtain by the residue theorem

(3.18) $$\Delta(x) = (2\pi i)^{-1} \sum_{n=1}^{\infty} d(n) \int_{-a-iT}^{-a+iT} \chi^2(s) n^{s-1} x^s s^{-1} \, ds$$

$$+ O(x^\varepsilon) + O(T^{2a} x^{-2a}) + O(x^{1+\varepsilon} T^{-1}),$$

where the functional equation (1.23) was used, and $\zeta^2(1-s)$ was replaced by the absolutely convergent series which may be integrated term by term, so that (3.18) follows on estimating the integrals over the horizontal segments. Using Lemma 2.1 and the asymptotic formula (1.25) for $\chi(s)$ it is seen that the contribution of $\sum_{n>N}$ in (3.18) is contained in the error terms, and writing

$$\int_{-a-iT}^{-a+iT} \chi^2(s) n^{s-1} x^s s^{-1} \, ds = \int_{-i\infty}^{i\infty} - \left(\int_{iT}^{i\infty} + \int_{-i\infty}^{-iT} + \int_{-iT}^{-a-iT} + \int_{-a+iT}^{iT} \right),$$

we obtain, after estimating the integrals in the bracket above either trivially or by Lemma 2.1,

(3.19)

$$\Delta(x) = (2\pi i)^{-1} \sum_{n \leqslant N} d(n) n^{-1} \int_{-i\infty}^{i\infty} 2^{2s} \pi^{2s-2} \sin^2(\pi s/2) \Gamma^2(1-s)(nx)^s s^{-1} \, ds$$

$$+ O(x^\varepsilon) + O(x^{1/2+\varepsilon} N^{-1/2}),$$

where we chose $a = \varepsilon$, N satisfies $1 \ll N \ll x^A$ ($A > 0$ arbitrary), and we used (1.24). The above transformations of $\Delta(x)$ were necessary, since the change of variable $s = 1 - w$ in (3.19) shows that (3.16) follows from

$$-(\pi^2 n)^{-1} \int_{1-i\infty}^{1+i\infty} \cos^2(\pi w/2) \Gamma(w) \Gamma(w-1)(2\pi\sqrt{nx})^{2-2w} \, dw$$

$$= -4i(x/n)^{1/2} \left(K_1(4\pi\sqrt{nx}) + (\pi/2) Y_1(4\pi\sqrt{nx}) \right),$$

or by writing $2\pi\sqrt{nx} = X$ we have to show that

(3.20)

$$(2\pi i)^{-1} \int_{1-i\infty}^{1+i\infty} \cos^2(\pi w/2) \Gamma(w) \Gamma(w-1) X^{2-2w} \, dw$$

$$= X(K_1(2X) + (\pi/2) Y_1(2X)).$$

To see that (3.20) holds note that

(3.21) $f(x) = x^{-1}Y_1(x),$ $F(s) = 2^{s-2}\pi^{-1}\Gamma\left(\dfrac{s}{2}\right)\Gamma\left(\dfrac{s}{2} - 1\right)\cos\dfrac{\pi s}{2}$

$$\left(2 < \sigma < \tfrac{5}{2}\right)$$

and

(3.22) $f(x) = x^{-1}K_1(x),$ $F(s) = 2^{s-3}\Gamma\left(\dfrac{s}{2}\right)\Gamma\left(\dfrac{s}{2} - 1\right)$ $(\sigma > 2)$

are respective Mellin transforms in the sense of (A.1) and (A.3). Thus by (A.3)

$$\frac{\pi}{2}X^{-1}Y_1(X) + X^{-1}K_1(X)$$

$$= (2\pi i)^{-1}\int_{3-i\infty}^{3+i\infty} 2^{s-3}\Gamma\left(\frac{s}{2}\right)\Gamma\left(\frac{s}{2} - 1\right)\left(\cos\frac{\pi s}{2} + 1\right)X^{-s}\,ds$$

$$= 2(2\pi i)^{-1}\int_{1-i\infty}^{1+i\infty} 2^{2w-3}\Gamma(w)\Gamma(w - 1)(\cos\pi w + 1)X^{-2w}\,dw$$

$$= (2\pi i)^{-1}\int_{1-i\infty}^{1+i\infty} 2^{2w-1}\cos^2(\pi w/2)\Gamma(w)\Gamma(w - 1)X^{-2w}\,dw,$$

since the integrand in the first integral above is regular for $\sigma > 0$. Replacing X by $2X$ we obtain

$$\frac{\pi}{2}X^{-1}Y_1(2X) + X^{-1}K_1(2X)$$

$$= (2\pi i)^{-1}\int_{1-i\infty}^{1+i\infty}\cos^2(\pi w/2)\Gamma(w)\Gamma(w - 1)X^{-2w}\,dw,$$

and finally multiplication by X^2 proves (3.20) and therefore (3.16) too. A similar formula may be derived by this method for $\Delta_k(x)$, the error term in the asymptotic formula for $\sum_{n \leqslant x} d_k(n)$ (see Chapter 13), and for $k \geqslant 2$ fixed one obtains

(3.23) $\Delta_k(x) \ll x^{(k-1)/2k}\left|\displaystyle\sum_{n \leqslant N} d_k(n)n^{-(k+1)/2k}e\left(k(xn)^{1/k}\right)\right|$

$$+ x^\varepsilon + x^{(k-1+\varepsilon)/k}N^{-1/k}.$$

3.3   THE WEIGHTED VORONOI FORMULA

An effective way to prove both (3.1) and (3.2) is to consider the weighted sum

(3.24) $D_{q-1}(x) = \dfrac{x^{q-1}}{\Gamma(q)}\displaystyle\sum_{n \leqslant x}(1 - n/x)^{q-1}d(n)$ $(q \geqslant 1),$

and find an asymptotic expansion of this sum (q is a fixed number, not necessarily an integer) in terms of some "generalized Bessel functions." This line of approach has been successfully used by several authors, and we shall follow here the work of A. L. Dixon and W. L. Ferrar (1931). By the inversion formula (A.14) we have

$$D_{q-1}(x) = (2\pi i)^{-1} \int_{2-i\infty}^{2+i\infty} \zeta^2(s) \frac{x^{s+q-1}\Gamma(s)}{\Gamma(s+q)} \, ds,$$

which is the starting point for the evaluation of $D_{q-1}(x)$. For $q > 2$ and $0 < c < \min(\frac{1}{2}, \frac{1}{2}q - 1)$ it is seen by Stirling's formula (A.34) that the line of integration in the above integral may be replaced by the line $\operatorname{Re} s = -c$, and hence by the residue theorem

$$(3.25) \quad D_{q-1}(x) = (2\pi i)^{-1} \int_{-c-i\infty}^{-c+i\infty} \zeta^2(s) \frac{x^{s+q-1}\Gamma(s)}{\Gamma(s+q)} \, ds + \frac{x^{q-1}}{4\Gamma(q)}$$

$$+ \frac{x^q}{\Gamma(q+1)} (\gamma + \log x - \psi(1+q)),$$

since the integrand has a simple pole at $s = 0$ and a double pole at $s = 1$. As in the proof of (3.16) we use now the functional equation for the zeta-function and replace $\zeta^2(1-s)$ by the absolutely covergent series which may be integrated termwise to give

$$(2\pi i)^{-1} \int_{-c-i\infty}^{-c+i\infty} \zeta^2(s) \frac{x^{s+q-1}\Gamma(s)}{\Gamma(s+q)} \, ds = (4^q \pi^{2q-2})^{-1} \sum_{n=1}^{\infty} d(n) n^{-q} (2\pi i)^{-1}$$

$$\times \int_{-c-i\infty}^{-c+i\infty} \frac{(4\pi^2 nx)^{s+q-1}}{\Gamma(s)\Gamma(s+q)\cos^2(\pi s/2)} \, ds.$$

For n and x fixed the second integral above is equal to minus $2\pi i$ times the sum of residues at its double poles $s = 2m + 1$ ($m = 0, 1, 2, \ldots$). To calculate these residues observe that $\cos z$ is an even function of z and for $z = s - (2m + 1) \to 0$

$$\cos^2 \frac{\pi s}{2} = \cos^2\left(\frac{\pi z}{2} + (2m+1)\frac{\pi}{2}\right) = \sin^2 \frac{\pi z}{2} = (1 + o(1)) \frac{\pi^2 z^2}{4},$$

while the linear part of the expansion of

$$\frac{(4\pi^2 nx)^{s+q-1}}{\Gamma(s)\Gamma(s+q)} = \frac{f(s)}{g(s)}$$

at $s = 2m + 1$ is equal by Taylor's formula to

$$\left(\frac{f}{g}\right)'\bigg|_{s=2m+1} = \frac{f}{g}\left(\frac{f'}{f} - \frac{g'}{g}\right)\bigg|_{s=2m+1}$$

$$= \frac{(4\pi^2 nx)^{2m+q}}{\Gamma(2m+1)\Gamma(2m+1+q)}$$

$$\times \left(\log(4\pi^2 nx) - \psi(2m+1) - \psi(2m+1+q)\right).$$

Therefore we have, when $q > 2$,

$$(3.26) \quad D_{q-1}(x) = \frac{x^{q-1}}{4\Gamma(q)} + \frac{x^q}{\Gamma(1+q)}(\gamma + \log x - \psi(q+1))$$

$$+ 2\pi x^q \sum_{n=1}^{\infty} d(n)\lambda_q(4\pi\sqrt{nx}),$$

where

$$(3.27) \quad \lambda_q(z) = -\frac{2}{\pi} \sum_{m=0}^{\infty} \frac{(z/2)^{4m}}{\Gamma(2m+1)\Gamma(2m+1+q)}$$

$$\times (2\log(z/2) - \psi(2m+1) - \psi(2m+1+q))$$

is the "generalized Bessel function." To see that this terminology is justified, note that from (3.6) and (3.7) it follows that

$$K_1(2z) + \frac{\pi}{2}Y_1(2z) = z \sum_{m=0}^{\infty} \frac{z^{4m}}{(2m)!\,(2m+1)!}(2\log z - \psi(2m+1)$$

$$-\psi(2m+2)),$$

so that a comparison with (3.27) shows that

$$(3.28) \quad \lambda_1(4\pi\sqrt{nx}) = -\pi^{-2}(xn)^{-1/2}\big(K_1(4\pi\sqrt{nx}) + (\pi/2)Y_1(4\pi\sqrt{nx})\big).$$

Therefore the main effort must be directed towards showing that (3.26) holds not only for $q > 2$, but for $q \geqslant 1$, since for $q = 1$ (3.26) reduces to (3.1) in view of (3.28), proving Voronoi's formula (3.1) for $\Delta(x)$ when x is not an integer.

The definition of λ_q as an integral shows that

$$(3.29) \quad \frac{2}{\pi}\lambda_q(z) = (2\pi i)^{-1} \int_{c-i\infty}^{c+i\infty} \frac{(z/2)^{2s-2}\,ds}{\Gamma(s)\Gamma(s+q)\cos^2(\pi s/2)}$$

for $\operatorname{Re} q > 0$, $-1 < c < 1$, and $\operatorname{Re} q + 2c > 2$. Using a technique similar to the one used in the proof of (3.26) it is seen that the integral in (3.29) may be asymptotically evaluated to yield, for $-\pi/2 < \arg z < 3\pi/2$,

(3.30)

$$\lambda_q(z) \sim (z/2)^{-q}\left(-Y_q(z) + \frac{2}{\pi}e^{q\pi i}K_q(z)\right) - \frac{2}{\pi}\sum_{r=1}^{\infty}\frac{\Gamma(2r)}{\Gamma(q-2r+1)}(2/z)^{4r}.$$

Hence by the asymptotic formulas (3.12) and (3.13)

(3.31)

$$\lambda_q(4\pi\sqrt{nx}) = (nx)^{-q/2-1/4}\left\{A_q\sin(4\pi\sqrt{nx} - \pi/4 - \pi q/2) + B_qe^{-4\pi\sqrt{nx}}\right\}$$

$$+ O\left((nx)^{-q/2-5/4}\right) + O\left((nx)^{-2}\right),$$

where A_q and B_q are uniformly bounded. Therefore the series in (3.26) converges absolutely for $\operatorname{Re} q > \frac{3}{2}$, and this establishes by analytic continuation the validity of (3.26) in the range $q > \frac{3}{2}$. To investigate (3.26) for $q \leqslant \frac{3}{2}$ one needs to know the behavior of the partial sums of the series in (3.26). Letting

$$r_0(x) = D_0(x) - x(\gamma + \log x - \psi(2)) - \tfrac{1}{4}$$

and

$$r_1(x) = \int_0^x r_0(t)\, dt = \sum_{n \leqslant x}(x-n)d(n) - \tfrac{1}{2}x^2(\gamma + \log x - \psi(3)) - \tfrac{1}{4}x,$$

we obtain for $a, b > 0$, $f(x) \in C^2[a, b]$

$$\sum_{a < n \leqslant b} f(n)d(n) = \int_a^b f(t)\, dD_0(t) = \int_a^b f(t)\, dr_0(t) + \int_a^b(2\gamma + \log t)f(t)\, dt,$$

and integrating twice by parts it follows that

(3.32) $$\sum_{a < n \leqslant b} f(n)d(n) = \left(r_0(t)f(t) - r_1(t)f'(t)\right)\Big|_a^b + \int_a^b r_1(t)f''(t)\, dt$$

$$+ \int_a^b(2\gamma + \log t)f(t)\, dt.$$

From (3.26) and (3.31) it is seen that $r_1(x) \ll x^{3/4}$, while trivially one has $r_0(x) = \Delta(x) \ll x^{1/2}$. To establish the convergence of (3.26) for $1 \leqslant q \leqslant \frac{3}{2}$ we use (3.32) with $f(t) = \lambda_q(4\pi\sqrt{xt})$ and note that from the series expansion

(3.27) we obtain for integral q

(3.33)
$$\frac{d}{dt}\left(t^q\lambda_q(4\pi\sqrt{nt})\right) = t^{q-1}\lambda_{q-1}(4\pi\sqrt{nt}),$$

which is analogous to (3.11). By some calculations it follows that

$$f''(t) \sim Ax^{3/4-q/2}t^{-q/2-5/4}\sin(4\pi\sqrt{xt} - q\pi/2 - 5\pi/4) + O(x^{-2}t^{-4}),$$

and using again $r_0(x) \ll x^{1/2}$, $r_1(x) \ll x^{3/4}$ it is seen from (3.32) that the series in (3.26) converges for $q \geqslant 1$. Moreover when x is not an integer the convergence is uniform for x lying in any closed interval free of integers for $q > \frac{1}{2}$. A more careful analysis, based on investigation of the function $f(t) = \lambda_1(4\pi\sqrt{xt}) - \lambda_1(4\pi\sqrt{mt})$, $m = [x]$, settles the case $q = 1$, x is an integer. The details may be found in the work of Dixon and Ferrar (1931).

Finally, it remains to discuss the proof of the summation formula (3.2). Note first that using (3.1) and (3.11) one obtains formally

$$\sum_{a<n\leqslant b} f(n)d(n) = \int_a^b f(x)\,dD(x)$$

$$= \int_a^b (\log x + 2\gamma)f(x)\,dx + \int_a^b f(x)\,d\Delta(x)$$

$$= \int_a^b (\log x + 2\gamma)f(x)\,dx + \sum_{n=1}^\infty d(n)\int_a^b f(x)\alpha(nx)\,dx,$$

that is, one obtains formally (3.2) from (3.1) by differentiating $\Delta(x)$ term by term, but this procedure is hard to justify. A rigorous proof may be based on (3.26) and the summation formula (3.32), when we substitute

$$r_1(t) = 2\pi t^2 \sum_{n=1}^\infty d(n)\lambda_2(4\pi\sqrt{nt}).$$

Since $f''(t)$ is bounded and the series for $r_1(t)$ is absolutely and uniformly convergent, the order of summation and integration may be inverted, and the first integral in (3.32) becomes

$$2\pi \sum_{n=1}^\infty d(n)\int_a^b t^2\lambda_2(4\pi\sqrt{nt})f''(t)\,dt.$$

Integrating twice by parts and using (3.33) we have

(3.34)

$$\int_a^b t^2\lambda_2(4\pi\sqrt{nt})f''(t)\,dt = \left.\left(t^2\lambda_2(4\pi\sqrt{nt})f'(t) - t\lambda_1(4\pi\sqrt{nt})f(t)\right)\right|_a^b$$

$$+ \int_a^b \lambda_0(4\pi\sqrt{nt})f(t)\,dt.$$

From (3.4), (3.5), and (3.27) it is seen that

$$2\pi\lambda_0(4\pi\sqrt{nx}) = 4K_0(4\pi\sqrt{nx}) - 2\pi Y_0(4\pi\sqrt{nx}),$$

and using (3.26) with $q = 1$ and $q = 2$ (in view of (3.1) and (3.24)), it follows that

$$r_0(t) = \tfrac{1}{2}d(t) + 2\pi t \sum_{n=1}^{\infty} d(n)\lambda_1(4\pi\sqrt{nt}), \quad r_1(t) = 2\pi t^2 \sum_{n=1}^{\infty} d(n)\lambda_2(4\pi\sqrt{nt}),$$

where we set $d(t) = 0$ if t is not an integer. Therefore if we multiply (3.34) by $2\pi d(n)$ and sum over n we obtain (3.2) from (3.32).

3.4 OTHER FORMULAS OF THE VORONOI TYPE

There exists an extensive literature concerning various generalizations of Voronoi's formulas (3.1) and (3.2) to other arithmetical functions, whose generating functions satisfy functional equations similar to the functional equation for the zeta-function of Riemann. This possibility of generalizations is one of the most important aspects of the research initiated by Voronoi, but since our main purpose is the investigation concerning the zeta-function, we shall mention here explicitly only one result which is similar to Voronoi's formulas. This concerns the classical lattice-point problem known as the circle problem, which is in many ways similar to the divisor problem (see Chapter 13) and consists of estimating the function

(3.35) $$P(x) = R(x) - \pi x - 1 = {\sum_{n \leqslant x}}' r(n) - \pi x - 1,$$

where $r(n)$ is the number of ways n may be written as a sum of two integer squares. In 1916 G. H. Hardy (1916b) proved the summation formula

$$(3.36) \quad \frac{1}{\Gamma(q)} \sum_{n \leqslant x}{}' (x-n)^{q-1} r(n) = \frac{\pi x^q}{\Gamma(q+1)} - \frac{x^{q-1}}{\Gamma(q)}$$

$$+ \pi^{1-q} x^{q/2} \sum_{n=1}^{\infty} n^{-q/2} r(n) J_q(2\pi\sqrt{nx})$$

for $q \geqslant 1$ [here $\sum'$ means that only for $q = 1$ and $n = x$ one should take $\frac{1}{2} r(n)$ instead of $r(n)$]. From Hardy's formula one may derive a summation formula for $\sum'_{a \leqslant n \leqslant b} f(n) r(n)$ analogous to (3.2). The expression on the right-hand side of (3.36) is simpler than the corresponding one in (3.26), and (3.36) may be deduced more simply than (3.26). This is so because the generating series of $r(n)$ has a simple pole at $s = 1$, while $\zeta^2(s)$ [which is the generating series of $d(n)$] has a second-order pole at $s = 1$.

Notes

G. F. Voronoi (1904a, 1904b) proved by a complicated analytic method the formulas (3.1) and (3.2), and a little later (1905) he succeeded in generalizing his method to certain other functions which are the number of representations of n by certain positive-definite quadratic forms. The methods introduced by Voronoi were deep and novel and inspired much subsequent research, of which one example was mentioned in Section 3.4. Modern developments of the theory of summation formulas for arithmetical functions may be found in the works of K. Chandrasekharan and R. Narasimhan (1961, 1962), B. C. Berndt (1969a, 1969b, 1971, 1975a) and J. L. Hafner (1981b).

The notation used in this chapter differs a little from the notation used in other chapters in two instances, but this should cause no confusion. First, the error term $\Delta(x) = \Delta_2(x)$ is defined somewhat differently than in Chapter 13, and second following the traditional notation we defined in (3.8) $\psi(z) = \Gamma'(z)/\Gamma(z)$. In the remaining chapters we use $\psi(x) = x - [x] - \frac{1}{2}$, except in Chapter 12 (The Distribution of Primes), where the (standard) notation $\psi(x) = \sum_{n \leqslant x} \Lambda(n)$ is used.

When p is an integer in (3.8) $\Gamma(p + k + 1)$ is undefined when $p + k + 1$ is a nonpositive integer, but for integer values of p it is clear that one should define

$$J_p(z) = \sum_{k=0}^{\infty} \frac{(-1)^k (z/2)^{p+2k}}{k!(p+k)!} \quad (p \geqslant 0), \qquad J_{-p}(z) = (-1)^p J_p(z),$$

and similarly for $I_p(z)$ when p is an integer.

There is a possibility of obtaining an explicit expression for $\Delta(x)$ which is completely different from (3.1). Namely, starting from the elementary expression

$$\sum_{n \leqslant x} d(n) = 2 \sum_{n \leqslant \sqrt{x}} \sum_{m \leqslant x/n} 1 - [\sqrt{x}]^2$$

and defining $\Delta(x)$ as in (3.1), a simple calculation gives at once

$$\Delta(x) = -2 \sum_{n \leqslant \sqrt{x}} \psi(x/n) + O(x^\varepsilon) \qquad [\psi(x) = x - [x] - \tfrac{1}{2}].$$

This is a useful formula, but for most purposes (3.1) and the flexible (3.17) are better.

All the facts used here about the Bessel functions may be found in the standard work of G. N. Watson (1944). Curiously enough, Watson mentions Voronoi's formula only once on p. 200, where he writes: "A novel application of these asymptotic expansions has been discovered in recent years; they are of some importance in the analytic theory of the divisors of numbers." In view of many important applications of (3.1) and (3.2) and all the research Voronoi's work has inspired, this remark seems a little unjust—Voronoi's formulas deserve more than a casual mention.

A detailed proof of (3.16) is given by E. C. Titchmarsh (1951) Chapter 12, and some details of the proof are for this reason suppressed here. However, his remark on p. 268, which amounts to saying that (3.20) holds, is rather casual. The reference to (7.9.8) and (7.9.11) of his book (1948) on Fourier integrals does not seem sufficient, and it is desirable to have a more detailed account of (3.20). To see that (3.21) and (3.22) hold, note that $f(x) = x^{a-1}$ $(0 < a < 1)$ and $F_c(x) = \sqrt{2/\pi}\,\Gamma(a)x^{-a}\cos(\pi a/2)$ are cosine transforms, that is,

$$F_c(x) = \sqrt{2/\pi}\int_0^\infty f(t)\cos(xt)\,dt.$$

By the definition of $J_p(x)$

$$\int_0^1 (1 - y^2)^{p-1/2}\cos(xy)\,dy = \sum_{n=0}^\infty \frac{(-1)^n x^{2n}}{(2n)!}\int_0^1 (1 - y^2)^{p-1/2} y^{2n}\,dy$$

$$= \frac{1}{2}\sum_{n=0}^\infty \frac{(-1)^n x^{2n}\Gamma\big(p + \tfrac{1}{2}\big)\Gamma\big(n + \tfrac{1}{2}\big)}{(2n)!\,\Gamma(p + n + 1)}$$

$$= 2^{p-1/2}\sqrt{\pi/2}\,\Gamma\big(p + \tfrac{1}{2}\big)x^{-p}J_p(x),$$

where we used (A.31) and $\Gamma(n + \tfrac{1}{2}) = 2^{-n}\pi^{1/2}(2n - 1)!!$. Thus

$$f(x) = \begin{cases} (1 - x^2)^{p-1/2} & (0 < x < 1) \\ 0 & (x \geqslant 1) \end{cases}, \qquad F_c(x) = 2^{p-1/2}\Gamma\big(p + \tfrac{1}{2}\big)x^{-p}J_p(x)$$

are also cosine transforms. Analogously to (A.6) one obtains

$$\int_0^\infty F_c(x)G_c(x)\,dx = \int_0^\infty f(x)g(x)\,dx, \qquad \int_0^\infty F_c(x)g(x)\,dx = \int_0^\infty G_c(x)f(x)\,dx,$$

for two pairs $f(x)$, $F_c(x)$ and $g(x)$, $G_c(x)$ of cosine transforms, which gives

(3.37)

$$\int_0^\infty J_p(x)x^{a-p-1}\,dx = \sqrt{2/\pi}\,\frac{\Gamma(a)\cos(\pi a/2)}{2^{p-1/2}\Gamma\big(p + \tfrac{1}{2}\big)}\int_0^1 (1 - x^2)^{p-1/2}x^{-a}\,dx$$

$$= \frac{\Gamma(a)\cos(\pi a/2)}{2^{p-1}\sqrt{\pi}\,\Gamma\big(p + \tfrac{1}{2}\big)}\,\frac{\Gamma\big(p + \tfrac{1}{2}\big)\Gamma\big(\tfrac{1}{2} - a/2\big)}{2\Gamma(p - a/2 + 1)} = \frac{2^{a-p-1}\Gamma(a/2)}{\Gamma(p - a/2 + 1)},$$

where we used (A.30). Taking $a = s$ to be complex in (3.37) it is seen that (3.37) holds for $0 < \sigma < p + \tfrac{3}{2}$ in view of (3.14), and we have the Mellin transforms

(3.38) $x^{-p}J_p(x)$, $\dfrac{2^{s-p-1}\Gamma(s/2)}{\Gamma(p - s/2 + 1)}$.

Finally, using the relations

$$Y_p(x) = \frac{J_p(x)\cos \pi p - J_{-p}(x)}{\sin \pi p}, \qquad K_p(x) = \frac{\pi i}{2} e^{p\pi i/2}\left(J_p(x) + iY_p(x)\right),$$

one obtains (3.21) and (3.22) easily from (3.38).

The method of considering the weighted divisor sum $D_{q-1}(x)$ in the proof of the Voronoi formula is due to A. L. Dixon and W. L. Ferrar (1931). Their paper (1932) contains an interesting investigation of a reciprocity relation connected with the Voronoi formula, which is motivated by the well-known reciprocity relation for Fourier transforms. The exposition presented in Section 3.3 concerning the proof of (3.1) and (3.2) follows Dixon and Ferrar (1931), where additional details [like the proof of (3.30), and especially the proof of (3.1) when x is an integer] may be found. The main idea in the proof of Dixon and Ferrar is to prove (3.26) for some q (specifically for $q > \frac{3}{2}$), and then to feed back (3.26) to itself again (in a certain sense) via the summation formula (3.32) to obtain (3.26) for values of q less than $\frac{3}{2}$ also. The crucial point in their proof of (3.2) is the fact that the expression for $r_1(t)$ allows one to invert the order of summation and integration—the rest is simply integration by parts. The paper of Dixon and Ferrar (1931) gives also an analysis of (3.2) when $a = 0$ and $b = \infty$, in which case there are some additional difficulties. A proof of (3.2) when $a = 0$, $b = \infty$ has been given recently by D. Hejhal (1979), who used a two-dimensional Poisson summation formula.

A nice generalization of the truncated formula (3.17) for $\Delta(x)$ to the error term in the asymptotic formula for $\sum'_{n \leqslant x} f(n)(x - n)^\kappa$, $\kappa \geqslant 0$ was made by H.-E. Richert (1957). Here $f(n)$ denotes an arithmetical function generated by a Dirichlet series which satisfies a certain type of functional equation involving gamma factors, similar to the ordinary functional equation (1.22) for the Riemann zeta-function. Application of Richert's results to the circle problem will be discussed in Chapter 13.

An interesting discussion of the sum $D_{q-1}(x)$ in (3.24) and related topics is given in Chapter 8 of K. Chandrasekharan (1970). Chandrasekharan's approach is different from the one used in this text, and utilizes results from the theory of Fourier series and integrals.

/ CHAPTER FOUR /
THE APPROXIMATE FUNCTIONAL EQUATIONS

4.1 THE APPROXIMATE FUNCTIONAL EQUATION FOR $\zeta(s)$

In this chapter we begin the study of more advanced topics of zeta-function theory. We shall be concerned with approximate functional equations, of which a simple variant was given already in Theorem 1.8. This result, however, has the feature that $\zeta(s)$ is approximated by a sum of length $\gg |t|$, and in many applications it is difficult to deal with such a "long" sum. Therefore it seems desirable to have other approximate functional equations at our disposal, both for $\zeta(s)$ and $\zeta^k(s)$ ($k \geqslant 2$), which will provide a considerable flexibility in many problems. Several results of this type will be presented in this chapter, and they all may be considered as "approximate functional equations," since they express $\zeta(s)$ as certain finite sums at the points s and $1 - s$, plus certain error terms. Perhaps the best known of all approximate functional equations is the following, which we state as

THEOREM 4.1. Let $0 \leqslant \sigma \leqslant 1$; $x, y, t > C > 0$; and $2\pi xy = t$. Then uniformly in σ

$$(4.1) \quad \zeta(s) = \sum_{n \leqslant x} n^{-s} + \chi(s) \sum_{n \leqslant y} n^{s-1} + O(x^{-\sigma}) + O(t^{1/2-\sigma}y^{\sigma-1}).$$

In this formula

$$\chi(s) = (2\pi)^s / (2\Gamma(s)\cos(\pi s/2)) = 2^s \pi^{s-1} \Gamma(1 - s)\sin(\pi s/2)$$

is the function which appears in the functional equation (1.23). Theorem 4.1 is a classical result due to G. H. Hardy and J. E. Littlewood, and a corresponding formula also holds [since $\zeta(\bar{s})1 = \overline{\zeta(s)}$] if $t < 0$, with t replaced by $|t|$ in the error term and $2\pi xy = t$. The approximate functional equation (4.1) possesses an interesting symmetric property for $\sigma = \frac{1}{2}$. Namely, from the functional equation (1.23) we obtain $\chi(\frac{1}{2} + it) = \chi^{-1}(\frac{1}{2} - it)$, so that $\zeta(\frac{1}{2} + it) \times \chi^{-1/2}(\frac{1}{2} + it)$ is real. Therefore (4.1) with $x = y = (t/2\pi)^{1/2}$ gives

(4.2)

$$\zeta(\tfrac{1}{2} + it)\chi^{-1/2}(\tfrac{1}{2} + it) = 2\,\mathrm{Re}\!\left(\chi^{1/2}(\tfrac{1}{2} + it)\sum_{n \leqslant (t/2\pi)^{1/2}} n^{-1/2+it}\right) + O(t^{-1/4}).$$

This is an important relation which will enable us to investigate in Chapter 10 the distribution of zeros of $\zeta(s)$ on the critical line $\sigma = \frac{1}{2}$.

There exist several methods of proving the approximate functional equation (4.1), and one such method, capable of dealing with $\zeta^k(s)$ too, will be developed in Section 4.3. The method of proof given below has the advantage that the error term is represented by a complex integral, whose asymptotic evaluation may be obtained by detailed analysis, which leads to an explicit asymptotic expansion of the error term. This type of result was first obtained by C. L. Siegel in 1932, who was inspired by looking at Riemann's notes on the zeta-function at the Göttingen library. This deep formula came to be known as the Riemann–Siegel formula, and it says that if $0 \leqslant \sigma \leqslant 1$, $m = [(t/2\pi)^{1/2}]$, $N < At$, where $A > 0$ is sufficiently small, then

$$(4.3) \quad \zeta(s) = \sum_{n \leqslant m} n^{-s} + \chi(s) \sum_{n \leqslant m} n^{s-1}$$

$$+ (-1)^{m-1} e\!\left(\frac{1-s}{4}\right)(2\pi t)^{(s-1)/2} \exp\!\left(-\frac{it}{2} - \frac{i\pi}{8}\right)\Gamma(1-s)$$

$$\times \left\{ S_N + O\big((ANt^{-1})^{N/6}\big)\right\} + O(e^{-At}),$$

where

$$(4.4) \quad S_N = \sum_{n=0}^{N-1} \sum_{\nu \leqslant n/2} \frac{n!\, i^{\nu-n}}{\nu!\,(n-2\nu)!\,2^n}(2/\pi)^{(n-2\nu)/2} a_n \Psi^{(n-2\nu)}\!\left(\frac{\eta}{\pi} - 2m\right).$$

Here the coefficients a_n are given by the recurrence relation

$$(n+1)t^{1/2}a_{n+1} = (\sigma - n - 1)a_n + ia_{n-2}$$

$$(a_{-2} = a_{-1} = 0,\ a_0 = 1;\ n = 0,1,\dots),$$

so that $a_n \ll ((5e/2n)t^{-1/2})^{n/3}$, and $\eta = (2\pi t)^{1/2}$, $\Psi(z) = (\cos \pi(\frac{1}{2}z^2 - z - \frac{1}{8}))/\cos \pi z$.

Taking $N = 3$ in the Riemann–Siegel formula and simplifying we obtain

(4.5)

$$\zeta(\tfrac{1}{2} + it)\chi^{-1/2}(\tfrac{1}{2} + it) = 2 \sum_{n=1}^{m} n^{-1/2}\cos(\vartheta - t\log n)$$

$$+ (-1)^{m-1}(2\pi t^{-1})^{1/4} \frac{\cos\{t - (2m + 1)\sqrt{2\pi t} - \pi/8\}}{\cos\sqrt{2\pi t}} + O(t^{-3/4}),$$

where $\vartheta = \vartheta(t) = -\tfrac{1}{2}\arg\chi(\tfrac{1}{2} + it)$. In particular a suitable choice of t shows that the second main term in (4.5) is $\gg t^{-1/4}$ for some arbitrarily large values of t, and this implies that the error term in (4.1) is best possible for $\sigma = \tfrac{1}{2}$.

Proof of Theorem 4.1. It is possible to extend the range for σ in Theorem 4.1 to any strip $-\sigma_0 < \sigma < \sigma_0$ by some minor changes in the argument. This is however not important, since it is certainly the critical strip $0 \leqslant \sigma \leqslant 1$ which is the most interesting range. For $\sigma > 0$

$$\int_0^\infty x^{s-1}e^{-nx}\,dx = n^{-s}\int_0^\infty y^{s-1}e^{-s}\,dy = n^{-s}\Gamma(s);$$

hence for $\sigma > 1$ and $m \geqslant 2$

$$\zeta(s) = \sum_{n\leqslant m} n^{-s} + \sum_{n>m} n^{-s} = \sum_{n\leqslant m} n^{-s} + \frac{1}{\Gamma(s)}\int_0^\infty x^{s-1}\left(\sum_{n>m} e^{-nx}\right)dx$$

$$= \sum_{n\leqslant m} n^{-s} + \frac{1}{\Gamma(s)}\int_0^\infty \frac{x^{s-1}e^{-mx}}{e^x - 1}\,dx,$$

where the inversion of the order of summation and integration is justified by absolute convergence. Now consider the integral

$$I(s) = \int_C \frac{z^{s-1}e^{-mz}}{e^z - 1}\,dz,$$

where C is the contour that starts at infinity on the positive real axis, encircles the origin once in the positive direction excluding the points $\pm 2\pi i, \pm 4\pi i, \ldots$, and returns to infinity. In $I(s)$ we have $z^{s-1} = \exp((s - 1)\log z)$, where the logarithm is real at the beginning of the contour, and its imaginary part varies from 0 to 2π around the contour. For $0 < \varepsilon < 1$ we may write

$$I(s) = -\int_\varepsilon^\infty \frac{x^{s-1}e^{-mx}}{e^x - 1}\,dx + \int_{|z|=\varepsilon} \frac{z^{s-1}e^{-mz}}{e^z - 1}\,dz + \int_\varepsilon^\infty \frac{(xe^{2\pi i})^{s-1}e^{-mx}}{e^x - 1}\,dx$$

$$= (e^{2\pi is} - 1)\int_0^\infty \frac{x^{s-1}e^{-mx}}{e^x - 1}\,dx,$$

on letting $\varepsilon \to 0$. The loop-integral $I(s)$ is uniformly convergent in any finite region of the s-plane, and so defines an integral function of s. Hence by analytic continuation we obtain, if s is not a positive integer,

$$(4.6) \qquad \zeta(s) = \sum_{n \leqslant m} n^{-s} + \frac{I(s)}{\Gamma(s)(e^{2\pi is} - 1)}$$

$$= \sum_{n \leqslant m} n^{-s} + \frac{e^{-\pi is}\Gamma(1 - s)}{2\pi i} \int_C \frac{w^{s-1}e^{-mw}}{e^w - 1} \, dw,$$

if we use the identity $\Gamma(s)\Gamma(1 - s) = \pi/\sin \pi s$.

Now let $x \leqslant y$, so that $x \leqslant (t/2\pi)^{1/2}$, $m = [x]$, $y = t/2\pi x$, $q = [y]$, and $\eta = 2\pi y$. The contour C in (4.6) is replaced by straight lines C_1, C_2, C_3, C_4 joining the points ∞, $c\eta + i\eta(1 + c)$, $-c\eta + i\eta(1 - c)$, $-c\eta - (2q + 1)\pi i$, ∞, where $0 < c \leqslant \frac{1}{2}$ is an absolute constant, and if y is an integer a small indentation is made above the pole $w = i\eta$ of the integrand in (4.6). The residue theorem gives

$$\zeta(s) = \sum_{n \leqslant m} n^{-s} - 2\pi i \sum \operatorname{Re} s + \frac{e^{-\pi is}\Gamma(1 - s)}{2\pi i}\left(\int_{C_1} + \int_{C_2} + \int_{C_3} + \int_{C_4}\right),$$

where summation is over the residues of

$$F(w) = \frac{e^{-\pi is}\Gamma(1 - s)e^{-mw}}{2\pi i(e^w - 1)} w^{s-1}$$

at the points $w = \pm 2n\pi i$ $(n = 1, 2, \ldots, q)$. The residues at $\pm 2n\pi i$ contribute together

$$\frac{e^{-\pi is}\Gamma(1 - s)}{2\pi i}\left\{(2ni\pi)^{s-1} + (-2ni\pi)^{s-1}\right\}$$

$$= \frac{(2n\pi)^{s-1}e^{-\pi is}\Gamma(1 - s)}{2\pi i}\left(e^{\pi i(s-1)/2} + e^{3\pi i(s-1)/2}\right)$$

$$= \frac{(2n\pi)^{s-1}\Gamma(1 - s)}{2\pi i}\left(-ie^{-\pi is/2} + ie^{\pi is/2}\right)$$

$$= -\frac{2^s(\pi n)^{s-1}\Gamma(1 - s)}{2\pi i}\sin\frac{\pi s}{2} = -(2\pi i)^{-1}n^{s-1}\chi(s).$$

Therefore we have

$$(4.7) \qquad \zeta(s) = \sum_{n \leqslant x} n^{-s} + \chi(s)\sum_{n \leqslant y} n^{s-1}$$

$$+ \frac{e^{-\pi is}\Gamma(1 - s)}{2\pi i}\left(\int_{C_1} + \int_{C_2} + \int_{C_3} + \int_{C_4}\right),$$

and the proof reduces to showing that the last expression in (4.7) is

$$\ll x^{-\sigma} + t^{1/2-\sigma}y^{\sigma-1}.$$

Let $w = u + iv = \rho e^{i\varphi}, 0 < \varphi < 2\pi$, so that $|w^{s-1}| = \rho^{\sigma-1}e^{-t\varphi}$. On C_4 we have $\varphi \geqslant \frac{5}{4}, \rho > A\eta, |e^w - 1| > A$, where $A > 0$ denotes (possibly different) absolute constants. Hence

$$\int_{C_4} \ll \eta^{\sigma-1}e^{-5\pi t/4}\int_{-c\eta}^{\infty} e^{-mu}\,du \ll \exp(mc\eta - 5\pi t/4) \ll \exp(tc - 5\pi t/4).$$

On C_3 we have

$$\varphi \geqslant \frac{\pi}{2} + \arctan\frac{c}{1-c} = \frac{\pi}{2} + \int_0^{c/(1-c)} \frac{dx}{1+x^2}$$

$$> \frac{\pi}{2} + A + \int_0^{c/(1-c)} \frac{dx}{(1+x)^2} = \frac{\pi}{2} + A + c.$$

Hence on C_3

$$w^{s-1}e^{-mw} \ll \eta^{\sigma-1}\exp\{-t(\pi/2 + A + c) + mc\eta\} \ll \eta^{\sigma-1}e^{-t(\pi/2+A)}.$$

Since $|e^w - 1| > A$ we obtain

$$\int_{C_3} \ll \eta^\sigma e^{-t(\pi/2+A)}.$$

On C_1 we have $|e^w - 1| > Ae^u$. Therefore

$$\frac{w^{s-1}e^{-mw}}{e^w - 1} \ll \eta^{\sigma-1}\exp\left(-t\arctan\frac{\eta + c\eta}{u} - mu - u\right).$$

Since $m + 1 \geqslant x = t/\eta$ and

$$\frac{d}{du}\left\{\arctan\frac{(1+c)\eta}{u} + \frac{u}{\eta}\right\} = -\frac{(1+c)\eta}{u^2 + (1+c)^2\eta^2} + \frac{1}{\eta} > 0,$$

we obtain

$$\arctan\frac{(1+c)\eta}{u} + \frac{u}{\eta} \geqslant \arctan\frac{1+c}{c} + c = \frac{\pi}{2} + c - \arctan\frac{c}{1+c}$$

$$= \frac{\pi}{2} + A,$$

since for $0 < \theta < 1$

$$\arctan \theta < \int_0^\theta \frac{dx}{(1-x)^2} = \frac{\theta}{1-\theta}.$$

Hence

$$\int_{C_1} \ll \eta^{\sigma-1} \int_0^{\pi\eta} e^{-(\pi/2+A)t}\, du + \eta^{\sigma-1} \int_{\pi\eta}^\infty e^{-xu}\, du$$

$$\ll \eta^\sigma e^{-(\pi/2+A)t} + \eta^{\sigma-1} e^{-\pi\eta x} \ll \eta^\sigma e^{-(\pi/2+A)t}.$$

Recalling that

(4.8)
$$e^{-\pi i s}\Gamma(1-s) \ll t^{1/2-\sigma} e^{\pi t/2}$$

we obtain

$$e^{-\pi i s}\Gamma(1-s)\left(\int_{C_1} + \int_{C_3} + \int_{C_4}\right) \ll e^{-At} \ll x^{-\sigma} + t^{1/2-\sigma} y^{\sigma-1},$$

and therefore it remains only to deal with $\int_{C_2}$. Here $w = i\eta + \lambda e(\frac{1}{8})$, where $-\sqrt{2}\, c\eta \leqslant \lambda \leqslant \sqrt{2}\, c\eta$. Hence

$$w^{s-1} = \exp\{(s-1)(\pi i/2 + \log(\eta + \lambda e^{-\pi i/4}))\}$$

$$= \exp\left\{(s-1)\left(\frac{\pi i}{2} + \log\eta + \frac{\lambda}{\eta} e^{-\pi i/4} - \frac{\lambda^2}{2\eta^2} e^{-\pi i/2} + O(\lambda^3\eta^{-3})\right)\right\}$$

$$\ll \eta^{\sigma-1}\exp\left\{\left(-\frac{\pi}{2} + 2^{-1/2}\frac{\lambda}{\eta} - \frac{\lambda^2}{2\eta^2} + O(\lambda^3\eta^{-3})\right)t\right\}.$$

Now for $u \geqslant 0$

$$\frac{e^{-mw+xw}}{e^w - 1} \ll \frac{e^{(x-m-1)u}}{1 - e^{-u}},$$

while for $u < 0$

$$\frac{e^{-mw+xw}}{e^w - 1} \ll \frac{e^{(x-m)u}}{e^u - 1},$$

which is bounded for $|u| > \pi/2$. Since $|e^{-xw}| = e^{-\lambda t/\eta\sqrt{2}}$ we see that the part of $\int_{C_2}$ with $|u| > \pi/2$ is

$$\ll \eta^{\sigma-1} e^{-\pi t/2} \int_{-c\eta\sqrt{2}}^{c\eta\sqrt{2}} \exp\left(-\frac{\lambda^2 t}{2\eta^2} + O(\lambda^3 t \eta^{-3})\right) d\lambda$$

$$\ll \eta^{\sigma-1} e^{-\pi t/2} \int_{-\infty}^{\infty} e^{-A\lambda^2 t/\eta^2} d\lambda$$

$$\ll \eta^{\sigma} t^{-1/2} e^{-\pi t/2}.$$

The arguments used above apply also if $|u| \leq \pi/2$ and $|e^w - 1| > A$. Otherwise suppose that the contour goes too near to the pole at $w = 2q\pi i$. Take it around the arc of the circle $|w - 2q\pi i| = \pi/2$. On this circle $w = 2q\pi i + (\pi/2)e^{i\theta}$ and

$$\log(w^{s-1} e^{-mw}) = -\frac{m}{2}\pi e^{i\theta} + (s-1)\left\{\frac{\pi i}{2} + \log\left(2q\pi + \frac{1}{2}\pi e^{i\theta}/i\right)\right\}$$

$$= -\frac{m}{2}\pi e^{i\theta} - \frac{\pi}{2}t + (s-1)\log(2q\pi) + \frac{te^{i\theta}}{4q} + O(1)$$

$$= -\frac{\pi}{2}t + (s-1)\log(2q\pi) + O(1),$$

since for $x \leq y$

$$m\pi - \frac{t}{2q} = \frac{2mq\pi - t}{2q}$$

$$= \frac{2\pi}{2q}\{[x][y] - ([x] + O(1))([y] + O(1))\} = O(1).$$

Therefore

$$w^{s-1} e^{-mw} \ll q^{\sigma-1} e^{-\pi t/2},$$

and the contribution of this part is

$$\ll \eta^{\sigma-1} e^{-\pi t/2}.$$

Using (4.8) again we finally have

$$e^{-\pi i s} \Gamma(1-s) \int_{C_2} \ll t^{1/2-\sigma}\left(\eta^{\sigma} t^{-1/2} + \eta^{\sigma-1}\right) \ll x^{-\sigma} + t^{-1/2} x^{1-\sigma},$$

which in view of (4.7) proves the theorem for $x \leq y$.

To deal with the case $x \geqslant y$ change s into $1 - s$ in (4.1). Then for $x \leqslant y$

$$\zeta(1 - s) = \sum_{n \leqslant x} n^{s-1} + \chi(1 - s) \sum_{n \leqslant y} n^{-s} + O(x^{\sigma-1}) + O(t^{\sigma - 1/2}y^{-\sigma});$$

hence by the functional equation for $\zeta(s)$ and (1.25)

$$\zeta(s) = \chi(s)\zeta(1 - s) = \chi(s) \sum_{n \leqslant x} n^{s-1} + \chi(s)\chi(1 - s) \sum_{n \leqslant y} n^{-s}$$

$$+ O(t^{1/2 - \sigma}x^{\sigma - 1}) + O(y^{-\sigma}).$$

Using $\chi(s)\chi(1 - s) = 1$ we have

(4.9)

$$\zeta(s) = \sum_{n \leqslant y} n^{-s} + \chi(s) \sum_{n \leqslant x} n^{s-1} + O(y^{-\sigma}) + O(t^{1/2 - \sigma}x^{\sigma - 1}) \qquad (x \leqslant y)$$

Interchanging x and y in (4.9) we obtain (4.1) with $x \geqslant y$.

4.2 THE APPROXIMATE FUNCTIONAL EQUATION FOR $\zeta^2(s)$

We proceed now to the analogue of Theorem 4.1 for $\zeta^2(s)$. This problem is naturally more difficult than the corresponding problem for $\zeta(s)$, and one cannot expect that the error term will be as small as the one in Theorem 4.1. The result is

THEOREM 4.2. Let $x, y, t > C > 0$ and $4\pi^2 xy = t^2$. Then for $0 < \sigma < 1$ uniformly in σ

$$(4.10) \quad \zeta^2(s) = \sum_{n \leqslant x} d(n)n^{-s} + \chi^2(s) \sum_{n \leqslant y} d(n)n^{s-1} + O(x^{1/2 - \sigma}\log t).$$

As in Theorem 4.1 a corresponding result holds if t is replaced by $-t$, and by an argument similar to the one used in (4.9) we may suppose in the proof of Theorem 4.2 that $x \geqslant y$. We also have the analogue of (4.2), namely,

$$(4.11) \quad \left|\zeta(\tfrac{1}{2} + it)\right|^2 = \zeta^2(\tfrac{1}{2} + it)\chi^{-1}(\tfrac{1}{2} + it)$$

$$= 2\,\mathrm{Re}\!\left(\chi(\tfrac{1}{2} + it) \sum_{n \leqslant t/2\pi} d(n)n^{-1/2 + it}\right) + O(\log t),$$

since $\zeta(\tfrac{1}{2} + it)\chi^{-1/2}(\tfrac{1}{2} + it)$ is real and $|\chi(\tfrac{1}{2} + it)| = 1$. The Riemann–Siegel formula (4.5) showed that the error term in (4.1) is best possible when $\sigma = \tfrac{1}{2}$, and this is also true of the error term in (4.10); however, the argument in this case is almost elementary. To obtain this let $t = 2\pi N$, $x = N - \delta t^{1/2}$, where N is an integer and $\delta > 0$ is sufficiently small. We shall suppose that the error term in (4.10) for $\sigma = \frac{1}{2}$ is $o(\log t)$ and derive a contradiction. Thus for $s = \frac{1}{2} + it$ we have

$$\zeta^2(s) = \sum_{n \leqslant x} d(n)n^{-s} + \chi^2(s) \sum_{n \leqslant y} d(n)n^{s-1} + o(\log t),$$

$$\zeta^2(s) = \sum_{n \leqslant y} d(n)n^{-s} + \chi^2(s) \sum_{n \leqslant x} d(n)n^{s-1} + o(\log t),$$

provided that $t \to \infty$. Multiplying by N^{it} and subtracting we obtain

(4.12)

$$\chi^2(\tfrac{1}{2} + it) \sum_{x < n \leqslant y} d(n)n^{-1/2}(Nn)^{it} - \sum_{x < n \leqslant y} d(n)n^{-1/2}(N/n)^{it} = o(\log t).$$

Our choice for t gives $e^{2it} = 1$, hence using (1.25) we have

$$\chi^2(\tfrac{1}{2} + it) = (2\pi/t)^{2it} e^{i(2t+\pi/2)}(1 + O(t^{-1})) = iN^{-2it}(1 + O(t^{-1})),$$

and so (4.12) becomes

(4.13) $$\sum_{x < n \leqslant y} d(n)n^{-1/2}\{-(n/N)^{-it} + i(n/N)^{it}\} = o(\log t).$$

Now for $x \leqslant n \leqslant y$ we have

$$(n/N)^{\pm it} = \exp\{\pm 2\pi iN \log(1 + (n-N)N^{-1})\}$$

$$= 1 + O((n-N)^2 N^{-1}) = 1 + O(\delta^2);$$

hence a suitable choice of δ gives

$$\left| \sum_{x < n \leqslant y} d(n)n^{-1/2}\{-(n/N)^{-it} + i(n/N)^{it}\} \right| \gg \log t,$$

since $y = N + \delta t^{1/2} + O(1)$ and we may use the asymptotic formula for $D(x) = \sum_{n \leqslant x} d(n)$ of Chapter 3. Inserting the above estimate in (4.13) we obtain a contradiction which shows that, in general, the error term in (4.10) cannot be $o(\log t)$ when $\sigma = \frac{1}{2}$.

Proof of Theorem 4.2. A complete proof of Theorem 4.2 may be obtained by the method of G.H. Hardy and J. E. Littlewood which will be presented in Section 4.3 in connection with the approximate functional equation for $\zeta^k(s)$. Here we shall treat the case $\sigma \geqslant \frac{1}{2}$, $x \geqslant y > t^\varepsilon$ by the Voronoi summation formula, which in view of the presence of $d(n)$ in the sums in

(4.10) seems a natural tool to use. With

$$D(x) + \sum_{n \leqslant x} d(n) = x(\log x + 2\gamma - 1) + \Delta(x)$$

and $\sigma > 1$ we have

$$\zeta^2(s) = \sum_{n \leqslant N} d(n)n^{-s} + \int_{N+0}^{\infty} x^{-s}dD(x)$$

$$= \sum_{n \leqslant N} d(n)n^{-s} + \int_{N}^{\infty} (\log x + 2\gamma)x^{-s}dx + \int_{N}^{\infty} x^{-s}d\Delta(x).$$

As discussed in Section 3.2 of Chapter 3, a trivial consequence of the truncated Voronoi formula is the order estimate $\Delta(x) \ll x^{1/3+\varepsilon}$. Thus an integration by part gives

(4.14)

$$\zeta^2(s) = \sum_{n \leqslant N} d(n)n^{-s} + (s-1)^{-1}N^{1-s}(\log N + 2\gamma) + (s-1)^{-2}N^{1-s}$$

$$+ O(N^{\varepsilon+1/3-\sigma}) + s\int_{N}^{\infty} x^{-s-1}\Delta(x)\,dx.$$

The integral in (4.14) is therefore seen to be absolutely convergent for $\sigma > \frac{1}{3}$ (and using Theorem 13.5 it is seen that the integral in question is actually convergent in a wider semiplane $\sigma > \frac{1}{4}$), so that (4.14) furnishes an analytic continuation of $\zeta^2(s)$ for $\sigma > \frac{1}{3}$. Our choice for N will be $N = t^c$, where $c > 0$ is fixed but sufficiently large. We use (3.2) and split the series involving $d(n)$ at $(1+\varepsilon)y$, where $xy = (t/2\pi)^2$. Integration by parts yields

(4.15)

$$\sum_{x < n \leqslant N} d(n)n^{-s} = \int_{x}^{N} (\log u + 2\gamma)u^{-s}\,du + O(x^{1/2-\sigma}\log t)$$

$$+ \sum_{n \leqslant (1+\varepsilon)y} d(n)\int_{x}^{N} \{4K_0(4\pi\sqrt{nu}) - 2\pi Y_0(4\pi\sqrt{nu})\} u^{-s}\,du$$

$$+ s \sum_{n > (1+\varepsilon)y} d(n)\int_{x}^{N} \left\{ -2\pi^{-1}(u/n)^{1/2} \right.$$

$$\times \left(K_1(4\pi\sqrt{nu}) + \frac{\pi}{2}Y_1(4\pi\sqrt{nu}) \right) \right\} u^{-s-1}\,du.$$

Noting that

$$\int_{x}^{N} (\log u + 2\gamma)u^{-s}\,du = (s-1)^{-1}x^{1-s}(\log x + 2\gamma) + (s-1)^{-2}x^{1-s}$$

$$-(s-1)^{-1}N^{1-s}(\log N + 2\gamma) - (s-1)^{-2}N^{1-s},$$

it follows on comparing (4.14) and (4.15) that the only difficulty lies in the estimation of the sums appearing on the right-hand sides of (4.15). Using (3.12) and (3.13) it is seen that the second sum on the right-hand side of (4.15) is equal to $c_1 s$ times

$$(4.16) \quad \sum_{n > (1+\varepsilon)y} d(n) \left\{ n^{-3/4} \int_x^N u^{-s-3/4} \sin(4\pi\sqrt{nu} - 3\pi/4) \, du \right.$$

$$+ O\left(\int_x^N n^{-5/4} u^{-\sigma-5/4} \, du \right) \right\}$$

$$= \pm (2i)^{-1} \sum_{n > (1+\varepsilon)y} d(n) n^{-3/4} \int_x^N u^{-\sigma-3/4}$$

$$\times \exp\left(-it \log u \pm 4\pi\sqrt{nu} \mp 3\pi/4 \right) du + O\left(x^{-\sigma-1/4} y^{-1/4} \log t \right),$$

and the error term above is $\ll t^{-1} x^{1/2-\sigma} \log t$. The integrals are of the form

$$\int_x^N u^{-\sigma-3/4} e(F(u)) \, du, \qquad F(u) = -\frac{t}{2\pi} \log u \pm 2\sqrt{nu} ,$$

so that F is monotonic and $|F'(u)| \gg (n/x)^{1/2}$ for $x \leqslant u \leqslant N$ in view of $n > (1 + \varepsilon)y$.

Using (2.3) it is then seen that the total contribution of the sum with $n > (1 + \varepsilon)y$ is

$$\ll x^{1/2-\sigma} \log t + t \sum_{n > (1+\varepsilon)y} d(n) n^{-3/4} n^{-1/2} x^{-\sigma-1/4}$$

$$\ll x^{1/2-\sigma} \log t + t y^{-1/4} x^{-\sigma-1/4} \log t \ll x^{1/2-\sigma} \log t,$$

since $t \ll x$ if $x \geqslant y$ and $xy = (t/2\pi)^2$. Setting for brevity $T = t/2\pi$ and using again (3.12) and (3.13) it is seen that the first sum on the right-hand side of (4.15) is equal to

$$-2^{1/2} \sum_{n \leqslant (1+\varepsilon)y} d(n) n^{-1/4} \int_x^N u^{-\sigma-1/4} \exp(-it \log u) \sin(4\pi\sqrt{nu} - \pi/4) \, du$$

$$+ O\left(\sum_{n \leqslant (1+\varepsilon)y} d(n) n^{-3/4} \int_x^N u^{-\sigma-3/4} \, du \right)$$

$$= \mp 2^{-1/2} i^{-1} \sum_{n \leqslant (1+\varepsilon)y} d(n) n^{-1/4}$$

$$\times \int_x^N u^{-\sigma-1/4} e\left(-T \log u \pm 2\sqrt{nu} \mp 1/8 \right) du + O\left(y^{1/4} x^{1/4-\sigma} \log t \right),$$

and clearly $y^{1/4}x^{1/4-\sigma} \ll x^{1/2-\sigma}$. The integral with $e(-T \log u - 2\sqrt{nu} + \frac{1}{8})$ is as in the previous case estimated by (2.3), the total contribution of these integrals being now

$$\ll \sum_{n \leqslant (1+\varepsilon)y} d(n)n^{-3/4}x^{1/4-\sigma} \ll y^{1/4}x^{1/4-\sigma}\log t \ll x^{1/2-\sigma}\log t.$$

Therefore we are left with

$$(4.17) \quad -2^{-1/2}i^{-1} \sum_{n \leqslant (1+\varepsilon)y} d(n)n^{-1/4}I_n, \qquad I_n = \int_x^N u^{-\sigma-1/4}e(f_n(u))\, du,$$

where

$$f_n(u) = -T \log u + 2\sqrt{nu} - \tfrac{1}{8}.$$

Thus we have

$$f_n'(u) = -Tu^{-1} + (n/u)^{1/2}, \qquad f_n''(u) = Tu^{-2} - \tfrac{1}{2}n^{1/2}u^{-3/2},$$

implying $f_n''(u) > 0$ for $u \leqslant u_0 = 4T^2n^{-1}$. But for $u \geqslant (1 - \varepsilon)u_0$ we see that $f_n'(u) \gg (n/u)^{1/2}$, and so using (2.3)

$$\sum_{n \leqslant (1+\varepsilon)y} d(n)n^{-1/4} \int_{(1-\varepsilon)u_0}^N u^{-\sigma-1/4}e(f_n(u))\, du$$

$$\ll \sum_{n \leqslant (1+\varepsilon)y} d(n)n^{-1/4}u_0^{1/4-\sigma}n^{-1/2}$$

$$\ll \sum_{n \leqslant (1+\varepsilon)y} t^{1/2-2\sigma}d(n)n^{\sigma-1} \ll t^{1/2-2\sigma}y^{\sigma}\log t \ll x^{1/2-\sigma}\log t,$$

if $\varepsilon > 0$ is a sufficiently small fixed number. For the integrals in the remaining sum

$$-2^{-1/2}i^{-1} \sum_{n \leqslant (1+\varepsilon)y} d(n)n^{-1/4} \int_x^{(1-\varepsilon)u_0} u^{-\sigma-1/4}e(f_n(u))\, du$$

we use Theorem 2.2 with $a = x$, $b = (1 - \varepsilon)u_0$, $\Phi(u) = u^{-\sigma-1/4}$, $f(u) = f_n(u)$, $k = 0$, and $F = \mu = t$, and the conditions of the theorem are readily checked. All the error terms in Theorem 2.2 are easily seen to contribute a total $\ll x^{1/2-\sigma}\log t$, except the error term $O(\Phi_a(|f_a' + k| + f_a''^{1/2})^{-1})$, which is discussed next. Observe that for a given n we have $f_n'(x) = 0$ if $n = T^2x^{-1} = y$ and y is an integer. Therefore making the substitution $n = [y] + m$, $|m| \leqslant \varepsilon y$, we have

$$f_n'(x) = f_n'(x) - f_y'(x) \asymp |m|t^{-1}$$

by the mean value theorem, so that

$$(4.18) \quad \left(|f_n'(x)| + (f_n''(x))^{1/2}\right)^{-1}$$

$$\ll \begin{cases} xt^{-1/2} & \text{for } |m| \leqslant T^2 x^{-2}, \\ |m|^{-1}t & \text{for } t^2 x^{-2} \ll |m| \leqslant \varepsilon y/2, \\ \max\left((n/x)^{-1/2}, x/t\right) & \text{for } |m| > \varepsilon y/2. \end{cases}$$

In view of $T^2 x^{-2} \ll 1$ the first estimate in (4.18) can hold for at most $O(1)$ values of m, and the total contribution of the error term

$$O\left(\Phi_a\left(|f_a' + k| + f_a''^{1/2}\right)^{-1}\right)$$

is then

$$(4.19) \qquad \sum_{|m| \ll t^2 x^{-2}} t^\varepsilon y^{-1/4} x t^{-1/2} x^{-\sigma-1/4}$$

$$+ \sum_{1 \ll |m| \leqslant \varepsilon y/2} t|m|^{-1} d([y]+m)([y]+m)^{-1/4} x^{-\sigma-1/4}$$

$$+ \sum_{\varepsilon y/2 \leqslant m \leqslant y} x^{1/2} d([y]+m)([y]+m)^{-3/4} x^{-\sigma-1/4}$$

$$+ \sum_{\varepsilon y/2 \leqslant m \leqslant y} xt^{-1} d([y]+m)([y]+m)^{-1/4} x^{-\sigma-1/4}$$

$$\ll t^{1/2+\varepsilon} x^{-\sigma} + x^{1/2-\sigma}\log t,$$

and where in the second sum in (4.19) we used the trivial $d(n) \ll n^\varepsilon$. Here the error term $t^{1/2+\varepsilon} x^{-\sigma}$ does not exceed $x^{1/2-\sigma}\log t$ if $x > t^{1+2\varepsilon}$, but it is possible by a more elaborate analysis of the error term $\Phi_a(|f_a' + k| + f_a''^{1/2})^{-1}$ in Atkinson's Theorem 2.2 to obtain that the contribution of the sums in (4.19) is indeed $\ll x^{1/2-\sigma}\log t$ for the whole range $x \geqslant y$.

Finally it remains to deal with the main terms, that is, the saddle-point terms coming from Theorem 2.2 and then to use (1.25). The only root of $f_n'(u) = 0$ for a fixed n is $x_0 = T^2 n^{-1}$, and $x_0 > x$ precisely if $n \leqslant y = T^2 x^{-1}$. Now

$$f_n''(x_0) = \tfrac{1}{2} T^{-3} n^2,$$

$$f_n(x_0) = -T\log(T^2 n^{-1}) + 2T - \tfrac{1}{8},$$

so that we have

$$-2^{-1/2}i^{-1}d(n)n^{-1/4}\varphi(x_0)f_n''(x_0)^{-1/2}\exp(2\pi i f_n(x_0) + \pi i/4)$$

$$= -2^{-1/2}i^{-1}d(n)n^{-1/4}(T^2n^{-1})^{-\sigma-1/4}2^{1/2}T^{3/2}n^{-1}$$

$$\times \exp(2it\log(2\pi/t) + it\log n + 2it + \pi i/2)\exp(-\pi i/2))$$

$$= d(n)n^{\sigma-1+it}\chi^2(\sigma + it) + O(d(n)n^{\sigma-1}t^{-2\sigma}).$$

Therefore

$$-2^{-1/2}i^{-1} \sum_{n \leqslant (1+\varepsilon)y} d(n)n^{-1/4}I_n = \sum_{n \leqslant y} d(n)n^{\sigma-1+it}\chi^2(\sigma + it)$$

$$+ O\left(t^{-2\sigma} \sum_{n \leqslant 2y} d(n)n^{\sigma-1}\right) + O(x^{1/2-\sigma}\log t)$$

$$= \chi^2(s) \sum_{n \leqslant y} d(n)n^{s-1} + O(x^{1/2-\sigma}\log t)$$

and in view of (4.17) this means that we have proved the approximate functional equation (4.10).

4.3 THE APPROXIMATE FUNCTIONAL EQUATION FOR HIGHER POWERS

We pass now to the analogues of (4.1) and (4.10) for $\zeta^k(s)$, where $k \geqslant 3$ is a fixed integer. The approach that will be used is that of R. Wiebelitz (1951–1952), and is based on Hardy and Littlewood's proof (1929) of the approximate functional equation for $\zeta^2(s)$, so that this method yields an alternative proof of (4.10) for $0 < \sigma < 1$; $4\pi^2 xy = t^2$; $x, y, t > C > 0$, but it seemed interesting to treat the important case $k = 2$ by Voronoi's formula also. The proof will use a certain differencing argument [essentially recovering $\sum_{n \leqslant x} d_k(n)n^{-s}$ from the weighted sum $\sum_{n \leqslant x} d_k(n)n^{-s}(\log x/n)^{k-1}$] and estimates for power moments of the zeta-function, which will be extensively discussed in later chapters (and which do not depend on results of this section). New estimates for power moments of the zeta-function on the critical line lead to overall improvements of Wiebelitz's results, but as k grows the order of the error terms in the approximate functional equation becomes rather large, which is to be only expected. Therefore for practical reasons the detailed analysis is carried out only for $k \leqslant 12$.

For simplicity of writing we shall use the notation

$$(4.20) \quad X(s) = \chi^k(s), \quad \log T = -\frac{X'(\frac{1}{2} + it)}{X(\frac{1}{2} + it)} = -k\frac{\chi'(\frac{1}{2} + it)}{\chi(\frac{1}{2} + it)},$$

and furthermore as in the proof of (4.10) we may suppose that $t > 0$ without loss of generality. From (1.22) and (1.23) it follows that

$$\frac{\chi'(s)}{\chi(s)} = \log \pi - \frac{1}{2}\frac{\Gamma'}{\Gamma}\left(\frac{1}{2} - \frac{s}{2}\right) - \frac{1}{2}\frac{\Gamma'}{\Gamma}(s);$$

hence for $s = \frac{1}{2} + it$ the above equation shows that T, as defined by (4.20), is real and moreover using (A.35) we have

$$-\frac{\chi'(\frac{1}{2} + it)}{\chi(\frac{1}{2} + it)} = -\log 2\pi + \log t + O(t^{-2}),$$

and this gives

$$(4.21) \qquad\qquad T = (t/2\pi)^k + O(t^{k-2}).$$

Further we suppose that $xy = (t/2\pi)^k, 0 < \sigma < 1$, and we define

$$(4.22) \quad R_x(s) = \frac{x^{1-s}}{1-s}\sum_{\nu=0}^{k-1}\sum_{\rho=\nu+1}^{k} a_{-\rho,k} r_{\rho,\nu}(1-s)^{1+\nu-\rho}\log^\nu x,$$

where $a_{j,k}$ is the coefficient of $(s-1)^j$ in the Laurent expansion of $\zeta^k(s)$ at $s = 1$ and

$$r_{j,m} = (m!)^{-1}\sum_{i=0}^{j-m-1}(-1)^i\binom{k+i-1}{i}\binom{k-1}{j-i-m-1}.$$

Therefore in general we have

$$(4.23) \quad R_x(s) + \chi^k(s)R_y(1-s) \ll x^{1/2-\sigma}t^{-1}(x+y)^{1/2}\log^{k-1}t,$$

while for $k = 3$ we may write for some absolute constant D

$$(4.24) \quad R_x(s) = \frac{x^{1-s}}{1-s}\left(\frac{1}{2}\log^2 x + 3\gamma\log x + D\right) + O(x^{1-\sigma}t^{-2}\log t).$$

We shall now give a proof of the approximate functional equation for $\zeta^k(s)$ for $3 \le k \le 12$, although it has been already remarked that the method of proof may be used both when $k = 2$ and when $k > 13$, but in the latter case the error terms tend to be large as k increases. The approximate functional equation that will be proved is contained in

THEOREM 4.3. With the notation introduced above we have

$$(4.25) \quad \zeta^k(s) = \sum_{n \leqslant x} d_k(n) n^{-s} + \chi^k(s) \sum_{n \leqslant y} d_k(n) n^{s-1} - R_x(s)$$

$$-\chi^k(s) R_y(1 - s) + \Delta_k(x, y),$$

where $xy = (t/2\pi)^k$, $0 < \sigma < 1$; $x, y, t > C > 0$ and $\Delta_k(x, y)$ may be considered as an error term which depends on k, x, and y. We have uniformly in σ

$$(4.26) \quad \Delta_3(x, y) \ll x^{1/2-\sigma} t^{1/8+\varepsilon}, \qquad \Delta_4(x, y) \ll x^{1/2-\sigma} t^{13/48+\varepsilon},$$

and

$$(4.27) \quad \Delta_k(x, y) \ll t^\varepsilon \{ x^{1/2-\sigma} \min(x^{1/2}, y^{1/2}) t^{-2}$$

$$+ (x + y)^{1/2} x^{1/2-\sigma} t^{-\beta} + x^{1/2-\sigma} t^{(31k-52)/216} \}$$

for $5 \leqslant k \leqslant 12$, where $\beta = (31k - 52)/(27k - 108)$.

In case $k = 3$ or $k = 4$ one may obtain special results from (4.25) and (4.26) analogous to (4.2) and (4.11). These are

(4.28)

$$\zeta^3(\tfrac{1}{2} + it) \chi^{-3/2}(\tfrac{1}{2} + it) = 2 \operatorname{Re} \left(\chi^{3/2}(\tfrac{1}{2} + it) \sum_{n \leqslant (t/2\pi)^{3/2}} d_3(n) n^{-1/2+it} \right)$$

$$+ O(t^{1/8+\varepsilon}),$$

and

$$(4.29) \quad |\zeta(\tfrac{1}{2} + it)|^4 = 2 \operatorname{Re} \left(\chi^2(\tfrac{1}{2} + it) \sum_{n \leqslant (t/2\pi)^2} d_4(n) n^{-1/2+it} \right)$$

$$+ O(t^{13/48+\varepsilon}),$$

which follows with $s = \tfrac{1}{2} + it$, $x = y = (t/2\pi)^{k/2}$ ($k = 3$ or 4), as the terms with R_x and R_y are by (4.23) absorbed into the O-term of (4.28) and (4.29), respectively.

Proof of Theorem 4.3. We begin the proof of (4.25) by remarking that for technical reasons the condition $xy = (t/2\pi)^k$ is replaced by $xy = T$

[see (4.20)]. The error that is made in this process seems to be $\ll x^{1/2-\sigma}\min(x^{1/2}, y^{1/2})t^{\varepsilon-2}$, which is negligible in (4.26) and present in (4.27). We shall begin the proof of the general (4.25), but at a suitable point we shall distinguish the cases $k = 3$, $k = 4$, and $k > 4$. For notational convenience we define

$$(4.30) \quad \Phi_1(u) = \frac{\Phi(u)}{(u-s)^2} = \frac{T^{u-s}X(u) - X(s)}{(u-s)^2} \qquad \left(-\tfrac{1}{2} \leqslant \sigma \leqslant \tfrac{3}{2}\right)$$

so that for $\operatorname{Re} u \leqslant \tfrac{1}{2}$ and also for $\operatorname{Re} u < \min(\sigma, 1)$ the function $\Phi_1(u)$ is seen to be regular and moreover uniformly in s for $\operatorname{Re} u = \tfrac{1}{2}$ we have

$$(4.31) \quad \Phi(u) \ll t^{k/2 - k\sigma}\min(1, t^{-1}|s-u|^2).$$

In the course of the proof we shall need the power moment estimates

$$(4.32) \quad \int_{T-G}^{T+G}\left|\zeta(\tfrac{1}{2} + it)\right|^k dt \ll GT^{(k-4)c+\varepsilon}, \qquad G \geqslant T^{2/3}, k \geqslant 4,$$

where $\zeta(\tfrac{1}{2} + it) < t^{c+\varepsilon}$ (so that in view of Corollary 7.1 one may take $c = 35/216$ and

$$(4.33) \quad \int_0^T\left|\zeta(\tfrac{1}{2} + it)\right|^k dt \ll T^{1+(k-4)/8+\varepsilon}, \qquad 4 \leqslant k \leqslant 12.$$

The estimate (4.32) is a trivial consequence of a result of H. Iwaniec (1979/1980) (see Section 7.6), while (4.33) is contained in Theorem 8.3.

The first step in the proof is to use the inversion formula (A.11) to obtain

$$I = (2\pi i)^{-1}\int_{2-i\infty}^{2+i\infty}\zeta^k(s + w)x^w w^{-k}dw$$

$$= \frac{1}{(k-1)!}\sum_{n\leqslant x}d_k(n)n^{-s}\log^{k-1}(x/n) = S_x.$$

The line of integration is moved to $\operatorname{Re} w = -\gamma$, where $0 < \gamma < \tfrac{3}{4}$, $\sigma - 1 < \gamma, \gamma \neq \sigma, \gamma \neq \sigma - \tfrac{1}{2}$. There are poles of the integrand at $w = 0$ and $w = 1 - s$ with respective residues

$$F_x = \sum_{m=0}^{k-1}\frac{(\zeta^k(s))^{(m)}}{m!\,(k-1-m)!}(\log x)^{k-1-m}$$

and

$$Q_x = \frac{x^{1-s}}{(k-1)!(1-s)^k}\sum_{m=0}^{k-1}\frac{(-1)^m(k+m-1)!}{m!\,(1-s)^m}\sum_{r=m+1}^{k}a_{-r,k}\frac{\log^{r-m-1}x}{(r-m-1)!}.$$

Hence by the residue theorem

(4.34)

$$J_0 = (2\pi i)^{-1} \int_{-\gamma-i\infty}^{-\gamma+i\infty} \zeta^k(s+w) x^w w^{-k} \, dw = I - F_x - Q_x = S_x - F_x - Q_x.$$

Setting $s + w = z$, substituting x by T/y, and using the functional equation (1.23) we obtain

(4.35) $$J_0 = X(s)(2\pi i)^{-1} \int_{\sigma-\gamma-i\infty}^{\sigma-\gamma+i\infty} \zeta^k(1-z) y^{s-z}(z-s)^{-k} \, dz$$

$$+ (2\pi i)^{-1} \int_{\sigma-\gamma-i\infty}^{\sigma-\gamma+i\infty} \zeta^k(1-z) \Phi(z) y^{s-z}(z-s)^{-k} \, dz$$

$$= X(s) J_1 + J_2,$$

say. For $\sigma < \gamma$ we have by (A.11)

(4.36) $$J_1 = \frac{(-1)^k}{(k-1)!} \sum_{n \le y} d_k(n) n^{s-1} (\log x/n)^{k-1} = S_y,$$

similarly to the notation already introduced in evaluating the integral I. For $\sigma > \gamma$ we must take into account the pole $z = 0$, where the integrand has a residue

$$Q_y = \frac{(-1)^k y^s}{s^k(k-1)!} \sum_{m=0}^{k-1} \frac{(-1)^m(k+m-1)!}{m! s^m} \sum_{r=m+1}^{k} a_{-r,k} \frac{\log^{r-m-1}}{(r-m-1)!},$$

so that altogether

$$J_1 = S_y - \varepsilon_\gamma Q_y,$$

where $\varepsilon_\gamma = 0$ if $\sigma < \gamma$ and $\varepsilon_\gamma = 1$ if $\sigma > \gamma$.

The line of integration in J_2 is moved to $\mathrm{Re}\, z = \frac{1}{4}$, and for $\sigma < \gamma$ the pole $z = 0$ of the integrand is passed. In calculating the residue note that $X(0) = X'(0) = \cdots = X^{(k-1)}(0) = 0$, since in $X(u)$ and its first $k-1$ derivatives the factor $\sin(\pi u/2)$ comes in. This leads to

(4.37)

$$J_2 = (2\pi i)^{-1} \int_{1/4-i\infty}^{1/4+i\infty} \zeta^k(1-z) \Phi(z)(z-s)^{-k} y^{s-z} \, dz - (1-\varepsilon_\gamma) X(s) Q_y$$

$$= J_y - (1-\varepsilon_\gamma) X(s) Q_y.$$

Inserting the expressions (4.36) and (4.37) in (4.35) we obtain

$$(4.38) \quad F_x - S_x + Q_x = -X(s)J_1 - J_2 = -X(s)(S_y - Q_y) - J_y.$$

At this stage of the proof we shall use a differencing argument. The underlying idea is that (4.38) remains true if x and y are replaced by $xe^{\nu h}$ and $xe^{-\nu h}$, respectively, where $0 < h \leqslant 1$ and ν is an integer for which $\nu \leqslant k - 1$, and moreover h will be suitably chosen later. Now we shall sum (4.38) with weight $(-1)^\nu \binom{k-1}{\nu}$ for $0 \leqslant \nu \leqslant k - 1$ to recover the approximate functional equation by means of the elementary identity

$$(4.39) \qquad \sum_{\nu=0}^{m} (-1)^\nu \binom{m}{\nu} \nu^p = \begin{cases} 0 & p < m \\ m! & p = m \end{cases}$$

and the estimate

$$(4.40) \qquad e^z = \sum_{n=0}^{m-1} \frac{z^n}{n!} + O(|z|^m), \qquad m \geqslant 2, \quad a \leqslant \mathrm{Re}\, z \leqslant b,$$

where a and b are fixed.

To distinguish better the sums which will arise in this process we introduce left indexes to obtain then from (4.38)

$$\sum_{\nu=0}^{k-1} (-1)^\nu \binom{k-1}{\nu} \left({}_\nu F_x - {}_\nu S_x + {}_\nu Q_x + X(s)\,{}_\nu S_y - X(s)\,{}_\nu Q_y + {}_\nu J_y \right) = 0,$$

or abbreviating,

$$(4.41) \qquad \overline{F}_x - \overline{S}_x + \overline{Q}_x + X(s)\overline{S}_y - X(s)\overline{Q}_x + \overline{J}_y = 0.$$

Each term in (4.41) will be evaluated now separately. We have

$$\overline{F}_x = \sum_{m=0}^{k-1} \frac{(\zeta^k(s))^{(m)}}{m!\,(k-1-m)!} A_m(x),$$

where we have set

$$A_m(x) = \sum_{\nu=0}^{k-1} (-1)^\nu \binom{k-1}{\nu} (\log x + \nu h)^{k-1-m}.$$

Using (4.39) it follows that

$$A_m(x) = \sum_{r=0}^{k-1} \binom{k-1-m}{r} h^r \log^{k-1-m-r} x \sum_{\nu=0}^{k-1} (-1)^\nu \binom{k-1}{\nu} \nu^r$$

$$= h^{k-1}(k-1)!$$

when $m = 0$, and $A_m(x) = 0$ when $m > 0$, so that we have

$$(4.42) \qquad \overline{F}_x = h^{k-1} \zeta^k(s),$$

and this is exactly what is needed for the approximate functional equation that will eventually follow on dividing (4.41) by h^{k-1} with a suitably chosen h. Consider next

(4.43)

$$\overline{S}_x = \frac{1}{(k-1)!} \sum_{n \leqslant x} d_k(n) n^{-s} \sum_{\nu=0}^{k-1} \binom{k-1}{\nu} (-1)^\nu (\nu h + \log x/n)^{k-1}$$

$$+ \frac{1}{(k-1)!} \sum_{\nu=0}^{k-1} (-1)^\nu \binom{k-1}{\nu} \sum_{x \leqslant n \leqslant xe^{\nu h}} d_k(n) n^{-s} (\nu h + \log x/n)^{k-1}$$

$$= \sum_1 + \sum_2,$$

say. Analogously to the evaluation of $\overline{F}_x$ it follows on using (4.39) again that

$$(4.44) \qquad \sum_1 = h^{k-1} \sum_{n \leqslant x} d_k(n) n^{-s},$$

and we estimate $\sum_2$ trivially [using $d_k(n) \ll n^\varepsilon$] as

$$(4.45) \quad \left| \sum_2 \right| \leqslant \frac{1}{(k-1)!} \sum_{\nu=0}^{k-1} \binom{k-1}{\nu} (\nu h)^{k-1} x^{-\sigma} \sum_{x < n \leqslant xe^{(k-1)h}} d_k(n)$$

$$\ll h^{k-1} x^{-\sigma} t^\varepsilon \big(1 + x(e^{(k-1)h} - 1)\big) \ll t^\varepsilon \big(h^{k-1} x^{-\sigma} + h^k x^{1-\sigma}\big).$$

Estimating analogously $\overline{S}_y$ we obtain

$$(4.46) \quad -X(s)\overline{S}_y = h^{k-1} \chi^k(s) \sum_{n \leqslant y} d_k(n) n^{s-1} + O\big(h^{k-1} t^\varepsilon x^{1/2-\sigma} y^{-1/2}\big)$$

$$+ O\big(h^k t^\varepsilon x^{1/2-\sigma} y^{1/2}\big).$$

Next we have

$$\overline{Q}_x = \frac{x^{1-s}}{(k-1)!(1-s)^k} \sum_{\mu=0}^{k-1} \frac{(-1)^\mu (k+\mu-1)!}{\mu!(1-s)^\mu} \sum_{\rho=\mu+1}^{k} \frac{a_{-\rho,k}}{(\rho-\mu-1)!} B_{\mu\rho},$$

where

$$B_{\mu\rho} = \sum_{\nu=0}^{k-1} (-1)^\nu \binom{k-1}{\nu} e^{\nu h(1-s)} (\nu h + \log x)^{\rho-\mu-1}.$$

Using (4.39) and (4.40) with $m = k$ it follows

$$B_{\mu\rho} = (k-1)! h^{k-1} \sum_{\substack{n+m=k-1 \\ 0<m<\rho-m-1}} (n!)^{-1}(1-s)^n \log^{\rho-\mu-1-m} x \binom{\rho-\mu-1}{m}$$

$$+ O(h^k t^k \log^{\rho-\mu-1} x).$$

If we set $\nu = \rho - \mu - 1 - m$, change the order of summation, and collect the constants we obtain

$$(4.47) \quad \overline{Q}_x = x^{1-s}(1-s)^{-1} h^{k-1} \sum_{\nu=0}^{k-1} \sum_{\rho=\nu+1}^{k} a_{-\rho,k} r_{\rho,\nu} (1-s)^{1+\nu-\rho} \log^\nu x$$

$$+ O(h^k x^{1+\varepsilon-\sigma})$$

$$= h^{k-1} R_x(s) + O(h^k x^{1+\varepsilon-\sigma}).$$

The same argument applies to $\overline{Q}_y$ and yields

$$(4.48) \qquad \overline{Q}_y = -h^{k-1} R_y(1-s) + O(h^k x^{1/2+\varepsilon-\sigma} y^{1/2}).$$

Therefore we are left with the evaluation of

$$\overline{J}_y = (2\pi i)^{-1} \int_{1/4-i\infty}^{1/4+i\infty} \zeta^k(1-z) \Phi(z) y^{s-z} (z-s)^{-k}$$

$$\times \left(\sum_{m=0}^{k-1} (-1)^m \binom{k-1}{m} e^{-hm(s-z)} \right) dz.$$

Observing that $\Phi(z)$ has a double zero at $z = s$ and that

$$\sum_{m=0}^{k-1} (-1)^m \binom{k-1}{m} e^{-hm(s-z)} = (1 - e^{-hs+hz})^{k-1}$$

has a zero of order $k - 1$ at $z = s$, we can move the line of integration in $\overline{J}_y$ to

$\text{Re}\, z = \frac{1}{2}$ to obtain with $w = u + iv$ (u, v real)

(4.49)

$$\bar{J}_y = (2\pi i)^{-1} \int_{1/2-i\infty}^{1/2+i\infty} \zeta^k(1-w)\Phi(w)(w-s)^{-k}y^{s-w}$$

$$\times \left(\sum_{m=0}^{k-1} (-1)^m \binom{k-1}{m} e^{-hm(s-w)} \right) dw$$

$$= (2\pi i)^{-1} \left(\int_{|v-t|\leqslant G} \cdots dw + \int_{|v-t|>G,\,|v|\leqslant 2t^\beta} \cdots dw + \int_{|v|\geqslant 2t^\beta} \cdots dw \right)$$

$$= j_1 + j_2 + j_3,$$

say. Here β ($\geqslant 1$) is the number appearing in (4.27), and $t^\varepsilon < G \leqslant t^{2/3}$ is a parameter that will be suitably chosen. We distinguish now the cases $3 \leqslant k \leqslant 4$ and $k > 4$, and treat first the latter case. For j_1 we use (4.31) in the form

$$\Phi(w) \ll t^{k/2-k\sigma-1}|s-w|^2,$$

and majorize the sum $\sum_{m=0}^{k-1}$ in (4.49) by $O(h^{k-1}|s-w|^{k-1})$, which follows when we combine (4.39) and (4.40). Therefore we have

(4.50)

$$j_1 \ll h^{k-1}y^{\sigma-1/2}t^{k(1/2-\sigma)-1}G\int_{t-G}^{t+G}\left|\zeta(\tfrac{1}{2}+iv)\right|^k dv$$

$$\ll h^{k-1}x^{1/2-\sigma}Gt^{-1}\int_{t-t^{2/3}}^{t+t^{2/3}}\left|\zeta(\tfrac{1}{2}+iv)\right|^k dv$$

$$\ll h^{k-1}x^{1/2-\sigma}Gt^{(k-4)c+\varepsilon-1/3},$$

where (4.32) was used. To estimate j_2 we use $\Phi(w) \ll t^{k/2-k\sigma}$ and the same majorization for $\sum_{m=0}^{k-1}$ as above to obtain

(4.51)

$$j_2 \ll h^{k-1}x^{1/2-\sigma}\int_{|v-t|>G,\,|v|\leqslant 2t^\beta}\left|\zeta(\tfrac{1}{2}+iv)\right|^k|v-t|^{-1} dv.$$

The integral in (4.51) is split into subintegrals $j_{21}, j_{22}, j_{23}, j_{24}, j_{25}$ over the intervals $[-2t^\beta, -2t]$, $[-2t, t/2]$, $[t/2, t-G]$, $[t+G, 2t]$, $[2t, 2t^\beta]$, respectively. Using (4.33) it follows at once that

$$j_{22} \ll t^{(k-4)/8+\varepsilon},$$

and the other integrals are integrated by parts and then estimated. For example, for j_{23} we have with

$$H(v) = -\int_v^{t-G}\left|\zeta(\tfrac{1}{2}+ix)\right|^k dx$$

and (4.33) that

$$(4.52) \qquad j_{23} = H(v)(t-v)^{-1}\Big|_{t/2}^{t-G} - \int_{t/2}^{t-G} H(v)(t-v)^{-2}\,dv$$

$$\ll t^{(k-4)/8+\varepsilon} + G^{-1}t^{1+(k-4)/8+\varepsilon} \ll G^{-1}t^{1+(k-4)/8+\varepsilon},$$

since $G \leqslant t^{2/3}$, and the same bound similarly holds for j_{24}. Using again (4.33) we have

$$j_{21} + j_{25} \ll t^{\beta((k-4)/8+\varepsilon)},$$

and so

$$(4.53) \quad j_1 + j_2 \ll t^{\varepsilon}\big(h^{k-1}x^{1/2-\sigma}Gt^{(k-4)c-1/3} + h^{k-1}x^{1/2-\sigma}G^{-1}t^{1+(k-4)/8}$$

$$+ h^{k-1}x^{1/2-\sigma}t^{\beta(k-4)/8}\big).$$

We choose now G in such a way that the first two terms on the right-hand side of (4.53) are equal. Thus with $c = \frac{35}{216}$ we let

$$G = t^{2/3+(k-4)(1/8-c)/2}.$$

This choice of G obviously satisfies the condition $t^{\varepsilon} < G \leqslant t^{2/3}$, and then we obtain

$$(4.54) \qquad j_1 + j_2 \ll t^{\varepsilon}h^{k-1}x^{1/2-\sigma}\big(t^{(31k-52)/216} + t^{\beta(k-4)/8}\big)$$

$$\ll t^{\varepsilon}h^{k-1}x^{1/2-\sigma}t^{(31k-52)/216},$$

if as in (4.27) we take

$$(4.55) \qquad\qquad\qquad \beta = \frac{31k-52}{27k-108} > 1.$$

Integration by parts and (4.33) give

$$(4.56) \qquad j_3 \ll x^{1/2-\sigma}\int_{2t^{\beta}}^{\infty}\Big|\zeta\big(\tfrac{1}{2}+iv\big)\Big|^{k}v^{-k}\,dv \ll x^{1/2-\sigma}t^{\beta(4-7k+\varepsilon)/8},$$

where we used $\Phi(w) \ll t^{k/2-k\sigma}$ and $\sum_{m=0}^{k-1} \ll 1$ for the sum appearing in (4.49), since $e^{-hm(s-w)} = e^{-hm(\sigma-1/2)} \ll 1$. Finally, combining all the estimates (4.41)–(4.56) we obtain

$$(4.57) \quad h^{k-1}\left(\zeta^k(s) - \sum_{n \leqslant x} d_k(n)n^{-s} - \chi^k(s) \sum_{n \leqslant y} d_k(n)n^{s-1} \right.$$

$$\left. + R_x(s) + \chi^k(s)R_y(1-s) \right)$$

$$\ll t^\varepsilon \left(h^k(x+y)^{1/2}x^{1/2-\sigma} + h^{k-1}x^{1/2-\sigma}t^{(31k-52)/216} \right.$$

$$\left. + x^{1/2-\sigma}t^{\beta(4-7k)/8} \right).$$

Choosing $h = t^{-\beta}$, where β is already defined by (4.55), it is seen that the last two terms in (4.57) are equal, and the approximate functional equation follows from (4.57) on dividing by h^{k-1}, if we recall that the error made by replacing the condition $xy = (t/2\pi)^k$ by $xy = T$ is $\ll x^{1/2-\sigma}\min(x^{1/2}, y^{1/2})t^{\varepsilon-2}$.

This settles the case $k > 4$, and we have still to consider the cases $k = 3$ and $k = 4$. The only changes in the proof will be in the estimation of the integrals j_1, j_2, j_3 appearing in (4.49), where sharper estimates than those used for the general case $k > 4$ are available.

For $k = 3$ we choose $G = 2t^{1/2}$ in (4.49) to obtain with the third moment estimate (7.75)

$$j_1 \ll h^2 x^{1/2-\sigma}t^{-1/2}\int_{t-2t^{1/2}}^{t+2t^{1/2}}\left|\zeta\left(\tfrac{1}{2}+iv\right)\right|^3 dv \ll h^2 x^{1/2-\sigma}t^{1/8+\varepsilon}.$$

For j_{23} we use (7.75) again to obtain

$$j_{23} \ll h^2 x^{1/2-\sigma}\int_{t/2}^{t-2t^{1/2}}\left|\zeta\left(\tfrac{1}{2}+iv\right)\right|^3 (t-v)^{-1} dv$$

$$\ll h^2 x^{1/2-\sigma}\sum_{n=1}^{O(t^{1/2})}\int_{t-2(n+1)t^{1/2}}^{t-2nt^{1/2}}\left|\zeta\left(\tfrac{1}{2}+iv\right)\right|^3 (t-v)^{-1} dv$$

$$\ll h^2 x^{1/2-\sigma}\sum_{n=1}^{O(t^{1/2})}t^{-1/2}n^{-1}\int_{t-2nt^{1/2}-2t^{1/2}}^{t-2nt^{1/2}}\left|\zeta\left(\tfrac{1}{2}+iv\right)\right|^3 dv$$

$$\ll h^2 x^{1/2-\sigma}t^{\varepsilon-1/2}\sum_{n=1}^{O(t^{1/2})}n^{-1}(t^{1/2}+t^{5/8}) \ll h^2 x^{1/2-\sigma}t^{1/8+\varepsilon}.$$

At last using $\int_0^T|\zeta(\tfrac{1}{2}+it)|^3\,dt \ll T^{1+\varepsilon}$ we obtain

$$j_1 + j_2 + j_3 \ll t^\varepsilon x^{1/2-\sigma}\left(h^2 t^{1/2} + t^{-2\beta}\right).$$

Thus for $k = 3$ we obtain (4.57), only the right-hand side will now be

$$\ll t^\varepsilon\left(h^3(x+y)^{1/2}x^{1/2-\sigma} + h^2 x^{1/2-\sigma}t^{1/8} + x^{1/2-\sigma}t^{-2\beta}\right).$$

Now we set $\beta = \frac{11}{8}$, $h = t^{-\beta}$. Then from $x + y \ll t^3$ we infer that

$$h^3(x+y)^{1/2}x^{1/2-\sigma} \ll h^2 x^{1/2-\sigma}t^{1/8},$$

and dividing (4.57) by h^2 we obtain the desired approximate functional equation

$$(4.58) \qquad \zeta^3(s) = \sum_{n \leqslant x} d_3(n)n^{-s} + \chi^3(s)\sum_{n \leqslant y} d_3(n)n^{s-1} - R_x(s)$$

$$-\chi^3(s)R_y(1-s) + O\left(x^{1/2-\sigma}t^{1/8+\varepsilon}\right).$$

It remains to consider yet the case $k = 4$, where the estimation is identical with the general case up to (4.51), only now for $H(v)$ we shall use

$$\int_0^T \left|\zeta(\tfrac{1}{2}+it)\right|^4 dt = T\sum_{j=0}^4 a_j\log^{4-j}T + O\left(T^{7/8+\varepsilon}\right),$$

which is a result of D. R. Heath-Brown (see Notes of Chapter 5). Therefore for j_{23} in (4.51) with $k = 4$ we have

$$j_{23} \ll t^\varepsilon\left(1 + \int_{t/2}^{t-G}(t - v - G + t^{7/8})(t-v)^{-2}\,dv\right) \ll G^{-1}t^{7/8+\varepsilon},$$

which means that we have saved a factor $t^{1/8}$ from the general estimate used in (4.51). We have than

$$j_1 + j_2 \ll t^\varepsilon\left(h^3 x^{1/2-\sigma}Gt^{-1/3} + h^3 x^{1/2-\sigma}G^{-1}t^{7/8}\right) \ll h^3 x^{1/2-\sigma}t^{13/48+\varepsilon}$$

for $G = t^{29/48} < t^{2/3}$. Since $j_3 \ll x^{1/2-\sigma}t^{\beta(\varepsilon-3)}$ we obtain (4.57), where the right-hand side will be

$$\ll t^\varepsilon\left(h^4(x+y)^{1/2}x^{1/2-\sigma} + h^3 x^{1/2-\sigma}t^{13/48+\varepsilon} + x^{1/2-\sigma}t^{-3\beta}\right).$$

The result given by (4.26) follows for $\beta = 2$, $h = t^{-\beta}$ on dividing (4.57) (with $k = 4$) by h^3, since

$$h(x+y)^{1/2}x^{1/2-\sigma} \ll t^{-2}(t^4)^{1/2}x^{1/2-\sigma} = x^{1/2-\sigma}.$$

4.4 THE REFLECTION PRINCIPLE

The approximate functional equations discussed in Sections 4.1–4.3 have a symmetric property if $x = y$, especially when $\sigma = \frac{1}{2}$, which allows one to obtain useful expressions like (4.2), (4.11), (4.28), and (4.29). However, when one seeks estimates for averages of powers of moduli of the zeta-function in the critical strip, it turns out that the approximate functional equations of the type just considered have two shortcomings. First, the lengths of the sums over n depend on t, and second, the error terms for $k \geqslant 3$ and $\sigma = \frac{1}{2}$ already are not small (i.e., they are not $\ll t^\varepsilon$). We shall proceed now to derive another type of approximate functional equation, which though lacking symmetry is in many problems concerning averages of the zeta-function quite adequate. The idea, which permeates the whole theory since the pioneering days of Riemann, is to use the functional equation in the form $\zeta^k(w) = \chi^k(w)\zeta^k(1 - w)$ for some w with $\operatorname{Re} w < 0$, to split the series $\zeta^k(1 - w) = \sum_{n=1}^{\infty} d_k(n)n^{w-1}$ at some suitable M and estimate the terms with $n > M$ trivially. This approach is very flexible, and the error terms that will arise will be small. The starting point is the Mellin integral (A.7) where we set $x = Y^h$, $s = w/h$ and suppose $Y, h > 0$. In view of $\Gamma(z + 1) = z\Gamma(z)$ we obtain by moving the line of integration

$$(4.59) \qquad e^{-Y^h} = (2\pi i)^{-1} \int_{2-i\infty}^{2+i\infty} Y^{-w}\Gamma(1 + w/h)w^{-1}\,dw.$$

Replacing now Y by n/Y and using $\sum_{n=1}^{\infty} d_k(n)n^{-z} = \zeta^k(z)$ $(\operatorname{Re} z > 1)$ it follows when $\sigma \geqslant 0$ and $k \geqslant 1$ is an integer that

$$(4.60)$$

$$\sum_{n=1}^{\infty} e^{-(n/Y)^h}d_k(n)n^{-s} = (2\pi i)^{-1} \int_{2-i\infty}^{2+i\infty} \zeta^k(s + w)Y^w\Gamma(1 + w/h)w^{-1}\,dw.$$

Now we suppose $s = \sigma + it, 0 \leqslant \sigma \leqslant 1, h^2 \leqslant t \leqslant T, h = \log^2 T, 1 \ll Y \ll T^c$ for some fixed $c > 0$, and we move the line of integration in (4.60) to $\operatorname{Re}(s + w) = -\frac{1}{2}$. Using Stirling's formula (A.34) it is seen that the residue coming from the pole $w = 1 - s$ is $o(1)$, while the residue at the pole $w = 0$ is $\zeta^k(s)$. Using the functional equation (1.23) we have then

$$(4.61) \qquad \sum_{n=1}^{\infty} d_k(n)e^{-(n/Y)^h}n^{-s} = \zeta^k(s) + o(1) + I_1 + I_2,$$

say, where for some $1 \ll M \ll T^c$

$$(4.62) \qquad I_1 = (2\pi i)^{-1} \int_{\operatorname{Re}(s+w)=-1/2} \chi^k(s + w)$$

$$\times \sum_{n \leqslant M} d_k(n)n^{w+s-1}Y^w\Gamma(1 + w/h)w^{-1}\,dw,$$

and

$$(4.63) \qquad I_2 = (2\pi i)^{-1} \int_{\text{Re}(s+w)=-1/2} \chi^k(s+w)$$

$$\times \sum_{n>M} d_k(n) n^{w+s-1} Y^w \Gamma(1+w/h) w^{-1} dw.$$

In I_2 we move the line of integration to $\text{Re}(s+w) = -h/2$, noting that the integrand is regular for $-h/2 \leqslant \text{Re}(s+w) \leqslant -\frac{1}{2}$, and aiming to choose M in such a way that $I_2 = o(1)$ as $T \to \infty$. With $w = u + iv$ (u, v real), we obtain using Stirling's formula

$$I_2 \ll \int_{-\infty}^{\infty} \left| \chi\left(-\frac{h}{2} + iv + it \right) \right|^k$$

$$\times \sum_{n>M} d_k(n) n^{-1-h/2} Y^{-h/2} |\Gamma(\tfrac{1}{2} - \sigma/h + iv/h)| \, dv$$

$$\ll (MY)^{-h/2} \log^k T \int_0^T (t+v)^{k(1+h)/2} \, dv + \int_T^{\infty} e^{-v/h} \, dv$$

$$\ll (MY)^{-h/2} T \log^k T (2T)^{k(1+h)/2} + o(1) = o(1)$$

if

$$(4.64) \qquad M \geqslant (3T)^k Y^{-1}.$$

The flexibility of this method is best seen in various possibilities for the estimation of I_1 in (4.62). The line of integration in I_1 may be moved to $\text{Re}(s+w) = \alpha, 0 < \alpha < 1$ fixed and $\alpha \neq \sigma$, so that the sum appearing in (4.62) will be "reflected", hence the name "reflection principle." Letting $\delta_\alpha = 1$ if $\alpha > \sigma$ and $\delta_\alpha = 0$ if $\alpha < \sigma$ we obtain by the residue theorem

$$(4.65) \qquad I_1 = -\delta_\alpha \chi^k(s) \sum_{n \leqslant M} d_k(n) n^{s-1}$$

$$+ (2\pi i)^{-1} \int_{\text{Re}(s+w)=\alpha} \chi^k(s+w)$$

$$\times \sum_{n \leqslant M} d_k(n) n^{w+s-1} Y^w \Gamma\left(1 + \frac{w}{h} \right) \frac{dw}{w}.$$

The terms with $n > 2Y$ in (4.61) are trivially $o(1)$, and the part of the integral in (4.65) with $|v| = |\text{Im} w| \geqslant h^2$ is also $o(1)$ by Stirling's formula, so that combining (4.61)–(4.65) we obtain

THEOREM 4.4. For $s = \sigma + it, 0 \leqslant \sigma \leqslant 1, k \geqslant 1$ a fixed integer, $h = \log^2 T$, $h^2 \leqslant t \leqslant T$, $1 \ll Y \ll T^c$, $0 < \alpha < 1$ ($\alpha \neq \sigma$), $\delta_\alpha = 1$ if $\alpha > \sigma$ and zero otherwise, $M \geqslant (3T)^k Y^{-1}$, we have uniformly in σ and t

$$(4.66) \quad \zeta^k(s) = \sum_{n \leqslant 2Y} d_k(n) e^{-(n/Y)^h} n^{-s} + \delta_\alpha \chi^k(s) \sum_{n \leqslant M} d_k(n) n^{s-1}$$

$$- (2\pi i)^{-1} \int_{\substack{\mathrm{Re}(\sigma+w)=\alpha \\ |\mathrm{Im}\, w| \leqslant h^2}} \chi^k(s+w)$$

$$\times \sum_{n \leqslant M} d_k(n) n^{w+s-1} Y^w \Gamma\left(1 + \frac{w}{h}\right) \frac{dw}{w} + o(1).$$

Therefore we have obtained the desired type of the (unsymmetrical) approximate functional equation, where the lengths of the sums do not depend on t, but on T, and where the error term is $o(1)$. Another useful variant follows from (4.66) with $k = 1$, $\alpha = \frac{1}{2}$ when we replace Y by $2Y$ and subtract the resulting expressions, or we may proceed directly and observe that there is now no pole at $w = 0$ because of the zero $w = 0$ of $(2Y)^w - Y^w$. We obtain then

(4.67)

$$\sum_{n=1}^{\infty} \left(e^{-(n/2Y)^h} - e^{-(n/Y)^h}\right) n^{-s} \ll 1 + Y^{1/2-\sigma} \int_{-h^2}^{h^2} \left| \sum_{n \leqslant M} n^{-1/2+it+iv} \right| dv,$$

which will be very useful later in Chapter 11 for zero-density estimates. In (4.67) as in (4.66) we have

$$0 \leqslant \sigma \leqslant 1, \qquad h^2 \leqslant t \leqslant T, \qquad 1 \ll Y \ll T^c, \qquad h = \log^2 T, \qquad M \geqslant 3TY^{-1}.$$

Notes

The approximate functional equation (4.1) was proved by G. H. Hardy and J. E. Littlewood (1922, 1929). The proof given in the text follows the one given in Chapter 4 of E. C. Titchmarsh (1951). As shown by Titchmarsh (1951), this approach leads to a proof of the complicated Riemann–Siegel formula (4.3), proved first by C. L. Siegel (1932). The recent dissertation of W. Gabcke (1979) contains a simplified proof of the Riemann–Siegel formula, together with explicit estimations of the error terms, which is useful in calculations relating to the zeros on the critical line.

E.C. Titchmarsh (1951) states Theorem 4.2 at the end of Chapter 4 without proof. However, the error term $O((x + y)^{1/2-\sigma} \log t)$ that he gives is incorrect, as was kindly pointed out to the author by M. Jutila. The function $f(x) = \sum_{n \leqslant x} d(n) n^{-s}$ is discontinuous at integers x with jumps $d(x) x^{-s}$, accordingly the error term in (4.10) must be at least $O(x^{-\sigma})$. But if x is small and σ is near unity, then $(x + y)^{1/2-\sigma} \log t$ is much smaller than $x^{-\sigma}$. The correct error term is $O(x^{1/2-\sigma} \log t)$, and this was obtained by Titchmarsh himself (1938). His method of proof there is an extension of the proof used by Hardy and Littlewood (1929) in the proof of the approximate functional equation for $\zeta(s)$. The first step in Titchmarsh's proof is to obtain an exact formula for

$\zeta^2(s)$, valid for $\sigma > -\frac{1}{4}$. This is

$$(4.68) \quad \zeta^2(s) = \sum_{n \leqslant x} d(n) n^{-s} - x^{-s} \sum_{n \leqslant x} d(n) + \frac{2s - s^2}{(s-1)^2} x^{1-s} + \frac{s}{s-1} x^{1-s}(2\gamma + \log x)$$

$$+ \frac{1}{4} x^{-s} - 2^{4s} \pi^{2s-2} s \sum_{n=1}^{\infty} d(n) n^{s-1} \int_{4\pi\sqrt{nx}}^{\infty} \left(K_1(v) + \frac{\pi}{2} Y_1(v) \right) v^{-2s} \, dv,$$

and using the asymptotic expansions for the Bessel functions K_1, Y_1 (see Chapter 3) one is led to the estimation of certain exponential integrals which eventually yield (4.10).

Although we have seen in Section 4.2 that the error term $O(x^{1/2-\sigma} \log t)$ in (4.10) is in general the best possible [the argument is due to M. Jutila (unpublished)], it has been found out recently that a better error term is possible for various ranges of σ, x, and t. Thus M. Jutila (unpublished) used Titchmarsh's formula (4.68), a subtle averaging process and the saddle-point method to obtain that the error term in (4.10) is

$$\ll x^{-\sigma} t^{1/2} \min(1, xt^{-1}) L \log(tx^{-1} + xt^{-1}) + x^{1-\sigma} t^{-1}(1 + tx^{-1}) \min(x^{\varepsilon} + L, y^{\varepsilon} + L),$$

where $L = \log t$. This is better than $O(x^{1/2-\sigma} L)$ except when x or y is bounded, or when these numbers are of the same order as t.

However, in the important case $x = y = t/(2\pi)$ much more is true, as shown by Y. Motohashi (1983). It occurred to Motohashi to use the standard splitting-up argument of Dirichlet, namely

$$(4.69) \quad \sum_{n \leqslant N} a_n d(n) = 2 \sum_{n \leqslant \sqrt{N}} \sum_{m \leqslant N/n} a_{nm} - \sum_{n \leqslant \sqrt{N}} \sum_{m \leqslant \sqrt{N}} a_{nm},$$

and to evaluate the sums involving $d(n)$ in (4.10) using (4.69) and the approximate functional equation for $\zeta(s)$. If $U = t/2\pi$, $u = U^{1/2}$, $E(s, x)$ is the error term in (4.1), then one obtains

$$(4.70) \quad \zeta^2(s) = \sum_{n \leqslant U} d(n) n^{-s} + \chi^2(s) \sum_{n \leqslant U} d(n) n^{s-1} + 2\chi(s) \sum_{n \leqslant u} n^{-1}$$

$$+ 2 \sum_{m \leqslant u} m^{-s} E(s, U/m) + 2\chi^2(s) \sum_{m \leqslant u} m^{s-1} E(1 - s, U/m) + E^2(s, u).$$

With the bound $E(s, x) \ll x^{-\sigma} + t^{1/2-\sigma} y^{\sigma-1}$ of Theorem 4.1 one obtains at once (4.10) for $x = y = t/2\pi$, but Motohashi carried the analysis much further than that. Namely, using for $E(s, x)$ the expression which arises in the proof of the Riemann–Siegel formula and some involved techniques, Motohashi deduces from (4.70) an analogue of the Riemann–Siegel formula (4.3) for $\zeta^2(s)$. The result is too complicated to be stated here in full, but in particular it gives

$$(4.71) \quad \zeta^2(s) = \sum_{n \leqslant t/2\pi} d(n) n^{-s} + \chi^2(s) \sum_{n \leqslant t/2\pi} d(n) n^{s-1} + O(t^{1/3-\sigma}),$$

which is not obtainable from Theorem 4.2.

The deduction of the approximate functional equation (4.10) for $\zeta^2(s)$ via the Voronoi summation formula was made by the author (1983b). However, the possibility of such an approach is mentioned by M. Jutila (1984a), who also indicated to the author how an averaging process and a modification of Atkinson's Theorem 2.2 dispose of the term $t^{1/2+\varepsilon} x^{-\sigma}$ in (4.19).

Note that the definition of $\Delta(x)$ given in Section 4.2 differs slightly from the one made in (3.1), but this can cause no confusion since the two expressions differ only by $O(x^{\varepsilon})$, which is absorbed in (4.15) in the error term $O(x^{1/2-\sigma} \log t)$.

To investigate more precisely the absolute convergence of the integral in (4.14) we proceed as follows. By Theorem 13.5 and the Cauchy–Schwarz inequality for integrals we have

$$\left| \int_M^{2M} x^{-s-1} \Delta(x)\, dx \right| \le \left(\int_M^{2M} x^{-2\sigma-2}\, dx \right)^{1/2} \left(\int_M^{2M} \Delta^2(x)\, dx \right)^{1/2}$$

$$\ll (M^{-1-2\sigma} M^{3/2})^{1/2} \ll M^{-\varepsilon}$$

for $\sigma > \frac{1}{4}$. Taking $M = N, 2N, 2^2 N, \ldots$ and adding up the estimates, it is seen that the integral in (4.14) is absolutely convergent for $\sigma > \frac{1}{4}$.

To see how the error term $O(x^{1/2-\sigma} \log t)$ appears in (4.15) we use (3.1) in evaluating $\int_x^N u^{-s} d\Delta(u)$. Writing

$$\Delta(x) = \sum_{n \le (1+\varepsilon)y} + \sum_{n > (1+\varepsilon)y}$$

the first sum in view of (3.1) and (3.11) gives the first sum with K_0 and Y_0 on the right-hand side of (4.15). Integration by parts gives

$$\int_x^N u^{-s} d\left(\sum_{n > (1+\varepsilon)y} \cdots \right)$$

$$= \left(\sum_{n > (1+\varepsilon)y} \cdots \right) u^{-s} \Big|_x^N + s \sum_{n > (1+\varepsilon)y} d(n) \int_x^N \left\{ -2\pi \left(\frac{u}{n} \right)^{1/2} (K_1(\cdots) \right\} du,$$

while (3.16) gives

$$x^{-s} \sum_{n > (1+\varepsilon)y} d(n) \left\{ -2(x/n)^{1/2} \left(K_1(4\pi\sqrt{nx}) + \frac{\pi}{2} Y_1(4\pi\sqrt{nx}) \right) \right\}$$

$$\ll x^{-\sigma} \left(|\Delta(x)| + (x/y)^{1/2} x^\varepsilon \right) \ll x^{1/2-\sigma} \log t,$$

since $y \gg t^\varepsilon$.

The approximate functional equation for $\zeta^k(s)$ in Theorem 4.3 is due to the author (1983b). The method of proof is based on R. Wiebelitz (1951–1952), who was guided by the work of Hardy and Littlewood (1929). Power moment estimates (4.32) and (4.33) were necessary for our proof, and Theorem 4.3 is one of the few results of this text whose proof is not self-contained. For comparison with our Theorem 4.3, we now present Wiebelitz's approximate functional equation for $\zeta^k(s)$.

Wiebelitz uses the estimate $\zeta(\frac{1}{2} + it) \ll t^{c+\varepsilon}$, $c = \frac{15}{92}$ [due to S.-H. Min (1949)], which was the best result available at the time of his writing, and assumes that $k \ge 3$ is a fixed integer, $xy = (|t|/2\pi)^k$; $x, y \gg 1$, $-\frac{1}{2} \le \sigma \le \frac{3}{2}$. Then (4.25) holds uniformly in σ with

$$\Delta_k(x, y) \ll (x+y)^{1/2} x^{1/2-\sigma} |t|^{-\beta} \log^{k-1} |t| + x^{1/2-\sigma} (x+y)^{2\beta c(k-4)/k+\varepsilon}$$

$$+ x^{1/2-\sigma} |t|^{(k-2)c} \log^2 |t| + x^{1/2-\sigma} \min(x^{1/2}, y^{1/2}) |t|^{-2},$$

where $\beta = \frac{3}{2}$ for $k = 3$ and $\beta = 23k/(15k + 32)$ for $k \ge 4$. The terms R_x and R_y in (4.25) for $\rho \ge \beta + \nu$ may be incorporated in the above error terms, which was done by Wiebelitz. Though the result is proved for $-\frac{1}{2} \le \sigma \le \frac{3}{2}$, it is really of interest for $0 \le \sigma \le 1$ (and especially for $\sigma \ge \frac{1}{2}$) in view of the functional equation for $\zeta(s)$.

When estimating Σ_2 in (4.45) Wiebelitz uses the asymptotic formula for $\sum_{n \le x} d_k(n)$, which will be extensively discussed in Chapter 13. This enabled him to obtain the term $h^{k-1} x^{-\sigma} \log^{k-1} |t|$, while in (4.45) we had $h^{k-1} x^{-\sigma} t^\varepsilon$. However, this approach introduces the error term $\Delta_k(x)$ in the general divisor problem, which ultimately affects Wiebelitz's estimate for $\Delta_k(x, y)$, as given above. Using the trivial estimate

$$\sum_{a < n \leqslant b} d_k(n) \ll b^\varepsilon (1 + b - a)$$

in (4.45) we managed to dispose of the error term $\Delta_k(x)$, and the estimates for power moments of the zeta-function that were used involve the factor t^ε, so that we would gain nothing in following Wiebelitz in the use of $\Delta_k(x)$ in (4.45). Estimates given in (4.26) and (4.27) for $\Delta_k(x, y)$ are clearly superior to the ones given by Wiebelitz.

Concerning (4.30), observe that by Taylor's formula

$$T^{u-s}X(u) = X(s) + (u - s)\left(X'(s) + X(s)\log T\right) + \frac{1}{2!}\left(X''(s) + \cdots\right)(u - s)^2 + \cdots,$$

and since by (4.20) we have

$$X'\left(\tfrac{1}{2} + it\right) + X\left(\tfrac{1}{2} + it\right)\log T = 0,$$

it is seen that $\Phi_1(u)$ is regular for $\operatorname{Re} u = \tfrac{1}{2}$, when the double zeros $u = s = \tfrac{1}{2} + it$ of the numerator and denominator cancel each other, and the other ranges for u are easy. This discussion also shows why the definition of T in (4.20), which may have looked a little mysterious, is a natural one to make.

To see that (4.31) holds observe that for $\operatorname{Re} u = \tfrac{1}{2}$ we have $|X(u)| = 1$, so that by (1.25)

$$|\Phi(u)| \leqslant |T^{u-s}| + |X(s)| \ll t^{k(1/2 - \sigma)}.$$

This proves the first estimate in (4.31), and for the second one note that if $|s - u|^2 \leqslant t$, then $\operatorname{Re} u \asymp t$, and since we have

$$\frac{d^2}{ds^2}(\log \chi(s)) \ll t^{-1},$$

we may write

$$\Phi(u) = T^{u-s}X(u)\left(1 - \frac{X(s)}{X(u)}T^{s-u}\right)$$

and use Taylor's formula and (4.20) with $u = \tfrac{1}{2} + iv$ (v real). This gives

$$1 - \frac{X(s)}{X(u)}T^{s-u} = 1 - \exp\left\{-k(s - u)\frac{\chi'}{\chi}\left(\tfrac{1}{2} + it\right) + k\log\chi(\sigma + it) - k\log\chi\left(\tfrac{1}{2} + it\right)\right\}$$

$$= 1 - \exp\left\{-k(s - u)\left(\frac{\chi'}{\chi}\left(\tfrac{1}{2} + it\right) - \frac{\chi'}{\chi}\left(\tfrac{1}{2} + iv\right)\right) + O\left(|s - u|^2 t^{-1}\right)\right\}$$

$$= 1 - \exp\left(k\left(\sigma + it - iv + \tfrac{1}{2}\right)^2 O(t^{-1})\right) \ll |s - u|^2 t^{-1},$$

and then (4.31) follows.

To prove (4.39) one may start from

$$(1 + x)^m = \binom{m}{0} + \binom{m}{1}x + \binom{m}{2}x^2 + \cdots + \binom{m}{m}x^m$$

and differentiate, taking eventually $x = -1$. In the first step we have

$$m(1 + x)^{m-1} = \binom{m}{1} + 2\binom{m}{2}x + \cdots + m\binom{m}{m}x^{m-1},$$

and the proof is finished if $p = 1$ (here p does not have to be a prime) by taking $x = -1$. If $p \neq 1$, then the above equation is multiplied by x and differentiated again and the process is repeated sufficiently many times. Finally, for $p = m$, we obtain

$$\sum_{\nu=0}^{m} (-1)^{\nu}\binom{m}{\nu}\nu^p = m!,$$

since we arrive at an expression whose left side is $m!$ plus a polynomial in $x + 1$, and taking $x = -1$ we obtain the above identity.

In (4.35) one uses $T^{u-s}X(u) = X(s) + \Phi(u)$, which is the main reason why $\Phi(u)$ was introduced by (4.30).

The discussion of the reflection principle in Section 4.4 is based primarily on M. Jutila (1976, 1977). This simple and powerful method was used in a similar form before Jutila's work by M. N. Huxley (1973b, 1975a, 1975b) and K. Ramachandra (1974b). A general principle in analytic number theory is to express a sum (or series) by a contour integral in the complex plane, and to attain flexibility by moving the contour of integration and applying the residue theorem. When coupled with the use of the functional equation (1.23) for $\zeta(s)$, this idea leads to the reflection principle.

CHAPTER FIVE

THE FOURTH POWER MOMENT

5.1 INTRODUCTION

Estimates of integrals of the type $I_k = \int_0^T |\zeta(\frac{1}{2} + it)|^k \, dt$ ($k \geq 1$ a fixed integer) play a prominent role in many parts of zeta-function theory. Applications of upper bounds for I_k to zero-density estimates and divisor problems will be considered in Chapter 11 and Chapter 13 respectively, while omega results for I_k will be discussed in Chapter 9. The asymptotic formula for I_2, known as Atkinson's formula, is one of the main topics of this text. A detailed account of Atkinson's formula and some of its applications will be presented in Chapter 15, while power moment estimates for $k > 4$ will be treated in Chapter 8. This chapter is concerned with I_k when $k = 4$, and we shall prove the following

THEOREM 5.1.

$$(5.1) \qquad \int_0^T \left| \zeta(\tfrac{1}{2} + it) \right|^4 dt = (2\pi^2)^{-1} T \log^4 T + O(T \log^3 T).$$

This is a classical result of zeta-function theory, proved first by A. E. Ingham in 1926. Ingham's proof was difficult, and the asymptotic formula (5.1) remained the best-known mean value estimate of the zeta-function for a very long time, though for the somewhat easier problem of estimating the integral $\int_0^\infty e^{-\delta t} |\zeta(\frac{1}{2} + it)|^4 \, dt$ ($\delta \to 0+$) a sharp formula was obtained by F. V. Atkinson. Recently, D. R. Heath-Brown improved substantially (5.1) by showing that

$$(5.2) \qquad \int_0^T \left| \zeta(\tfrac{1}{2} + it) \right|^4 dt = T \sum_{k=0}^{4} c_k (\log T)^{4-k} + O(T^{7/8+\varepsilon}),$$

where $c_0 = (2\pi^2)^{-1}$, and the other constants c_k are computable. As is to be expected, the proof of (5.2) is long and difficult and for these reasons will not be given here. However, the proof of the classical result (5.1) may be given now in a relatively simple way by combining the reflection principle of Chapter 4 with the mean value theorem for Dirichlet polynomials. The mean value theorem for Dirichlet polynomials is a very useful tool in analytic number theory, and may be put into two forms: discrete and integral. The integral variant of the theorem may be formulated as

THEOREM 5.2. Let $a_1, \ldots, a_N$ be arbitrary complex numbers. Then

$$(5.3) \qquad \int_0^T \left| \sum_{n \leqslant N} a_n n^{it} \right|^2 dt = T \sum_{n \leqslant N} |a_n|^2 + O\left(\sum_{n \leqslant N} n |a_n|^2 \right),$$

and the above formula remains also valid if $N = \infty$, provided that the series on the right-hand side of (5.3) converge.

In Section 5.2 a proof of Theorem 5.2 and its discrete variant, Theorem 5.3, will be given, while in Section 5.3 a proof of (5.1) is presented.

5.2 THE MEAN VALUE THEOREM FOR DIRICHLET POLYNOMIALS

We begin now the proof of Theorem 5.2. Squaring and integrating, it is seen that the left-hand side of (5.3) equals

$$(5.4) \qquad T \sum_{n \leqslant N} |a_n|^2 + \sum_{m \neq n \leqslant N} a_m \bar{a}_n \frac{(m/n)^{iT} - 1}{\log m - \log n},$$

so that (5.3) is a consequence of

$$(5.5) \qquad \sum_{m \neq n \leqslant N} \frac{a_m \bar{a}_n}{\log m - \log n} \ll \sum_{n \leqslant N} n |a_n|^2,$$

applied once directly and once with a_m replaced by $a_m m^{iT}$.

To obtain (5.5) we shall first prove

$$(5.6) \qquad \left| \sum_{m \neq n} \frac{a_m \bar{a}_n}{m - n} \right| \leqslant \pi \sum |a_n|^2,$$

which is known in literature as Hilbert's inequality, and then deduce (5.5) from (5.6). In (5.6) the a's are arbitrary complex numbers, and m, n run over the same (possibly infinite) subset of the integers, subject only to the condition $m \neq n$. To see that (5.6) holds let

$$E = \sum_{m \neq n} \frac{a_m \bar{a}_n}{m - n}.$$

Then obviously $E = -\bar{E}$, which means that E is purely imaginary, hence E/i is real. Recalling that for integral k we have

(5.7)
$$\int_0^1 e(kx)\, dx = \int_0^1 e^{2\pi i kx}\, dx = \begin{cases} 1 & k = 0, \\ 0 & k \neq 0, \end{cases}$$

it follows that

(5.8)

$$0 \leqslant \int_0^1 \int_0^y \left| \sum a_n e(nx) \right|^2 dx\, dy$$

$$= \int_0^1 \int_0^y \left(\sum |a_n|^2 + \sum_{m \neq n} a_m \bar{a}_n e((m-n)x) \right) dx\, dy$$

$$= \frac{1}{2} \sum |a_n|^2 + \int_0^1 \sum_{m \neq n} a_m \bar{a}_n \frac{e((m-n)y) - 1}{2\pi i (m-n)}\, dy = \frac{1}{2} \sum |a_n|^2 - \frac{E}{2\pi i}.$$

From (5.8) it follows that (5.6) holds if $E/i \geqslant 0$, and if $E/i < 0$ then the result follows if we repeat the above reasoning with $|\sum a_n e(-nx)|^2$ in place of $|\sum a_n e(nx)|^2$. The proof actually shows that we obtain

(5.9)
$$\left| \sum_{m \neq n} \frac{a_m \bar{a}_n}{q_m - q_n} \right| \leqslant \pi \sum |a_n|^2$$

if $\{q_n\}_{n=1}^\infty$ is any sequence of integers such that $q_m \neq q_n$ if $m \neq n$. Moreover, one also has

(5.10)
$$\left| \sum_{m \neq n} \frac{a_m \bar{b}_n}{q_m - q_n} \right| \leqslant 3\pi \left(\sum |a_n|^2 \right)^{1/2} \left(\sum |b_n|^2 \right)^{1/2},$$

which follows from

$$(2\pi i)^{-1} \sum_{m \neq n} \frac{a_m \bar{b}_n}{q_m - q_n} = -\int_0^1 \int_0^y \sum a_m e(q_m x) \sum \bar{b}_n e(-q_n x)\, dx\, dy + \frac{1}{2} \sum a_n \bar{b}_n$$

if one uses the Cauchy–Schwarz inequality, (5.8) and (5.9), since

$$\left| \int_0^1 \int_0^y \sum a_m e(q_m x) \sum \bar{b}_n e(-q_n x)\, dx\, dy \right|^2$$

$$\leqslant \int_0^1 \int_0^y \left| \sum a_n e(q_n x) \right|^2 dx\, dy \int_0^1 \int_0^y \left| \sum \bar{b}_n e(-q_n x) \right|^2 dx\, dy$$

$$\leqslant \sum |a_n|^2 \sum |b_n|^2.$$

For simplicity of writing, now let $L_n = \log n$ and

$$G = \sum_{m \neq n} \frac{a_m \bar{a}_n}{L_m - L_n},$$

so that G is purely imaginary and as in the proof of (5.6) it will be sufficient to assume that $G/i \geqslant 0$. From

$$0 \leqslant \int_0^1 \int_0^y \left| \sum a_n e(L_n x) \right|^2 dx\, dy$$

$$= \frac{1}{2} \sum |a_n|^2 + \int_0^1 \sum_{m \neq n} a_m \bar{a}_n \frac{e((L_m - L_n)y) - 1}{2\pi i (L_m - L_n)}\, dy$$

we then obtain

(5.11)

$$\frac{G}{2\pi i} \leqslant \frac{1}{2} \sum |a_n|^2 + \frac{1}{2\pi} \sum_{k,l \geqslant 1} \left| \sum_{(m,n) \in I_k \times I_l} \int_0^1 a_m \bar{a}_n \frac{e((L_m - L_n)y)}{L_m - L_n}\, dy \right|,$$

where the range of summation $1 \leqslant n \leqslant N$ in (5.5) has been divided into intervals $I_j = (N2^{-j}, N2^{1-j}]$, $j = 1, 2, \ldots$. Since

$$\left| \int_0^1 e^{ixt}\, dx \right| = \left| \frac{e^{it} - 1}{t} \right| \leqslant 2|t|^{-1} \qquad (0 \neq t \in \text{Re}),$$

we have, for $|k - l| \geqslant 2$ in (5.11),

$$\sum_{(m,n) \in I_k \times I_l} \int_0^1 a_m \bar{a}_n \frac{e((L_m - L_n)y)}{L_m - L_n}\, dy$$

$$\ll \sum_{(m,n) \in I_k \times I_l} |a_m a_n| \max_{(m,n) \in I_k \times I_l} (L_m - L_n)^{-2}$$

$$\ll (k - l)^2 \left(\sum_{(m,n) \in I_k \times I_l} |a_m|^2 \right)^{1/2} \left(\sum_{(m,n) \in I_k \times I_l} |a_n|^2 \right)^{1/2}$$

$$\ll (k - l)^{-2} N 2^{-(k+l)/2} \left(\sum_{m \in I_k} |a_m|^2 \right)^{1/2} \left(\sum_{n \in I_l} |a_n|^2 \right)^{1/2}$$

$$\ll (k - l)^{-2} (S_k S_l)^{1/2},$$

where

$$S_j = \sum_{n \in I_j} n |a_n|^2.$$

Here we used the Cauchy–Schwarz inequality and

$$|L_m - L_n| \geqslant \log(N2^{-l}) - \log(N2^{1-k}) = (\log 2)(k - l - 1) \geqslant \tfrac{1}{4}(k - l),$$

which holds if $k - l \geqslant 2$, while the case $l - k \geqslant 2$ is analogous. A further application of the Cauchy–Schwarz inequality gives

$$\sum_{|k-l| \geqslant 2} (k - l)^{-2} (S_k S_l)^{1/2}$$

$$\leqslant \left(\sum_{k \neq l} S_k (k - l)^{-2} \right)^{1/2} \left(\sum_{k \neq l} S_l (k - l)^{-2} \right)^{1/2} \ll \sum_{n \leqslant N} n |a_n|^2,$$

since $\sum_{k, k \neq l} (k - l)^{-2}$ converges, so that the contribution of $\sum_{|k-l| \geqslant 2}$ to (5.11) is of the desired order of magnitude. The terms in (5.11) for which $|k - l| \leqslant 1$ may be written as

$$\int_0^1 \sum_{\substack{(m, n) \in I_k \times I_l \\ m \neq n}} a_m \bar{a}_n \frac{e((L_m - L_n) y)}{L_m - L_n} \, dy$$

$$= M \int_0^1 \left(\sum_{\substack{(m, n) \in I_k \times I_l \\ m \neq n}} \frac{a'_m \bar{a}'_n}{ML_m - ML_n} \right) dy,$$

where $a'_m = a_m e(L_m y)$, $M = N2^{6-k}$, and it will be sufficient to majorize the last sum. The reason for introducing M is that if $|k - l| \leqslant 1$, then for $m > n$ and $(m, n) \in I_k \times I_l$

(5.12)

$$[ML_m] - [ML_n] \geqslant M(L_m - L_n) - 2 = M \log\left(1 + \frac{m - n}{n}\right) - 2$$

$$\geqslant \frac{2^{6-k} N(m - n)}{3N2^{2-k}} - 2 \geqslant \frac{16}{3}(m - n) - 2 \geqslant m - n,$$

since $m - n \geqslant 1$, $0 \leqslant (m - n)/n \leqslant 3$, and $\log(1 + x) \geqslant x/3$ for $0 \leqslant x \leqslant 3$. Therefore we have, for $|k - l| \leqslant 1$,

(5.13)

$$M \sum_{\substack{(m, n) \in I_k \times I_l \\ m \neq n}} \frac{a'_m \bar{a}'_n}{ML_m - ML_n} = M \sum_{\substack{(m, n) \in I_k \times I_l \\ m \neq n}} \frac{a'_m \bar{a}'_n}{[ML_m] - [ML_n]}$$

$$+ O\left(M \sum_{\substack{(m, n) \in I_k \times I_l \\ m \neq n}} \frac{|a_m a_n|}{(m - n)^2} \right),$$

and the O-term above is easily seen to be $\ll (S_k S_l)^{1/2}$. The other term on the right-hand side of (5.13) is estimated by (5.10) with $q_m = [ML_m]$, and the resulting estimate is multiplied by M to yield also $\ll (S_k S_l)^{1/2}$. Using once again the Cauchy–Schwarz inequality and $|k - l| \leqslant 1$ we obtain (5.5), and consequently (5.3). If $N = \infty$ the reasoning is the same, only we define $I_j = (2^{j-1}, 2^j]$ this time.

We proceed now to the discrete form of the mean value theorem for Dirichlet polynomials, which turns out to be more useful in certain applications than the integral form of the theorem. This may be formulated as

THEOREM 5.3. Let $1 \leqslant t_1 < \cdots < t_R \leqslant T$ be real numbers such that $|t_r - t_s| \geqslant 1$ for $r \neq s \leqslant R$ and let $a_1, \ldots, a_N$ be arbitrary complex numbers. Then

$$(5.14) \qquad \sum_{r \leqslant R} \left| \sum_{n \leqslant N} a_n n^{-it_r} \right|^2 \ll \left(T \sum_{n \leqslant N} |a_n|^2 + \sum_{n \leqslant N} n|a_n|^2 \right) \log N.$$

Proof of Theorem 5.3. If $0 \leqslant x \leqslant 1$ and $f(x) \in C^1[0,1]$, then an integration by parts shows that

$$f(x) = \int_0^1 f(t)\, dt + \int_0^x tf'(t)\, dt + \int_x^1 (t-1)f'(t)\, dt;$$

hence

$$(5.15) \qquad |f(1/2)| \leqslant \int_0^1 \left(|f(x)| + \tfrac{1}{2}|f'(x)| \right) dx.$$

Taking $f(x) = F(x - \tfrac{1}{2} + t_r)$ we have from (5.15)

$$(5.16) \qquad |F(t_r)| \leqslant \int_{t_r - 1/2}^{t_r + 1/2} |F(t)|\, dt + \frac{1}{2} \int_{t_r - 1/2}^{t_r + 1/2} |F'(t)|\, dt.$$

Now we use (5.16) with $F(t) = (\sum_{n \leqslant N} a_n n^{-it})^2$. By the spacing condition imposed on the t_r's it is seen that the intervals $(t_r - \tfrac{1}{2}, t_r + \tfrac{1}{2})$ $(r \leqslant R)$ are disjoint; hence the left-hand side of (5.14) is not greater than

$$(5.17) \qquad \int_0^{T+1} |F(t)|\, dt + \frac{1}{2} \int_0^{T+1} |F'(t)|\, dt.$$

The first integral in (5.17) is estimated directly by Theorem 5.2 and makes a contribution which is $\ll T\sum_{n \leqslant N}|a_n|^2 + \sum_{n \leqslant N}|a_n|^2 n$. For the other integral in (5.17) note that

$$F'(t) = -2 \sum_{n \leqslant N} a_n n^{-it} \sum_{n \leqslant N} a_n (\log n) n^{-it};$$

hence by the Cauchy–Schwarz inequality and Theorem 5.2 we obtain

(5.18)

$$\frac{1}{2}\int_0^{T+1}|F'(t)|\,dt$$

$$\leqslant \left(\int_0^{T+1}\left|\sum_{n\leqslant N}a_n n^{it}\right|^2 dt\right)^{1/2}\left(\int_0^{T+1}\left|\sum_{n\leqslant N}a_n(\log n)n^{it}\right|^2 dt\right)^{1/2}$$

$$\ll \left(T\sum_{n\leqslant N}|a_n|^2 + \sum_{n\leqslant N}n|a_n|^2\right)^{1/2}\left(T\sum_{n\leqslant N}|a_n|^2\log^2 n\right.$$

$$\left. + \sum_{n\leqslant N}n|a_n|^2\log^2 n\right)^{1/2}$$

$$\ll \left(T\sum_{n\leqslant N}|a_n|^2 + \sum_{n\leqslant N}n|a_n|^2\right)\log N.$$

This completes the proof of Theorem 5.3, but it may be remarked that in the case $N = \infty$ the above proof gives

(5.19)

$$\sum_{r\leqslant R}\left|\sum_{n=1}^{\infty}a_n n^{-it_r}\right|^2 \ll T\sum_{n=1}^{\infty}|a_n|^2\log^2(n+1) + \sum_{n=1}^{\infty}n|a_n|^2\log^2(n+1),$$

provided that the series on the right-hand side converge.

Another conclusion is that from $|a_n| = |a_n n^{iT_0}|$, T_0 an arbitrary, fixed, real number, it follows that Theorem 5.2 remains valid if $\int_0^T$ is replaced by $\int_{T_0}^{T_0+T}$, and similarly in Theorem 5.3, we may suppose that $T_0 + 1 \leqslant t_1 < \cdots < t_R \leqslant T_0 + T$.

5.3 PROOF OF THE FOURTH POWER MOMENT ESTIMATE

As an application of Theorem 5.2 we shall now present a proof of Theorem 5.1 by using a variant of the reflection principle, which was discussed in Chapter 4. With $w = u + iv$, $s = \sigma + it$, u and v real, $0 < \sigma < \frac{3}{4}$, $T/2 \leqslant t \leqslant T$, we obtain from (A.7), on applying the residue theorem,

(5.20)

$$\sum_{n=1}^{\infty} d(n)e^{-n/T}n^{-s} = (2\pi i)^{-1}\int_{u=2} \zeta^2(s+w)\Gamma(w)T^w dw$$

$$= \zeta^2(s) + O(T^{-c}) + (2\pi i)^{-1}\int_{u=-3/4} \chi^2(s+w)\zeta^2(1-s-w)\Gamma(w)T^w dw$$

$$= \zeta^2(s) + O(T^{-c}) + (2\pi i)^{-1}\int_{u=-3/4} \chi^2(s+w)\sum_{n=1}^{\infty} d(n)n^{w+s-1}\Gamma(w)T^w dw$$

$$= \zeta^2(s) + O(T^{-c}) + (2\pi i)^{-1}\int_{u=-3/4} \chi^2(s+w)\sum_{n>T} d(n)n^{w+s-1}\Gamma(w)T^w dw$$

$$-\chi^2(s)\sum_{n\leqslant T} d(n)n^{s-1} + (2\pi i)^{-1}\int_{u=1/4} \chi^2(s+w)\sum_{n\leqslant T} d(n)n^{w+s-1}\Gamma(w)T^w dw.$$

Here we used the functional equation (1.23) and Stirling's formula (A.34) to obtain the error term $O(T^{-c})$ (here $c > 0$ is arbitrary, but fixed) which majorizes the residue of $\zeta^2(s+w)\Gamma(w)T^w$ at the double pole $w = 1 - s$. Now we set $s = \frac{1}{2} + it$ and use again Stirling's formula together with

$$\zeta^2(\tfrac{1}{2}+it)\chi^{-1}(\tfrac{1}{2}+it) = |\zeta(\tfrac{1}{2}+it)|^2$$

to deduce from (5.20)

(5.21)
$$|\zeta(\tfrac{1}{2}+it)|^2 = \sum_{k=1}^{6} J_k(s) + O(T^{-c})$$

for any fixed $c > 0$, where $J_k = J_k(s)$, and for $s = \frac{1}{2} + it$

$$J_2 = \bar{J}_1 = \chi(\tfrac{1}{2}+it)\sum_{n\leqslant T} d(n)n^{-1/2+it},$$

$$J_3 = \chi^{-1}(\tfrac{1}{2}+it)\sum_{n>T} d(n)e^{-n/T}n^{-1/2-it},$$

$$J_4 = \chi^{-1}(\tfrac{1}{2}+it)\sum_{n\leqslant T} d(n)(e^{-n/T}-1)n^{-1/2-it},$$

$$J_5 = -(2\pi i)^{-1}\chi^{-1}(\tfrac{1}{2}+it)\int_{u=-3/4,|v|\leqslant \log^2 T} \chi^2(\tfrac{1}{2}+it+w)$$
$$\times \sum_{n>T} d(n)n^{w-1/2+it}\Gamma(w)T^w dw,$$

$$J_6 = -(2\pi i)^{-1}\chi^{-1}(\tfrac{1}{2}+it)\int_{u=1/4,|v|\leqslant \log^2 T} \chi^2(\tfrac{1}{2}+it+w)$$
$$\times \sum_{n\leqslant T} d(n)n^{w-1/2+it}\Gamma(w)T^w dw.$$

Theorem 5.1 will follow then from

$$(5.22) \qquad \int_{T/2}^{T} |\zeta(\tfrac{1}{2} + it)|^4 \, dt = (4\pi^2)^{-1} T \log^4 T + O(T \log^3 T),$$

when one replaces T by $T/2, T/2^2, \ldots,$ and adds all the results. Observe that trivially $J_k \ll T^{1/2} \log T$ for each k. Therefore squaring and integrating (5.21) we have

$$(5.23) \quad \int_{T/2}^{T} |\zeta(\tfrac{1}{2} + it)|^4 \, dt = 2 \int_{T/2}^{T} |J_1|^2 \, dt + \int_{T/2}^{T} \left(J_1^2 + J_2^2 \right) dt$$

$$+ O\left(\sum_{k=3}^{6} \int_{T/2}^{T} |J_k|^2 \, dt \right)$$

$$+ O\left(\sum_{k=3}^{6} \left| \int_{T/2}^{T} (J_1 + J_2) J_k \, dt \right| \right) + O(1),$$

and the main contribution in (5.22) will come from the first integral on the right-hand side of (5.23). To see this note that from the Dirichlet-series representations

$$\sum_{n=1}^{\infty} d^2(n) n^{-s} = \frac{\zeta^4(s)}{\zeta(2s)}, \qquad \sum_{n=1}^{\infty} \mu(n) n^{-2s} = \frac{1}{\zeta(2s)},$$

which are valid for $\operatorname{Re} s > 1$ and $\operatorname{Re} s > \tfrac{1}{2}$, respectively, one obtains by an easy convolution argument

$$(5.24) \qquad \sum_{n \leqslant x} d^2(n) = \pi^{-2} x \log^3 x + O(x \log^2 x).$$

Thus by partial summation it follows that

$$(5.25) \quad \sum_{n \leqslant x} d^2(n) n^a = \begin{cases} c(a) x^{1+a} \log^3 x + O(x^{1+a} \log^2 x) & (a > -1), \\ (4\pi^2)^{-1} \log^4 x + O(\log^3 x) & (a = -1). \end{cases}$$

Now we shall apply Theorem 5.2 and (5.25), obtaining first

$$(5.26) \qquad 2 \int_{T/2}^{T} |J_1|^2 \, dt = T \sum_{n \leqslant T} d^2(n) n^{-1} + O\left(\sum_{n \leqslant T} d^2(n) \right)$$

$$= (4\pi^2)^{-1} T \log^4 T + O(T \log^3 T),$$

since $|\chi(\tfrac{1}{2} \pm it)| = 1$. Therefore (5.26) does contribute the main term in

(5.22), and in fact the main idea of the proof is to apply Theorem 5.2 to the remaining integrals in (5.23) using (5.25) with $a > -1$. Thus we have

$$\int_{T/2}^{T} |J_3|^2 \, dt \ll T \sum_{n > T} d^2(n) e^{-2n/T} n^{-1} + \sum_{n > T} d^2(n) e^{-2n/T} \ll 1,$$

$$\int_{T/2}^{T} |J_4|^2 \, dt \ll T \sum_{n \leqslant T} d^2(n)(e^{-n/T} - 1)^2 n^{-1} + \sum_{n \leqslant T} d^2(n)(e^{-n/T} - 1)^2$$

$$\ll T^{-1} \sum_{n \leqslant T} d^2(n) n + T^{-2} \sum_{n \leqslant T} d^2(n) n^2 \ll T \log^3 T,$$

$$\int_{T/2}^{T} |J_5|^2 \, dt \ll T^{5/2} \sum_{n > T} d^2(n) n^{-5/2} + T^{3/2} \sum_{n > T} d^2(n) n^{-3/2} \ll T \log^3 T,$$

$$\int_{T/2}^{T} |J_6|^2 \, dt \ll T^{1/2} \sum_{n \leqslant T} d^2(n) n^{-1/2} + T^{-1/2} \sum_{n \leqslant T} d^2(n) n^{1/2} \ll T \log^3 T.$$

Next, we write

$$(5.27) \quad i \int_{T/2}^{T} J_1^2 \, dt = \int_{1/2 + iT/2}^{1/2 + iT} J_1^2(s) \, ds, \qquad J_1(s) = \chi^{-1}(s) \sum_{n \leqslant T} d(n) n^{-s},$$

and consider the last integral as an integral of the complex variable s. To avoid (5.25) with $a = -1$ we replace by Cauchy's theorem the segment of integration in (5.27) by segments joining the points $\frac{1}{2} + \frac{1}{2}iT, \frac{1}{4} + \frac{1}{2}iT, \frac{1}{4} + iT, \frac{1}{2} + iT$. Using (1.25) it is seen that the integrals over horizontal segments are $\ll T \log^3 T$, while using Theorem 5.2 and (5.25) we obtain

$$\int_{1/4 + iT/2}^{1/4 + iT} J_1^2(s) \, ds \ll T^{-1/2} \int_{T/2}^{T} \left| \sum_{n \leqslant T} d(n) n^{-1/4 + it} \right|^2 dt \ll T \log^3 T.$$

The same procedure may be applied to the integral of J_2^2 to yield

$$(5.28) \qquad \int_{T/2}^{T} (J_1^2 + J_2^2) \, dt \ll T \log^3 T.$$

The remaining integrals in (5.23) are written as

$$i \int_{1/2 + iT/2}^{1/2 + iT} J_1(s) J_k(s) \, ds + i \int_{1/2 + iT/2}^{1/2 + iT} J_2(s) J_k(s) \, ds \qquad (k = 3, 4, 5, 6),$$

and are treated similarly. In integrals with $J_1(s)$, the segment of integration $[\frac{1}{2} + \frac{1}{2}iT, \frac{1}{2} + iT]$ is being replaced by the segment $[\frac{3}{8} + \frac{1}{2}iT, \frac{3}{8} + iT]$ with an error $\ll T \log^3 T$, while in integrals containing $J_2(s)$ this segment is replaced by the segment $[\frac{5}{8} + \frac{1}{2}iT, \frac{5}{8} + iT]$, again with an error $\ll T \log^3 T$. Applying the Cauchy–Schwarz inequality, Theorem 5.2, and collecting all the estimates we finally obtain

$$(5.29) \qquad \int_{T/2}^{T} |\zeta(\tfrac{1}{2} + it)|^4 \, dt = (4\pi^2)^{-1} T \log^4 T + O(T \log^3 T).$$

Theorem 5.1 follows then from (5.29) on replacing T by $T/2, T/2^2, \ldots,$ and adding all the results.

Notes

A. E. Ingham (1926) proved (5.1). Various mean value estimates for integrals

$$\int_0^T |\zeta(\sigma + it)|^k \, dt \qquad (\tfrac{1}{2} \leqslant \sigma < 1, k \geqslant 1)$$

are discussed in Chapter 7 of E. C. Titchmarsh (1951), but (5.1) is not proved there, only the weaker formula

$$(5.30) \qquad \int_0^T |\zeta(\tfrac{1}{2} + it)|^4 \, dt = (1 + o(1))(2\pi^2)^{-1} T \log^4 T \qquad (T \to \infty).$$

This follows from the investigation of the integrals

$$I(T) = \int_0^T |\zeta(\sigma + it)|^{2k} \, dt, \qquad J(\delta) = \int_0^\infty |\zeta(\sigma + it)|^{2k} e^{-\delta t} \, dt,$$

where $k \geqslant 1$ is a fixed integer, $T \to \infty$ and $\delta \to 0 +$, $\tfrac{1}{2} \leqslant \sigma < 1$ is fixed. A simple Tauberian argument shows that, for $C, D > 0, I(T) \sim CT \log^D T$ is equivalent to $J(\delta) \sim C\delta^{-1}(\log \delta^{-1})^D$, and Titchmarsh deduces then (5.30) from

$$\int_0^\infty |\zeta(\tfrac{1}{2} + it)|^4 e^{-\delta t} \, dt = (1 + o(1))(2\pi^2)^{-1} \delta^{-1}(\log \delta^{-1})^4 \qquad (\delta \to 0 +).$$

As mentioned in Section 5.1, a sharper result than the one above was obtained by F. V. Atkinson (1941b), who proved for $\delta \to 0 +$

$$(5.31) \qquad \int_0^\infty |\zeta(\tfrac{1}{2} + it)|^4 e^{-\delta t} \, dt = \sum_{j=0}^{4} A_j \delta^{-1}(\log \delta^{-1})^{4-j} + O\big((\delta^{-1})^{13/14+\varepsilon}\big),$$

where $A_0 = (2\pi^2)^{-1}$ and the other constants A_j are also computable. A method is also indicated in Atkinson's paper by which the exponent $13/14$ may be reduced to $8/9$. However, (5.31) does not seem to apply (5.1), but only the weaker (5.30). Similarly it may be mentioned that one has [Theorem 7.15 (A) of E. C. Titchmarsh (1951)], as $\delta \to 0 +$

$$\int_0^\infty |\zeta(\tfrac{1}{2} + it)|^2 e^{-2\delta t} \, dt = \frac{\gamma - \log(4\pi\delta)}{2 \sin \delta} + \sum_{n=0}^{N} c_n \delta^n + O(\delta^{N+1}),$$

for any fixed integer $N \geqslant 1$, but this sharp result does not seem to imply anything like Atkinson's formula for $\int_0^T |\zeta(\tfrac{1}{2} + it)|^2 \, dt$, which is extensively discussed in Chapter 15.

D. R. Heath-Brown's proof (1979a) of (5.2) is based on several new ideas. The first is the use of an approximate functional equation which may be written as

$$(5.32) \quad |\zeta(\tfrac{1}{2} + it)|^{2k} = \sum_{mn \leqslant cT^k} d_k(m) d_k(n) (mn)^{-1/2} (m/n)^{it} K(mn, t) + O(T^{-2}).$$

Here $k \geqslant 1$ is a fixed integer, $c \geqslant 1$ is a constant depending on k, $T \leqslant t \leqslant 2T$, and

$$(5.33) \quad K(x, t) = (\pi i)^{-1} \int_{1-i\infty}^{1+i\infty} t^{kz} (2\pi)^{-kz} x^{-z} e^{z^2 T^{-1}} \left(1 + \sum_{u=1}^{U} \sum_{v} \alpha(u, v) z^u t^{-2v} \right) z^{-1} dz,$$

where $\alpha(u, v)$ is a constant and U is an integer depending on k, while $\sum_v$ denotes summation for $\max(1, u/3) \leqslant v \leqslant U$.

The function $K(x, t)$ in (5.33) is fairly complicated, but in contrast with the approximate functional equations for $\zeta^k(s)$ of Chapter 4, the error term $O(T^{-2})$ in (5.32) is remarkably sharp. This enabled Heath-Brown to integrate (5.32) termwise when $k = 2$, but there were difficulties which arose from the dependence of $K(x, t)$ on t. A further feature of the proof of (5.2) is the use of an exponential averaging technique, which permits one to evaluate $\int_{2T}^{4T} |\zeta(\tfrac{1}{2} + it)|^4 \, dt$ from the weighted integral $\int_T^{5T} w(t) |\zeta(\tfrac{1}{2} + it)|^4 \, dt$, where the function $w(t)$ is precisely defined in Section 3 of Heath-Brown (1979a). The proof in its last stage requires an asymptotic formula for the divisor sum

$$D(x, r) = \sum_{n \leqslant x} d(n) d(n + r) = m(x, r) + E(x, r),$$

which in turn depends on estimates for Kloosterman sums. Here $m(x, r)$ is the main term of the form

$$m(x, r) = \sum_{j=0}^{2} c_j(r) x \log^j x,$$

and it should be stressed that r may be increasing with x. The exponent $7/8$ in (5.2) is primarily limited by the mean value estimate

$$\int_X^{2X} E^2(x, r) \, dx = O(X^{5/2 + \varepsilon}),$$

which as present is known to hold uniformly in r for $r \leqslant X^{3/4}$.

The proof of Theorem 5.2 is based on K. Ramachandra (1980b), and the crucial estimate (5.5) is a special case of a more general inequality due to H. L. Montgomery and R. C. Vaughan (1974): Suppose that $R \geqslant 2$ and $\lambda_1, \lambda_2, \ldots, \lambda_R$ are distinct real numbers which satisfy $0 < \delta_n = \min_{m \neq n} |\lambda_n - \lambda_m|$. If $a_1, a_2, \ldots, a_R$ are arbitrary complex numbers, then

$$(5.34) \quad \left| \sum_{m \neq n} a_m \bar{a}_n (\lambda_m - \lambda_n)^{-1} \right| \leqslant \frac{3}{2} \pi \sum_n |a_n|^2 \delta_n^{-1}.$$

This inequality is closely connected with large sieve-type inequalities for which the reader may consult the expository paper of H. L. Montgomery (1978) and his article (1982).

In presenting the proof of Theorem 5.1 we have followed the work of K. Ramachandra (1975b). Ramachandra's method does not seem to extend to give anything sharper than (5.1), yet it is considerably simpler than the method used by A. E. Ingham (1926) in proving (5.1). For other mean value theorems for Dirichlet polynomials besides Theorems 5.2 and 5.3 the reader may consult H. L. Montgomery (1971), Chapters 6 and 7.

In estimating the first sum on the right-hand side of (5.11) by (5.10) we take $a_m = a'_m$ for $m \in I_k$ and zero otherwise, and $b_n = a'_n$ for $n \in I_l$ and zero otherwise. Then the sum in question is

$$\ll M \left(\sum_{m \in I_k} |a_m|^2 \right)^{1/2} \left(\sum_{n \in I_l} |a_n|^2 \right)^{1/2}$$

$$\ll \left(\sum_{m \in I_k} m|a_m|^2 \right)^{1/2} \left(\sum_{n \in I_l} n|a_n|^2 \right)^{1/2} = (S_k S_l)^{1/2},$$

as asserted.

Concerning the convolution argument that leads to (5.24) observe that from the Dirichlet-series representation (1.106) one has

$$d^2(n) = \sum_{kl^2 = n} d_4(k)\mu(l);$$

hence

$$\sum_{n \leqslant x} d^2(n) = \sum_{kl^2 \leqslant x} d_4(k)\mu(l) = \sum_{l \leqslant x^{1/2}} \mu(l) \sum_{k \leqslant xl^{-2}} d_4(k)$$

$$= \sum_{l \leqslant x^{1/2}} \mu(l) \left(\tfrac{1}{6}xl^{-2}\log^3(xl^{-2}) + O(xl^{-2}\log^2 x) \right) = \pi^{-2}x\log^3 x + O(x\log^2 x).$$

Here we used a weak asymptotic formula for $\sum_{n \leqslant x} d_4(n)$ and $\sum_{n=1}^{\infty} \mu(n)n^{-2} = 1/\zeta(2) = 6/\pi^2$. For an asymptotic formula which supersedes (5.24), see (14.30).

/ CHAPTER SIX /
THE ZERO-FREE REGION

6.1 A SURVEY OF RESULTS

By a zero-free region for $\zeta(s)$ we mean a region in the s-plane where $\zeta(s) \neq 0$. It follows from (1.2) and Theorem 1.5 that $\sigma \geq 1$ is a zero-free region, and later in Chapter 1 [see (1.55)] it was indicated that $\sigma \geq 1 - C/\log t$ ($C > 0$, $t \geq 2$) is also a zero-free region. The determination of the zero-free region is of fundamental importance in zeta-function theory, both from a theoretical and practical viewpoint. A classical application of the zero-free region is the estimation of the error term in the prime number theorem, a problem which will be discussed in Chapter 12. Although all attempts to obtain a zero-free region of the form $\sigma > \sigma_0$ for some $\frac{1}{2} \leq \sigma_0 < 1$ have failed so far, it is nevertheless possible to improve substantially the zero-free region (1.55). The sharpest known zero-free region is given by

THEOREM 6.1. There is an absolute constant $C > 0$ such that $\zeta(s) \neq 0$ for

$$(6.1) \qquad \sigma \geq 1 - C(\log t)^{-2/3}(\log \log t)^{-1/3} \qquad (t \geq t_0).$$

It will turn out in the course of the proof that the quality of a result like Theorem 6.1 depends on the region where the estimate $\zeta(s) \ll \log^A |t|$ holds for some $A > 0$. Since trivially $\zeta(1 + it) \ll \log|t|$, this hints that one should consider the order of $\zeta(s)$ near the line $\sigma = 1$. Taking $x = t$ in (1.56) we may write

$$(6.2) \qquad \zeta(s) = \sum_{n \leq t} n^{-\sigma - it} + O(t^{-\sigma}) \qquad (t \geq t_0)$$

and after removing $n^{-\sigma}$ from the sum above by partial summation, it is seen that the estimation of $\zeta(s)$ near $\sigma = 1$ ultimately depends on the estimation of exponential sums of the form

$$(6.3) \qquad S(t; N, N_1) = \sum_{N < n \leqslant N_1 \leqslant 2N} n^{it} \qquad (N_0 \leqslant N \leqslant t/2, t \geqslant t_0).$$

The last sums may be quite appropriately called "zeta sums," and the proof of Theorem 6.1 will depend on a uniform estimate for these sums, given by

THEOREM 6.2. If $S(t; N, N_1)$ is the zeta sum defined by (6.3), then uniformly in N, N_1 and t we have

$$(6.4) \qquad S(t; N, N_1) \ll N\exp\left(-\frac{\log^3 N}{100{,}000\log^2 t}\right).$$

Theorem 6.2 will be proved by an application of the Vinogradov–Korobov method for the estimation of exponential integrals of the form

$$(6.5)$$

$$J_{k,n}(P) = \int_0^1 \cdots \int_0^1 \left| \sum_{x \leqslant P} e\left(\alpha_1 x + \alpha_2 x^2 + \cdots + \alpha_n x^n\right) \right|^{2k} d\alpha_1\, d\alpha_2 \cdots d\alpha_n.$$

Here, in the usual notation, $k, n \geqslant 1$ are integers, $x \geqslant 1$ is an integer variable, and P is a large integer. The method of Vinogradov–Korobov supersedes the earlier methods of Weyl, Hardy–Littlewood, and van der Corput, and may be applied to various problems from analytic number theory. In Section 6.2 we shall develop the essentials of this method, and in Section 6.3 we shall prove Theorem 6.2. Finally, in Section 6.4 we shall derive a suitable order estimate for $\zeta(s)$, which by the lemmas of Section 6.5 then lead to (6.1). Our order result for $\zeta(s)$ will be

THEOREM 6.3. There is an absolute constant $\eta > 0$ such that for $1 - \eta \leqslant \sigma \leqslant 1$ uniformly in σ

$$(6.6) \qquad \zeta(\sigma + it) \ll t^{122(1-\sigma)^{3/2}}\log^{2/3} t \qquad (t \geqslant t_0).$$

This result implies at once

$$(6.7) \qquad \zeta(1 + it) \ll \log^{2/3} t \qquad (t \geqslant t_0),$$

which is the best known order result for $\zeta(1 + it)$, although the Riemann hypothesis implies the stronger bound $\zeta(1 + it) \ll \log\log t$. Apart from the numerical constant 122, Theorem 6.3 is also the strongest known result of its type, and its significance lies in being able to provide good bounds when σ is "very close" to 1. Namely, it follows from the bounds of $\mu(\sigma)$ in Chapter 7 that there exists $0 < \eta_0 < \frac{1}{2}$ such that the bound given by (6.6) may be improved for $\sigma \leqslant 1 - \eta_0$. Also, the value of the constant η in the formulation of Theorem 6.3 (this constant may be evaluated explicitly) does not matter much in many applications.

6.2 THE METHOD OF VINOGRADOV – KOROBOV

The basis of this method is the elementary identity (5.7), that is,

$$(6.8) \qquad \int_0^1 e(kx)\, dx = \int_0^1 e^{2\pi i k x}\, dx = \begin{cases} 1 & \text{if } k = 0, \\ 0 & \text{if } k \text{ is a nonzero integer.} \end{cases}$$

Using $|z|^2 = z\bar{z}$ and (6.8) it is seen that $J_{k,n}(P)$, as defined by (6.5), represents in fact the number of integer solutions of the system of equations

$$x_1 + \cdots + x_k - x_{k+1} - \cdots - x_{2k} = 0,$$
$$\vdots$$
$$x_1^n + \cdots + x_k^n - x_{k+1}^n - \cdots - x_{2k}^n = 0,$$

where $1 \leqslant x_1, \ldots, x_{2k} \leqslant P$. More generally, let us define $J_{k,n}(\lambda_1, \ldots, \lambda_n)$ for given integers $\lambda_1, \ldots, \lambda_n$ as the number of solutions of the system

$$(6.9) \qquad x_1 + \cdots + x_k - x_{k+1} - \cdots - x_{2k} = \lambda_1,$$
$$\vdots$$
$$x_1^n + \cdots + x_k^n - x_{k+1}^n - \cdots - x_{2k}^n = \lambda_n,$$

in integers $1 \leqslant x_1, \ldots, x_{2k} \leqslant P$, so that in this notation $J_{k,n}(P) = J_{k,n}(0, \ldots, 0)$. Using (6.8) we have

$$J_{k,n}(\lambda_1, \ldots, \lambda_n) = \int_0^1 \cdots \int_0^1 \left| \sum_{x \leqslant P} e(\alpha_1 x + \cdots + \alpha_n x^n) \right|^{2k}$$
$$\times e(-\alpha_1\lambda_1 - \cdots - \alpha_n\lambda_n)\, d\alpha_1 \cdots d\alpha_n$$
$$= \left| \int_0^1 \cdots \int_0^1 \left| \sum \right|^{2k} e(-\cdots)\, d\alpha_1 \cdots d\alpha_n \right|$$
$$\leqslant \int_0^1 \cdots \int_0^1 \left| \sum_{x \leqslant P} e(\alpha_1 x + \cdots + \alpha_n x^n) \right|^{2k} d\alpha_1 \cdots d\alpha_n$$
$$= J_{k,n}(P);$$

that is, for any integers $\lambda_1, \ldots, \lambda_n$,

$$(6.10) \qquad J_{k,n}(\lambda_1, \ldots, \lambda_n) \leqslant J_{k,n}(P).$$

When $x_1, \ldots, x_{2k}$ run over all possible P^{2k} values, then the left-hand side of the system (6.9) assumes all possible values $\lambda_1, \ldots, \lambda_n$, which may be interpreted formally as

$$(6.11) \qquad \sum_{\lambda_1, \ldots, \lambda_n} J_{k,n}(\lambda_1, \ldots, \lambda_n) = P^{2k}.$$

The number of summands in (6.11) is certainly finite, since (6.9) immediately gives

$$(6.12) \qquad |\lambda_1| < kP, \quad |\lambda_2| < kP^2, \ldots, |\lambda_n| < kP^n.$$

Further, we have trivially $J_{k,n}(P) \leqslant P^{2k}$, and moreover $J_{k,n}(P)$ is clearly nondecreasing as a function of k or P. Our interest will be primarily in the upper bounds for $J_{k,n}(P)$, but we may note here that a lower bound may be obtained as follows. From (6.10)–(6.12) we have

$$P^{2k} = \sum_{\lambda_1, \ldots, \lambda_n} J_{k,n}(\lambda_1, \ldots, \lambda_n)$$

$$\leqslant J_{k,n}(P) \sum_{\lambda_1, \ldots, \lambda_n} 1 \leqslant J_{k,n}(P)(2k)P \cdots (2k)P^n$$

$$= J_{k,n}(P)(2k)^n P^{n(n+1)/2},$$

which gives

$$(6.13) \qquad J_{k,n}(P) \geqslant (2k)^{-n} P^{2k - n(n+1)/2},$$

and this is a nontrivial bound if $k > \frac{1}{4}(n^2 + n)$.

Further, we have

$$(6.14) \qquad \left| \sum_{x \leqslant P} e(\alpha_1 x + \cdots + \alpha_n x^n) \right|^{2k}$$

$$= \sum_{\lambda_1, \ldots, \lambda_n} J_{k,n}(\lambda_1, \ldots, \lambda_n) e(-\alpha_1 \lambda_1 - \cdots - \alpha_n \lambda_n),$$

if we raise the left-hand side of (6.14) to the power $2k$ and collect sums with various $\lambda_1, \ldots, \lambda_n$ defined by (6.9). Replacing x by xy in (6.14) and summing, we also obtain

$$(6.15) \qquad \sum_{x \leqslant P} \left| \sum_{y \leqslant P} e(\alpha_1 xy + \cdots + \alpha_n x^n y^n) \right|^{2k}$$

$$\leqslant \sum_{\lambda_1, \ldots, \lambda_n} J_{k,n}(\lambda_1, \ldots, \lambda_n) \left| \sum_{x \leqslant P} e(-\alpha_1 \lambda_1 x - \cdots - \alpha_n \lambda_n x^n) \right|.$$

If we consider only the first $n - 1$ equations in (6.9), then the number of their solutions is $J_{k,n-1}(\lambda_1, \ldots, \lambda_{n-1})$, and if we let $|\lambda_n|$ take all possible values $(< kP^n)$ in the last equation in (6.9), then we obtain

$$(6.16) \qquad \sum_{|\lambda_n| < kP^n} J_{k,n}(\lambda_1, \ldots, \lambda_n) = J_{k,n-1}(\lambda_1, \ldots, \lambda_{n-1}).$$

We now proceed to obtain less elementary estimates of $J_{k,n}(P)$, and to accomplish this we shall need a lemma which counts the number of solutions of a certain system of congruences modulo a prime power. This is

LEMMA 6.1. Let $m \geqslant 1$, $p > n$ be a prime, and let T denote the number of solutions of the system of congruences

$$(6.17) \qquad \begin{aligned} x_1 + \cdots + x_n &\equiv \mu_1 \pmod{p}, \\ x_1^2 + \cdots + x_n^2 &\equiv \mu_2 \pmod{p^2}, \\ &\vdots \qquad\qquad \vdots \\ x_1^n + \cdots + x_n^n &\equiv \mu_n \pmod{p^n}, \end{aligned}$$

where $A \leqslant x_r \leqslant A + mp^n - 1$ $(r = 1, \ldots, n)$, A is a given integer, and $x_i \not\equiv x_j$ $(\bmod\ p)$ for $i \neq j$. Then for any integers $\mu_1, \ldots, \mu_n$

$$(6.18) \qquad T \leqslant n! \, m^n p^{n(n-1)/2}.$$

Proof of Lemma 6.1. We may suppose $A = 0$ without loss of generality, since if $x_1, \ldots, x_n$ satisfy (6.17) and $x_j \equiv y_j \pmod{p^n}$ $(j = 1, \ldots, n)$, then $y_1, \ldots, y_n$ also satisfy (6.17). We write x_r $(r = 1, \ldots, n)$ as

$$x_r = x_{r,1} + p x_{r,2} + \cdots + p^{n-1} x_{r,n} + p^n x_{r,n+1},$$

where $0 \leqslant x_{r,j} < p$ $(j = 1, \ldots, n)$, $0 \leqslant x_{r,n+1} < m$, and $x_{i,1} \neq x_{j,1}$ if $i \neq j$. The numbers $x_{1,1}, \ldots, x_{n,1}$ satisfy then the system of congruences

$$(6.19) \qquad \begin{aligned} x_{1,1} + \cdots + x_{n,1} &\equiv \mu_1 \pmod{p}, \\ x_{1,1}^2 + \cdots + x_{n,1}^2 &\equiv \mu_2 \pmod{p}, \\ &\vdots \qquad\qquad \vdots \\ x_{1,1}^n + \cdots + x_{n,1}^n &\equiv \mu_n \pmod{p}. \end{aligned}$$

If $s_1, \ldots, s_n$ are elementary symmetric functions in $x_{1,1}, \ldots, x_{n,1}$, then $s_1, \ldots, s_n$ are uniquely determined mod p by $\mu_1, \ldots, \mu_n$. The congruence

$$x^n - s_1 x^{n-1} + \cdots + (-1)^n s_n \equiv 0 \pmod{p}$$

has at most n $(< p)$ roots, and it obviously has $x_{1,1}, \ldots, x_{n,1}$ as roots for any solution $x_{1,1}, \ldots, x_{n,1}$ of (6.19). Thus all solutions of (6.19) can be permutations of one solution in distinct numbers $x_{1,1}, \ldots, x_{n,1}$, hence $T_1 \leqslant n!$ if T_1 denotes the number of solutions of (6.19). If we fix a solution of (6.19), then (6.17) gives the system

$$(6.20) \qquad (x_{1,1} + px_{1,2})^2 + \cdots + (x_{n,1} + px_{n,2})^2 \equiv \mu_2 \pmod{p^2},$$
$$\vdots \qquad\qquad\qquad \vdots$$
$$(x_{1,1} + px_{1,2})^n + \cdots + (x_{n,1} + px_{n,2})^n \equiv \mu_n \pmod{p^2},$$

which reduces to the following system of $n - 1$ congruences in unknowns $x_{1,2}, \ldots, x_{n,2} \pmod{p}$:

$$(6.21) \qquad x_{1,1}x_{1,2} + \cdots + x_{n,1}x_{n,2} \equiv \mu'_2 \pmod{p},$$
$$\vdots \qquad\qquad\qquad \vdots$$
$$x_{1,1}^{n-1}x_{1,2} + \cdots + x_{n,1}^{n-1}x_{n,2} \equiv \mu'_n \pmod{p}.$$

Since the numbers $x_{1,1}, \ldots, x_{n,1}$ are distinct we may suppose that the numbers $x_{1,1}, \ldots, x_{n-1,1}$ are nonzero. Then

$$\begin{vmatrix} x_{1,1} & \cdots & x_{n-1,1} \\ \vdots & & \vdots \\ x_{1,1}^{n-1} & \cdots & x_{n-1,1}^{n-1} \end{vmatrix} = x_{1,1} \cdots x_{n-1,1} \prod_{1 \leqslant i < j \leqslant n-1} (x_{i,1} - x_{j,1}) \not\equiv 0 \pmod{p}.$$

Therefore for each fixed $x_{n,2}$ (which can take up to p values) the numbers $x_{1,2}, \ldots, x_{n-1,2}$ are uniquely determined from (6.21), and if T_2 is the number of solutions of (6.21), then $T_2 = p$. Considering the analogous system mod p^3, we obtain $n - 2$ congruences in $x_{1,3}, \ldots, x_{n,3}$ with the number of solutions $T_3 = p^2$, and continuing in this way $T_4 = p^3, \ldots, T_n = p^{n-1}$. Since each $x_{j,n+1}$ can take up to m values we have $T_{n+1} = m^n$, and we finally obtain

$$T \leqslant T_1 T_2 \cdots T_n T_{n+1} \leqslant n! \, m^n p^{1+2+\cdots+(n-1)} = n! \, m^n p^{n(n-1)/2},$$

as asserted.

We are now ready to give a recurrent estimate for $J_{k,n}(P)$, which will enable us to bound it explicitly. This is the crucial part of the Vinogradov–Korobov method, and the result is

LEMMA 6.2. Let $n \geqslant 2$, $P \geqslant (2n)^{3n}$, and $k \geqslant n^2 + n$. Then

$$(6.22) \qquad J_{k,n}(P) \leqslant 4^{2k} P^{2k/n + (3n-5)/2} J_{k-n,n}(P_1),$$

where P_1 is a number which satisfies $P^{(n-1)/n} \leqslant P_1 \leqslant 4P^{(n-1)/n}$.

Proof of Lemma 6.2. Let p be a prime from $[\frac{1}{2}P^{1/n}, P^{1/n}]$ (such a prime exists e.g., by the prime number theorem). Thus $p > n$, and if we set $P_1 = [Pp^{-1}] + 1$, then $P^{(n-1)/n} \leqslant P_1 \leqslant 4P^{(n-1)/n}$, $P < pP_1$. This gives $J_{k,n}(P) \leqslant J_{k,n}(pP_1)$, where $J_{k,n}(pP_1)$ represents the number of solutions $1 \leqslant x_j \leqslant p$, $0 \leqslant y_j < P_1$ $(j = 1, \ldots, 2k)$ of the system

(6.23)

$$(x_1 + py_1) + \cdots + (x_k + py_k) - (x_{k+1} + py_{k+1}) - \cdots - (x_{2k} + py_{2k}) = 0,$$
$$\vdots$$
$$(x_1 + py_1)^n + \cdots + (x_k + py_k)^n - (x_{k+1} + py_{k+1})^n - \cdots - (x_{2k} + py_{2k})^n = 0.$$

The solutions of (6.23) are divided now into two classes: the first class consists of solutions such that both $x_1, \ldots, x_k$ and $x_{k+1}, \ldots, x_{2k}$ contain n different numbers, while the second class contains all other solutions. If the number of solutions belonging to the first and second class are denoted, respectively, by J_1 and J_2, then

(6.24)
$$J_{k,n}(P) \leqslant J_{k,n}(pP_1) = J_1 + J_2;$$

hence the problem reduces to the estimation of J_1 and J_2. Noting that n numbers may be placed in $k(k-1)\cdots(k-n+1)$ ways in k places and relabeling the solutions counted by J_1 we have $J_1 \leqslant k^{2n}J_1'$, where J_1' is the number of solutions of (6.23) where $x_1, \ldots, x_n$ and $x_{k+1}, \ldots, x_{k+n}$ are distinct numbers. Putting

$$S(x) = \sum_{y=0}^{P_1-1} e(f(x + py)), \qquad f(t) = \alpha_1 t + \cdots + \alpha_n t^n,$$

we have

$$J_1' = \int_0^1 \cdots \int_0^1 \left| \sum_{x_1, \ldots, x_n} S(x_1) \cdots S(x_n) \right|^2 \left| \sum_{x \leqslant p} S(x) \right|^{2k-2n} d\alpha_1 \cdots d\alpha_n,$$

where $\sum_{x_1, \ldots, x_n}$ denotes summation over distinct numbers $x_1, \ldots, x_n$. Using Hölder's inequality (Section A.4) we obtain

$$J_1' \leqslant p^{2k-2n-1} \int_0^1 \cdots \int_0^1 \left| \sum_{x_1, \ldots, x_n} S(x_1) \cdots S(x_n) \right|^2 \sum_{x \leqslant p} |S(x)|^{2k-2n} d\alpha_1 \cdots d\alpha_n$$

$$\leqslant p^{2k-2n} \max_{1 \leqslant x \leqslant p} \int_0^1 \cdots \int_0^1 \left| \sum_{x_1, \ldots, x_n} S(x_1) \cdots S(x_n) \right|^2 |S(x)|^{2k-2n} d\alpha_1 \cdots d\alpha_n.$$

Thus x is fixed in the last integral, which represents the number of solutions of the system

$$(x_1 + py_1) + \cdots + (x_n + py_n) + (x + py_{n+1}) + \cdots + (x + py_k)$$
$$= (x_{k+1} + py_{k+1}) + \cdots + (x_{k+n} + py_{k+n})$$
$$+ (x + py_{k+n+1}) + \cdots + (x + py_{2k}),$$
$$\vdots \qquad\qquad\qquad\qquad \vdots$$
$$(x_1 + py_1)^n + \cdots + (x_n + py_n)^n + (x + py_{n+1})^n + \cdots + (x + py_k)^n$$
$$= (x_{k+1} + py_{k+1})^n + \cdots + (x_{k+n} + py_{k+n})^n$$
$$+ (x + py_{k+n+1})^n + \cdots + (x + py_{2k})^n.$$

Since the number of solutions counted by $J_{k,n}(P)$ does not change if x is subtracted from the unknowns, it follows that the number of solutions of the last system is the same as the number of solutions of the system

$$(x_1 - x + py_1) + \cdots + (x_n - x + py_n) - (x_{k+1} - x + py_{k+1}) - \cdots$$
$$- (x_{k+n} - x + py_{k+n}) = -p(y_{n+1} + \cdots - y_{2k}),$$
$$\vdots \qquad\qquad\qquad\qquad \vdots$$
$$(x_1 - x + py_1)^n + \cdots + (x_n - x + py_n)^n - (x_{k+1} - x + py_{k+1})^n - \cdots$$
$$- (x_{k+n} - x + py_{k+n})^n = -p^n(y_{n+1}^n + \cdots - y_{2k}^n),$$

which we denote by J_0. Setting

$$y_{n+1} + \cdots - y_{2k} = \lambda_1, \qquad y_{n+1}^n + \cdots - y_{2k}^n = \lambda_n,$$

we see by (6.10) that the number of solutions of the last system for any fixed $\lambda_1, \ldots, \lambda_n$ does not exceed $J_{k-n,n}(P_1)$. Further, fixing $0 \leqslant x_{k+1} + py_{k+1} \leqslant pP_1, \ldots, 0 \leqslant x_{k+n} + py_{k+n} \leqslant pP_1$ we see that $x_1 + py_1, \ldots, x_n + py_n$ satisfy the conditions of Lemma 6.1 with $m \geqslant pP_1 p^{-n} > Pp^{-n}$, so that we may take $m = [Pp^{-n}] + 1$. Thus

$$J_0 \leqslant (pP_1)^n TJ_{k-n,n}(P_1),$$

and using (6.18) we obtain

(6.26)

$$J_1 \leqslant k^{2n} p^{2k-2n} n! \, m^n p^{n(n-1)/2} (pP_1)^n J_{k-n,n}(P_1).$$

Since $p \geqslant \frac{1}{2} P^{1/n}$ we have $Pp^{-n} \leqslant 2^n$, implying $m^n \leqslant 2^{n^2+n} \leqslant 2^k$. Using further $p \leqslant P^{1/n}$, $P_1 \leqslant 4P^{(n-1)/n}$, we obtain from (6.25)

(6.26)

$$J_1 \leqslant n! \, 2^k k^{2n} p^{2k/n+(3n-5)/2} 4^n J_{k-n,n}(P_1) \leqslant \frac{1}{4} 2^{2k} p^{2k/n+(3n-5)/2} J_{k-n,n}(P_1),$$

because $k \geqslant n^2 + n$, $n \geqslant 2$.

Thus it remains to estimate J_2. In this case there are at most $n - 1$ distinct numbers between $x_1, \ldots, x_k$, and each x_i can take p values, so there are at most $n^k p^{n-1}$ choices for the numbers $x_1, \ldots, x_k$. Hence if $x_1, \ldots, x_{2k}$ belong to the second class, an application of Hölder's inequality gives

$$\left| \sum_{x_1, \ldots, x_{2k}} S(x_1) \cdots \overline{S(x_{2k})} \right|$$

$$\leqslant \left(\sum_{x_1, \ldots, x_{2k}} |S(x_1)|^{2k} \right)^{1/2k} \cdots \left(\sum_{x_1, \ldots, x_{2k}} |S(x_{2k})|^{2k} \right)^{1/2k}$$

$$\leqslant \sum_{x \leqslant p} |S(x)|^{2k} \sum_{x_1, \ldots, x_{2k}} 1 \leqslant p^{k+n-1} n^k \sum_{x \leqslant p} |S(x)|^{2k}$$

$$\leqslant p^{k+n-1} n^k P_1^{2n} \sum_{x \leqslant p} |S(x)|^{2(k-n)},$$

since trivially $|S(x)| \leqslant P_1$. Therefore

$$J_2 = \int_0^1 \cdots \int_0^1 \sum_{x_1, \ldots, x_k, x_{k+1}, \ldots, x_{2k}} S(x_1) \cdots$$

$$\times S(x_k) \overline{S(x_{k+1})} \cdots \overline{S(x_{2k})} \, d\alpha_1 \cdots d\alpha_n$$

$$\leqslant p^{k+n-1} n^k P_1^{2n} \int_0^1 \cdots \int_0^1 \sum_{x \leqslant P_1} |S(x)|^{2(k-n)} \, d\alpha_1 \cdots d\alpha_n$$

$$= p^{k+n-1} n^k P_1^{2n} J_{k-n, n}(P_1) \leqslant \tfrac{1}{2} 4^{2k} P^{2k/n + (3n-5)/2} J_{k-n, n}(P_1),$$

which combined with (6.26) proves (6.22) in view of (6.24).

Lemma 6.2 enables us to prove an upper bound for $J_{k, n}(P)$ which depends explicitly on k, n, and P. This is

LEMMA 6.3. Let $r \geqslant 0$ be an integer, $k \geqslant n^2 + nr$, $P \geqslant P_0$ and $c_r = \tfrac{1}{2}(n^2 + n)(1 - 1/n)^r$. Then

(6.27) $$J_{k, n}(P) \leqslant (4n)^{4kr} P^{2k - (n^2+n)/2 + c_r}.$$

Proof of 6.3. We use induction on r. For $r = 0$ (6.27) is true, since trivially $J_{k, n}(P) \leqslant P^{2k}$. Suppose now that (6.27) is true for $r = m \geqslant 0$ and consider $r = m + 1$. If

(6.28) $$P \geqslant (2n)^{3n(1 + 1/(n-1))^r},$$

then $k \geqslant n^2 + n(m+1)$, $P \geqslant (2n)^{3n(1+1/(n-1))^{m+1}}$, and an application of Lemma 6.2 gives

(6.29) $$J_{k,n}(P) \leqslant 4^{2k} P^{2k/n+(3n-5)/2} J_{k-n,n}(P_1).$$

To bound $J_{k-n,n}(P_1)$ we may use the the induction hypothesis, since

$$k - n \geqslant n^2 + nm, \qquad P_1 \geqslant P^{1-1/n} \geqslant (2n)^{3n(1+1/(n-1))^m}.$$

It follows that

$$J_{k-n,n}(P_1) \leqslant (4n)^{4(k-n)m} P_1^{2(k-n)-(n^2+n)/2+c_m}$$

$$\leqslant (4n)^{4(k-n)m} 4^{2(k-n)} P^{(1-1/n)(2k-2n-(n^2+n)/2+c_m)},$$

which combined with (6.29) gives (6.27). Let now

(6.30) $$P < (2n)^{3n(1+1/(n-1))^r}$$

and use induction again, supposing

(6.31) $$P < (2n)^{3n(1+1/(n-1))^{m+1}}, \qquad k \geqslant n^2 + n(m+1).$$

Then we have trivially

(6.32) $$J_{k,n}(P) \leqslant P^{2n} J_{k-n,n}(P),$$

and if $P \geqslant (2n)^{3n(1+1/(n-1))^m}$ then by the first part of the proof $J_{k-n,n}(P)$ may be estimated by (6.27). Otherwise, we use the induction hypothesis to estimate $J_{k-n,n}(P)$, so that we obtain in any case

$$J_{k-n,n}(P) \leqslant (4n)^{4km} P^{2(k-n)-(1/2)(n^2+n)+c_m},$$

and (6.32) gives

$$J_{k,n}(P) \leqslant (4n)^{4k(m+1)} P^{2k-(1/2)(n^2+n)+c_{m+1}},$$

since (6.31) implies

$$P^{c_m - c_{m+1}} \leqslant (4n)^{4k}.$$

6.3 ESTIMATION OF THE ZETA SUM

We shall now use Lemma 6.3 to prove Theorem 6.2. First, we need two simple, technical lemmas which will enable us to estimate suitably a certain sum which arises in the proof of Theorem 6.2.

LEMMA 6.4. For $N \geqslant 1$ and α real

(6.33)
$$\left| \sum_{n \leqslant N} e(\alpha n) \right| \leqslant \min(N, 1/(2\|\alpha\|)),$$

where $\|\alpha\|$ is the distance of α to the nearest integer, that is,

$$\|\alpha\| = \min(\alpha - [\alpha], 1 + [\alpha] - \alpha).$$

Proof of 6.4. Obvious for $\alpha = 0$, hence by periodicity of the exponential function suppose $0 < \alpha < 1$. Then

$$\left| \sum_{n \leqslant N} e(\alpha n) \right| = \left| \frac{e(\alpha N) - 1}{e(\alpha) - 1} \right| \leqslant \frac{1}{|\sin \pi \alpha|} \leqslant \frac{1}{2\|\alpha\|},$$

since $\sin x \geqslant (2/\pi)x$ for $0 \leqslant x \leqslant \pi/2$.

LEMMA 6.5. Let $\alpha = a/q + \theta/q^2$, where a and q are integers such that $(a, q) = 1$, $q \geqslant 1$, and $|\theta| \leqslant 1$. Then for arbitrary β, $U > 0$ and $N \geqslant 1$

(6.34)
$$\sum_{n \leqslant N} \min\left(U, \frac{1}{\|\alpha n + \beta\|}\right) \leqslant 6\left(\frac{N}{q} + 1\right)(U + q \log q).$$

Proof of Lemma 6.5. Writing

$$\sum_{n \leqslant N} = \sum_{n=1}^{q} + \sum_{n=q+1}^{2q} + \cdots$$

and changing indexes we obtain at most $N/q + 1$ sums

$$S = \sum_{n \leqslant q} \min\left(U, (\|\alpha n + \beta_1\|)^{-1}\right)$$

with perhaps different β_1's, so it will be sufficient to prove

(6.35)
$$S \leqslant 6(U + q \log q).$$

For $1 \leqslant n \leqslant q$ we have

$$\alpha n + \beta_1 = (an + [q\beta_1])q^{-1} + \theta'(n)q^{-2},$$

$$\theta'(n) = \theta n + (q\beta_1 - [q\beta_1])q, \qquad |\theta'(n)| < 2q.$$

We make the change of variable $m = an + [q\beta_1]$, so that by periodicity of $\|x\|$ and $(a, q) = 1$ we may assume that m runs over a complete system of residues mod q, and $|m| \leqslant q/2$. Thus

$$S = \sum_{|m| \leqslant q/2} \min\left(U, \left\| \frac{m}{q} + \frac{\theta''(m)}{q} \right\|^{-1} \right),$$

where $|\theta''(m)| = |\theta'(m)/q| < 2$. For $m = 0, \pm 1, \pm 2$, we estimate the min by U in the sum above, and for $2 < |m| \leqslant q/2$ we have

$$\left\| \frac{m}{q} + \frac{\theta''(m)}{q} \right\| > \frac{|m| - 2}{q};$$

hence

$$S \leqslant 5U + \sum_{2 < |m| \leqslant q/2} \frac{q}{|m| - 2} < 6(U + q \log q).$$

Proof of Theorem 6.2. The most suitable way to proceed seems to write $n^{it} = \exp(it \log n)$ and to use sufficiently many terms of the Taylor expansion for the log function. To accomplish this let x, y be integer variables such that

$$1 \leqslant x \leqslant a, \qquad 1 \leqslant y \leqslant a, \qquad a = [N^{2/5}].$$

If we define

$$S_1 = \sum_{n=N+1}^{N_1} \exp(it \log(n + xy)) = \sum_{n=N+xy+1}^{N_1 + xy} n^{it},$$

then we may write

$$S(t; N, N_1) = \sum_{N < n \leqslant N_1} n^{it} = S_1 + \sum_{N < n \leqslant N+xy} n^{it} - \sum_{N_1 < n \leqslant N_1 + xy} n^{it},$$

so that

$$S(t; N, N_1) = S_1 + 2\theta a^2 \qquad (|\theta| \leqslant 1),$$

$$a^2 S(t; N, N_1) = \sum_{n=N+1}^{N_1} \sum_{x \leqslant a} \sum_{y \leqslant a} \exp(it \log(1 + xy/n)) \exp(it \log n) + 2\theta a^4,$$

which gives

$$(6.36) \qquad S(t; N, N_1) \ll N a^{-2} \max_{N \leqslant n \leqslant 2N} |T(n)| + a^2,$$

where

(6.37) $$T(n) = \sum_{x \leqslant a} \sum_{y \leqslant a} \exp(it \log(1 + xy/n)).$$

Here $0 < xy/n < 1$, so that using the inequality $|e^{iu} - e^{iv}| \leqslant |u - v|$ $(u, v$ real) and the first r terms in the Taylor-series expansion for $\log(1 + x)$, we obtain

$$\exp(it \log(1 + xy/n)) = \exp(itF_r(x, y)) + t\theta_1(a^2 n^{-1})^{r+1} \qquad (|\theta_1| \leqslant 1),$$

$$F_r(x, y) = \sum_{k=1}^{r} (-1)^{k-1} k^{-1} (xyn^{-1})^{k}.$$

We choose now

(6.38) $$r = \left[5.01 \frac{\log t}{\log N} \right],$$

so that $r \geqslant 5$. This gives

$$Nt(a^2 n^{-1})^{r+1} \leqslant NtN^{-(r+1)/5}$$

$$\leqslant Nt \exp\left\{ \left(-\frac{1}{5} \log N \right) \left(5.01 \frac{\log t}{\log N} \right) \right\} = Nt^{-1/500};$$

hence

(6.39) $$S(t; N, N_1) \ll Na^{-2} \max_{N \leqslant n \leqslant 2N} U(n) + N^{4/5} + Nt^{-1/500},$$

where

(6.40) $$U(n) = \left| \sum_{x \leqslant a} \sum_{y \leqslant a} e(\alpha_1 xy + \cdots + \alpha_r x^r y^r) \right|$$

$$\left(\alpha_j = (-1)^{j-1} t/(2\pi j n^j), \ j = 1, \ldots, r \right).$$

Therefore (6.4) will follow from

(6.41) $$U(n) \ll a^2 \exp\left(-\frac{\log^3 N}{10^5 \log^2 t} \right),$$

and the exponential sum $U(n)$ is of such a form that the method of

Vinogradov–Korobov may be applied. By Hölder's inequality and (6.15)

$$U^{2k}(n) \leqslant a^{2k-1} \sum_{x \leqslant a} \left| \sum_{y \leqslant a} e(\alpha_1 xy + \cdots + \alpha_r x^r y^r) \right|^{2k}$$

$$\leqslant a^{2k-1} \sum_{\lambda_1, \ldots, \lambda_r} J_{k,r}(\lambda_1, \ldots, \lambda_r) \left| \sum_{x \leqslant a} e(\alpha_1 \lambda_1 x + \cdots + \alpha_r \lambda_r x^r) \right|.$$

Another application of Hölder's inequality gives

$$U^{4k^2}(n) \leqslant a^{4k^2 - 2k} \left(\sum_{\lambda_1, \ldots, \lambda_r} J_{k,r}(\lambda_1, \ldots, \lambda_r) \right)^{2k-1}$$

$$\times \sum_{\lambda_1, \ldots, \lambda_r} J_{k,r}(\lambda_1, \ldots, \lambda_r) \left| \sum_{x \leqslant a} e(\alpha_1 \lambda_1 x + \cdots + \alpha_r \lambda_r x^r) \right|^{2k}$$

$$\leqslant J_{k,r}(a) a^{8k^2 - 4k} \sum_{\lambda_1, \ldots, \lambda_r} \left| \sum_{x \leqslant a} e(\alpha_1 \lambda_1 x + \cdots + \alpha_r \lambda_r x^r) \right|^{2k},$$

where we used (6.10) and (6.11). To bound the last sum we use (6.14) and Lemma 6.4. This gives

$$\sum_{\lambda_1, \ldots, \lambda_r} \left| \sum_{x \leqslant a} e(\alpha_1 \lambda_1 x + \cdots + \alpha_r \lambda_r x^r) \right|^{2k}$$

$$= \sum_{\lambda_1, \ldots, \lambda_r, \mu_1, \ldots, \mu_r} J_{k,r}(\mu_1, \ldots, \mu_r) e(-\alpha_1 \lambda_1 \mu_1 - \cdots - \alpha_r \lambda_r \mu_r)$$

$$\leqslant \sum_{\mu_1, \ldots, \mu_r} J_{k,r}(a) \left| \sum_{\lambda_1} e(\alpha_1 \lambda_1 \mu_1) \right| \cdots \left| \sum_{\lambda_r} e(\alpha_r \lambda_r \mu_r) \right|$$

$$\leqslant J_{k,r}(a) \sum_{\mu_1, \ldots, \mu_r} \min(2A_1, \|\alpha_1 \mu_1\|^{-1}) \cdots \min(2A_r, \|\alpha_r \mu_r\|^{-1})$$

$$\leqslant J_{k,r}(a) \prod_{m=1}^{r} \sum_{|\mu_m| < A_m} \min(2A_m, \|\alpha_m \mu_m\|^{-1}),$$

where $A_m = 2ka^m$ $(m = 1, \ldots, r)$. The last sum above is trivially $\leqslant (2A_m)^2$, but for certain values of m one can do considerably better. Namely, for

$$1 + \left[\frac{2 \log t}{\log N} \right] \leqslant m \leqslant \left[\frac{5 \log t}{\log N} \right]$$

we shall use Lemma 6.5 to obtain

$$\sum_{\mu_m} \min\left(2A_m, \|\alpha_m \mu_m\|^{-1}\right) \leqslant 6\left(\frac{2A_m}{q_m} + 1\right)\left(2A_m + q_m \log q_m\right)$$

$$\leqslant 6(2A_m)^2\left(\frac{1}{q_m} + \frac{1}{A_m} + \frac{q_m}{4A_m^2}\right)\log q_m,$$

where

$$\alpha_m = \frac{(-1)^{m-1}t}{2\pi m n^m} = \frac{a_m}{q_m} + \frac{\theta_m}{q_m^2} \qquad (|\theta_m| \leqslant 1),$$

$$a_m = (-1)^{m-1}, \qquad q_m = [2\pi m n^m t^{-1}] > 1.$$

We have $\log q_m \leqslant m \log(2N) \leqslant 2m \log N$, and for $k \ll r^2, t \geqslant t_0$, we obtain using $m \geqslant 2\log t/\log N$

$$q_m \geqslant m n^m t^{-1} \geqslant m N^m t^{-1} \geqslant 8ka^m = 4A_m;$$

hence

$$\frac{1}{q_m} + \frac{1}{A_m} + \frac{q_m}{4A_m^2} \leqslant \frac{q_m}{A_m^2} \leqslant \frac{2\pi m 2^m N^m}{4t a^{2m}} \leqslant m2^{2m+1}N^{m/5}t^{-1}.$$

This gives

$$\prod_{m=1}^{r} \sum_{|\mu_m| < A_m} \min\left(2A_m, \|\alpha_m \mu_m\|^{-1}\right)$$

$$\leqslant \prod_{m=1}^{r} (2A_m)^2 \prod_{1+[(2\log t)/(\log N)] \leqslant m \leqslant [(5\log t)/(\log N)]} \left(m^2 2^{2m+5}N^{m/5}t^{-1}\log N\right)$$

$$\leqslant W(4k)^{2r}2^{8r^2}a^{r^2+r}\log^r N,$$

where

$$W = \prod_{1+[(2\log t)/(\log N)] \leqslant m \leqslant [(5\log t)/(\log N)]} N^{m/5}t^{-1}.$$

If $b > a \geqslant 1$ are integers, then

$$\sum_{a+1 \leqslant n \leqslant b} n = \tfrac{1}{2}(b-a)(b+a+1).$$

Therefore writing $Y = \log t/\log N$ we obtain

$$\log W = \frac{\log N}{10}\{([5Y] - [2Y])([5Y] + [2Y] + 1) - 10Y([5Y] - [2Y])\}$$

$$\leqslant \frac{\log N}{10}([5Y] - [2Y])(1 - 3Y) \leqslant -\frac{\log N}{10}(3Y - 1)^2,$$

where we used $[x] - [y] \geqslant x - y - 1$. Now observe that we may suppose $Y = \log t/\log N \geqslant 10$, that is, we may restrict ourselves to the range

$$\exp(\log^{2/3} t) \leqslant N \leqslant t^{1/10}.$$

If N does not satisfy the lower bound of this inequality then (6.4) is trivial, while if it does not satisfy the upper bound we use the theory of exponent pairs as in Section 7.5 of Chapter 7 to estimate $S(t; N, N_1)$. If (κ, λ) is an exponent pair, then by (7.54) we have

$$S(t; N, N_1) \ll t^\kappa N^{\lambda - \kappa} \leqslant N \exp\left(-\frac{\log^3 N}{10^5 \log^2 t}\right)$$

for $t^{1/10} \leqslant N \leqslant t$ if

$$(6.42) \qquad\qquad \kappa + 10^{-5} \leqslant (1 + \kappa - \lambda)/10.$$

But using induction and Lemma 2.8 we see that starting from $(\kappa, \lambda) = (\frac{1}{6}, \frac{4}{6})$, we obtain that

$$(6.43) \qquad\qquad (\kappa, \lambda) = \left(\frac{1}{2^l - 2}, \frac{2^l - l - 1}{2^l - 2}\right) \qquad (l \geqslant 3)$$

is also an exponent pair, and (6.42) holds with $l = 13$.

Thus the condition $Y \geqslant 10$ gives

$$\log W \leqslant -\frac{\log N}{10}\left(\frac{29}{10}Y\right)^2 = -\frac{841 \log^2 t}{1000 \log N};$$

hence

$$\prod_{m=1}^r \sum_{|\mu_m| < A_m} \min\left(2A_m, \|\alpha_m\mu_m\|^{-1}\right) \leqslant (4k)^{2^r} 2^{8r^2} a^{r^2 + r} \exp\left(-\frac{841 \log^2 t}{1000 \log N}\right) \log^r N,$$

and

$$U^{4k^2}(n) \leqslant J_{k,r}^2(a)(4k)^{2^r} 2^{8r^2} a^{8k^2 - 4k + r^2 + r} \exp\left(-\frac{841 \log^2 t}{1000 \log N}\right) \log^r N.$$

To estimate $J_{k,r}(a)$ we apply Lemma 6.3 with $k = r^2 + (Rr)r$, where $R \geqslant 0$ is an integer to be determined in a moment. We then have

$$(6.44) \quad U(n) \leqslant \left((4k)^{2r} 2^{8r^2} (4r)^{8krR} \right)^{(2k)^{-2}}$$

$$\times a^{(r^2+r)(2k)^{-2}(1-r^{-1})^{Rr}} a^2 (\log N)^{r(2k)^{-2}} \exp\left(- \frac{841 \log^2 t}{4000 k^2 \log N} \right).$$

The first factor on the right-hand side of (6.44) is clearly bounded, while for any $\varepsilon > 0$, and $N > N_0(\varepsilon)$

$$(\log N)^{r(2k)^{-2}} < \exp\left(\frac{\varepsilon \log^2 t}{4000 k^2 \log N} \right).$$

The condition $N \leqslant t^{1/10}$ ensures by (6.38) that $r \geqslant 50$, hence $r^2 + r \leqslant (51/50)r^2$, so that using $(1 - r^{-1})^{rR} \leqslant e^{-R}$ we have

$$a^{(r^2+r)(2k)^{-2}(1-r^{-1})^{rR}} \leqslant \exp\left(\frac{2}{5} (\log N) \frac{51}{50} r^2 e^{-R} (2k)^{-2} \right)$$

$$= \exp\left(\frac{51 e^{-R} \log N}{500(R+1)^2 r^2} \right).$$

Inserting this bound in (6.44) we obtain

$$(6.45) \qquad U(n) \ll a^2 \exp\left(\left(\frac{51 e^{-R}}{500} + \frac{\varepsilon - 841}{100,401} \right) \frac{\log N}{r^2 (R+1)^2} \right).$$

We choose now $R \geqslant 0$ in such a way that

$$f(R) = \left(\frac{51 e^{-R}}{500} - \frac{841}{100,401} \right)(R+1)^{-2}$$

is a minimum. Since R has to be an integer, a calculation shows that the optimal value is $R = 4$, which gives $f(R) = -1/3841.2987\ldots$. Therefore (6.45) yields for ε sufficiently small

$$(6.46) \qquad U(n) \ll a^2 \exp\left(- \frac{\log N}{3842 r^2} \right) \leqslant a^2 \exp\left(- \frac{\log^3 N}{96,500 \log^2 t} \right),$$

and Theorem 6.2 follows. In fact (6.46) is even sharper than (6.41), and some further improvements could be obtained by more elaborate calculations. This we shall not do, since our aim was to prove Theorem 6.2 with the reasonable value 10^5 of the constant in $\exp(\,\cdots\,)$, and the optimal value of the constant in question is not easy to determine.

6.4 THE ORDER ESTIMATE OF $\zeta(s)$ NEAR $\sigma = 1$

We proceed now to deduce Theorem 6.3 from Theorem 6.2. We shall first use Theorem 6.2 to estimate the "long" sum

$$\Sigma(t, N) = \sum_{n \leqslant N} n^{it} \qquad (N_0 \leqslant N \leqslant t).$$

Replacing N in (6.4) by $N/2, N/2^2, \ldots$ we obtain

$$\Sigma(t, N) \ll N \overset{O(\log N)}{\underset{k=0}{\sum}} 2^{-k}\exp\left(-\frac{\log^3(N2^{-k})}{10^5\log^2 t}\right)$$

$$= N\left(\sum_{k < \delta \log N} + \sum_{k \geqslant \delta \log N}\right) = N\left(\sum_1 + \sum_2\right)$$

say, where $\delta > 0$ is a fixed, small number. In Σ_1 we have $N2^{-k} \geqslant N^{1-\delta\log 2}$, hence

$$\sum_1 \leqslant \sum_{k=0}^{\infty} 2^{-k}\exp\left(\frac{-(1 - \delta\log 2)^3\log^3 N}{10^5\log^2 t}\right) \leqslant 2\exp\left((\varepsilon - 10^{-5})\frac{\log^3 N}{\log^2 t}\right)$$

for any $\varepsilon > 0$ if $\delta = \delta(\varepsilon)$ is sufficiently small. Since trivially

$$\sum_2 \leqslant \sum_{k \geqslant \delta \log N} 2^{-k} \leqslant 2^{-\delta \log N} = N^{-\delta \log 2},$$

we obtain, after changing δ to $\delta/\log 2$,

$$(6.47) \qquad \Sigma(t, N) \ll N\exp\left((\varepsilon - 10^{-5})\frac{\log^3 N}{\log^2 t}\right) + N^{1-\delta},$$

where $\delta > 0$ is a fixed, small number.

We now set $N = [\exp(\log^{2/3}t)]$ and using (6.2) we write

$$\zeta(\sigma + it) = \sum_{n \leqslant N} n^{-\sigma - it} + \sum_{N < n \leqslant t} n^{-\sigma - it} + O(t^{-\sigma}).$$

The first sum is trivially

$$\ll N^{1-\sigma} \sum_{n \leqslant N} n^{-1} \ll e^{(1-\sigma)\log N}\log^{2/3}t \leqslant t^{(1-\sigma)^{3/2}}\log^{2/3}t$$

if $(1 - \sigma)\log^{2/3} t \leqslant 1$. If $(1 - \sigma)\log^{2/3} t > 1$, then it follows that $(1 - \sigma)\log^{2/3} t \leqslant (1 - \sigma)^{3/2}\log t$, so that in any case

$$\sum_{n \leqslant N} n^{-\sigma - it} \ll t^{(1-\sigma)^{3/2}}\log^{2/3} t.$$

Partial summation and (6.47) give for some $\delta_0 > 0$

$$\sum_{N < n \leqslant t} n^{-s} \ll t^{-\sigma}\left|\sum(t, t)\right| + \int_N^t u^{-\sigma - 1}\left|\sum(t, u)\right| du$$

$$\ll t^{1 - \sigma - \delta_0} + \int_N^t u^{-\sigma}\exp\left((\varepsilon - 10^{-5})\frac{\log^3 u}{\log^2 t}\right) du$$

$$\ll t^{(1-\sigma)^{3/2}}\log^{2/3} t + \int_{\log N}^{\log t}\exp\left(v(1 - \sigma) + (\varepsilon - 10^{-5})\frac{v^3}{\log^2 t}\right) dv,$$

provided that $1 - \delta_1 \leqslant \sigma \leqslant 1$ for fixed, small $\delta_1 > 0$.

Writing for brevity $D = 10^{-5} - 2\varepsilon$ we see that the last integral above does not exceed

$$(6.48) \quad \left\{ \max_{\log N \leqslant v \leqslant \log t} \exp\left(v(1 - \sigma) - \frac{Dv^3}{\log^2 t}\right) \right\} \int_{\log N}^{\log t}\exp\left(-\frac{\varepsilon v^3}{\log^2 t}\right) dv.$$

Change of the variable $v^3 = y \log^2 t$ gives here

$$\int_{\log N}^{\log t}\exp\left(-\frac{\varepsilon v^3}{\log^2 t}\right) dv = \frac{1}{3}\log^{2/3} t \int_{1 + o(1)}^{\log t} y^{-2/3}e^{-\varepsilon y} dy \ll \log^{2/3} t,$$

while if $f(v) = v(1 - \sigma) - Dv^3\log^{-2} t$, then $f'(v) = 0$ for

$$v = v_0 = (\log t)\left(\frac{1 - \sigma}{3D}\right)^{1/2},$$

and

$$f(v_0) = \tfrac{2}{3}(3D)^{-1/2}(1 - \sigma)^{3/2}\log t,$$

so that the expression in (6.48) is

$$\ll t^{(2/3)(3D)^{-1/2}(1-\sigma)^{3/2}}\log^{2/3} t.$$

Therefore we finally obtain for some $\eta > 0$

$$\zeta(s) \ll t^{(2/3)(3D)^{-1/2}(1-\sigma)^{3/2}}\log^{2/3} t \quad (1 - \eta \leqslant \sigma \leqslant 1),$$

and (6.6) follows for $D = 10^{-5} - 2\varepsilon$ if ε is sufficiently small.

If we suppose that instead of (6.4) the more general inequality

(6.49)

$$S(t; N, N_1) \ll N \exp\left(-\gamma \frac{\log^{a+1}N}{\log^a t}\right) \qquad (N_0 \leqslant N \leqslant t/2, t \geqslant t_0, a > 0)$$

holds for some fixed $a > 0$ and an absolute constant $\gamma > 0$, then the foregoing argument would yield

$$(6.50) \qquad \zeta(\sigma + it) \ll t^{C(1-\sigma)^{(a+1)/a}} \log^{a/(a+1)} t \qquad (1 - \eta \leqslant \sigma \leqslant 1)$$

for some small $\eta > 0$ and some absolute $C > 0$. This obviously generalizes (6.6), but at present the Vinogradov–Korobov method does not seem to be able to yield (6.49) with any $a < 2$.

6.5 THE DEDUCTION OF THE ZERO-FREE REGION

At last we are able to deduce the zero-free region of Theorem 6.1. We shall base our arguments on the general estimate (6.50), so that in view of (6.4) Theorem 6.1 will follow in the special case $a = 2$. Since

$$t^{C(1-\sigma)^{(a+1)/a}} = \exp\left(C(1-\sigma)^{(a+1)/a}\log t\right) \leqslant \exp(\varepsilon \log \log t) = \log^\varepsilon t$$

for

$$(6.51) \qquad \sigma \geqslant 1 - \left(\varepsilon C^{-1}\log\log t/\log t\right)^{a/(a+1)} \qquad (t \geqslant t_0),$$

we have for any $\varepsilon > 0$

$$(6.52) \qquad \zeta(\sigma + it) \ll \log^{a/(a+1)+\varepsilon} t,$$

provided that (6.51) holds. From (6.52) we shall deduce

$$(6.53) \qquad \zeta(s) \neq 0 \quad \text{for} \quad \sigma \geqslant 1 - C(\log t)^{-a/(a+1)}(\log\log t)^{-1/(a+1)}$$

$$(C > 0, t \geqslant t_0)$$

by a method which is well known and primarily technical in nature. To accomplish this we shall now formulate and prove three lemmas, of which the first two are general, while the third one is specialized to furnish zero-free regions for the zeta-function.

LEMMA 6.6. Suppose $f(z) = \sum_{n=0}^{\infty} c_n(z - z_0)^n$ is regular for $|z - z_0| < R$ and $\operatorname{Re} f(z) \leqslant U$ for $|z - z_0| < R$. Then

$$(6.54) \quad |c_n| = |f^{(n)}(z_0)/n!| \leqslant 2(U - \operatorname{Re} c_0)R^{-n} \quad (n = 1, 2, \dots),$$

while for $|z - z_0| \leqslant r < R$ we have

$$(6.55) \qquad |f(z) - f(z_0)| \leqslant \frac{2r}{R - r}(U - \operatorname{Re} f(z_0)),$$

$$(6.56) \quad \left| \frac{f^{(k)}(z)}{k!} \right| \leqslant \frac{2R}{(R - r)^{k+1}}(U - \operatorname{Re} f(z_0)) \quad (k = 1, 2, \dots).$$

Proof of Lemma 6.6. By the change of variable $z' = z - z_0$ we may suppose without loss of generality that $z_0 = 0$. For $|z| < R$ let

$$F(z) = U - f(z) = U - c_0 - \sum_{n=1}^{\infty} c_n z^n = \sum_{n=0}^{\infty} b_n z^n.$$

If $0 < r < R$, then

$$(6.57) \qquad b_n = (2\pi i)^{-1} \int_{|z|=r} F(z) z^{-n-1} dz$$

$$= (2\pi r^n)^{-1} \int_{-\pi}^{\pi} (P + iQ) e^{-n\theta} d\theta \quad (n \geqslant 0),$$

where

$$F(re^{i\theta}) = P(r, \theta) + iQ(r, \theta) = P + iQ \quad (P, Q \text{ real}).$$

Since $F(z)z^{n-1}$ $(n \geqslant 1)$ is regular in $|z| \leqslant r$ we have

$$0 = r^n(2\pi)^{-1} \int_{-\pi}^{\pi} (P + iQ) e^{in\theta} d\theta;$$

hence taking the conjugate and adding the resulting expression to (6.57) we obtain

$$b_n r^n = \pi^{-1} \int_{-\pi}^{\pi} P e^{-in\theta} d\theta \quad (n \geqslant 1).$$

Since $P = U - \operatorname{Re} f(z) \geqslant 0$ for $|z| < R$, we have

$$|b_n| r^n \leqslant \pi^{-1} \int_{-\pi}^{\pi} |P e^{-in\theta}| d\theta = \pi^{-1} \int_{-\pi}^{\pi} P \, d\theta = 2 \operatorname{Re} b_0 \quad (n \geqslant 1)$$

in view of (6.57). Letting $r \to R$ it follows that

$$(6.58) \quad |b_n| \leqslant 2(\operatorname{Re} b_0) R^{-n} \quad (b_0 = U - c_0, b_n = -c_n \text{ for } n \geqslant 1),$$

which implies (6.54). Therefore for $|z| \leqslant r < R$

$$|F(z) - F(0)| = \left| \sum_{n=1}^{\infty} b_n z^n \right| \leqslant \sum_{n=1}^{\infty} 2(\operatorname{Re} b_0) r^n R^{-n} = \frac{2r \operatorname{Re} b_0}{R - r},$$

which gives (6.55). Finally, using (6.58) we obtain

$$|F^{(k)}(z)| \leqslant \sum_{n=k}^{\infty} n(n-1) \cdots (n-k+1) \frac{2(\operatorname{Re} b_0) r^{n-k}}{R^n}$$

$$= \frac{2(\operatorname{Re} b_0) R k!}{(R-r)^{k+1}} \quad (k = 1, 2, \ldots),$$

which yields (6.56).

LEMMA 6.7. Let $f(z)$ be regular for $|z - z_0| \leqslant r$, $f(z_0) \neq 0$ and $|f(z)/f(z_0)| \leqslant M$ for $|z - z_0| \leqslant r$. If $f(z) \neq 0$ for $|z - z_0| \leqslant r/2$, $\operatorname{Re}(z - z_0) \geqslant 0$, then

$$(6.59) \qquad \operatorname{Re} f'(z_0)/f(z_0) \geqslant -\frac{4}{r} \log M,$$

$$(6.60) \qquad \operatorname{Re} f'(z_0)/f(z_0) \geqslant -\frac{4}{r} \log M + \operatorname{Re}(z_0 - \rho)^{-1},$$

where ρ is an arbitrary zero of $f(z)$ in the region $|z - z_0| \leqslant r/2$, $\operatorname{Re}(z - z_0) < 0$.

Proof of Lemma 6.7. Let

$$g(z) = f(z) \prod_\rho (z - \rho)^{-1}, \qquad z \neq \rho, \qquad g(\rho) = \lim_{z \to \rho} g(z),$$

where ρ denotes zeros of $f(z)$ in the circle $|z - z_0| \leqslant r/2$ counted with respective multiplicities. Then $g(z)$ is regular for $|z - z_0| \leqslant r/2$, and for $|z - z_0| = r$

$$\left| \frac{g(z)}{g(z_0)} \right| = \left| \frac{f(z)}{f(z_0)} \prod_\rho \frac{z_0 - \rho}{z - \rho} \right| \leqslant M,$$

and by the maximum modulus principle this holds also for $|z - z_0| \leqslant r$. Since

$g(z) \neq 0$ for $|z - z_0| \leqslant r/2$, then taking the principal branch of the logarithm we see that $F(z) = \log g(z) - \log g(z_0)$ is regular for $|z - z_0| \leqslant r/2$ and

$$\operatorname{Re} F(z) = \log|g(z)/g(z_0)| \leqslant \log M,$$

and $M \geqslant 1$, since $g(z)/g(z_0) = 1$ when $z = z_0$. Moreover $\operatorname{Re} F(z_0) = 0$, so that applying (6.54) with $R = r/2$ we obtain

$$|F'(z_0)| = |g'(z_0)/g(z_0)| \leqslant \frac{4}{r} \log M,$$

while by logarithmic differentiation we have

$$\left| \frac{g'(z_0)}{g(z_0)} \right| = \left| \frac{f'(z_0)}{f(z_0)} - \sum_{\rho} \frac{1}{z_0 - \rho} \right| \leqslant \frac{4}{r} \log M,$$

which implies

(6.61) $$\operatorname{Re}\left(\frac{f'(z_0)}{f(z_0)} - \sum_{\rho} \frac{1}{z_0 - \rho} \right) \geqslant -\frac{4}{r} \log M.$$

Finally, the condition $\operatorname{Re}(z_0 - \rho) > 0$ ensures that the conclusion of the lemma will follow from (6.61).

LEMMA 6.8. Let $\varphi(t)$ and $1/\theta(t)$ be two positive, nondecreasing functions of t for $t \geqslant t_0$ such that $\theta(t) \leqslant 1$, $\varphi(t) \to \infty$ as $t \to \infty$ and

(6.62) $$\varphi(t)/\theta(t) = o(e^{\varphi(t)}) \qquad (t \to \infty).$$

If $\zeta(s) \ll e^{\varphi(t)}$ for $1 - \theta(t) \leqslant \sigma \leqslant 2, t \geqslant t_0$, then $\zeta(s) \neq 0$ for

(6.63) $$\sigma \geqslant 1 - C \frac{\theta(2t + 1)}{\varphi(2t + 1)} \qquad (t \geqslant t_0),$$

where $C > 0$ is an absolute constant.

Proof of Lemma 6.8. Let $\zeta(\beta + i\gamma) = 0, \beta \leqslant 1, \gamma \geqslant \gamma_0$ and let σ_0 satisfy $1 + e^{-\varphi(2\gamma+1)} \leqslant \sigma_0 \leqslant 2$. Let further $s_0 = \sigma_0 + i\gamma, s_0' = \sigma_0 + 2i\gamma$, and $r = \theta(2\gamma + 1)$. Then both the circles $|s - s_0| \leqslant r$ and $|s - s_0'| \leqslant r$ lie in the region $\sigma \geqslant 1 - \theta(r), t \geqslant t_0$, since

$$\sigma_0 - r = \sigma_0 - \theta(2\gamma + 1) \geqslant 1 + e^{-\varphi(2\gamma+1)} - \theta(2\gamma + 1)$$

$$\geqslant 1 - \theta(2\gamma + 1) \geqslant 1 - \theta(r).$$

For $\sigma > 1$ and some $A > 0$ we have $|1/\zeta(s)| \leqslant \zeta(\sigma) < A(\sigma - 1)^{-1}$, hence

$$|1/\zeta(s_0)| \leqslant Ae^{\varphi(2\gamma+1)}, \qquad |1/\zeta(s_0')| \leqslant Ae^{\varphi(2\gamma+1)}.$$

By hypothesis $\zeta(s) \ll e^{\varphi(t)}$ for $1 - \theta(t) \leqslant \sigma \leqslant 2$, so that there must exist $A_2 > 0$ such that

$$|\zeta(s)/\zeta(s_0)| < e^{A_2\varphi(2\gamma+1)} \quad \text{for} \quad |s - s_0| \leqslant r,$$

$$|\zeta(s)/\zeta(s_0')| < e^{A_2\varphi(2\gamma+1)} \quad \text{for} \quad |s - s_0'| \leqslant r.$$

Using (6.59) of Lemma 6.7 we obtain with $M = e^{A_2\varphi(2\gamma+1)}$

$$(6.64) \qquad -\operatorname{Re} \frac{\zeta'(\sigma_0 + 2i\gamma)}{\zeta(\sigma_0 + 2i\gamma)} < A_3 \frac{\varphi(2\gamma + 1)}{\theta(2\gamma + 1)} \qquad (A_3 > 0).$$

We have $\beta \leqslant 1 < \sigma_0$, while for $\beta > \sigma_0 - r/2$, (6.60) of Lemma 6.7 gives

$$(6.65) \qquad -\operatorname{Re} \frac{\zeta'(\sigma_0 + i\gamma)}{\zeta(\sigma_0 + i\gamma)} < A_3 \frac{\varphi(2\gamma + 1)}{\theta(2\gamma + 1)} - \frac{1}{\sigma_0 - \beta}.$$

Also we have

$$(6.66) \qquad -\zeta'(\sigma_0)/\zeta(\sigma_0) < B/(\sigma_0 - 1),$$

where $B \to 1 + 0$ as $\sigma_0 \to 1 + 0$, since $s = 1$ is a pole of first order of $\zeta(s)$ with residue 1. As in the deduction of $\zeta(1 + it) \neq 0$ in Theorem 1.5 we use the elementary inequality (1.20) to deduce from (6.64)–(6.66)

$$\frac{3B}{\sigma_0 - 1} + 5A_3 \frac{\varphi(2\gamma + 1)}{\theta(2\gamma + 1)} - \frac{4}{\sigma_0 - \beta} \geqslant 0,$$

or

$$\sigma_0 - \beta \geqslant \left(\frac{3B}{4(\sigma_0 - 1)} + \frac{5}{4}A_3 \frac{\varphi(2\gamma + 1)}{\theta(2\gamma + 1)} \right)^{-1},$$

which gives then

$$(6.67) \qquad 1 - \beta \geqslant \frac{1 - \frac{3}{4}B - \frac{5}{4}A_3(\varphi(2\gamma + 1)/\theta(2\gamma + 1))(\sigma_0 - 1)}{(3B/4(\sigma_0 - 1)) + \frac{5}{4}A_3(\varphi(2\gamma + 1)/\theta(2\gamma + 1))}.$$

Now we choose $B = \frac{5}{4}$ and $\sigma_0 = 1 + (40A_3)^{-1}\theta(2\gamma + 1)/\varphi(2\gamma + 1)$, and then regardless of A_3 the condition $\sigma_0 \geqslant 1 + \exp(-\varphi(2\gamma + 1))$ holds in view of (6.62). With this choice of B and σ_0 (6.67) reduces to

$$1 - \beta \geqslant \frac{\theta(2\gamma + 1)}{1240 A_3 \varphi(2\gamma + 1)},$$

which gives (6.63). It remains to consider the case $\beta \leqslant \sigma_0 - r/2$, when

$$\beta \leqslant \sigma_0 - r/2 = 1 + (40 A_3)^{-1} \frac{\theta(2\gamma + 1)}{\varphi(2\gamma + 1)} - \frac{1}{2}\theta(2\gamma + 1)$$

$$\leqslant 1 - (1240 A_3)^{-1} \frac{\theta(2\gamma + 1)}{\varphi(2\gamma + 1)},$$

since $\lim_{t \to \infty} \varphi(t) = \infty$.

Having established Lemma 6.8 it is now a simple matter to prove Theorem 6.1, once we have (6.51) and (6.52) with $a = 2$. It is readily checked that the conditions of Lemma 6.8 hold with

$$\varphi(t) = \left(a/(a+1) + \varepsilon\right)\log\log t,$$

$$\theta(t) = C_1(\log t)^{-a/(a+1)}(\log\log t)^{a/(a+1)},$$

for example,

$$\varphi(t)/\theta(t) \ll (\log t)^{a/(a+1)}(\log\log t)^{1/(a+1)}$$

$$= o\left((\log t)^{(a/(a+1)+\varepsilon)}\right) = o\left(e^{\varphi(t)}\right)$$

as $t \to \infty$. Therefore Lemma 6.8 yields (6.53) as the zero-free region for $\zeta(s)$, and $a = 2$ gives finally Theorem 6.1.

Notes

A complete proof of Theorem 6.1 is given in A. Walfisz's book (1963). As told there on p. 226, this result is due to H.-E. Richert, who never published a proof. Richert (1967), however, proved Theorem 6.3 with the constant 122 replaced by the better constant 100, and his result holds uniformly for $\frac{1}{2} \leqslant \sigma \leqslant 1$. Richert's proof is based on the estimate [Walfisz (1963), Satz 2, p. 57]

$$\sum_{M < n \leqslant M' \leqslant 2M} (n + w)^{it} \ll M^{1-\theta},$$

where $t^{1/r} \leqslant M \leqslant t^{1/(r-1)}$, $r \geqslant 19$, $t \geqslant 1$, $0 \leqslant w \leqslant 1$, $\theta = (60{,}000 r^2)^{-1}$. The proof of this bound does not seem to be simpler than our proof of (6.4).

Incidentally, both N. M. Korobov (1958a) and I. M. Vinogradov (1958) claimed a stronger zero-free region than (6.1), namely, $\sigma \geqslant 1 - C \log^{-2/3} t$, $C > 0$, $t \geqslant t_0$. Unfortunately their claim was not substantiated, and one does not see at present how their method can yield anything better than (6.1).

Theorem 6.2 provides a simple form of the estimate of the zeta sum (6.3). A result of this type was proved in Chapter 5 of A. A. Karacuba (1975) with an ineffective constant γ in place of our 10^{-5}. To obtain the explicit value 10^{-5} in Theorem 6.3 we had to make more careful estimations in several places than was done by Karacuba. Karacuba in (1975) uses Theorem 6.2 to obtain a zero-free region for $\zeta(s)$, but instead of (6.1) he manages to obtain (Theorem 2 of Chapter 6)

$$\sigma \geqslant 1 - C \log^{-2/3} t (\log \log t)^{-1} \qquad (C > 0, t > t_0),$$

which is somewhat weaker than (6.1). In connection with this, two questions seem to remain unsettled yet: Does the zero-free region (6.1) imply the order result of the type (6.4) (i.e., with perhaps another numerical constant)? Is (6.49) with $a = 2$ the limit of the present form of the Vinogradov–Korobov method (i.e., Lemmas 6.2 and 6.3)?

It was mentioned in Section 6.1 that $\zeta(1 + it) \ll \log \log t$ if the Riemann hypothesis is true. For this result, and other consequences of the Riemann hypothesis, see Chapter 14 of E. C. Titchmarsh (1951).

Developed by I. M. Vinogradov in the 1930s [see his monographs (1976) and (1980)], the powerful methods for the estimation of exponential sums have found many applications in number theory, and a classical example is Vinogradov's result that every sufficiently large odd integer is a sum of three primes. However, the particular method outlined in Section 6.2 of this chapter has undergone several simplifications and improvements before assuming its present form. These are due to several people, most notably to N. M. Korobov (1958b, 1958c), and especially Appendix II of the Russian translation of K. Prachar (1967), so that the method justly deserves to be called the Vinogradov–Korobov method. Strangely enough, Vinogradov (1976, 1980) makes no mention of Korobov's works.

Before the Korobov works of 1958 it was known only that Vinogradov's method led to the zero-free region

$$(6.68) \qquad \sigma \geqslant 1 - C \log^{-3/4} t (\log \log t)^{-3/4} \qquad (C > 0, t \geqslant t_0),$$

which is weaker than (6.1). [For a proof of (6.68) see, e.g., Chapter 4 of K. Chandrasekharan (1970)].

The methods of Weyl, Hardy–Littlewood, and van der Corput that were developed in the 1920s for the estimation of exponential sums and the derivation of the zero-free region are extensively discussed in Chapter 5 of E. C. Titchmarsh (1951). The sharpest zero-free region obtainable by these methods seems to be

$$\sigma \geqslant 1 - C \log^{-1} t (\log \log t) \qquad (C \geqslant 0, t \geqslant t_0),$$

which is much poorer than (6.1) and only slightly sharper than (1.55). Thus one may acquire a historical perspective of how the zero-free region expanded from $\sigma \geqslant 1$ to the present $\sigma \geqslant 1 - C(\log t)^{-2/3}(\log \log t)^{-1/3}$. Even if the Riemann hypothesis is not true, the somewhat awkward form of the last function makes it very probable that the zero-free region (6.1) will be further improved in the foreseeable future.

The lower bound (6.13) for $J_{k,n}(P)$ may be compared with the upper bound of Lemma 6.3. When r is much larger than n in Lemma 6.3, say when $r \geqslant n \log n$, then c_r is small and the bound in (6.27) is very good. For a further discussion of upper and lower bounds for $J_{k,n}(p)$ see G. I. Arhipov and A. A. Karacuba (1978).

The proofs of Lemmas 6.1 (due to Yu. V. Linnik), 6.2, and 6.3 are to be found in A. A. Karacuba's book (1975).

Lemmas 6.6, 6.7, and 6.8 have been around in zeta-function theory for several decades. Proofs and other variants of these results may be found in competent works such as K. Chandrasekharan (1970), K. Prachar (1967), and E. C. Titchmarsh (1951).

For a derivation of Theorem 6.1 from Theorem 6.2 without function-theoretic results like Lemmas 6.6, 6.7, and 6.8, see Y. Motohashi (1978, 1981), where the Selberg sieve method is used.

Although (6.49) at present is known to hold only for $a = 2$, it seemed of interest to consider the more general case, since it presents no additional difficulties and gives an insight into how Lemma 6.8 provides easily a zero-function region for $\zeta(s)$. Similarly, a general expression for the zero-free region and the prime number theorem (but of a different form than ours) was considered by A. Walfisz (1963).

/ *CHAPTER SEVEN* /
MEAN VALUE ESTIMATES
OVER SHORT INTERVALS

7.1 INTRODUCTION

The proof of the fourth power moment estimate (5.1) depended on an approximate functional equation for $\zeta^2(s)$ [see (4.66)], and even the stronger asymptotic formula (5.2) of D. R. Heath-Brown depended on another approximate functional equation, namely, (5.32) and (5.33). Thus the natural line of approach in estimating integrals of the type

$$\int_0^T |\zeta(\sigma + it)|^k \, dt \qquad \left(\tfrac{1}{2} \leqslant \sigma < 1, \ k > 4 \text{ fixed}\right)$$

would be to use a suitable approximate equation for $\zeta^k(s)$ and then to integrate it termwise. However, to this day no satisfactory result based on this idea has been obtained, and in this chapter we focus our attention on integrals of the type

$$\int_{T-G}^{T+G} |\zeta(\sigma + it)|^2 \, dt \qquad \left[\tfrac{1}{2} \leqslant \sigma < 1, \ G = o(T) \text{ as } T \to \infty\right],$$

where σ is fixed, and the interval of integration is "short" because $G = o(T)$ as $T \to \infty$. The purpose of estimating this type of integral will be seen best in Chapter 8, where we shall use Theorem 7.2 to obtain general power moment estimates for $\zeta(s)$. This idea was first used by D. R. Heath-Brown in his proof of the twelfth power moment estimate (8.35). His crucial estimate in the proof of (8.35) (this is essentially our Theorem 7.2) depended on F. V. Atkinson's formula for the mean square of $\zeta(s)$, which will be discussed in detail in Chapter 15. The proof of Theorem 7.2 given in Section 7.4 is different and dispenses completely with Atkinson's formula. Besides the approximate functional equation (4.10) for $\zeta^2(s)$ it uses the Voronoi summation formula and Atkinson's saddle-point result (Theorem 2.2).

Mean value estimates of $\zeta(s)$ over short intervals may be used to provide order results for $\zeta(s)$ when σ lies in the critical strip, and this will be done in Section 7.5. We shall employ the theory of exponent pairs, and in the important case $\sigma = \frac{1}{2}$ we shall see how G. Kolesnik's method of the estimation of two-dimensional exponential sums leads to the bound $\mu(\frac{1}{2}) \leqslant \frac{35}{216}$. The chapter concludes with a brief discussion (based on the work of H. Iwaniec) of the third and fourth power moments in short intervals.

7.2 AN AUXILIARY ESTIMATE

To facilitate subsequent estimates we shall start with a technical lemma. Its significance lies in the fact that it estimates the moduli of the zeta-function by an integral of nearly the same function. This is

LEMMA 7.1. Let $k \geqslant 1$ be a fixed integer and let $T/2 \leqslant t \leqslant 2T$. Then for $\frac{1}{2} \leqslant \sigma < 1$ fixed we have uniformly in t

$$(7.1) \quad |\zeta(\sigma + it)|^k \ll 1 + \log T \int_{-\log^2 T}^{\log^2 T} \left| \zeta\left(\sigma - \frac{1}{\log T} + it + iv \right) \right|^k e^{-|v|} \, dv,$$

and

$$(7.2) \quad \left| \zeta\left(\tfrac{1}{2} + it \right) \right|^k \ll \log T \left(1 + \int_{-\log^2 T}^{\log^2 T} \left| \zeta\left(\tfrac{1}{2} + it + iv \right) \right|^k e^{-|v|} \, dv \right).$$

Proof of Lemma 7.1. Let $\frac{1}{2} \leqslant \sigma < 1$ be fixed, $s' = \sigma + c + it$, and $c = 1/(\log T)$. From (A.7) we obtain by termwise integration

$$(7.3) \quad (2\pi i)^{-1} \int_{1-i\infty}^{1+i\infty} \zeta^k(s' + w) \, dw = \sum_{n=1}^{\infty} d_k(n) e^{-n} n^{-s'} \ll 1.$$

Moving the line of integration in (7.3) to $\operatorname{Re} w = -c$ we encounter poles at $w = 1 - s'$ (of order k) and $w = 0$ with residues $O(1)$ [in view of (A.34)] and $\zeta^k(s')$, respectively. Since $s = 0$ is a simple pole of $\Gamma(s)$, then also in view of (A.34) we have then for any real v

$$(7.4) \quad \Gamma(\pm c \pm iv) \ll e^{-|v|}(c + |v|)^{-1},$$

so that (7.3) yields, for $T/3 \leqslant t \leqslant 3T$,

$$(7.5) \qquad \zeta^k(s') \ll 1 + \int_{-\infty}^{\infty} |\zeta(\sigma + it + iv)|^k e^{-|v|} (c + |v|)^{-1} \, dv.$$

To obtain (7.1) from (7.5) we only have to note that $c^{-1} = \log T$ and that for any fixed $A > 0$

$$\int_{\pm \log^2 T}^{\pm \infty} |\zeta(\sigma + it + iv)|^k e^{-|v|} (c + |v|)^{-1} \, dv \ll \int_{\log^2 T}^{\infty} e^{-v/2} \, dv \ll T^{-A},$$

and finally we have to replace σ by $\sigma - c$ in (7.5).

Now suppose that $\sigma = \frac{1}{2}$ and note that by (1.23) and (1.25)

$$\left| \zeta \left(\tfrac{1}{2} - c + it \right) \right| \ll \left| \zeta \left(\tfrac{1}{2} + c + it \right) \right| T^c \ll \left| \zeta \left(\tfrac{1}{2} + c + it \right) \right|,$$

so that (7.5) remains true if $\sigma = \frac{1}{2}$, and $s' = \frac{1}{2} - c + it$. On the other hand by the residue theorem we have, for $s = \frac{1}{2} + it$,

$$(7.6) \qquad \zeta^k(s) = (2\pi i)^{-1} \int_{\mathcal{D}} \zeta^k(s + z) \Gamma(z) \, dz,$$

where $\mathcal{D}$ is the rectangle with vertices $\pm c \pm i \log^2 T$. Using Stirling's formula (A.34) again, it is seen that the integrals over horizontal sides of $\mathcal{D}$ are $o(1)$ as $T \to \infty$, and using (7.5) with $\sigma = \frac{1}{2}$, $s' = \frac{1}{2} \pm c \pm i(t + u)$, $|u| \leqslant \log^2 T$, we obtain

$$(7.7) \qquad \zeta^k(s) \ll 1 + \int_{-\log^2 T}^{\log^2 T} e^{-|u|} \left(1 + \int_{-\infty}^{\infty} \left| \zeta \left(\tfrac{1}{2} + it + iu + iv \right) \right|^k \right.$$

$$\left. \times (c + |v|)^{-1} e^{-|v|} \, dv \right) (c + |u|)^{-1} \, du.$$

To estimate the above expression first note that trivially

$$\int_{-\log^2 T}^{\log^2 T} e^{-|u|} (c + |u|)^{-1} \, du \ll c^{-1} = \log T,$$

and in the remaining integral we make the substitution $v = x - u$ and invert the order of integration. This gives

$$(7.8) \quad \zeta^k \left(\tfrac{1}{2} + it \right) \ll \log T + \int_{-\infty}^{\infty} \left| \zeta \left(\tfrac{1}{2} + it + ix \right) \right|^k$$

$$\times \left(\int_{-\infty}^{\infty} e^{-|u| - |x - u|} (c + |u|)^{-1} (c + |x - u|)^{-1} \, du \right) dx,$$

and the proof of (7.2) will be finished if we can show

(7.9) $$\int_{-\infty}^{\infty} e^{-|u|-|x-u|}(c + |u|)^{-1}(c + |x - u|)^{-1} du \ll c^{-1}e^{-|x|}.$$

This is obvious when $x = 0$, and since the cases $x > 0$ and $x < 0$ are treated analogously, we shall consider $x > 0$ only. Write

$$\int_{-\infty}^{\infty} e^{-|u|-|x-u|}(c + |u|)^{-1}(c + |x - u|)^{-1} du$$

$$= \int_{-\infty}^{0} + \int_{0}^{x} + \int_{x}^{\infty} = I_1 + I_2 + I_3,$$

say. Then

$$I_1 = \int_{0}^{\infty} e^{-x}(c + v)^{-1}(c + x + v)^{-1} dv$$

$$\ll e^{-x}\left(\int_{0}^{c} c^{-2} dv + \int_{c}^{\infty} v^{-2} dv\right) \ll e^{-x}c^{-1},$$

$$I_2 = \int_{0}^{x} e^{-x}(c + u)^{-1}(c + x - u)^{-1} du \ll e^{-x}\int_{0}^{x/2}(c + u)^{-2} du \ll e^{-x}c^{-1},$$

since $c + u = c + x - u$ for $u = x/2$, and finally

$$I_3 = \int_{x}^{\infty} e^{-2u+x}(c + u)^{-1}(c + u - x)^{-1} du$$

$$\leqslant \tfrac{1}{2}e^{-x}\int_{x}^{\infty}\left((c + u)^{-2} + (c + u - x)^{-2}\right) du \ll e^{-x}c^{-1}.$$

7.3   THE MEAN SQUARE WHEN σ IS IN THE CRITICAL STRIP

We consider now the integral

$$\int_{T-G}^{T+G} |\zeta(\sigma + it)|^2 dt \qquad (\tfrac{1}{2} < \sigma < 1, G = o(T), T \to \infty),$$

where σ is fixed, leaving aside the most important case $\sigma = \tfrac{1}{2}$ for the next section. We need the following lemma, whose proof is typical of several proofs in the sequel and uses the exponential integral (A.38) to "shorten" exponential sums under consideration.

LEMMA 7.2. For $N < N'$ $2N \ll T^A$ ($A > 0$ fixed), $\log T < G \leqslant T$, we have uniformly in G

(7.10) $\displaystyle\int_{T-G}^{T+G} \left| \sum_{N < n \leqslant N'} n^{-it} \right|^2 dt \ll NG \log T$

$$+ G \sum_{r \leqslant NG^{-1}\log T} \max_{N \leqslant N'' \leqslant N' - r} \left| \sum_{N < n \leqslant N''} \exp(iT \log(1 + r/n)) \right|.$$

Proof of Lemma 7.2.

(7.11)

$$\int_{T-G}^{T+G} \left| \sum_{N < n \leqslant N'} n^{-it} \right|^2 dt \leqslant e \int_{-G \log T}^{G \log T} \left| \sum_{N < n \leqslant N'} n^{-it-iT} \right|^2 \exp(-t^2 G^{-2}) \, dt$$

$$\ll NG \log T + \left| \sum_{N < m \neq n \leqslant N'} (m/n)^{-iT} \int_{-\infty}^{\infty} \exp\left(-it(\log m/n) - t^2 G^{-2}\right) dt \right|$$

$$+ o(1),$$

since

$$\int_{\pm G \log T}^{\pm \infty} \exp(-t^2 G^{-2}) \, dt \ll \exp\left(-\tfrac{1}{2}\log^2 T\right) \ll T^{-C}$$

for any fixed $C > 0$, and N is bounded by a fixed power of T. Because of symmetry we may suppose that $m > n$ in the last sum in (7.11) and use the exponential integral (A.38) to obtain

(7.12) $\displaystyle\sum_{N < n < m \leqslant N'} (m/n)^{-iT} \int_{-\infty}^{\infty} \exp\left(-it(\log m/n) - t^2 G^{-2}\right) dt$

$$= \pi^{1/2} G \sum_{N < n < m \leqslant N'} \exp(-iT(\log m/n)) \exp\left(-\tfrac{1}{4} G^2 \log^2 m/n\right)$$

$$= \pi^{1/2} G \sum_{r \leqslant NG^{-1}\log T} \sum_{N < n \leqslant N' - r} \exp(-iT \log(1 + r/n))$$

$$\times \exp\left(-\tfrac{1}{4} G^2 \log^2(1 + r/n)\right) + o(1).$$

Namely, writing $m = n + r$ we see that for $r > NG^{-1}\log T$

$$\exp\left(-\tfrac{1}{4}G^2\log^2 m/n\right) \leqslant \exp\left(-G^2 r^2/16N^2\right) \leqslant \exp\left(-\tfrac{1}{16}\log^2 T\right) \leqslant T^{-C}$$

for T sufficiently large and any fixed $C > 0$, because $\log(1 + x) \geqslant x/2$ for $0 \leqslant x \leqslant 1$. The lemma now follows from (7.11), (7.12), and partial summation. If the sums on the right-hand side of (7.10) are empty they shall be of course counted as zero.

We proceed now with the main result of this section, whose proof will follow easily from Lemma 7.2 and the approximate functional equation (4.1) for $\zeta(s)$.

THEOREM 7.1. Let $(\varkappa, \lambda)$ be an exponent pair and $\tfrac{1}{2} < \sigma < 1$ fixed. Then for $T^{(\varkappa+\lambda+1-2\sigma)/2(\varkappa+1)}(\log T)^{(2+\varkappa)/(\varkappa+1)} \leqslant G \leqslant T,\ 1 + \lambda - \varkappa \geqslant 2\sigma$, we have uniformly in G

(7.13)
$$\int_{T-G}^{T+G}|\zeta(\sigma + it)|^2\,dt \ll G.$$

Proof of Theorem 7.1. From the approximate functional equation (4.1) we have with $x = y = (t/2\pi)^{1/2}$

(7.14) $\quad \zeta(\sigma + it) \ll 1 + \left|\displaystyle\sum_{n \leqslant (T/2\pi)^{1/2}} n^{-\sigma-it}\right| + T^{1/2-\sigma}\left|\displaystyle\sum_{n \leqslant (T/2\pi)^{1/2}} n^{\sigma-1-it}\right|,$

where the error made by replacing $(t/2\pi)^{1/2}$ by $(T/2\pi)^{1/2}$ in the range of summation is clearly $O(1)$ if $T - G \leqslant t \leqslant T + G$ and $G \leqslant T^{(1+\sigma)/2}$. For the less interesting range $T^{(1+\sigma)/2} \leqslant G \leqslant T$ the theorem follows from the approximate functional equation (4.66) (where the lengths of the sums involved do not depend on t) and the mean value theorem 5.2 for Dirichlet polynomials. The intervals of summation in (7.14) are split into $O(\log T)$ subintervals of the form $(N, 2N]$, $N = [(T/2\pi)^{1/2}]2^{-j}$, $j = 1, 2, \ldots$, and thus by partial summation [see (A.19)]

(7.15)

$$\int_{T-G}^{T+G}|\zeta(\sigma + it)|^2\,dt \ll G + \sum_N \max_{N < N' \leqslant 2N} N^{-2\sigma}\int_{T-G}^{T+G}\left|\sum_{N < n \leqslant N'} n^{-it}\right|^2\,dt,$$

since $\sigma > \tfrac{1}{2}$, $N \ll T^{1/2}$. For $N \leqslant T^\varepsilon$ we again use the mean value theorem for Dirichlet polynomials, and for $N \geqslant T^\varepsilon$ we use Lemma 7.2, which leads to the estimation of the exponential sum

(7.16)

$$S = \sum_{N<n\leqslant N'} \exp(iT\log(1+r/n)) = \sum_{N<n\leqslant N'} \exp(if(n)) \qquad (N \leqslant N' \leqslant 2N),$$

where for r, T fixed we have set

$$f(x) = T\log(1+r/x) \qquad (N \leqslant x \leqslant 2N).$$

The condition $N \ll T^{1/2}$ ensures that $f'(x) \gg 1$ for $N \leqslant x \leqslant 2N$, and in the same range we have also

$$f^{(k)}(x) \asymp TrN^{-k-1} \qquad (k = 1, 2, \ldots)$$

so that we may use the theory of exponent pairs, as presented in Section 2.3 of Chapter 2, to estimate S. We obtain

(7.17)
$$S \ll \max_{N\leqslant x\leqslant 2N} |f'(x)|^{\varkappa}N^{\lambda} \ll T^{\varkappa}r^{\varkappa}N^{\lambda-2\varkappa}$$

for any exponent pair $(\varkappa, \lambda)$. Therefore combining (7.14)–(7.17) and Lemma 7.2, we obtain for $T^{\varepsilon} < N \ll T^{1/2}$, $1 + \lambda - \varkappa \geqslant 2\sigma$,

(7.18)

$$\int_{T-G}^{T+G} |\zeta(\sigma + it)|^2 \, dt \ll G + GT^{\varepsilon(1-2\sigma)}\log T + G\sum_N \sum_{r\leqslant NG^{-1}\log T} T^{\varkappa}r^{\varkappa}N^{\lambda-2\varkappa-2\sigma}$$

$$\ll G + G\sum_N T^{\varkappa}(NG^{-1}\log T)^{1+\varkappa}N^{\lambda-2\varkappa-2\sigma}$$

$$\ll G + \log T \max_N G^{-\varkappa}T^{\varkappa}N^{1+\lambda-\varkappa-2\sigma}\log^{1+\varkappa}T \ll G,$$

for $G \geqslant T^{(\varkappa+\lambda+1-2\sigma)/2(\varkappa+1)}(\log T)^{(2+\varkappa)/(1+\varkappa)}$, proving Theorem 7.1.

Following the foregoing argument in the case $\sigma = \frac{1}{2}$, we arrive at

(7.19)
$$\int_{T-G}^{T+G} \left|\zeta(\tfrac{1}{2} + it)\right|^2 dt \ll G\log T$$

$$\left[T^{(\varkappa+\lambda)/2(\varkappa+1)}(\log T)^{(\varkappa+2)/(\varkappa+1)} \leqslant G \leqslant T\right],$$

but in the next section we shall obtain an estimate which improves (7.19).

7.4 THE MEAN SQUARE WHEN $\sigma = \frac{1}{2}$

The theorem which will be stated and proved in this section is one of the fundamental results of this text, since it serves as a basis for the derivation of higher power moments of the zeta-function and provides a technically simple way of estimating $|\zeta(\frac{1}{2} + iT)|$. The result is due to D. R. Heath-Brown, but our proof is different and does not depend on Atkinson's formula for the mean square (see Chapter 15), while Heath-Brown's proof (1978a) does.

THEOREM 7.2. For $T^{\varepsilon} \leqslant G \leqslant T^{1/2-\varepsilon}$ uniformly in G

$$(7.20) \quad \int_{T-G}^{T+G} \left|\zeta(\tfrac{1}{2} + it)\right|^2 dt \ll G \log T$$

$$+ G\sum_{K} (TK)^{-1/4}\left(|S(K)| + K^{-1}\int_0^K |S(x)|\, dx\right) e^{-G^2 K/T},$$

where

$$(7.21) \quad S(x) = S(x, K, T) = \sum_{K \leqslant n \leqslant K+x} (-1)^n d(n)\exp(if(T, n)),$$

$$(7.22) \quad f(T, n) = 2T \operatorname{ar} \sinh\left((\pi n/2T)^{1/2}\right) + (\pi^2 n^2 + 2\pi nT)^{1/2},$$

and summation is over $K = 2^k$ such that $T^{1/3} \leqslant K \leqslant N$, where for $\delta > 0$ fixed

$$(7.23) \quad N = B^2/(T/2\pi - B), \qquad B = T(2\pi G)^{-1}\log^{(1+\delta)/2}T.$$

Proof of Theorem 7.2. From (7.23) we have $K \ll TG^{-2}\log^{1+\delta}T$, and the interesting range for K is $K \geqslant T^{1/3}$, since the trivial bound $S(x) \ll K \log T$ gives

$$\sum_{j}\sum_{2^{-j}N = K \leqslant T^{1/3}} (TK)^{-1/4}\left(|S(K)| + K^{-1}\int_0^K |S(x)|\, dx\right) e^{-G^2 K/T}$$

$$\ll \sum_{j}\sum_{2^{-j}N = K \leqslant T^{1/3}} (\log T)T^{-1/4}K^{3/4}e^{-G^2 K/T} \ll \log T.$$

Therefore it is seen that the relevant range for G is $G \leqslant T^{1/3}\log T$, since we may take $T^{1/3} \leqslant K \leqslant TG^{-2}\log^{1+\delta}T$. Another remark is that in view of the exponential factor $\exp(-G^2 K/T)$ in (7.20), the proof that will be given actually shows that, uniformly for $T/2 \leqslant \tau \leqslant T$ and $T^{\varepsilon} \leqslant G \leqslant T^{1/2-\varepsilon}$, we obtain

(7.24) $\displaystyle\int_{\tau-G}^{\tau+G}|\zeta(\tfrac{1}{2}+it)|^2\,dt \ll G\log T$

$$+ G\sum_{K}(TK)^{-1/4}\left(|S(K)| + K^{-1}\int_0^K|S(x)|\,dx\right)e^{-G^2K/(2T)},$$

where $S(x) = S(x, K, \tau)$ is given by (7.21), $T^{1/3} \leqslant K = 2^k \leqslant N$, and N is given by (7.23). This form of the mean value estimate will be particularly useful for higher power moments in Chapter 8.

Since $\operatorname{ar\,sinh} x = (1 + o(1))x$ as $x \to 0$, it is seen that (7.20) will follow readily by partial summation from

(7.25) $\displaystyle\int_{T-G}^{T+G}|\zeta(\tfrac{1}{2}+it)|^2\,dt$

$$\ll G\log T + G\left|\sum_{n\leqslant N}(-1)^n d(n)n^{-1/2}\left(\frac{1}{4} + \frac{T}{2\pi n}\right)^{-1/4}\right.$$

$$\left. \times \exp\left\{-\left(G\operatorname{ar\,sinh}\left(\left(\frac{\pi n}{2T}\right)^{1/2}\right)\right)^2\right\}\exp(if(T, n))\right|,$$

and so we set out to prove (7.25). To facilitate the notation we introduce the abbreviations

(7.26) $\quad T' = T/2\pi, \qquad L = \log T, \qquad I = \displaystyle\int_{T-G}^{T+G}|\zeta(\tfrac{1}{2}+it)|^2\,dt.$

The first step is similar to the one made in the proof of Theorem 7.1, and consists of majorizing I by a "short" exponential sum (of length $\ll TL/G$). We start from the approximate functional equation (4.11), which gives

(7.27) $\qquad I \ll \displaystyle\int_{-GL}^{GL}|\zeta(\tfrac{1}{2} + iT + it)|^2\exp(-t^2 G^{-2})\,dt$

$$\ll GL + \int_{-GL}^{GL}(S_1 + \bar{S}_1)\exp(-t^2 G^{-2})\,dt,$$

Here

(7.28) $\qquad S_1 = \chi(\tfrac{1}{2} + it + iT)\displaystyle\sum_{n\leqslant(T+t)/2\pi}d(n)n^{-1/2+it+iT}$

$$= \sum_{n\leqslant T'}d(n)n^{-1/2}e^{iq(T+t)} + O(L),$$

and we have set

(7.29) $$q(x) = x\log(2\pi n/x) + x + \pi/4$$

and used the asymptotic formula (1.25). For $-GL \leqslant t \leqslant GL$ we have by Taylor's formula

(7.30) $$q(T + t) = q(T) + t\log(2\pi n/T) - t^2/(2T) + O(G^3L^3T^{-2}),$$

so that from (7.27)–(7.30) we obtain

(7.31) $$I \ll GL + I_1 + \bar{I}_1,$$

where

(7.32)

$$I_1 = \int_{-GL}^{GL} \sum_{n \leqslant T'} d(n)n^{-1/2}e^{iq(T)}\exp\left(it(\log n/T') - it^2/2T - t^2G^{-2}\right)dt$$

$$+ O(G^4L^5T^{-3/2})$$

$$= \sum_{n \leqslant T'} d(n)n^{-1/2}e^{iq(T)}\int_{-\infty}^{\infty} \exp\left(it(\log n/T') - it^2/2T - t^2G^{-2}\right)dt$$

$$+ O(GL),$$

since $G \leqslant T^{1/2-\varepsilon}$ and for any fixed $c > 0$

$$\int_{\pm GL}^{\pm\infty} \exp\left(it(\log n/T') - it^2/2T - t^2G^{-2}\right)dt \ll T^{-c}.$$

The second integral in (7.32) is evaluated using (A.38), and setting

$$Y = -(2iT)^{-1} + G^{-2}$$

we obtain from (7.31) and (7.32)

(7.33) $$I \ll GL + \left| \sum_{n \leqslant T'} d(n)n^{-1/2}e^{iq(T)}Y^{-1/2}\exp\left(-(\log n/T')^2/4Y\right) \right|$$

$$\ll GL + G\left| \sum_{n \leqslant T'} d(n)n^{-1/2}e^{iq(T)}\exp\left(-\frac{G^2}{4}(\log n/T')^2\right) \right|.$$

The presence of the negative exponential factor in (7.33) will make the contribution of many summands negligible. To see this, let $n = [T'] - m$ and suppose $m \geqslant T'G^{-1}L^{(1+\delta)/2}$, where $\delta > 0$ is arbitrarily small but fixed. For these m we have

$$G^2(\log n/T')^2 \gg G^2 m^2/(T')^2 \gg L^{1+\delta},$$

and since $\exp(-c_1 L^{1+\delta}) \ll T^{-c_2}$ for any fixed $c_1, c_2 > 0$, we have

(7.34) $$I \ll GL + G|S|,$$

where S is the "short" exponential sum

(7.35) $$S = \sum_{T'-T'G^{-1}L^{(1+\delta)/2} \leqslant n \leqslant T'} d(n)n^{-1/2-iT}\exp\left(-\frac{G^2}{4}(\log n/T')^2\right).$$

Thus we have reduced the problem to the estimation of the short exponential sum S in (7.35), which will be transformed now by the Voronoi summation formula (3.2) to yield eventually (7.25). Since the summation formula (3.2) involves an infinite series whose tails are not easy to estimate, we shall use Lemma 2.3, and instead of S in (7.35) it will be more suitable to consider the averaged sum

(7.36)
$$S' = U^{-1}\int_0^U S(u)\, du, \quad S(u)$$

$$= \sum_{T'-T'G^{-1}L^{(1+\delta)/2}+u \leqslant n \leqslant T'-u} d(n)n^{-1/2-iT}\exp\left(2\pi in - \frac{G^2}{4}(\log n/T')^2\right).$$

The factor $\exp(2\pi in) = 1$ that is introduced in $S(u)$ does not affect the value of the sum, but it is inserted to regulate the distribution of saddle points of exponential integrals which arise after Voronoi's formula is applied. We shall choose

(7.37) $$U = G^{1/2},$$

and then trivially $S - S' \ll UT^{-1/2+\varepsilon}$. Now we apply (3.2) to $S(u)$ in (7.36), setting for convenience of notation

$$M_1 = T' - T'G^{-1}L^{(1+\delta)/2}, \qquad M_2 = T' = T/(2\pi).$$

Then (3.2) gives

(7.38)

$$S(u) = \int_{M_1+u}^{M_2-u}(\log x + 2\gamma)x^{-1/2-iT}\exp\left(2\pi ix - \frac{G^2}{4}(\log x/T')^2\right) dx + O(1)$$

$$+ \sum_{n=1}^{\infty} d(n)\int_{M_1+u}^{M_2-u}x^{-1/2-iT}\exp\left(2\pi ix - \frac{G^2}{4}(\log x/T')^2\right)\alpha(nx)\, dx,$$

where $\alpha(x)$ is defined by (3.3). We recall the asymptotic formula (3.15), which we write here as

(7.39)

$$\alpha(nx) = -2^{1/2}x^{-1/4}n^{-1/4}$$

$$\times \left\{ \sin(4\pi\sqrt{nx} - \pi/4) - (32\pi)^{-1}(nx)^{-1/2}\cos(4\pi\sqrt{nx} - \pi/4) \right\}$$

$$+ O(n^{-5/4}x^{-5/4}),$$

noting first that the contribution of the O-term above to the sum in (7.38) is certainly $O(1)$. The first integral in (7.38) is estimated by (2.5) as

$$\ll M_1^{-1/2}L \max_{M_1 \leqslant x \leqslant M_2} (Tx^{-2})^{-1/2} \ll L,$$

and therefore its contribution to S' in (7.36) is again $\ll L$ and so it gives $\ll GL$ in (7.34). To treat the terms containing sines and cosines in (7.39) let N be defined by (7.23), and write the series in (7.38) as

(7.40)

$$\sum_{n=1}^{\infty} d(n) \int_{M_1+u}^{M_2-u} \cdots \alpha(nx) \, dx = R_1 + R_2 + O(1)$$

say, where

(7.41)

$$R_1 = -2^{1/2} \sum_{n \leqslant (1+\varepsilon)N} d(n)n^{-1/4} \int_{M_1+u}^{M_2-u} x^{-3/4-iT}$$

$$\times \left\{ \sin(4\pi\sqrt{nx} - \pi/4) - (32\pi)^{-1}(nx)^{-1/2}\cos(4\pi\sqrt{nx} - \pi/4) \right\}$$

$$\times \exp\left(2\pi ix - \frac{G^2}{4}(\log x/T')^2 \right) dx,$$

and

(7.42)

$$R_2 = -2^{1/2} \sum_{n > (1+\varepsilon)N} d(n)n^{-1/4} \int_{M_1+u}^{M_2-u} \cdots .$$

[In (7.42), $\cdots$ stands for the same terms as in (7.41)]. The sum R_2 will be estimated as $\ll L$, and this can be seen at once for terms coming from $\cos(4\pi\sqrt{nx} - \pi/4)$. Namely, using (2.3) with $f(z) = z - T'\log z \pm 2\sqrt{nz}$ we have

$$\int_{M_1+u}^{M_2-u}\exp\left(-\frac{G^2}{4}(\log x/T')^2\right)x^{-5/4}e\left(x-T'\log x\pm 2\sqrt{nx}\right)dx$$

$$\ll M_1^{-5/4}\max_{M_1\leqslant x\leqslant M_2}|f'(x)|^{-1}\ll M_1^{-5/4}(M_1/n)^{1/2},$$

since for $M_1\leqslant x\leqslant M_2$ and $n>(1+\varepsilon)N$ we have $|f'(x)|\gg(n/x)^{1/2}$. Therefore the cosine terms in (7.42) contribute a total of

$$\sum_{n>(1+\varepsilon)N}d(n)n^{-5/4}M_1^{-3/4}=o(1).$$

To estimate the contribution of the sine terms in (7.42) we shall make use of S', as defined by (7.36). By the properties of the function $\alpha(nx)$ we may integrate termwise, and we are left with the estimation of

$$(7.43)\qquad \sum_{n>(1+\varepsilon)N}d(n)n^{-1/4}\left|U^{-1}\int_0^U\int_{M_1+u}^{M_2-u}x^{-3/4-iT}\right.$$

$$\left.\times\exp\left(2\pi ix-\frac{G^2}{4}(\log x/T')^2\pm 4\pi i\sqrt{nx}\right)dx\,du\right|,$$

which will be carried out with the use of Lemma 2.3, where we take

$$f(z)=z\pm 2\sqrt{nz}-T'\log z,\qquad g(z)=z^{-3/4}\exp\left(-\frac{G^2}{4}(\log z/T')^2\right),$$

$a=M_1, b=M_2, G=M_1^{-3/4}, \mu\asymp M_1, M=(n/M_1)^{1/2}$. This is the only point in the proof where the parameter $U=G^{1/2}$ is needed, and we can easily verify that the hypotheses of Lemma 2.3 are satisfied. Taking into account that $TG^{-2}L^{1+\delta}\asymp N$, we therefore obtain that the expression given by (7.43) is

$$(7.44)\qquad \ll M_1^{-3/4}\sum_{n>N}d(n)\left\{n^{-1/4}M_1\exp\left(-A(nT)^{1/2}\right)+n^{-5/4}M_1U^{-1}\right\}$$

$$\ll M_1^{1/4}U^{-1}N^{-1/4}L\ll G^{1/2}U^{-1}L=L.$$

We have therefore terminated the estimation of R_2 in (7.42), and now we turn to R_1 in (7.41). First, we observe that the contribution of terms with $\cos(4\pi\sqrt{nx}-\pi/4)$ is trivially

$$\ll\sum_{n\leqslant(1+\varepsilon)N}d(n)n^{-3/4}\int_{M_1}^{M_2}x^{-5/4}dx\ll N^{1/4}LT^{-1/4}\ll L.$$

Since U will not be needed anymore in the proof we replace the limits of integration in (7.41) by M_1 and M_2 respectively, creating an error which is

$$\ll \sum_{n \leqslant (1+\varepsilon)N} d(n)n^{-1/4}\left(\left|\int_{M_1}^{M_1+u}\right| + \left|\int_{M_2-u}^{M_2}\right|\right) \ll \sum_{n \leqslant (1+\varepsilon)N} d(n)n^{-1/4}UT^{-3/4}$$

$$\ll N^{3/4}LUT^{-3/4} \ll UG^{-3/2}L \ll 1.$$

Thus we have yet to consider

$$(7.45) \qquad \qquad \pm i2^{-1/2} \sum_{n \leqslant (1+\varepsilon)N} d(n)n^{-1/4}I_n^{\pm},$$

where

$$(7.46) \qquad I_n^{\pm} = \int_{M_1}^{M_2} x^{-3/4}\exp\left(-\frac{G^2}{4}(\log x/T')^2\right)$$

$$\times \exp\left(2\pi i x - iT \log x \pm i\left(4\pi\sqrt{nx} - \pi/4\right)\right) dx,$$

which means that in I_n^+ the $+$ sign is to be taken in $\exp(2\pi i x - \cdots)$, while in I_n^- the minus sign is to be taken.

The sum in (7.45) will give the sum in (7.25). To see this we shall estimate $I_n^{\pm}$ by applying Theorem 2.2 with $a = M_1, b = M_2, k = 1$,

$$f(z) = f_n^{\pm}(z) = -T'\log z \pm 2(nz)^{1/2},$$

$$\varphi(z) = z^{-3/4}\exp\left(-\frac{G^2}{4}(\log z/T')^2\right),$$

$$\Phi(x) = x^{-3/4}, \qquad F(x) = T, \qquad \mu(x) = T.$$

The conditions of Theorem 2.2 are readily verified, for example, $N \ll TG^{-2}\log^{1+\delta}T$ implies

$$\left|f_n^{\pm}(z)''\right|^{-1} \ll \mu^2(x)F^{-1}(x).$$

The saddle points are the roots of the equation $(f_n^{\pm}(x))' = -1$, which is

$$(7.47) \qquad \qquad 1 - T/2\pi x \pm (n/x)^{1/2} = 0,$$

and these roots must lie in $[M_1, M_2]$ for the main terms in Theorem 2.2 to exist. Of the two roots of Eq. (7.47) only

$$(7.48) \qquad \qquad x_0 = T/(2\pi) + n/2 - (n^2/4 + nT/2\pi)^{1/2}$$

need be considered, since the other root always exceeds M_2, and x_0 corresponds to the integral I_n^+. Trivially, $x_0 \leq M_2 = T' = T/(2\pi)$, and $x_0 \geq M_1$ holds for $n \leq n_0$, where

$$T' + \tfrac{1}{2}n_0 - \left(\tfrac{1}{4}n_0^2 + n_0 T'\right)^{1/2} = T' - B, \qquad B = T'G^{-1}L^{(1+\delta)/2};$$

hence

$$-n_0 T' / \left(\tfrac{1}{2}n_0 + \left(\tfrac{1}{4}n_0^2 + n_0 T'\right)^{1/2}\right) = -B.$$

Solving for n_0 we obtain

$$n_0 = N = B^2/(T' - B),$$

where N is given by (7.23).

The error terms arising from I_n^+ and I_n^- after Theorem 2.2 is applied are treated analogously, and thus only the error terms coming from I_n^+ will be considered (no main terms come from I_n^-). Alternatively, one can estimate the sum with I_n^- by applying (2.3). To calculate the main terms coming from the saddle points of I_n^+ note that

$$f_n'(z) = f_n'^+(z) = -T'/z + (n/z)^{1/2},$$

$$f_n''(z) = T'z^{-2} - \tfrac{1}{2}n^{1/2}z^{-3/2} = z^{-3/2}\left(T'z^{-1/2} - \tfrac{1}{2}n^{1/2}\right),$$

and in view of $(n/x_0)^{1/2} = T'/x_0 - 1$, it follows that

$$f_n''(x_0) = x_0^{-3/2}\left(T'x_0^{-1/2} - \tfrac{1}{2}n^{1/2}\right),$$

$$T'x_0^{-1/2} - \tfrac{1}{2}n^{1/2} = T'n^{-1/2}(T'/x_0 - 1) - \tfrac{1}{2}n^{1/2} = n^{1/2}\left(\tfrac{1}{4} + T'/n\right)^{1/2},$$

since by rationalizing the right-hand side of (7.48) we obtain

$$T'^2/(nx_0) = T'/n + \tfrac{1}{2} + \left(\tfrac{1}{4} + T'/n\right)^{1/2}.$$

Hence

$$\left(f_n''(x_0)\right)^{-1/2} = x_0^{3/4}n^{-1/4}\left(\tfrac{1}{4} + T'/n\right)^{-1/4},$$

and likewise

$$T'x_0^{-1} = 1 + n/2T' + \left(n^2/(2T')^2 + n/T'\right)^{1/2}$$

$$= \left((n/4T')^{1/2} + \left(1 + n/4T'\right)^{1/2}\right)^2,$$

which gives

$$\log(T'/x_0) = 2\log\left((n/4T')^{1/2} + (1 + n/4T')^{1/2}\right) = 2 \text{ ar } \sinh\left((n/4T')^{1/2}\right).$$

Therefore

$$\varphi(x_0)f_n''(x_0)^{-1/2}e\left(f(x_0) + kx_0 + \tfrac{1}{8}\right)$$

$$= n^{-1/4}\left(\tfrac{1}{4} + T'/n\right)^{-1/4}\exp\left(-\left\{G \text{ ar } \sinh\left((n/4T')^{1/2}\right)\right\}^2\right)$$

$$\times e\left(-T'\log x_0 + 2(nx_0)^{1/2} + x_0 + \tfrac{1}{8}\right),$$

and since $e(-n/2) = e^{n\pi i} = (-1)^n$ it is seen using (7.48) and $(nx_0)^{1/2} = T' - x_0$ that

$$\left|\sum_{n \leqslant N} d(n)n^{-1/4}\varphi(x_0)f_n''(x_0)^{-1/2}e\left(f(x_0) + x_0 + \tfrac{1}{8}\right)\right|$$

$$\leqslant \left|\sum_{n \leqslant N} (-1)^n d(n)n^{-1/2}\left(\tfrac{1}{4} + T'/n\right)^{-1/4}\exp(if(T, n))\right.$$

$$\left. \times \exp\left(-\left\{G \text{ ar } \sinh\left((n/4T')^{1/2}\right)\right\}^2\right)\right|.$$

Here $f(T, n)$ is given by (7.22), and the last sum is exactly the one that appears in (7.25).

Thus it remains to show that the contribution of the error terms of I_n^+ to the sum in (7.45) is $\ll L$. This is analogous to the corresponding proof of the approximate functional equation (4.10) in Chapter 4 by the use of Voronoï's formula, and the only terms which are nontrivial to estimate are

$$\sum_{n \leqslant (1+\varepsilon)N} d(n)n^{-1/4}T^{-3/4}\left\{\left(|f_n'(M_1)| + 1| + T^{-1/2}\right)^{-1}\right.$$

$$\left. + \left(|f_n'(M_2)| + 1| + T^{-1/2}\right)^{-1}\right\},$$

since $f_n''(x) \asymp T^{-1}$ for $M_1 \leqslant x \leqslant M_2$. Next, $f_n'(M_2) + 1 = (n/T')^{1/2}$, and this gives

$$\sum_{n \leqslant (1+\varepsilon)N} d(n)n^{-1/4}T^{-3/4}\left(|f_n'(M_2)| + 1| + T^{-1/2}\right)^{-1}$$

$$\ll T^{-3/4}\sum_{n \leqslant 2N} d(n)n^{-3/4}T^{1/2} \ll L.$$

The equation $1 + f_n'(M_1) = 0$ has only one solution in n, namely $n = N$, so it is convenient to write $n = [N] + k$. Then

$$1 + f_n'(M_1) = f_{[N]+k}'(M_1) - f_N'(M_1) \asymp |k| N^{-1/2} T^{-1/2},$$

since $(N + k)^{1/2} - N^{1/2} \asymp |k| N^{-1/2}$ for $|k| \leqslant N/2$. Also, for $\varepsilon > 0$ fixed and sufficiently small, for $|k| > (\varepsilon/2)N$ in $f_n'(M_1) + 1$ either the term

$$1 - T'/M_1 \asymp G^{-1} L^{(1+\delta)/2}$$

or the term $(n/M_1)^{-1/2}$ dominates, and in either case we have

$$\left(|f_n'(M_1) + 1| + T^{-1/2} \right)^{-1}$$

$$\ll \begin{cases} T^{1/2}, & \text{if} \quad |k| < T^{1/2} G^{-1} L^{(1+\delta)/2}, \\ |k|^{-1} T G^{-1} L^{(1+\delta)/2}, & \text{if} \quad T^{1/2} G^{-1} L^{(1+\delta)/2} \leqslant |k| \leqslant (\varepsilon/2)N, \\ \max(G, T^{1/2} n^{-1/2}), & \text{if} \quad |k| > (\varepsilon/2)N. \end{cases}$$

Therefore, finally, we obtain that the contribution of the error terms coming from I_n^+ is

$$\ll T^{-3/4} \sum_{n \leqslant (1+\varepsilon)N} d(n) n^{-1/4} \left(|f_n'(M_1) + 1| + T^{-1/2} \right)^{-1} + L$$

$$\ll T^{-1/4} \sum_{|k| \leqslant T^{1/2} G^{-1} L^{(1+\delta)/2}} d([N] + k)([N] + k)^{-1/4} + T^{1/4+\varepsilon} G^{-1} N^{-1/4}$$

$$+ T^{\varepsilon - 1/4} \sum_{(\varepsilon/2)N \leqslant |k| \leqslant \varepsilon N} d([N] + k)([N] + k)^{-3/4}$$

$$+ G T^{-3/4} \sum_{(\varepsilon/2)N \leqslant |k| \leqslant \varepsilon N} d([N] + k)([N] + k)^{-1/4} + L \ll L,$$

which completes the proof of Theorem 7.2. Actually, our arguments show that (7.25) holds without the absolute value sign if $\exp(if(T, n))$ is replaced by $\sin f(T, n)$; namely, we obtain for some absolute C_1, C_2

(7.49)

$$\int_{T-G}^{T+G} |\zeta(\tfrac{1}{2} + it)|^2 \, dt < C_1 G L + C_2 G \sum_{n \leqslant N} (-1)^n d(n) n^{-1/2} \left(\frac{1}{4} + \frac{T}{2\pi n} \right)^{-1/4}$$

$$\times \exp\left(-\left\{ G \operatorname{ar} \sinh\left((\pi n/2T)^{1/2} \right) \right\}^2 \right) \sin f(T, n).$$

In concluding this section let us mention that the proof of Theorem 7.2 depended on the use of the approximate functional equation (4.11) for $|\zeta(\frac{1}{2} + it)|^2$, whose proof is not easy. Instead of (4.11) one may use the reflection principle (Theorem 4.4), which is simpler than the approximate functional equation for $\zeta^2(s)$, and obtain a result of the same strength as Theorem 7.2 (the unimportant exponential factors which will appear in the course of the proof may be easily removed by partial summation). Namely, from (4.66) with $h = \log^2 T$, $k = 2$, $T - GL \leqslant t \leqslant T + GL$, $L = \log T$, $s = \frac{1}{2} + it$, $M = 9T^2/Y$, $w = u + iv$, $\alpha = 1/2 + \varepsilon$ we have

$$\zeta^2(\tfrac{1}{2} + it) = \sum_{n \leqslant 3Y} d(n)e^{-(n/Y)^h}n^{-1/2-it} + \chi^2(\tfrac{1}{2} + it) \sum_{n \leqslant 9T^2/Y} d(n)n^{-1/2+it}$$

$$+ O(1) - (2\pi i)^{-1}\int_{u=\varepsilon,|v|\leqslant h^2}\chi^2(\tfrac{1}{2} + it + w)$$

$$\times \sum_{n \leqslant 9T^2/Y} d(n)n^{-1/2+it+w}Y^w\Gamma\left(1 + \frac{w}{h}\right)\frac{dw}{w}.$$

Here we choose $Y = 3T$ to equalize the lengths of the sums and multiply by $\chi^{-1}(\frac{1}{2} + it)$ to obtain

$$\chi^{-1}(\tfrac{1}{2} + it)\zeta^2(\tfrac{1}{2} + it) = |\zeta(\tfrac{1}{2} + it)|^2.$$

From this point on the proof would be quite similar to the one already given for Theorem 7.2. The reflected sum (with $-\frac{1}{2} + it + w$) will give the same upper bound as the other two, namely (7.20).

7.5 THE ORDER OF $\zeta(s)$ IN THE CRITICAL STRIP

The problem of determining the precise order of $\zeta(\sigma + it)$ in the "critical strip" $0 \leqslant \sigma \leqslant 1$ is one of the deepest problems of analytic number theory, with many different applications. The present state of knowledge is far from satisfactory, unless one is willing to accept the truth of unproved conjectures like Lindelöf's or Riemann's. The order of $\zeta(s)$ was already considered in Chapter 1, where Theorem 1.9 was proved, and later in Chapter 6, where we proved Theorem 6.3, which gives sharp results when σ is close to 1. Here we shall be concerned with the order of $\zeta(s)$ when $\frac{1}{2} \leqslant \sigma < 1$, since the range $0 < \sigma < \frac{1}{2}$ is covered by the functional equation for $\zeta(s)$ and the asymptotic formula (1.25). The problem may be formulated as finding upper bounds for the function $\mu(\sigma)$, defined in (1.65) by

$$\mu(\sigma) = \limsup_{t \to \infty} \frac{\log|\zeta(\sigma + it)|}{\log t}.$$

Lemma 7.1 and Theorem 7.1 provide at once the means for the estimation of $\mu(\sigma)$, and one can easily improve (1.66) for $\frac{1}{2} \leqslant \sigma < 1$. Namely, for $\frac{1}{2} < \sigma < 1$ fixed we have from (7.1) and (7.13)

$$|\zeta(\sigma + iT)|^2 \ll 1 + \log T \int_{T-\log^2 T}^{T+\log^2 T} |\zeta(\sigma - 1/(\log T) + iv)|^2 \, dv$$

$$\ll 1 + \log T \int_{T-G}^{T+G} |\zeta(\sigma - 1/(\log T) + iv)|^2 \, dv \ll G \log T$$

for

$$G \geqslant T^{(\varkappa + \lambda + 1 - 2\sigma)/(2\varkappa + 2)} (\log T)^{(2+\varkappa)/(1+\varkappa)},$$

if $(\varkappa, \lambda)$ is an exponent pair for which $1 + \lambda - \varkappa \geqslant 2\sigma$. This gives

(7.50)

$$\mu(\sigma) \leqslant (\varkappa + \lambda + 1 - 2\sigma)/(4\varkappa + 4) \qquad (\tfrac{1}{2} < \sigma < 1, 1 + \lambda - \varkappa \geqslant 2\sigma),$$

and likewise for $\sigma = \frac{1}{2}$ we obtain

(7.51)
$$\mu(\tfrac{1}{2}) \leqslant (\varkappa + \lambda)/(4\varkappa + 4).$$

The bound furnished by (7.50) provides fairly good estimates with an adequate choice of $(\varkappa, \lambda)$, while (7.51) yields $\mu(\tfrac{1}{2}) \leqslant \frac{27}{164} = 0.16463 \cdots$ with $(\varkappa, \lambda) = (\frac{11}{30}, \frac{16}{30})$. Another possibility is to use the theory of exponent pairs and the approximate functional equation (4.1) for $\zeta(s)$. Namely, choosing $x = y = (t/2\pi)^{1/2}$ in (4.1) we obtain by partial summation

(7.52)
$$\zeta(\sigma + it) \ll 1 + \sum_N N^{-\sigma} \max_{N < N' \leqslant 2N} \left| \sum_{N < n \leqslant N'} n^{it} \right|$$

$$+ \sum_N N^{\sigma-1} t^{1/2-\sigma} \max_{N < N' \leqslant 2N} \left| \sum_{N < n \leqslant N'} n^{it} \right|,$$

where N takes $O(\log t)$ values of the form $(t/2\pi)^{1/2} 2^{-j}$, $j = 1, 2, \ldots$. Since $N \ll t^{1/2}$ it is possible to use the theory of exponent pairs of Chapter 2 to estimate

(7.53)

$$S(t; N, N') = \sum_{N < n \leqslant N'} n^{it} = \sum_{N < n \leqslant N'} \exp(iF(n)), \qquad F(x) = t \log x.$$

Here $F'(x) = t/x \gg 1$ for $N \leqslant x \leqslant 2N$, and also

$$F^{(k)}(x) \asymp tN^{-k} \qquad (k = 1, 2, \ldots).$$

Therefore we obtain

(7.54)
$$S(t; N, N') \ll (tN^{-1})^{\varkappa} N^{\lambda},$$

and from (7.52) and (7.54) we infer that

(7.55) $$\zeta(\sigma + it) \ll 1 + \sum_N \left(t^{\varkappa} N^{\lambda - \varkappa - \sigma} + t^{1/2 - \sigma + \varkappa} N^{\lambda - \varkappa + \sigma - 1} \right).$$

If further we have

(7.56)
$$\sigma \geqslant \tfrac{1}{2}, \qquad \lambda - \varkappa \geqslant \sigma,$$

then (7.55) gives

$$\zeta(\sigma + it) \ll 1 + \left(t^{\varkappa + (\lambda - \varkappa - \sigma)/2} + t^{1/2 - \sigma + \varkappa + (\lambda - \varkappa + \sigma - 1)/2} \right) \log t,$$

or

(7.57) $$\zeta(\sigma + it) \ll t^{(\varkappa + \lambda - \sigma)/2} \log t \qquad (\sigma \geqslant 1/2, \lambda - \varkappa \geqslant \sigma).$$

In case $\sigma = \tfrac{1}{2}$ (when $\lambda - \varkappa \geqslant \tfrac{1}{2}$ has to be observed) we obtain a small improvement of $\mu(\tfrac{1}{2}) \leqslant 27/164$, namely,

(7.58)
$$\mu\left(\tfrac{1}{2}\right) \leqslant 0.164510678\ldots,$$

by taking the exponent pair $(\varkappa, \lambda) = (\alpha/2 + \varepsilon, \tfrac{1}{2} + \alpha/2 + \varepsilon)$, $\alpha = 0.3290213568\ldots$, and this seems to be the present limit obtainable by the method of exponent pairs in the one-dimensional case. The bound (7.57) contains the classical estimate of van der Corput

(7.59) $$\mu(\sigma) \leqslant 1/(2L - 2), \qquad \sigma = 1 - l/(2L - 2),$$

where $L = 2^{l-1}, l \geqslant 3$ an integer, which follows from (7.57) when one uses the exponent pair in (6.43). Similarly, one has

(7.60) $$\mu(\sigma) \leqslant 1/L(l + 1) \quad \text{for} \quad \sigma = 1 - 1/L \qquad (L = 2^{l-1}, l \geqslant 3),$$

which is an estimate of G. H. Hardy and J. E. Littlewood.

We proceed now to improve on (7.58), using Theorem 7.2 and Lemma 7.1. Apart from the unimportant factor $(-1)^n$, the exponential sum in (7.21) is a two-dimensional exponential sum, which follows on writing $d(n) = \sum_{kl=n} 1$. Thus two-dimensional techniques for the estimation of exponential sums can be put at once to our advantage, while in the general case it is not always easy to transform a one-dimensional exponential sum into a suitable two-dimensional sum, although a general method is provided by a variant of Lemma 2.5, since the sum on the left-hand side of (2.42) is a double sum. For the estimation of $\mu(\frac{1}{2})$ and $\int_{T-G}^{T+G}|\zeta(\frac{1}{2} + it)|^2 dt$, we could use the theory of two-dimensional exponent pairs of Section 2.4 of Chapter 2, but a sharper result is obtained by using a result of G. Kolesnik, which we state here as

LEMMA 7.3. Let $\mathcal{D}$ be the domain $X \leqslant x \leqslant X_1 \leqslant 2X$, $Y \leqslant y \leqslant Y_1 \leqslant 2Y$, $XY = N$, where x and y are positive integers. Then

$$(7.61) \qquad \sum_{(x,y)\in\mathcal{D}} e(f(x, y)) \ll N \log N (N^{61/38}F^{-1} + N^{-85/38}F)^{1/8},$$

if $f(x, y) \ll F$, $f_{x^k y^l}(x, y) = C_{k,l} f(x, y) x^{-k} y^{-l} + O(\delta F x^{-k} y^{-l})$ for $(x, y) \in \mathcal{D}$, $\delta \ll N^{-1/3}$, $X \geqslant Y$, and a certain system of equations involving partial derivatives of $f(x, y)$ must be satisfied.

The last condition mentioned above, involving a certain system of equations, is rather technical and lengthy, and therefore is not stated in detail for the sake of brevity. For the details of the difficult proof of (7.61) the reader is referred to G. Kolesnik (1982), since a complete proof of this result falls beyond the scope of this text. Having at our disposal an estimate such as (7.61), one can slightly improve (7.19) and also (7.58) by proving

THEOREM 7.3. For $G \geqslant T^{35/108}$ and $T \geqslant T_0$

$$(7.62) \qquad \int_{T-G}^{T+G}|\zeta(\tfrac{1}{2} + it)|^2 dt \ll G \log^2 T.$$

There is an extra log factor in (7.62), which comes from the log factor in (7.61). In view of Lemma 7.1 the above estimate gives for $G = T^{35/108}$

$$|\zeta(\tfrac{1}{2} + iT)|^2 \ll \log T + \log T \int_{-\log^2 T}^{\log^2 T} e^{-|u|}|\zeta(\tfrac{1}{2} + iT + iu)|^2 du$$

$$\ll \left(1 + \int_{T-G}^{T+G}|\zeta(\tfrac{1}{2} + it)|^2 dt\right)\log T \ll T^{35/108}\log^3 T,$$

giving at once

COROLLARY 7.1.

$$(7.63) \qquad \zeta(\tfrac{1}{2} + iT) \ll T^{35/216}\log^{3/2} T.$$

This improves (7.58) since $\frac{35}{216} = 0.162037037\ldots$. The estimate (7.63) is the best published result, due to G. Kolesnik (1982), who only had the weaker T^ε instead of our $\log^{3/2} T$. Further refinements of Kolesnik's method could slightly decrease the exponent $\frac{35}{216}$.

Proof of Theorem 7.3.   The interesting range in (7.62) is $G \leqslant T^{1/3}$, since the larger values of G are covered by the proof of Theorem 7.1. We use Theorem 7.2 with $\delta = 1$, which leads to the estimation of the exponential sum

$$(7.64) \quad S(x, K, T) = \sum_{K < n \leqslant K + x} (-1)^n d(n) \exp(if(T, n)) \qquad (1 \leqslant x \leqslant K),$$

where $f(T, n)$ is given by (7.22). The factor $(-1)^n$ in (7.64) is harmless. Indeed, if n, k, k_1, m, m_1 are positive integers, $0 < a < b$, and $g(n)$ is any arithmetical function, then

$$\sum_{a < n \leqslant b} (-1)^n d(n) g(n) = \sum_{a < km \leqslant b} (-1)^{km} g(km) = \sum_{a/2 < k_1 m \leqslant b/2} g(2k_1 m)$$

$$+ \sum_{a/2 < km_1 \leqslant b/2} g(2km_1) - \sum_{a/4 < k_1 m_1 \leqslant b/4} g(4k_1 m_1)$$

$$- \sum_{a < (2k_1 + 1)(2m_1 + 1) \leqslant b} g((2k_1 + 1)(2m_1 + 1)).$$

But the last sum above is equal to

$$\sum_{a < km \leqslant b} g(km) - \sum_{a/2 < k_1 m \leqslant b/2} g(2k_1 m) - \sum_{a/2 < km_1 \leqslant b/2} g(2km_1)$$

$$+ \sum_{a/4 < k_1 m_1 \leqslant b/4} g(4k_1 m_1),$$

which shows that the estimation of $S(x, K, T)$ in (7.64) reduces to the estimation of several sums of the type

$$(7.65) \qquad S_1(x, K, T) = \sum_{K/C < mn \leqslant (K+x)/C} \exp(if(T, Cmn)),$$

where $C = 1$, 2, or 4. All these sums are estimated analogously, and thus henceforth we shall assume that $C = 1$. Writing

$$g(z) = \operatorname{ar\,sinh} z + z(z^2 + 1)^{1/2}$$

it is seen that $g'(z) = 2(z^2 + 1)^{1/2}$; hence for $|z| < 1$

$$(7.66) \qquad g(z) = \sum_{m=1}^{\infty} a_m z^{2m-1}, \qquad a_m = \frac{2}{2m - 1} \binom{\frac{1}{2}}{m - 1},$$

and thus $a_1 = 2$ and $|a_m| \leqslant 1/(2m - 1)$ for $m \geqslant 2$. Therefore with some suitable constants $b_1, b_2, \ldots$ we may write

$$(7.67) \quad f(T, u) = \sum_{j=1}^{\infty} b_j T^{3/2-j} u^{j-1/2} = F(T, u) + \sum_{j=3}^{\infty} b_j T^{3/2-j} u^{j-1/2},$$

where $b_1 \neq 0$,

$$(7.68) \qquad\qquad F(T, u) = b_1 T^{1/2} u^{1/2} + b_2 T^{-1/2} u^{3/2},$$

and the function F will be easier to estimate by Kolesnik's Lemma 7.3 than f itself. For $\delta = 1$ in Theorem 7.2 we have $K \ll TG^{-2} \log^2 T$, and if we suppose $G \geqslant T^{1/4}$, then we obtain

(7.69)

$$\sum_{K < mn \leqslant K+x} \{\exp(if(T, mn)) - \exp(iF(T, mn))\}$$

$$\ll \sum_{K < n \leqslant K+x} d(n)|f(T, n) - F(T, n)|$$

$$\ll T^\varepsilon K \max_{K < n \leqslant K+x} |f(T, n) - F(T, n)| \ll T^\varepsilon K^{7/2} T^{-3/2} \ll T^\varepsilon K^{1/2}.$$

Therefore for $G \geqslant T^{1/4}$ in Theorem 7.2 we have reduced the problem to the estimation of the sum

$$(7.70) \qquad S_2(x, K, T) = \sum_{K < mn \leqslant K+x} \exp(iF(T, mn)),$$

$$F(T, u) = b_1 T^{1/2} u^{1/2} + b_2 T^{-1/2} u^{3/2}.$$

To estimate $S_2(x, K, T)$ we apply Lemma 7.3, taking $f(x, y) = F(T, xy)$ [here F refers to (7.68)], $F = (TK)^{1/2}$, $N = K$, and dividing the domain $K < mn \leqslant K + x$ into $O(\log T)$ subdomains of the form $X \leqslant x \leqslant X_1 \leqslant 2X, Y \leqslant y \leqslant Y_1 \leqslant 2Y, X \geqslant Y$. It may be readily verified that the conditions of Lemma 7.3 hold, and we obtain

$$(7.71) \quad S_2(x, K, T) \ll \log^2 T(K^{173/152} T^{-1/16} + K^{119/152} T^{1/16}).$$

Further observe that for any fixed $C > 0$ we have

$$(7.72) \qquad \sum_{K=2^k \leqslant TG^{-2} \log^2 T} K^C \exp(-G^2 K/T)$$

$$= \sum_{K=2^k \leqslant TG^{-2}} + \sum_{K=2^k > TG^{-2}} \ll (TG^{-2})^C.$$

Therefore for $G \geqslant T^{35/108}$ it follows from Theorem 7.2 and the above estimates that

$$\int_{T-G}^{T+G} |\zeta(\tfrac{1}{2} + it)|^2 \, dt$$

$$\ll G \log^2 T \left(1 + \sum_{K=2^k \leqslant TG^{-2}\log^2 T} (K^{135/152}T^{-5/16} + K^{81/152}T^{-3/16} \right.$$

$$\left. + K^{1/4}T^{\varepsilon-1/4}) e^{-G^2 K/T} \right)$$

$$\ll G \log^2 T (1 + T^{175/304}G^{-540/304} + T^{105/304}G^{-324/304} + T^\varepsilon G^{-1/2})$$

$$\ll G \log^2 T,$$

which completes the proof of Theorem 7.3.

7.6 THIRD AND FOURTH POWER MOMENTS IN SHORT INTERVALS

Having developed a method based on the use of Fourier coefficients of cusp forms and Kloosterman sums, H. Iwaniec (1979/1980) recently obtained a deep estimate for the fourth power moment in short intervals. His Theorem 4 states the following: If $T \geqslant 2$, $T^{1/2} < G \leqslant T$, and $T \leqslant t_1 < t_2 < \cdots < t_R \leqslant 2T$, $t_{r+1} - t_r \geqslant G$ for $r = 1, 2, \ldots, R - 1$, then

$$(7.73) \qquad \sum_{r \leqslant R} \int_{t_r}^{t_r + G} |\zeta(\tfrac{1}{2} + it)|^4 \, dt \ll T^\varepsilon (RG + R^{1/2}G^{-1/2}T).$$

The proof of this result is too complex to be included here, and (7.73) is used in this text only for the derivation of the approximate functional equation for $\zeta^k(s)$ ($k > 2$) in Chapter 4, where the special case

$$(7.74) \qquad \int_{T-G}^{T+G} |\zeta(\tfrac{1}{2} + it)|^4 \, dt \ll T^\varepsilon G, \qquad G \geqslant T^{2/3}$$

of (7.73) is needed. It was mentioned by Iwaniec (1979/1980) that (7.73) may be used for the proof of the twelfth power moment estimate (8.35) of Heath-Brown, which will be proved here with the aid of Theorem 7.2. It may be also remarked that in the case of the twelfth power moment, Iwaniec's method yields T^ε instead of $\log^{17}T$ in (8.35). As an application of (7.73) for $R = 1$ note that for $G > T^{1/2}$ we have by Theorem 7.2 and the

Cauchy–Schwarz inequality

(7.75)

$$\int_{T-G}^{T+G}|\zeta(\tfrac{1}{2}+it)|^3\,dt \leqslant \left(\int_{T-G}^{T+G}|\zeta(\tfrac{1}{2}+it)|^2\,dt\right)^{1/2}\left(\int_{T-G}^{T+G}|\zeta(\tfrac{1}{2}+it)|^4\,dt\right)^{1/2}$$

$$\ll T^\varepsilon G^{1/2}(G+TG^{-1/2})^{1/2} \ll T^\varepsilon(G+T^{1/2}G^{1/4})$$

$$(G>T^{1/2}),$$

which is used in Chapter 4 in the proof of the approximate functional equation for $\zeta^3(s)$ (with $G=2T^{1/2}$).

Other interesting mean values involving the zeta-function were investigated by H. Iwaniec (1980a) and Deshouillers and Iwaniec (1982). A natural way to attack the sixth power moment for the zeta-function, namely, the estimate $\int_0^T|\zeta(\tfrac{1}{2}+it)|^6\,dt \ll T^{1+\varepsilon}$, is to try to prove

(7.76)
$$\int_0^T|\zeta(\tfrac{1}{2}+it)|^4\left|\sum_{n\leqslant N}a_n n^{-it}\right|^2\,dt \ll T^{1+\varepsilon}\sum_{n\leqslant N}|a_n|^2$$

for $N \ll T^{1/2}$, where $a_1,\ldots,a_N$ are arbitrary complex numbers, since by the approximate functional equation (4.1) (or the reflection principle) it is seen that $\zeta(\tfrac{1}{2}+it)$ may be majorized by two Dirichlet polynomials of length $\ll T^{1/2}$. Proving (7.76) for the range $N \ll T^{1/2}$ seems to be out of reach at present, but using intricate techniques involving Kloosterman sums, H. Iwaniec (1980a) obtained (7.76) for the range $N \leqslant T^{1/10}$, while J.-M. Deshouillers and H. Iwaniec (1982) improved this range to $N \leqslant T^{1/5}$. In their work they also mention that under the truth of a certain conjecture involving the lower bound of eigenvalues of the non-Euclidean Laplacian of Hecke congruence subgroups, their method would give $N \leqslant T^{1/4}$ as the range for which (7.76) holds.

Notes

Lemmas 7.1 and 7.2 are from the author's work (1983b). The former is a straightforward generalization of an estimate of D. R. Heath-Brown (1978a). D. R. Heath-Brown (1981c) uses an inequality due to P. X. Gallagher [Lemma 1.10 of H. L. Montgomery (1971)] to obtain an estimate very similar to Lemma 7.2, and then he proves (7.19) for the range $G \geqslant T^{1/3}$. There are several ways to estimate the integral in (7.10), and in the proof of Lemma 7.2 we employed the exponential integral (A.38), which is systematically used on many occasions in this text. An alternative approach consists of using the integral [see also (12.126)]

$$\int_{-\infty}^{\infty} e^{2\pi ixy}\left(\frac{\sin\pi x}{\pi x}\right)^2\,dx = \begin{cases} 1-|y|, & \text{if } |y|\leqslant 1, \\ 0, & \text{if } |y|\geqslant 1, \end{cases}$$

where y is real. Since for $|t| \leqslant G$ we have

$$1 \leqslant \frac{\pi^2}{4}\left(\frac{\sin \pi t/2G}{\pi t/2G}\right)^2,$$

we may write for an arbitrary complex sequence $\{a_n\}_{n=1}^{\infty}$

$$(7.77) \quad I(T,G) = \int_{-G}^{G}\left|\sum_{n=1}^{\infty} a_n n^{it+iT}\right|^2 dt \leqslant \frac{\pi^2}{4}\int_{-G}^{G}\left(\frac{\sin \pi t/2G}{\pi t/2G}\right)^2\left|\sum_{n=1}^{\infty} a_n n^{it+iT}\right|^2 dt$$

$$\leqslant \frac{\pi^2}{4}(2G)\int_{-\infty}^{\infty}\left(\frac{\sin \pi x}{\pi x}\right)^2\left|\sum_{n=1}^{\infty} a_n n^{iT}\exp\left(2\pi ix\left(G\pi^{-1}\right)\log n\right)\right|^2 dx$$

$$= \frac{1}{2}\pi^2 G\left(\sum_{n=1}^{\infty}|a_n|^2 + \sum_{m \neq n,\, |\log n/m| \leqslant \pi G^{-1}} a_n \bar{a}_m\left(1 - \frac{G}{\pi}\left|\log \frac{n}{m}\right|\right)\left(\frac{n}{m}\right)^{iT}\right).$$

For a fixed m the last double sum is empty unless $|n - m| \leqslant 2\pi mG^{-1}$; hence using the elementary inequality $|ab| \leqslant \frac{1}{2}(|a|^2 + |b|^2)$ we infer from (7.77)

$$I(T,G) = \int_{T-G}^{T+G}\left|\sum_{n=1}^{\infty} a_n n^{it}\right|^2 dt \leqslant CG\sum_{n=1}^{\infty}|a_n|^2 + C\sum_{n=1}^{\infty} n|a_n|^2$$

for $C = 5\pi^3$ say, which is a weakened form of the mean value theorem for Dirichlet polynomials (Theorem 5.2) for the short interval $[T - G, T + G]$.

In the important case when $a_n = 1$ if $N < n \leqslant N'$ and zero otherwise, in (7.77) we may remove the factor $1 - G\pi^{-1}|\log nm^{-1}|$ by partial summation and obtain essentially (7.10) by writing $m = n + r$.

The bound (7.57) is from the author's text (1983b), while for a detailed discussion of (7.59) and (7.60) the reader is referred to E. C. Titchmarsh's book (1951).

The proof of Theorem 7.2 is given by the author (1983b) and is based on M. Jutila's paper (1984a) and the exponential averaging technique which gives (7.35). Indeed it would have been shorter to apply Jutila's Theorem 1 (1984a) to (4.11) and then to integrate the resulting expression, but the proof that we have given is self-contained. Also several details in the proof are simpler than the corresponding ones in Jutila's proof, since we are dealing here with a special Dirichlet polynomial, while Jutila considers the general case of transforming

$$S_1(M_1, M_2; t) = \sum_{M_1 \leqslant n \leqslant M_2} d(n) n^{-1/2 - it}$$

by Voronoi's summation formula. This approach, based on the use of Voronoi's formula, seems very natural and the aforementioned paper of Jutila contains generalizations to other Dirichlet polynomials (e.g., whose coefficients are generated by certain cusp forms).

Theorem 7.3 and Corollary 7.1 are given in the author's paper (1980b). The final version of G. Kolesnik's estimate (7.61), as published in his paper (1982), had $N^{1+\varepsilon}(N^{61/38}F^{-1} + \dots$ instead of $N \log N(N^{61/38}F^{-1} + \dots$, but the slightly sharper version used in our text may be obtained by refining his argument a little.

As mentioned in Notes of Chapter 2, G. Kolesnik's paper (in press) presents a theory of n-dimensional exponent pairs. Therein it is stated (without details of proof) that Kolesnik's method of two-dimensional exponent pairs would lead to the value $\mu(\frac{1}{2}) \leqslant \frac{139}{858} = 0.162004\dots$ for the order of $\zeta(\frac{1}{2} + it)$, which is just slightly better than $\mu(\frac{1}{2}) \leqslant \frac{35}{216} = 0.162037\ldots$ of Corollary 7.1. The bounds for $\mu(\frac{1}{2})$ obtained historically by various authors are as follows:

$\mu(\frac{1}{2}) \leqslant \frac{1}{6} = 0.166\ldots$	G. H. Hardy and J. E. Littlewood (1921)
$\mu(\frac{1}{2}) \leqslant \frac{163}{988} = 0.1649797\ldots$	A. Walfisz (1924)
$\mu(\frac{1}{2}) \leqslant \frac{27}{164} = 0.1646341\ldots$	E. C. Titchmarsh (1931a, 1931b)
$\mu(\frac{1}{2}) \leqslant \frac{229}{1392} = 0.1645114\ldots$	E. Phillips (1933)
$\mu(\frac{1}{2}) \leqslant \frac{19}{116} = 0.1637931\ldots$	E. C. Titchmarsh (1942)
$\mu(\frac{1}{2}) \leqslant \frac{15}{92} = 0.1630434\ldots$	S. H. Min (1949)
$\mu(\frac{1}{2}) \leqslant \frac{6}{37} = 0.1621621\ldots$	W. Haneke (1962–1963) and Chen Jing-run (1965a)
$\mu(\frac{1}{2}) \leqslant \frac{173}{1067} = 0.1621368\ldots$	G. Kolesnik (1973)
$\mu(\frac{1}{2}) \leqslant \frac{35}{216} = 0.162037037\ldots$	G. Kolesnik (1982)
$\mu(\frac{1}{2}) \leqslant \frac{139}{858} = 0.1620046\ldots$	G. Kolesnik (in press)

Omega results for $\zeta(s)$ are discussed in Chapter 9. Concerning order results for $|\zeta(\frac{1}{2} + it)|$, it may be mentioned that M. Jutila (1983b) has proved the following interesting result: There exist positive constants a_1, a_2, and a_3 such that for $T \geqslant 10$

$$\exp\left(a_1 (\log\log T)^{1/2} \right) \leqslant \left| \zeta\left(\tfrac{1}{2} + it \right) \right| \leqslant \exp\left(a_2 (\log\log T)^{1/2} \right)$$

on a subset of measure at least $a_3 T$ of the interval $[0, T]$.

To assess the strength of H. Iwaniec's estimate (7.74), note that by Lemma 7.1 with $k = 4$

$$\left| \zeta\left(\tfrac{1}{2} + iT \right) \right|^4 \ll \log T \left(1 + \int_{T-G}^{T+G} \left| \zeta\left(\tfrac{1}{2} + it \right) \right|^4 dt \right) \ll GT^\varepsilon$$

for $G \geqslant T^{2/3}$; hence for $G = T^{2/3}$ one obtains $\zeta(\frac{1}{2} + iT) \ll T^{1/6+\varepsilon}$, which is the classical result of G. H. Hardy and J. E. Littlewood (1921).

CHAPTER EIGHT
HIGHER POWER MOMENTS

8.1 INTRODUCTION

In this chapter we focus our attention on higher power moments for $\zeta(s)$ (i.e., higher than the fourth, which was discussed in Chapter 5) when $\frac{1}{2} \leqslant \sigma < 1$ is fixed. As in the case of mean values over short intervals in Chapter 7, it will be expedient to distinguish the cases $\sigma = \frac{1}{2}$ and $\frac{1}{2} < \sigma < 1$. Accordingly we may define two numbers, $M(A)$ and $m(\sigma)$, which characterize power moments when $\sigma = \frac{1}{2}$ and $\frac{1}{2} < \sigma < 1$, respectively. For any fixed number $A \geqslant 4$ the number $M(A)$ ($\geqslant 1$) is defined as the infimum of all numbers M ($\geqslant 1$) such that

$$(8.1) \qquad \int_1^T \left| \zeta\left(\tfrac{1}{2} + it\right) \right|^A dt \ll T^{M+\varepsilon}$$

for any $\varepsilon > 0$. Similarly, for $\frac{1}{2} < \sigma < 1$ fixed we define $m(\sigma)$ ($\geqslant 4$) as the supremum of all numbers m ($\geqslant 4$) such that

$$(8.2) \qquad \int_1^T \left| \zeta(\sigma + it) \right|^m dt \ll T^{1+\varepsilon}$$

for any $\varepsilon > 0$, and naturally we seek upper bounds for $M(A)$ and lower bounds for $m(\sigma)$. The difference between the definitions of $M(A)$ and $m(\sigma)$ seems in place, since $M(A) = 1$ is not known to hold for any $A > 4$ at the time of writing of this text, while for any fixed $\frac{1}{2} < \sigma < 1$ it is possible to find a number $m > 4$ such that (8.2) holds. From the discussion made in Section 1.9 of Chapter 1 it follows that $M(A) = 1$ for $A \geqslant 1$ is equivalent to the Lindelöf hypothesis.

Power moments are an important part of zeta-function theory. They also occur frequently in applications, and the results of this chapter will be used in proving zero-density estimates in Chapter 11 and in bounding the error term in the Dirichlet divisor problem in Chapter 13.

An important feature of power moments for $\zeta(s)$ is that (8.1) and (8.2) with $M = M(A)$, $m = m(\sigma)$ are respectively equivalent to

(8.3)
$$\sum_{r \leqslant R} \left|\zeta\left(\tfrac{1}{2} + it_r\right)\right|^A \ll T^{M(A)+\varepsilon}$$

and

(8.4)
$$\sum_{r \leqslant R} \left|\zeta(\sigma + it_r)\right|^{m(\sigma)} \ll T^{1+\varepsilon},$$

where

(8.5)

$$T \leqslant t_r \leqslant 2T \quad \text{for} \quad r = 1, \ldots, R; \qquad |t_r - t_s| \geqslant \log^C T \quad \text{for} \quad 1 \leqslant r \neq s \leqslant R$$

and $C \geqslant 0$ is fixed. To see this, let for $m = 1, 2, \ldots, [T \log^{-C} T]$

$$t_m = \max_{t \in I_m} \left|\zeta\left(\tfrac{1}{2} + it\right)\right|, \qquad I_m = \left[2T - m \log^C T, 2T - (m-1)\log^C T\right],$$

and denote by $\{t_r\}$ either of the sets $\{t_{2m}\}$ or $\{t_{2m-1}\}$. Then the t_r's satisfy (8.5) and

$$\int_T^{2T} \left|\zeta\left(\tfrac{1}{2} + it\right)\right|^A dt \ll T^{1/6} + \sum_r \left|\zeta\left(\tfrac{1}{2} + it_r\right)\right|^A \ll T^{M(A)+\varepsilon}.$$

Replacing T by $T/2$, $T/2^2$, $\ldots$ and adding we obtain (8.1), and a similar discussion shows that (8.4) implies (8.2). On the other hand, using Lemma 7.1 we have

$$\sum_{r \leqslant R} \left|\zeta\left(\tfrac{1}{2} + it_r\right)\right|^A \ll R \log T + \log T \sum_{r \leqslant R} \int_{-\log^2 T}^{\log^2 T} \left|\zeta\left(\tfrac{1}{2} + it_r + iv\right)\right|^A dv$$

$$\ll T^{M(A)+\varepsilon} + \log T \int_1^{2T + \log^2 T} \left|\zeta\left(\tfrac{1}{2} + it\right)\right|^A dt \ll T^{M(A)+\varepsilon},$$

provided that (8.1) holds, and similarly (8.2) implies (8.4).

More generally we can eschew the terminology of both discrete sums and integrals and use large values of $|\zeta(\sigma + it)|$ instead. Namely, let $a(\sigma)$ ($\geqslant 1$) and $b(\sigma)$ ($\geqslant 4$) be such functions of σ ($\tfrac{1}{2} \leqslant \sigma < 1$ fixed) for which

(8.6)
$$R \ll T^{a(\sigma)+\varepsilon} V^{-b(\sigma)},$$

where for t_r defined by (8.5) we have

(8.7) $$|\zeta(\sigma + it_r)| \geqslant V \geqslant T^\varepsilon \qquad (r = 1, 2, \ldots, R).$$

Then this is equivalent to

(8.8) $$\sum_{r \leqslant R} |\zeta(\sigma + it_r)|^{b(\sigma)} \ll T^{a(\sigma) + \varepsilon}$$

or

(8.9) $$\int_1^T |\zeta(\sigma + it)|^{b(\sigma)} \, dt \ll T^{a(\sigma) + \varepsilon}.$$

The equivalence of (8.8) and (8.9) is established analogously as the equivalence of (8.1) and (8.3). For the other equivalence suppose that (8.8) holds and let $\{t_{V,1}, \ldots, t_{V,R_1}\}$ be the subset of $\{t_r\}$ such that

$$|\zeta(\sigma + it_{V,j})| \geqslant V \qquad (j = 1, \ldots, R_1).$$

Then (8.8) implies

$$R_1 V^{b(\sigma)} \leqslant \sum_{r \leqslant R} |\zeta(\sigma + it_r)|^{b(\sigma)} \ll T^{a(\sigma) + \varepsilon},$$

and consequently (8.6) holds with $R_1 = R$. Conversely, let (8.6) hold and denote by $t_{V,1}, \ldots, t_{V,R(V)}$ those of the points $t_1, \ldots, t_R$ for which

$$V \leqslant |\zeta(\sigma + it_{V,j})| \leqslant 2V \qquad [j = 1, \ldots, R(V)].$$

There are $O(\log T)$ choices for each V, and we set $V = T^{1/6}$, $V = 2^{-1}T^{1/6}$, $V = 2^{-2}T^{1/6}, \ldots$. Then we obtain

$$\sum_{r \leqslant R} |\zeta(\sigma + it_r)|^{b(\sigma)} \ll T^\varepsilon + \sum_V \sum_{j \leqslant R(V)} (2V)^{b(\sigma)}$$

$$\ll T^\varepsilon + \sum_V T^{a(\sigma) + \varepsilon} \ll T^{a(\sigma) + \varepsilon}.$$

In Section 8.2 we shall derive some convexity estimates for $a(\sigma)$, $b(\sigma)$ by an argument which gives also the convexity of the function $\mu(\sigma)$, defined by (1.65). Upper bounds for $M(A)$ will be dealt with in Section 8.3, and our main tool will be Theorem 7.2. In Section 8.4 a method will be developed to cope with $m(\sigma)$ for $\frac{1}{2} < \sigma < 1$, and finally in Section 8.5 we shall turn our attention to asymptotic formulas for $\int_1^T |\zeta(\sigma + it)|^{2k} \, dt$ ($k \geqslant 2$ a fixed integer, $\frac{1}{2} < \sigma < 1$ fixed), and not only to upper bounds of the type (8.2).

8.2 SOME CONVEXITY ESTIMATES

In this section we shall obtain a convexity estimate for the functions $a(\sigma)$ and $b(\sigma)$, defined by (8.6) and (8.7), and as a corollary the convexity property of the function $\mu(\sigma)$, defined by (1.65).

Let $s = \sigma + it$, $T \leqslant t \leqslant 2T$, $0 < \sigma < 1$. From Theorem 1.8 we have

$$(8.10) \qquad \zeta(s) = \sum_{n \leqslant 2T} n^{-s} + O(T^{-\sigma}),$$

which means that $\zeta(s)$ is approximated by $O(\log T)$ sums of the form $\sum_{N < n \leqslant 2N} n^{-s}$, where $N \leqslant T$. For $\sigma_0 > 0$ partial summation gives

$$(8.11)$$

$$\sum_{N < n \leqslant 2N} n^{-\sigma - it} \ll N^{\sigma_0 - \sigma} \left| \sum_{N < n \leqslant 2N} n^{-\sigma_0 - it} \right| + N^{\sigma_0 - \sigma - 1} \int_N^{2N} \left| \sum_{N < n \leqslant x} n^{-\sigma_0 - it} \right| dx.$$

Take now $N \leqslant N_1 < N_2 \leqslant 2N \leqslant 2T$. Then by the inversion formula (A.10) we have

$$(8.12)$$

$$\sum_{N_1 < n \leqslant N_2} n^{-s} = (2\pi i)^{-1} \int_{1 - \sigma + \varepsilon - iT}^{1 - \sigma + \varepsilon + iT} \zeta(s + w)(N_2^w - N_1^w) w^{-1} dw$$

$$+ O(N^{1 - \sigma + \varepsilon} T^{-1})$$

$$= (2\pi i)^{-1} \int_{-iT}^{iT} \zeta(s + w)(N_2^w - N_1^w) w^{-1} dw$$

$$+ O\left(\int_0^{1 - \sigma + \varepsilon} |\zeta(\sigma + u + iT + it)| N^u T^{-1} du \right) + O(T^{\varepsilon - \sigma}).$$

Now observe that

$$(N_2^w - N_1^w) w^{-1} = \int_{N_1}^{N_2} z^{w-1} dw \ll \min(1, |v|^{-1}) \qquad (w = iv)$$

and

$$\int_0^{1 - \sigma + \varepsilon} |\zeta(\sigma + u + iT + it)| N^u T^{-1} du \ll T^{\varepsilon - \sigma} \int_0^{1 - \sigma + \varepsilon} (N/T)^u du \ll T^{\varepsilon - \sigma},$$

since trivially $\zeta(\sigma + u + iT + it) \ll T^{1 - \sigma - u + \varepsilon}$. Therefore, we have uniformly in t, N_1, N_2 and $0 < \sigma_0 < 1$

$$(8.13) \qquad \sum_{N_1 < n \leqslant N_2} n^{-\sigma_0 - it} \ll \int_0^T |\zeta(\sigma_0 + it + iv)| \frac{dv}{v + 1} + T^{\varepsilon - \sigma_0},$$

and (8.11) gives

$$(8.14) \quad \sum_{N < n \leqslant 2N} n^{-\sigma - it} \ll N^{\sigma_0 - \sigma} T^{\varepsilon - \sigma_0} + N^{\sigma_0 - \sigma} \int_0^T |\zeta(\sigma_0 + it + iv)| \frac{dv}{v + 1}.$$

As our first application of (8.14) consider, as in (1.65),

$$\mu(\sigma) = \limsup_{t \to \infty} \frac{\log|\zeta(\sigma + it)|}{\log t},$$

and suppose $0 < \sigma_1 < \sigma_2 < 1$, $\sigma_1 \leqslant \sigma \leqslant \sigma_2$. Applying (8.14) with $\sigma_0 = \sigma_1$ and $\sigma_0 = \sigma_2$ we obtain

$$(8.15) \quad \sum = \sum_{N < n \leqslant 2N} n^{-\sigma - it} \ll N^{\sigma_1 - \sigma} T^{\mu(\sigma_1) + \varepsilon}, \qquad \sum \ll N^{\sigma_2 - \sigma} T^{\mu(\sigma_2) + \varepsilon}.$$

For an arbitrary $0 \leqslant \alpha \leqslant 1$ (8.15) gives

$$(8.16) \quad \sum = \sum^{\alpha} \sum^{1 - \alpha} \ll N^{\alpha(\sigma_1 - \sigma) + (1 - \alpha)(\sigma_2 - \sigma)} T^{\alpha\mu(\sigma_1) + (1 - \alpha)\mu(\sigma_2) + \varepsilon}.$$

Now we shall choose α in such a way that the exponent of N in (8.16) vanishes. Thus we take

$$\alpha = (\sigma_2 - \sigma)/(\sigma_2 - \sigma_1)$$

and obtain

$$\sum \ll T^{\mu(\sigma_1)(\sigma_2 - \sigma)/(\sigma_2 - \sigma_1) + \mu(\sigma_2)(\sigma - \sigma_1)/(\sigma_2 - \sigma_1) + \varepsilon}.$$

By (8.10) $\zeta(s)$ is approximated by $O(\log T)$ sums $\sum$, which means that we have proved (1.67), namely,

$$\mu(\sigma) \leqslant \frac{\mu(\sigma_1)(\sigma_2 - \sigma)}{(\sigma_2 - \sigma_1)} + \frac{\mu(\sigma_2)(\sigma - \sigma_1)}{(\sigma_2 - \sigma_1)}$$

for $0 < \sigma_1 < \sigma < \sigma_2 < 1$, which represents the convexity of the function $\mu(\sigma)$.

We proceed now to a convexity estimate for $a(\sigma)$ and $b(\sigma)$, where the notation is the same as in Section 8.1. Suppose that for $N \leqslant T$ we have

$$T^{\varepsilon} \leqslant V \leqslant \left| \sum_{N < n \leqslant 2N} n^{-\sigma - it_r} \right| \qquad (r = 1, 2, \ldots, R),$$

where the t_r's satisfy (8.5). Using (8.14) with $\sigma_0 = \sigma_1$ we obtain

$$RV \leqslant \sum_{r \leqslant R} \left| \sum_{N < n \leqslant 2N} n^{-\sigma - it_r} \right|$$

$$\ll N^{\sigma_1 - \sigma} \sum_{r \leqslant R} \left(T^{\varepsilon - \sigma_1} + \int_0^T |\zeta(\sigma_1 + it_r + iv)| \frac{dv}{v + 1} \right),$$

whence

$$RV \ll N^{\sigma_1 - \sigma} \int_0^T \sum_{r \leqslant R} |\zeta(\sigma_1 + it_r + iv)| \frac{dv}{v+1}$$

$$\ll N^{\sigma_1 - \sigma} R^{1 - 1/b(\sigma_1)} \left(T^{a(\sigma_1) + \varepsilon}\right)^{1/b(\sigma_1)} \log T,$$

where we used Hölder's inequality (see Section A.4). Hence

(8.17) $$R \ll T^{a(\sigma_1) + \varepsilon} V^{-b(\sigma_1)} N^{b(\sigma_1)(\sigma_1 - \sigma)},$$

and analogously it follows

(8.18) $$R \ll T^{a(\sigma_2) + \varepsilon} V^{-b(\sigma_2)} N^{b(\sigma_2)(\sigma_2 - \sigma)}.$$

However if (8.7) holds, then in view of (8.10) there must exist a subset $\{t_{r_1}, t_{r_2}, \dots\}$ of $\{t_r\}$ which contains at least $CR/\log T$ of all R points, so that for some $N \leqslant T$ and points of this subset

$$V \leqslant \left| \sum_{N < n \leqslant 2N} n^{-\sigma - it_r} \right| \qquad (r = r_1, r_2, \dots).$$

Therefore using (8.17) and (8.18) we have, for $0 \leqslant \alpha \leqslant 1$,

(8.19)

$$R = R^\alpha R^{1-\alpha}$$

$$\ll T^{\alpha a(\sigma_1) + (1-\alpha)a(\sigma_2) + \varepsilon} V^{-\alpha b(\sigma_1) - (1-\alpha)b(\sigma_2)} N^{\alpha b(\sigma_1)(\sigma_1 - \sigma) + (1-\alpha)b(\sigma_2)(\sigma_2 - \sigma)}.$$

If we choose

$$\alpha = \frac{b(\sigma_2)(\sigma_2 - \sigma)}{b(\sigma_2)(\sigma_2 - \sigma) + b(\sigma_1)(\sigma - \sigma_1)},$$

then the exponent of N in (8.19) vanishes, and we obtain

(8.20) $$RT^{-\varepsilon} \ll T^{\frac{a(\sigma_1)b(\sigma_2)(\sigma_2 - \sigma) + a(\sigma_2)b(\sigma_1)(\sigma - \sigma_1)}{b(\sigma_2)(\sigma_2 - \sigma) + b(\sigma_1)(\sigma - \sigma_1)}} V^{-\frac{b(\sigma_1)b(\sigma_2)(\sigma_2 - \sigma_1)}{b(\sigma_2)(\sigma_2 - \sigma) + b(\sigma_1)(\sigma - \sigma_1)}}.$$

If we consider $a(\sigma)$ and $b(\sigma)$ as lower and upper bounds of the numbers in (8.6) respectively, then (8.20) gives

(8.21) $$a(\sigma) \leqslant \frac{a(\sigma_1)b(\sigma_2)(\sigma_2 - \sigma) + a(\sigma_2)b(\sigma_1)(\sigma - \sigma_1)}{b(\sigma_2)(\sigma_2 - \sigma) + b(\sigma_1)(\sigma - \sigma_1)},$$

(8.22) $$b(\sigma) \geqslant \frac{b(\sigma_1)b(\sigma_2)(\sigma_2 - \sigma_1)}{b(\sigma_2)(\sigma_2 - \sigma) + b(\sigma_1)(\sigma - \sigma_1)},$$

if $\frac{1}{2} \leqslant \sigma_1 < \sigma < \sigma_2 < 1$. Taking $a(\sigma_1) = a(\sigma_2) = 1$, we obtain as a special case of (8.22) a sort of a convexity property of the function $m(\sigma)$, which we state as

THEOREM 8.1. Let $m(\sigma)$ be defined by (8.2). Then for $\frac{1}{2} \leqslant \sigma_1 < \sigma < \sigma_2 < 1$

$$(8.23) \qquad m(\sigma) \geqslant \frac{m(\sigma_1)m(\sigma_2)(\sigma_2 - \sigma_1)}{m(\sigma_2)(\sigma_2 - \sigma) + m(\sigma_1)(\sigma - \sigma_1)}.$$

8.3 POWER MOMENTS FOR $\sigma = \frac{1}{2}$

In this section we shall suppose that $t_1 < t_2 < \cdots < t_R$ are real numbers which satisfy

$$(8.24)$$
$$|t_r| \leqslant T \quad \text{for} \quad r = 1, \ldots, R; \qquad |t_r - t_s| \geqslant 1 \quad \text{for} \quad 1 \leqslant r \neq s \leqslant R,$$

and

$$(8.25) \qquad |\zeta(\tfrac{1}{2} + it_r)| \geqslant V > 0 \qquad (r = 1, 2, \ldots, R).$$

The spacing condition (8.24) is similar to the one required by (8.5), and indeed we can assume (8.5) to hold with $C = 0$. Our aim is to derive upper bounds for $R = R(V)$ of the type (8.6), which will then lead to estimates of $M(A)$, as described in Section 8.1. We begin by noting that

$$(8.26) \qquad R \ll TV^{-4}\log^5 T,$$

which is the analogue of the fourth power moment, and it will turn out that (8.26) is the best available bound for large values of moduli of $\zeta(\frac{1}{2} + it)$ when V is small. To see that (8.26) holds, it is sufficient to suppose that $T/2 \leqslant t_r \leqslant T$, and then to replace T by $T/2, T/2^2, \ldots$ and to sum all the results. From the reflection principle estimate (4.66) with $k = 2$, $h = \log^2 T$, $Y = 3T$, $M = 3T$, $s = \frac{1}{2} + it_r$, and $\alpha = \frac{3}{4}$ we obtain

$$(8.27)$$

$$RV^4 \leqslant \sum_{r \leqslant R} \left| \zeta(\tfrac{1}{2} + it_r) \right|^4$$

$$\ll \sum_{r \leqslant R} \left| \sum_{n \leqslant 3T} d(n) e^{-(n/3T)^h} n^{-1/2 - it_r} \right|^2$$

$$+ \sum_{r \leqslant R} \left| \sum_{n \leqslant 3T} d(n) n^{-1/2 - it_r} \right|^2 + R$$

$$+ T^{-1/2} \left(\max_{|v| \leqslant h^2} \sum_{r \leqslant R} \left| \sum_{n \leqslant 3T} d(n) n^{-1/4 - it_r - iv} \right|^2 \right) \left(\int_{-h^2}^{h^2} e^{-|v|h^{-1}} |\tfrac{1}{4} + iv|^{-1} dv \right)^2,$$

where the Cauchy–Schwarz inequality was used, (1.25), and Stirling's formula (A.34). The integral above is clearly

$$\ll 1 + \int_1^{h^2} v^{-1}\, dv \ll \log\log T.$$

The sums over $r \le R$ are estimated by the mean value estimate given by Theorem 5.3, when one uses (5.25). We obtain

$$RV^4 \ll T\log^5 T + R + T^{-1/2}(\log\log T)^2$$

$$\times\left(T\sum_{n\le 3T} d^2(n)n^{-1/2} + \sum_{n\le 3T} d^2(n)n^{1/2} \right)\log T$$

$$\ll T\log^5 T,$$

if $V \ge \log T$, and (8.26) follows. For $V < \log T$, (8.26) follows again from the trivial estimate $R \le T$.

We proceed now to the main result of this section, which is

THEOREM 8.2. Let $(\varkappa,\lambda)$ be any exponent pair with $\varkappa > 0$, and let $t_1 < \cdots < t_R$ satisfy (8.24) and (8.25). Then

$$(8.28) \quad R \ll TV^{-6}\log^8 T + T^{(\varkappa+\lambda)/\varkappa}V^{-2(1+2\varkappa+2\lambda)/\varkappa}(\log T)^{(3+6\varkappa+4\lambda)/\varkappa}.$$

For special choices of the exponent pair $(\varkappa,\lambda)$ we obtain from (8.28)

COROLLARY 8.1. Under the hypotheses of Theorem 8.2 we have

$$(8.29) \qquad R \ll TV^{-6}\log^8 T + T^{29/13}V^{-178/13}\log^{235/13} T,$$

$$(8.30) \qquad R \ll TV^{-6}\log^8 T + T^{5/2}V^{-31/2}\log^{81/4} T,$$

$$(8.31) \qquad R \ll TV^{-6}\log^8 T + T^3 V^{-19}\log^{49/2} T,$$

$$(8.32) \qquad R \ll TV^{-6}\log^8 T + T^4 V^{-128/5}\log^{162/5} T,$$

$$(8.33) \qquad R \ll TV^{-6}\log^8 T + T^{15/4}V^{-24}\log^{61/2} T.$$

The exponent $(\varkappa,\lambda) = (\tfrac{1}{2},\tfrac{1}{2})$ in Theorem 8.2 leads to an important result, proved first by D. R. Heath-Brown. This is

COROLLARY 8.2. Under the hypotheses of Theorem 8.2 we have

$$(8.34) \qquad\qquad R \ll T^2 V^{-12}\log^{16} T.$$

From this result it follows that $M(12) \leqslant 2$, or more precisely

(8.35)
$$\int_1^T \left| \zeta(\tfrac{1}{2} + it) \right|^{12} dt \ll T^2 \log^{17} T,$$

which by Lemma 7.1 gives at once $\mu(\tfrac{1}{2}) \leqslant \tfrac{1}{6}$. In fact (8.35) may be considered as an integral analogue of the classical estimate $\mu(\tfrac{1}{2}) \leqslant \tfrac{1}{6}$, although the stronger result $M(6) = 1$ is still hitherto unproved.

Before the proof of Theorem 8.2 we shall give a lemma which provides estimates for moduli of $S(x, K, t_r)$ over well-spaced points t_r, where $S(x, K, t_r)$ is defined by (7.21). The result is contained in

LEMMA 8.1. Let $\mathcal{A}$ be a set of real numbers t_r such that $T/2 \leqslant t_r \leqslant T$ and $\log^2 T \leqslant G \leqslant |t_r - t_s| \leqslant J$ for $r \neq s$. If $S(x, K, T)$ is defined by (7.21) and $|\mathcal{A}|$ denotes the cardinality of $\mathcal{A}$, then for $K \leqslant T/\log T$, $T \geqslant T_0$ and any exponent pair $(\varkappa, \lambda)$ we have

(8.36)
$$\sum_{t_r \in \mathcal{A}} |S(x, K, t_r)| \ll \left((K + K^{3/4} T^{1/4} G^{-1/2} \log^{1/2} T) |\mathcal{A}|^{1/2} \right.$$

$$\left. + J^{\varkappa/2} T^{-\varkappa/4} |\mathcal{A}| K^{(2\lambda - \varkappa + 2)/4} \right) \log^{3/2} T.$$

Proof of Lemma 8.1. We start from (A.39) and choose $\xi = \{\xi_n\}_{n=1}^{\infty}$ with $\xi_n = (-1)^n d(n)$ for $K \leqslant n \leqslant K + x$ and zero otherwise, $\varphi_r = \{\varphi_{r,n}\}_{n=1}^{\infty}$ with $\varphi_{r,n} = \exp(if(t_r, n))$ for $K \leqslant n \leqslant 2K$ and zero otherwise, where $f(t_r, n)$ is defined by (7.22). Then by (A.43) we have uniformly for $x \leqslant K$

$$\|\xi\|^2 = (\xi, \xi) = \sum_{K < n \leqslant K + x} d^2(n) \ll K \log^3 T,$$

(8.37)
$$\sum_{t_r \in \mathcal{A}} |S(x, K, t_r)|$$

$$\ll K^{1/2} \log^{3/2} T \left\{ \sum_{t_r, t_s \in \mathcal{A}} \left| \sum_{K < n \leqslant 2K} \exp(if(t_r, n) - if(t_s, n)) \right| \right\}^{1/2}.$$

The inner sum on the right-hand side of (8.37) is $O(K)$ if $r = s$, and if $r \neq s$ we shall use the theory of exponent pairs (Section 2.3 of Chapter 2) to estimate

(8.38) $S = \sum_{K < n \leqslant 2K} \exp(if(n)), \quad f(u) = f(t_r, u) - f(t_s, u) \quad (r \neq s).$

Defining $g(z) = \operatorname{ar\,sinh} z + z(z^2 + 1)^{1/2}$, we recall that (7.66) holds and since

$$f(u) = 2t_r g\left((\pi u / 2 t_r)^{1/2} \right) - 2t_s g\left((\pi u / 2 t_s)^{1/2} \right) \quad (r \neq s),$$

it is therefore seen that for r, s fixed, $K \leqslant u \leqslant 2K$, and $j = 1, 2, \ldots$ we have

$$f^{(j)}(u) \asymp |t_r - t_s| K^{1/2 - j} T^{-1/2}.$$

Here we used (7.66), the mean value theorem and the condition $K \leqslant T/\log T$. This implies that if $F = |t_r - t_s| K^{-1/2} T^{-1/2} \gg 1$ we may use the theory of exponent pairs to estimate S in (8.38), and if this condition is not satisfied we use Lemmas 1.2 and 2.1 to obtain in any case

$$(8.39) \qquad S = \sum_{K < n \leqslant 2K} \exp(if(n)) \ll F^\varkappa K^\lambda + \max_{K \leqslant u \leqslant 2K} |f'(u)|^{-1}$$

$$\ll J^\varkappa T^{-\varkappa/2} K^{\lambda - \varkappa/2} + (KT)^{1/2} |t_r - t_s|^{-1}.$$

The spacing condition imposed on the t_r's gives

$$\sum_{t_r, t_s \in \mathscr{A}, r \neq s} |t_r - t_s|^{-1} \ll G^{-1} |\mathscr{A}| \sum_{n \leqslant |\mathscr{A}|} n^{-1} \ll G^{-1} |\mathscr{A}| \log T,$$

and thus substituting (8.38) and (8.39) in (8.37) we obtain (8.36).

Proof of Theorem 8.2. Having at our disposal Theorem 7.2 and Lemma 8.1 it will be a fairly simple matter to derive Theorem 8.2. First, let $\mathscr{A}_3$ denote a set of points t_r satisfying the spacing condition (8.24), but with $2T/3 \leqslant t_r \leqslant 5T/6$. We shall divide the interval $[2T/3, 5T/6]$ into N subintervals of length at most $J = T/6N$, and we shall denote by $\mathscr{A}_{1,k}$ $(k = 1, \ldots, N)$ the set of points in the kth of these intervals. Then the points of each $\mathscr{A}_{1,k}$ lie in the interval $[T_0, T_0 + J]$ for some T_0 which satisfies $2T/3 \leqslant T_0 \leqslant 5T/6 - J$. We shall estimate first $|\mathscr{A}_{1,k}|$ by taking

$$(8.40) \qquad\qquad BG \log^2 T = V^2$$

for some suitable $B > 0$ and defining

$$\mathscr{A}'_{1,k} = \mathscr{A}'_{1,k}(\tau) = \mathscr{A}_{1,k} \cap [\tau - G/2, \tau + G/2]$$

for $7T/12 \leqslant \tau \leqslant 11T/12$. By (7.63) the relevant range for V in Theorem 8.2 is $V \ll T^{35/216} \log^{3/2} T$; hence $G \leqslant T^{1/3}$, which will enable us to use Theorem 7.2, where one requires $T^\varepsilon \leqslant G \leqslant T^{1/2 - \varepsilon}$. By Lemma 7.1 for each t_r we have

$$\zeta(\tfrac{1}{2} + it_r) \ll \log^{1/2} T$$

or

$$(8.41) \qquad |\zeta(\tfrac{1}{2} + it_r)|^2 \ll \log T \int_{-\log^2 t_r}^{\log^2 t_r} e^{-|u|} |\zeta(\tfrac{1}{2} + iu + it_r)|^2 \, du.$$

We may suppose that $V \geq T^\varepsilon$, for otherwise the trivial $R \leq T$ is better than the second term in (8.28) for $\varepsilon < \frac{1}{12}$, and henceforth we suppose that (8.41) holds. Summation of (8.41) over $t_r \in \mathscr{A}'_{1,k}$ gives for some absolute $C_1 > 0$

$$(8.42) \qquad |\mathscr{A}'_{1,k}|V^2 \leq C_1 \log T \int_{\tau-G}^{\tau+G} \left|\zeta(\tfrac{1}{2} + it)\right|^2 \sum_{t_r \in \mathscr{A}'_{1,k}} e^{-|t-t_r|} dt,$$

provided that

$$\left[t_r - \log^2 t_r, \, t_r + \log^2 t_r\right] \subseteq [\tau - G, \tau + G]$$

for $t_r \in [\tau - G/2, \tau + G/2]$, which is certainly satisfied for $V \geq T^\varepsilon$. The spacing condition $|t_r - t_s| \geq 1$ ($r \neq s$) implies that the sum in (8.42) is bounded, and for the integral in (8.42) we use Theorem 7.2 with $\delta = 1$. For our range of τ we may actually use (7.24), which gives

(8.43)

$$|\mathscr{A}'_{1,k}|BG \log^2 T \leq C_2 G \log^2 T$$

$$+ C_2 G \log T \sum_{T^{1/3} \leq K = 2^k \leq TG^{-2}\log^2 T} (TK)^{-1/4} e^{-G^2 K/(2T)}$$

$$\times \left(|S(K, K, \tau)| + K^{-1}\int_0^K |S(x, K, \tau)| \, dx\right).$$

Choosing $B = 2C_2$ (8.43) simplifies to

$$(8.44) \qquad |\mathscr{A}'_{1,k}| \ll \log^{-1} T \sum_K (TK)^{-1/4} e^{-G^2 K/(2T)}$$

$$\times \left(|S(K, K, \tau)| + K^{-1}\int_0^K |S(x, K, \tau)| \, dx\right).$$

Let now $\mathscr{A}_{2,k}$ denote the set of numbers $\tau = T_0 + G/2 + nG$ such that $\mathscr{A}'_{1,k}(\tau) \neq \varnothing$ and n is such an integer for which $T_0 \leq \tau \leq T_0 + J + G/2$. If τ_r and τ_s are two different elements of $\mathscr{A}_{2,k}$ we have $G \leq |\tau_r - \tau_s| \leq J$, and we may apply Lemma 8.1 to obtain

(8.45)

$$\sum_{\tau \in \mathscr{A}_{2,k}} |A'_{1,k}(\tau)| \ll \log^{1/2} T \sum_K (TK)^{-1/4} e^{-G^2 K/(2T)}$$

$$\times \left\{\left(K + K^{3/4}T^{1/4}G^{-1/2}\log^{1/2}T\right)|\mathscr{A}_{2,k}|^{1/2} + |\mathscr{A}_{2,k}|J^{\varkappa/2}T^{-\varkappa/4}K^{(2\lambda-\varkappa+2)/4}\right\}.$$

Using (7.72), the obvious inequalities

$$(8.46) \qquad |\mathscr{A}_{1,k}| \leqslant \sum_{\tau \in \mathscr{A}_{2,k}} |\mathscr{A}'_{1,k}(\tau)|, \; |\mathscr{A}_{2,k}| \leqslant \sum_{\tau \in \mathscr{A}_{2,k}} |\mathscr{A}'_{1,k}(\tau)|$$

and summing over $K = 2^k$ it follows that

(8.47)

$$\sum_{\tau \in \mathscr{A}_{2,k}} |\mathscr{A}'_{1,k}(\tau)| \ll T^{-1/2}\log T \sum_K K^{3/2}e^{-G^2K/T} + G^{-1}\log^2 T \sum_K Ke^{-G^2K/T}$$

$$+ J^{\varkappa/2}T^{-(\varkappa+1)/4}\log^{1/2}T|\mathscr{A}_{2,k}| \sum_K K^{(2\lambda-\varkappa+1)/4}e^{-G^2K/(2T)}$$

$$\ll TG^{-3}\log^2 T + |\mathscr{A}_{2,k}|J^{\varkappa/2}G^{(\varkappa-1-2\lambda)/2}T^{(\lambda-\varkappa)/2}\log^{1/2}T.$$

Therefore using (8.46) we obtain

$$(8.48) \qquad |\mathscr{A}_{1,k}| \leqslant \sum_{\tau \in \mathscr{A}_{2,k}} |\mathscr{A}'_{1,k}(\tau)| \ll TG^{-3}\log^2 T,$$

provided that for some suitable $C_3 > 0$ we have

$$(8.49) \qquad J \leqslant C_3 G^{(2\lambda-\varkappa+1)/\varkappa}T^{(\varkappa-\lambda)/\varkappa}\log^{-1/\varkappa}T.$$

Now we choose N in such a way that

$$(8.50) \quad J = T/(6N) \leqslant C_3 T^{(\varkappa-\lambda)/\varkappa}G^{(2\lambda-\varkappa+1)/\varkappa}\log^{-1/\varkappa}T < T/(6N-6).$$

This gives

$$(8.51) \qquad\qquad N \ll 1 + T^{\lambda/\varkappa}G^{-(2\lambda-\varkappa+1)/\varkappa}\log^{1/\varkappa}T$$

and

$$(8.52) \quad |\mathscr{A}_3| = \sum_{k \leqslant N} |\mathscr{A}_{1,k}| \ll NTG^{-3}\log^2 T$$

$$\ll TV^{-6}\log^8 T + T^{(\varkappa+\lambda)/\varkappa}V^{-2(1+2\varkappa+2\lambda)/\varkappa}(\log T)^{(3+6\varkappa+4\lambda)/\varkappa}$$

if $G \leqslant J$. This condition is certainly satisfied for

$$(8.53) \qquad\qquad G \leqslant C_4 T^{(\varkappa-\lambda)/\varkappa}G^{(2\lambda-\varkappa+1)/\varkappa}\log^{-1/\varkappa}T,$$

or in view of (8.40) for

$$(8.54) \qquad V > T_1 = C_5 T^{(\lambda-\varkappa)/(2+4\lambda-4\varkappa)}(\log T)^{(3-4\varkappa+4\lambda)/(2+4\lambda-4\varkappa)},$$

where $C_4, C_5 > 0$. Summing over intervals of the form $[T(5/4)^{-j-1}, T(5/4)^{-j}]$ it is seen from (8.52) that Theorem 8.2 follows if (8.54) is satisfied. If (8.54) does not hold, then (8.28) follows from (8.26), since

$$R \ll TV^{-4}\log^5 T \ll T^{(\varkappa+\lambda)/\varkappa}V^{-2(1+2\varkappa+2\lambda)/\varkappa}(\log T)^{(3+6\varkappa+4\lambda)/\varkappa}$$

for $V < T^{\lambda/(2+4\lambda)}\log^{C_6}T = T_2$. But for T_1 given by (8.54) we have $T_1 < T_2$ for any fixed $C_6 > 0$, $\varkappa > 0$ and T sufficiently large, which completes the proof of Theorem 8.2.

The proof of Theorem 8.2 relied on Lemma 8.1, whose proof necessitated the use of the Halász–Montgomery inequality (A.39) (or some of its equivalents). However Theorem 8.2 can be proved without the use of (A.39), but only with the Cauchy–Schwarz inequality. The crucial step in the proof of Theorem 8.2 consists of bounding

$$(8.55) \qquad \sum_{\tau_r \in \mathscr{A}_{2,k}} \int_{\tau_r - G}^{\tau_r + G} \left|\zeta\left(\tfrac{1}{2} + it\right)\right|^2 dt.$$

The integral in (8.55) can be majorized by (7.49), which is very suitable since it does not contain absolute value signs. Thus using (7.49) and removing $\exp(-(G \text{ ar sinh}\sqrt{\pi n/2t_r})^2)$ by partial summation, we see in view of (7.72) that the expression in (8.55) is majorized by $C|\mathscr{A}_{2,k}|G\log T$ plus the largest of sums of the type

$$S = CG\left| \sum_{\tau_r \in \mathscr{A}_{2,k}} \sum_{K < n \leqslant 2K} (-1)^n d(n)(n\tau_r)^{-1/4}\sin f(\tau_r, n)\right|$$

$$\ll G\left(\sum_{K < n \leqslant 2K} d^2(n)n^{-1/2}\right)^{1/2}\left(\sum_{K < n \leqslant 2K}\left| \sum_{\tau_r \in \mathscr{A}_{2,k}} \tau_r^{-1/4}\exp(if(\tau_r, n))\right|^2\right)^{1/2}$$

$$\ll GK^{1/4}\log^{3/2}T\left(\sum_{K < n \leqslant 2K} \sum_{\tau_r, \tau_s} (\tau_r\tau_s)^{-1/4}\exp(if(\tau_r, n) - if(\tau_s, n))\right)^{1/2}$$

$$\ll GT^{-1/4}K^{1/4}\log^{3/2}T$$

$$\times\left(|\mathscr{A}_{2,k}|K + \sum_{\tau_r \neq \tau_s \in \mathscr{A}_{2,k}}\left| \sum_{K < n \leqslant 2K}\exp(if(\tau_r, n) - if(\tau_s, n))\right|\right)^{1/2}.$$

Here $K \leqslant TG^{-2}\log^2 T$, and the last expression corresponds to the right-hand side of (8.37) and henceforth the estimation would be identical with the one already carried out in the proof of Lemma 8.1. Note that in the foregoing argument we could invert the order of summation before applying the Cauchy–Schwarz inequality because there were no absolute value signs in (7.49).

Corollary 8.1 follows from Theorem 8.2 with the exponent pairs $(\varkappa, \lambda) = (\frac{13}{31}, \frac{16}{31}), (\frac{4}{11}, \frac{6}{11}), (\frac{2}{7}, \frac{4}{7}), (\frac{5}{24}, \frac{15}{24}), (\frac{4}{18}, \frac{11}{18})$, respectively, while Corollary 8.2 follows with $(\varkappa, \lambda) = (\frac{1}{2}, \frac{1}{2})$. Actually (8.33) gives

$$(8.56) \qquad R \ll \begin{cases} TV^{-6}\log^8 T & \text{if } V \geqslant T^{11/72}\log^{5/4}T, \\ T^{15/4}V^{-24}\log^{61/2}T & \text{if } V \leqslant T^{11/72}\log^{5/4}T. \end{cases}$$

Here $\frac{11}{72} = 0.1527\ldots$, and therefore the sixth moment estimate $R \ll TV^{-6}\log^8 T$ holds for relatively large values of V, and (8.28) tells us that in a certain sense either $M(6) = 1$ or $M((2 + 4\varkappa + 4\lambda)/\varkappa) \leqslant (\varkappa + \lambda)/\varkappa$ holds, which will be used in proving zero-density estimates in Chapter 11.

Theorem 8.2 enables us to obtain power moment estimates when $\sigma = \frac{1}{2}$, and we shall prove

THEOREM 8.3. If $A \geqslant 4$ is a fixed number and $M(A)$ is defined by (8.1), then

$$(8.57) \quad M(A) \leqslant \begin{cases} 1 + (A - 4)/8, & 4 \leqslant A \leqslant 12, \\ 2 + 3(A - 12)/22, & 12 \leqslant A \leqslant \frac{178}{13} = 13.6923\ldots, \\ 1 + 35(A - 6)/216, & A \geqslant \frac{178}{13}. \end{cases}$$

Proof of Theorem 8.3. If the spacing condition (8.24) is satisfied, then by (8.26) and (8.34) we have, for $0 \leqslant \alpha \leqslant 1$,

$$R = R^\alpha R^{1-\alpha} \ll (TV^{-4}\log^5 T)^\alpha (T^2 V^{-12}\log^{16} T)^{1-\alpha}$$

$$= T^{2-\alpha}V^{-(12-8\alpha)}(\log T)^{16-11\alpha} \leqslant T^{(4+A)/8+\varepsilon}V^{-A}$$

for $\alpha = (12 - A)/8$. Hence the first estimate in (8.57) follows from the equivalence of (8.6) and (8.9).

The second estimate in (8.57) is obtained analogously by using (8.34) and (8.29). Namely, we have

$$R = R^\alpha R^{1-\alpha} \ll T^\varepsilon (T^2 V^{-12})^\alpha (T^{29/13}V^{-178/13})^{1-\alpha}$$

$$= T^{(29-3\alpha)/13+\varepsilon}V^{-(178-22\alpha)/13} = T^{2+3(A-12)/22+\varepsilon}V^{-A}$$

for $\alpha = (178 - 13A)/22$.

Finally, the third estimate in (8.57) will follow from a more general result, namely

$$(8.58) \quad S = \sum_{r \leqslant R} \left| \zeta(\tfrac{1}{2} + it_r) \right|^A \ll T^{c(A-6)+1+\varepsilon}, \qquad A \geqslant \tfrac{178}{13}, \qquad c \geqslant \tfrac{4}{25},$$

where the t_r's satisfy (8.24) and $\zeta(\frac{1}{2} + it) \ll |t|^{c+\varepsilon}$, so that by (7.63) the value $c = \frac{35}{216}$ leads to $M(A) \leqslant 1 + 35(A - 6)/216$. To see that (8.58) holds note that each t_r satisfies

$$(8.59) \qquad V \leqslant |\zeta(\tfrac{1}{2} + it_r)| \leqslant 2V$$

for some $0 < V = 2^k \ll T^{35/216}\log^{3/2}T$. Write now $S = S_1 + S_2$, where in S_1 we consider those t_r's for which (8.59) holds with $V \geqslant T^{4/25}$, and in S_2 we consider those t_r's for which (8.59) holds with $V < T^{4/25}$. Denoting by R_V the number of corresponding t_r's in each case, then for S_1 we have by (8.56) $R_V \ll T^{1+\varepsilon}V^{-6}$, so that summing over $O(\log T)$ values of V we obtain

$$(8.60) \quad S_1 \ll \sum_{V=2^k \geqslant T^{4/25}} R_V V^A \ll T^{1+\varepsilon} \sum_{V=2^k \geqslant T^{4/25}} V^{(A-6)} \ll T^{c(A-6)+1+\varepsilon}.$$

In S_2 we have $R_V \ll T^{29/13+\varepsilon}V^{-178/13}$ by (8.29), hence

$$(8.61) \quad S_2 \ll \sum_{V=2^k < T^{4/25}} R_V V^A \ll T^{29/13+\varepsilon} \sum_{V=2^k < T^{4/25}} V^{(A-178/13)}$$

$$\ll T^{(A-178/13)c+29/13+\varepsilon} \ll T^{c(A-6)+1+\varepsilon},$$

provided that $c \geqslant 4/25$ holds. Combining (8.60) and (8.61) we obtain (8.58), which completes the proof of Theorem 8.3.

8.4 POWER MOMENTS FOR $\frac{1}{2} < \sigma < 1$

In this section we suppose that $\frac{1}{2} < \sigma < 1$ is fixed, and we consider power moments of the type

$$\int_1^T |\zeta(\sigma + it)|^{m(\sigma)} \, dt \ll T^{1+\varepsilon},$$

which in view of the discussion made in Section 8.1 will follow from large value estimates of the type

$$(8.62) \qquad R \ll T^{1+\varepsilon}V^{-m(\sigma)},$$

where $t_1 < \cdots < t_R$ satisfy (8.5) with $C = 4$ and (8.7) holds.
    Our starting point is the relation

$$(8.63) \quad \sum_{n=1}^{\infty} d_k(n)e^{-n/Y}n^{-s} = (2\pi i)^{-1}\int_{2-i\infty}^{2+i\infty} Y^w\Gamma(w)\zeta^k(s + w)\, dw,$$

which is just (4.60) with $h = 1$. We shall need (8.63) with $k = 1$ or $k = 2$, and

$Y = Y(r)$ will be a real number (to be suitably chosen) which satisfies $1 \ll Y \ll T^C$. For s we take $s = \sigma + it_r$, where t_r satisfies (8.5) and (8.7). Moving the line integration in (8.63) to $\operatorname{Re} w = \frac{1}{2} - \sigma$, we encounter a pole of order k at $w = 1 - s$ with residue $O(1)$ in view of (A.34), and a simple pole at $w = 0$ with residue $\zeta^k(s)$. Therefore

$$(8.64) \quad \sum_{n \leqslant Y} d_k(n) e^{-n/Y} n^{-s} = \zeta^k(s) + O(1)$$

$$+ (2\pi i)^{-1} \int_{\operatorname{Re} w = 1/2 - \sigma} \zeta^k(s + w) \Gamma(w) Y^w dw.$$

The portion of the integral in (8.64) for which $|\operatorname{Im} w| \geqslant \log^2 T$ is $o(1)$ as $T \to \infty$ by (A.34), and so for each t_r under consideration we have

$$(8.65) \quad \zeta^k(\sigma + it_r) \ll 1 + \left| \sum_{n \leqslant Y} d_k(n) e^{-n/Y} n^{-\sigma - it_r} \right|$$

$$+ \int_{-\log^2 T}^{\log^2 T} \left| \zeta\left(\tfrac{1}{2} + it_r + iv\right) \right|^k Y^{1/2 - \sigma} e^{-|v|} dv.$$

Taking into account (8.7) it is seen that (8.65) implies either

$$(8.66) \quad V^k \ll \left| \sum_{n \leqslant Y} d_k(n) e^{-n/Y} n^{-\sigma - it_r} \right|$$

$$\ll \log T \max_{M \leqslant Y/2} \left| \sum_{M < n \leqslant 2M} d_k(n) e^{-n/Y} n^{-\sigma - it_r} \right|$$

or

$$(8.67) \quad V^k \ll Y^{1/2 - \sigma} \left| \zeta\left(\tfrac{1}{2} + it_r'\right) \right|^k,$$

where $V \geqslant T^\varepsilon$ and t_r' is defined by

$$(8.68) \quad \left| \zeta\left(\tfrac{1}{2} + it_r'\right) \right| = \max_{-\log^2 T \leqslant v \leqslant \log^2 T} \left| \zeta\left(\tfrac{1}{2} + it_r + iv\right) \right|.$$

This discussion shows that the estimation of R in (8.62) may be reduced to a large values estimate for Dirichlet polynomials which satisfy (8.66), and a large values estimate for (8.68), which is in fact furnished by Theorem 8.2 and its corollaries. Therefore before we formulate our results concerning bounds for $m(\sigma)$, it will be useful to derive a large values estimate for Dirichlet polynomials capable of dealing with the t_r's satisfying (8.66). This estimate is contained in

LEMMA 8.2. Let $t_1 < \cdots < t_R$ be real numbers such that $T \leqslant t_r \leqslant 2T$ for $r = 1, \ldots, R$ and $|t_r - t_s| \geqslant \log^4 T$ for $1 \leqslant r \neq s \leqslant R$. If

$$(8.69) \qquad T^\varepsilon < V \leqslant \left| \sum_{M < n \leqslant 2M} a(n) n^{-\sigma - it_r} \right|,$$

where $a(n) \ll M^\varepsilon$ for $M \leqslant n \leqslant 2M$, $1 \ll M \ll T^C$ $(C > 0)$, then

$$(8.70) \qquad R \ll T^\varepsilon (M^{2-2\sigma} V^{-2} + TV^{-f(\sigma)}),$$

where

$$(8.71) \qquad f(\sigma) = 2/(3 - 4\sigma) \qquad \text{for } \tfrac{1}{2} < \sigma \leqslant \tfrac{2}{3},$$

$$f(\sigma) = 10/(7 - 8\sigma) \qquad \text{for } \tfrac{2}{3} \leqslant \sigma \leqslant \tfrac{11}{14},$$

$$f(\sigma) = 34/(15 - 16\sigma) \qquad \text{for } \tfrac{11}{14} \leqslant \sigma \leqslant \tfrac{13}{15},$$

$$f(\sigma) = 98/(31 - 32\sigma) \qquad \text{for } \tfrac{13}{15} \leqslant \sigma \leqslant \tfrac{57}{62},$$

$$f(\sigma) = 5/(1 - \sigma) \qquad \text{for } \tfrac{57}{62} \leqslant \sigma \leqslant 1 - \varepsilon.$$

Proof of Lemma 8.2. The expected bound in (8.70) is $R \ll T^\varepsilon M^{2-2\sigma} V^{-2}$, and $TV^{-f(\sigma)}$ is the extra term which may be thought of as an error term. We start from the inequality (A.39), taking $\xi = \{\xi_n\}_{n=1}^\infty$, where $\xi_n = a(n)b^{-1/2}(n)n^{-\sigma}$ for $M < n \leqslant 2M$ and zero otherwise, and $\varphi_r = \{\varphi_{r,n}\}_{n=1}^\infty$, where $\varphi_{r,n} = b^{1/2}(n)n^{-it_r}$, $b(n) = e^{-(n/2M)^h} - e^{-(n/M)^h}$, and $h = \log^2 T$. Then by (A.43) $(\varphi_r, \varphi_s) = H(it_r - it_s)$, where

$$(8.72) \quad H(it) = \sum_{n=1}^\infty b(n) n^{-it}$$

$$= (2\pi i)^{-1} \int_{2-i\infty}^{2+i\infty} \zeta(w + it) \Gamma\left(1 + \frac{w}{h}\right) ((2M)^w - M^w) \frac{dw}{w},$$

which follows from (4.60) with $k = 1$, $Y = 2M$, and $Y = M$, respectively on subtracting. Note that for $M < n \leqslant 2M$ we have $1 \ll b(n) \ll 1$, $H(0) \ll M$, $\|\xi\|^2 \ll T^\varepsilon M^{1-2\sigma}$, and the integrand in (8.72) is regular for $\operatorname{Re} w > -h$, except for a simple pole at $w = 1 - it$ with residue $O(T^{-c})$ for any fixed $c > 0$ if $|t| \gg \log^3 T$.

From estimates given in Section 7.5 of Chapter 7 we have $\mu(\tfrac{1}{2}) < \tfrac{1}{6}$, $\mu(\tfrac{5}{7}) \leqslant \tfrac{1}{14}$ [(7.59) with $l = 4$], $\mu(\tfrac{5}{6}) \leqslant \tfrac{1}{30}$ [(7.59) with $l = 5$], so that by convexity of

$\mu(\sigma)$ this implies that we may take

$$(8.73) \qquad c(\theta) = \tfrac{1}{2} - \theta \qquad\qquad \text{for} \quad \theta \leqslant 0,$$

$$c(\theta) = (3 - 4\theta)/6 \qquad \text{for} \quad 0 \leqslant \theta \leqslant \tfrac{1}{2},$$

$$c(\theta) = (7 - 8\theta)/18 \qquad \text{for} \quad \tfrac{1}{2} \leqslant \theta \leqslant \tfrac{5}{7},$$

$$c(\theta) = (15 - 16\theta)/50 \quad \text{for} \quad \tfrac{5}{7} \leqslant \theta \leqslant \tfrac{5}{6},$$

$$c(\theta) = (1 - \theta)/5 \qquad \text{for} \quad \tfrac{5}{6} \leqslant \theta \leqslant 1,$$

where $c(\sigma)$ is an upper bound for $\mu(\sigma)$ in $(-\infty, \infty)$.

To estimate $H(it)$ in (8.72) we move the line of integration to $\operatorname{Re} w = \theta$, where

$$(8.74) \qquad \theta = (3\sigma - 2)/(2\sigma - 1) \qquad \text{for} \quad \tfrac{1}{2} < \sigma_0 \leqslant \sigma \leqslant \tfrac{2}{3},$$

$$\theta = (9\sigma - 6)/(4\sigma - 1) \qquad \text{for} \quad \tfrac{2}{3} \leqslant \sigma \leqslant \tfrac{11}{14},$$

$$\theta = (25\sigma - 16)/(8\sigma + 1) \qquad \text{for} \quad \tfrac{11}{14} \leqslant \sigma \leqslant \tfrac{13}{15},$$

$$\theta = (65\sigma - 40)/(16\sigma + 9) \qquad \text{for} \quad \tfrac{13}{15} \leqslant \sigma \leqslant \tfrac{57}{62},$$

$$\theta = (12\sigma - 7)/(2\sigma + 3) \qquad \text{for} \quad \tfrac{57}{62} \leqslant \sigma \leqslant 1 - \varepsilon,$$

so that the values of θ lie in the ranges $\theta \leqslant 0$, $0 \leqslant \theta \leqslant \tfrac{1}{2}$, $\tfrac{1}{2} \leqslant \theta \leqslant \tfrac{5}{7}$, $\tfrac{5}{7} \leqslant \theta \leqslant \tfrac{5}{6}$, $\tfrac{5}{6} \leqslant \theta \leqslant 1$, respectively and therefore (8.73) may be used. Hence with (8.73) and (A.34) we obtain, for $r \neq s$,

$$(8.75) \quad H(it_r - it_s) \ll T^\varepsilon \int_{-h^2}^{h^2} |\zeta(\theta + iv + it_r - it_s)| e^{-|v|/h} M^\theta \, dv + o(1)$$

$$\ll T^{c(\theta)+\varepsilon} M^\theta + o(1).$$

Therefore (A.39) gives

$$(8.76) \qquad R^2 V^2 \ll \sum_{M < n \leqslant 2M} |a(n)|^2 n^{-2\sigma} \left(\sum_{r,s} |H(it_r - it_s)| \right)$$

$$\ll T^\varepsilon M^{1-2\sigma} \left(RM + \sum_{r \neq s} |H(it_r - it_s)| \right),$$

and (8.75) leads to

$$(8.77) \quad R \ll T^\varepsilon (M^{2-2\sigma} V^{-2} + RM^{\theta+1-2\sigma} T^{c(\theta)} V^{-2}) \ll T^\varepsilon M^{2-2\sigma} V^{-2},$$

provided that

$$(8.78) \qquad T = T_0 = V^{(2-\varepsilon)/c(\theta)} M^{(2\sigma-1-\theta)/c(\theta)},$$

since $V > T^\varepsilon$ by hypothesis. If (8.77) is not satisfied it may be observed that if in (8.69) t_r is replaced by $t_r + T_0$ for any fixed T_0, then $a(n)$ is replaced by $a_0(n) = a(n)n^{-iT_0}$, and $|a_0(n)| = |a(n)| \ll M^\varepsilon$. Hence if the t_r's lie in an interval of length not exceeding T_0, then $R \ll T^\varepsilon M^{2-2\sigma}V^{-2}$, and dividing T into subintervals of length at most T_0 [where T_0 is given by (8.78)] we obtain

$$(8.79) \quad R \ll T^\varepsilon M^{2-2\sigma}V^{-2}(1 + T/T_0)$$

$$\ll T^\varepsilon\left(M^{2-2\sigma}V^{-2} + TM^{\{2c(\theta)+1+\theta-2\sigma(1+c(\theta))\}/c(\theta)}V^{-2(1+c(\theta))/c(\theta)}\right).$$

With $c(\theta)$ and θ given by (8.73) and (8.74) it is readily checked that

$$2c(\theta) + 1 + \theta - 2(1 + c(\theta))\sigma = 0, \quad 2(1 + c(\theta))/c(\theta) = f(\sigma),$$

where $f(\sigma)$ is given by (8.71), and thus (8.70) follows.

Having Theorem 8.2 and Lemma 8.2 at our disposal we are now ready to state and prove the main result of this section, which is

THEOREM 8.4. Let $m(\sigma)$ for each fixed $\frac{1}{2} < \sigma < 1$ be defined by (8.2). Then

(8.80)

$$
\begin{array}{lll}
m(\sigma) \geqslant 4/(3 - 4\sigma) & \text{for} & \frac{1}{2} < \sigma \leqslant \frac{5}{8}, \\[4pt]
m(\sigma) \geqslant 10/(5 - 6\sigma) & \text{for} & \frac{5}{8} \leqslant \sigma \leqslant \frac{35}{54}, \\[4pt]
m(\sigma) \geqslant 19/(6 - 6\sigma) & \text{for} & \frac{35}{54} \leqslant \sigma \leqslant \frac{41}{60}, \\[4pt]
m(\sigma) \geqslant 2112/(859 - 948\sigma) & \text{for} & \frac{41}{60} \leqslant \sigma \leqslant \frac{3}{4}, \\[4pt]
m(\sigma) \geqslant 12408/(4537 - 4890\sigma) & \text{for} & \frac{3}{4} \leqslant \sigma \leqslant \frac{5}{6}, \\[4pt]
m(\sigma) \geqslant 4324/(1031 - 1044\sigma) & \text{for} & \frac{5}{6} \leqslant \sigma \leqslant \frac{7}{8}, \\[4pt]
m(\sigma) \geqslant 98/(31 - 32\sigma) & \text{for} & \frac{7}{8} \leqslant \sigma \leqslant 0.91591\ldots, \\[4pt]
m(\sigma) \geqslant (24\sigma - 9)/(4\sigma - 1)(1 - \sigma) & \text{for} & 0.91591\cdots \leqslant \sigma \leqslant 1 - \varepsilon.
\end{array}
$$

In addition to this we have $m(\frac{2}{3}) \geqslant 9.6187\ldots$, $m(\frac{7}{10}) \geqslant 11$, and $m(\frac{5}{7}) \geqslant 12$.

Proof of Theorem 8.4. We begin the proof by considering first the range $\frac{1}{2} < \sigma \leqslant \frac{5}{8}$, when it is sufficient to prove

$$(8.81) \quad R \ll T^{1+\varepsilon}V^{-4/(3-4\sigma)}$$

for the number of points t_r which satisfy (8.5) and (8.7) [by (8.26) the bound in (8.81) holds for $\sigma = \frac{1}{2}$ also]. To simplify writing we shall omit in the rest of this proof factors like $T^\varepsilon \log^c T$ on right-hand sides of inequalities implied by $\ll$.

To obtain (8.81) we consider separately subsets $\mathscr{A}$ and $\mathscr{B}$ of $\{t_r\}$ such that $t_r \in \mathscr{A}$ if V in (8.7) satisfies $V \leqslant T^{(3-4\sigma)/8}$ and $t_r \in \mathscr{B}$ if $V > T^{(3-4\sigma)/8}$. If $R_1 = |\mathscr{A}|$, $R_2 = |\mathscr{B}|$, then $R = R_1 + R_2$ and

$$(8.82) \quad R_1 \ll Y_1^{2-2\sigma}V^{-4} + TV^{-4/(3-4\sigma)} + Y_1^{1/2-\sigma}V^{-2} \sum_{t_r \in \mathscr{A}} \left| \varsigma\left(\tfrac{1}{2} + it_r'\right)\right|^2,$$

which follows from (8.66) and (8.67) with $k = 2$ when one applies Lemma 8.2. Here M ($M \leqslant Y/2 = Y_1/2$) is chosen in such a way that $\gg R_1/\log T$ numbers $t_r \in \mathscr{A}$ satisfy (8.66) with that particular M. Using $M(4) = 1$ and the Cauchy–Schwarz inequality we have

$$(8.83) \qquad R_1 \ll Y_1^{2-2\sigma}V^{-4} + TV^{-4/(3-4\sigma)} + TV^{-4}Y_1^{1-2\sigma},$$

and in view of $V \leqslant T^{(3-4\sigma)/8}$ the choice $Y_1 = T$ gives

$$(8.84) \qquad R_1 \ll TV^{-4/(3-4\sigma)} + T^{2-2\sigma}V^{-4} \ll TV^{-4/(3-4\sigma)}.$$

To bound R_2 we reason analogously, only now we use Hölder's inequality to obtain

$$R_2 \ll Y_2^{2-2\sigma}V^{-4} + TV^{-4/(3-4\sigma)} + Y_2^{1/2-\sigma}R_2^{5/6}V^{-2}\left(\sum_{t_r \in \mathscr{B}} \left| \varsigma\left(\tfrac{1}{2} + it_r'\right)\right|^{12} \right)^{1/6},$$

whence using $M(12) \leqslant 2$ and simplifying we have

$$(8.85) \qquad R_2 \ll Y_2^{2-2\sigma}V^{-4} + TV^{-4/(3-4\sigma)} + Y_2^{3-6\sigma}T^2V^{-12}.$$

Choosing $Y_2 = T^{2/(4\sigma-1)}V^{-8/(4\sigma-1)} \gg 1$ it follows from (8.85) that

$$(8.86) \qquad R_2 \ll TV^{-4/(3-4\sigma)} + T^{(4-4\sigma)/(4\sigma-1)}V^{-12/(4\sigma-1)}.$$

The second term on the right-hand side of (8.86) does not exceed the first if

$$T^{(5-8\sigma)/(4\sigma-1)} \leqslant V^{8(5-8\sigma)/(4\sigma-1)(3-4\sigma)},$$

and this condition is satisfied since $\tfrac{1}{2} < \sigma \leqslant \tfrac{5}{8}$ and $V > T^{(3-4\sigma)/8}$. Thus from (8.84) and (8.86) we obtain (8.81), which implies the first estimate in (8.80).

Except for the last two bounds in (8.80), all the other bounds will follow from Theorem 8.1 and the bounds

$$(8.87) \quad m\left(\tfrac{35}{54}\right) \geqslant 9, \qquad m\left(\tfrac{41}{60}\right) \geqslant 10, \qquad m\left(\tfrac{3}{4}\right) \geqslant \tfrac{528}{37} = 14.27027\ldots,$$

$$m\left(\tfrac{5}{6}\right) \geqslant \tfrac{188}{7} = 26.85714\ldots, \qquad m\left(\tfrac{7}{8}\right) \geqslant \tfrac{184}{5} = 36.8.$$

The bounds in (8.87), and consequently the corresponding bounds in Theorem 8.4, can be somewhat improved by a more elaborate choice of the exponent pairs which will appear in the course of the proof.

We consider now the range $\frac{5}{8} \leqslant \sigma \leqslant \frac{2}{3}$, and this time let $\mathscr{A}$ and $\mathscr{B}$ denote subsets of $\{t'_r\}$ [see (8.62) and (8.68)] such that in (8.28)

$$(8.88) \qquad R \ll TV^{-6}$$

and

$$(8.89) \qquad R \ll T^{(\varkappa+\lambda)/\varkappa} V^{-2(1+2\varkappa+2\lambda)/\varkappa}$$

hold respectively for $R = R_1 = |\mathscr{A}|$ and $R = R_2 = |\mathscr{B}|$. In applying (8.88) and (8.89) we have to replace V by $VY^{(2\sigma-1)/4}$ in view of (8.67) with $k = 2$. Therefore we have

$$R_1 \ll Y_1^{2-2\sigma} V^{-4} + TV^{-4/(3-4\sigma)} + Y_1^{(3-6\sigma)/2} TV^{-6},$$

where Lemma 8.2 was used again. With the choice $Y_1 = (TV^{-2})^{2/(1+2\sigma)} \gg 1$ the above estimate becomes

$$(8.90) \qquad R_1 \ll TV^{-4/(3-4\sigma)} + T^{4(1-\sigma)/(1+2\sigma)} V^{-12/(1+2\sigma)}.$$

Analogously using (8.89) it follows that

$$(8.91) \qquad R_2 \ll Y_2^{2-2\sigma} V^{-4} + TV^{-4/(3-4\sigma)}$$

$$+ T^{(\varkappa+\lambda)/\varkappa} V^{-2(1+2\varkappa+2\lambda)/\varkappa} Y_2^{(1/2-\sigma)(1+2\varkappa+2\lambda)/\varkappa},$$

and we shall choose Y_2 to satisfy

$$Y_2 = T^{2(\varkappa+\lambda)/((2+4\lambda)\sigma-1+2\varkappa-2\lambda)} V^{-4(1+2\lambda)/((2+4\lambda)\sigma-1+2\varkappa-2\lambda)},$$

so that the first and the third term on the right-hand side of (8.91) are equal. Since by (8.73) we have $c(\theta) = \frac{1}{8}$ for $\theta \geqslant \frac{5}{8}$ and $2(\varkappa + \lambda)/4(1 + 2\lambda) \geqslant \frac{1}{8}$ the condition $Y_2 \gg 1$ will be satisfied, and hence from (8.90) and (8.91) we have

$$(8.92) \quad R \ll TV^{-4/(3-4\sigma)} + T^{(4-4\sigma)/(1+2\sigma)} V^{-12/(1+2\sigma)}$$

$$+ T^{4(1-\sigma)(\varkappa+\lambda)/((2+4\lambda)\sigma-1+2\varkappa-2\lambda)} V^{-4(1+2\varkappa+2\lambda)/((2+4\lambda)\sigma-1+2\varkappa-2\lambda)}.$$

The exponent of T of the last term in (8.92) equals unity for

$$(8.93) \qquad \sigma = \frac{1 + 2\varkappa + 6\lambda}{2 + 4\varkappa + 8\lambda},$$

which gives

(8.94) $R \ll TV^{-4/(3-4\sigma)} + T^{(4-4\sigma)/(1+2\sigma)}V^{-12/(1+2\sigma)} + TV^{-F}$,

where

$$F = \frac{2(1 + 2\varkappa + 2\lambda)(1 + 2\varkappa + 4\lambda)}{\varkappa + \lambda + 4\varkappa\lambda + 2\varkappa^2 + 2\lambda^2}.$$

In (8.94) the term TV^{-F} is the largest, which will be shown now for $\sigma = \frac{2}{3}$. In that case (8.93) reduces to $\lambda - \varkappa = \frac{1}{2}$, and therefore with the exponent pair $(\varkappa, \lambda) = (\alpha/2 + \varepsilon, \; \alpha/2 + \frac{1}{2} + \varepsilon)$, $\alpha = 0.3290213568\ldots$ one obtains from (8.94)

(8.95) $R \ll TV^{-9.61872\cdots} + (TV^{-9})^{4/7}$,

and one has $(TV^{-9})^{4/7} \leqslant TV^{-\varkappa}$ for

(8.96) $V \leqslant T^{3/(7\varkappa - 36)}$.

Since by (8.73) one has $c(\frac{2}{3}) = \frac{5}{54}$, it is seen that (8.96) is certainly satisfied for $\frac{5}{54} \leqslant 3/(7\varkappa - 36)$, or $\varkappa \leqslant \frac{342}{35} = 9.7714\ldots$. This proves $m(\frac{2}{3}) \geqslant 9.61872\ldots$, which is actually the optimal bound this method allows. With $(\varkappa, \lambda) = (\frac{2}{7}, \frac{4}{7})$ in (8.93) we obtain $\sigma = \frac{35}{54} = 0.6481481\ldots$, and a calculation similar to the one above gives $m(\frac{35}{54}) \geqslant 9$. The foregoing procedure may be also used when $\sigma \geqslant \frac{2}{3}$, only in view of Lemma 8.2 the first term in (8.92) is to be replaced by $TV^{-2f(\sigma)}$, namely, we obtain

(8.97) $R \ll TV^{-2f(\sigma)} + T^{(4-4\sigma)/(1+2\sigma)}V^{-12/(1+2\sigma)}$

$$+ \; T^{4(1-\sigma)(\varkappa+\lambda)/((2+4\lambda)\sigma-1+2\varkappa-2\lambda)}V^{-4(1+2\varkappa+2\lambda)/((2+4\lambda)\sigma-1+2\varkappa-2\lambda)}.$$

Calculations for $\sigma \geqslant \frac{2}{3}$ are carried out in the manner described above. The term $TV^{-2f(\sigma)}$ is always the smallest one, and the second and third term in (8.97) do not exceed $TV^{-\varkappa}$ and TV^{-y}, respectively, for values of $\varkappa$ and y which will depend on $c(\theta)$, where for $c(\theta)$ we use the values given by (8.73). With the exponent pair $(\varkappa, \lambda) = (\frac{1}{14}, \frac{11}{14})$ the last term in (8.97) is TV^{-10} for $\sigma = \frac{41}{60} = 0.68333\ldots$, and since then the other two terms in (8.97) are smaller, we obtain $m(\frac{41}{60}) \geqslant 10$. Using $(\varkappa, \lambda) = (\frac{2}{7}, \frac{4}{7})$ and $\sigma = \frac{7}{10}$, $\sigma = \frac{5}{7}$ we obtain likewise $m(\frac{7}{10}) \geqslant 11$ and $m(\frac{5}{7}) \geqslant 12$, respectively. For $\sigma \geqslant \frac{3}{4}$ we have from (8.97) that the first and the third term are $\ll TV^{-\varkappa}$ for

(8.98) $$\varkappa \leqslant \frac{8(3 + 6\varkappa + 2\lambda)}{1 + 4\varkappa + 2\lambda},$$

where we used $c(\frac{3}{4}) = \frac{1}{16}$. The choice $(\varkappa, \lambda) = (\frac{5}{24}, \frac{15}{24})$ gives $\varkappa \leqslant \frac{528}{37} =$

$14.270270\ldots$, so that $m(\frac{3}{4}) \geqslant \frac{528}{37}$, since the middle term in (8.97) turns out to be $T^{2/5}V^{-24/5} \leqslant TV^{-y}$ for $y \leqslant \frac{72}{5} = 14.4$. Similarly, one obtains $m(\frac{5}{6}) \geqslant \frac{188}{7}$ and $m(\frac{7}{8}) \geqslant \frac{184}{5}$ for $(\varkappa, \lambda) = (\frac{2}{7}, \frac{4}{7})$.

To finish the proof of Theorem 8.3 it remains yet to obtain the last two bounds in (8.80). In this case we shall use (8.65) with $k = 1$, since for $\sigma \geqslant \frac{7}{8}$ the values of $f(\sigma)$ given by Lemma 8.2 are large enough for our purposes and the estimate $TV^{-f(\sigma)}$ suffices, whereas for smaller values of σ it was necessary to use $k = 2$ in (8.65), with the effect that V in Lemma 8.2 is replaced by V^2. To avoid tedious calculations we choose $(\varkappa, \lambda) = (\frac{2}{7}, \frac{4}{7})$ in (8.89), and let similarly as before $\mathscr{A}$ and $\mathscr{B}$ denote subsets of $\{t_r'\}$ for which $R \ll TV^{-6}$ and $R \ll T^3V^{-19}$, respectively, hold with $R_1 = |\mathscr{A}|$ and $R_2 = |\mathscr{B}|$. Then applying Lemma 8.2 we obtain

$$R_1 \ll Y_1^{2-2\sigma}V^{-2} + Y_1^{3-6\sigma}TV^{-6} + TV^{-f(\sigma)},$$

$$R_2 \ll Y_2^{2-2\sigma}V^{-2} + Y_2^{19(1/2-\sigma)}T^3V^{-19} + TV^{-f(\sigma)}.$$

If we choose $Y_1 = (TV^{-4})^{1/(4\sigma-1)} \gg 1$ and $Y_2 = (T^6V^{-34})^{1/(34\sigma-15)} \gg 1$, then

$$(8.99) \qquad R_1 \ll T^{(2-2\sigma)/(4\sigma-1)}V^{-6/(4\sigma-1)} + TV^{-f(\sigma)},$$

$$(8.100) \qquad R_2 \ll T^{(12-12\sigma)/(34\sigma-15)}V^{-38/(34\sigma-15)} + TV^{-f(\sigma)}.$$

With $c(\theta) = (1 - \theta)/5$ for $\theta \geqslant \frac{5}{6}$ we obtain $R_1 + R_2 \ll TV^{-x}$ for

$$(8.101) \quad x = \min\left(f(\sigma), \frac{24\sigma - 9}{(4\sigma - 1)(1 - \sigma)}, \frac{192\sigma - 97}{(34\sigma - 15)(1 - \sigma)} \right),$$

where $f(\sigma) = 98/(31 - 32\sigma)$ for $\frac{13}{15} \leqslant \sigma \leqslant \frac{57}{62} = 0.91935\ldots$ and $f(\sigma) = 5/(1 - \sigma)$ for $\frac{57}{62} \leqslant \sigma \leqslant 1 - \varepsilon$. Now for $\frac{13}{15} \leqslant \sigma \leqslant 1$ we have $(24\sigma - 9)/(4\sigma - 1) \leqslant 5$ and also the second term in (8.101) does not exceed the third. For $\frac{57}{62} \leqslant \sigma \leqslant 0.91591\ldots$ we have $(24\sigma - 9)/(4\sigma - 1)(1 - \sigma) \leqslant 98/(31 - 32\sigma) = f(\sigma)$, hence the last two bounds of the theorem follow. In particular we have

$$(8.102) \quad m(\sigma) \geqslant 4.873/(1 - \sigma), \qquad 0.91591\cdots \leqslant \sigma \leqslant 1 - \varepsilon.$$

8.5 ASYMPTOTIC FORMULAS FOR POWER MOMENTS WHEN $\frac{1}{2} < \sigma < 1$

We conclude this chapter by considering the asymptotic formula

$$(8.103) \qquad \int_1^T |\zeta(\sigma + it)|^{2k}\, dt = T \sum_{n=1}^{\infty} d_k^2(n)n^{-2\sigma} + R(k, \sigma; T),$$

where $k \geqslant 1$ is a fixed integer and $R(k, \sigma; T)$ is supposed to be an error term, that is, $R(k, \sigma; T) = o(T)$ as $T \to \infty$. This problem was considered in Theorem 1.10 for $\sigma > 1$, and now we suppose that $\frac{1}{2} < \sigma < 1$ is fixed. In Section 8.4 we consider only upper bounds of the type $T^{1+\varepsilon}$ for the integral in (8.103), and although the present problem seems much more difficult than the problem of bounding $m(\sigma)$, this is only superficially so. Indeed the bounds for $m(\sigma)$ of Section 8.4 will enable us to obtain a result of the type (8.103), namely, Theorem 8.5. To obtain this result we shall need two lemmas: a general convexity result contained in Lemma 8.3, and Lemma 8.4, which provides the range for σ (for k fixed) for which (8.103) holds. These are

LEMMA 8.3. Let $F(s)$ be regular in the region $\mathscr{D}: \alpha \leqslant \sigma \leqslant \beta$, $t \geqslant 1$ and let, for $s \in \mathscr{D}$, $F(s) \ll e^{Ct^2}$. Then for any fixed $q > 0$ and $\alpha < \gamma < \beta$ we have

$$(8.104) \quad \int_2^T |F(\gamma + it)|^q \, dt \ll \left(\int_1^{2T} |F(\alpha + it)|^q \, dt + 1 \right)^{(\beta - \gamma)/(\beta - \alpha)}$$

$$\times \left(\int_1^{2T} |F(\beta + it)|^q \, dt + 1 \right)^{(\gamma - \alpha)/(\beta - \alpha)}.$$

LEMMA 8.4. For $k \geqslant 1$ a fixed integer let $\frac{1}{2} \leqslant \sigma_k^* < 1$ denote the infimum of all numbers σ^* for which

$$\int_1^T |\zeta(\sigma^* + it)|^{2k} \, dt \ll T^{1+\varepsilon}$$

holds for any $\varepsilon > 0$. Then the asymptotic formula (8.103) holds for $\sigma_k^* < \sigma < 1$ fixed with

$$(8.105) \quad R(k, \sigma; T) \ll T^{(2 - \sigma - \sigma_k^*)/(2 - 2\sigma_k^*) + \varepsilon}.$$

Proof of Lemma 8.3. Let $\mathscr{R}$ be the rectangle with vertices $\alpha + i$, $\alpha + 2it$, $\beta + i$, $\beta + 2it$, where $2 \leqslant t \leqslant T$. If $s = \sigma + it$, $\alpha < \sigma < \beta$ and $X > 0$ is a parameter, then by the residue theorem we have

$$(8.106) \quad F(s) = (2\pi i)^{-1} \int_{\mathscr{R}} \frac{X^{z-s} e^{8C(z-s)^2} F(z)}{z - s} \, dz.$$

Since $F(s) \ll e^{Ct^2}$ it is seen that the contribution of the integrals over the horizontal sides in (8.106) is

$$\ll (X^{\alpha - \sigma} + X^{\beta - \sigma}) e^{-Ct^2}$$

uniformly in X; hence (8.106) gives

$$F(\sigma + it) \ll X^{\alpha - \sigma}\left(\int_1^{2T} e^{-8C(y-t)^2}|F(\alpha + iy)|\, dy + e^{-Ct^2}\right)$$

$$+ X^{\beta - \sigma}\left(\int_1^{2T} e^{-8C(y-t)^2}|F(\beta + iy)|\, dy + e^{-Ct^2}\right)$$

$$= X^{\alpha - \sigma}I_1 + X^{\beta - \sigma}I_2,$$

say. Therefore choosing $X = (I_1/I_2)^{1/(\beta - \alpha)}$ we obtain

(8.107) $\quad F(\sigma + it) \ll \left(\int_1^{2T} e^{-8C(y-t)^2}|F(\alpha + iy)|\, dy + e^{-Ct^2}\right)^{(\beta - \sigma)/(\beta - \alpha)}$

$$\times \left(\int_1^{2T} e^{-8C(y-t)^2}|F(\beta + iy)|\, dy + e^{-Ct^2}\right)^{(\sigma - \alpha)/(\beta - \alpha)};$$

hence raising (8.107) to the power q and applying Hölder's inequality we have

(8.108) $\quad |F(\sigma + it)|^q$

$$\ll \left(\int_1^{2T} e^{-8C(y-t)^2}|F(\alpha + iy)|^q\, dy + e^{-qCt^2}\right)^{(\beta - \sigma)/(\beta - \alpha)}$$

$$\times \left(\int_1^{2T} e^{-8C(y-t)^2}|F(\beta + iy)|^q\, dy + e^{-qCt^2}\right)^{(\sigma - \alpha)/(\beta - \alpha)}.$$

Integrating (8.108) over t for $2 \leqslant t \leqslant T$, using again Hölder's inequality and noting that

$$\int_2^T e^{-8C(y-t)^2}\, dt \ll 1, \qquad \int_2^T e^{-qCt^2}\, dt \ll 1,$$

we then obtain (8.104) with $\gamma = \sigma$.

Proof of Lemma 8.4. With $s = \sigma + it$ we have

(8.109) $\quad \displaystyle\int_1^T |\zeta(s)|^{2k}\, dt = \int_1^T \left|\sum_{n \leqslant T} d_k(n)n^{-s}\right|^2 dt$

$$+ O\left(\int_1^T \left|\zeta^{2k}(s) - \left(\sum_{n \leqslant T} d_k(n)n^{-s}\right)\right|^2 dt\right).$$

Using Theorem 5.2 it follows that

$$(8.110) \quad \int_1^T \left| \sum_{n \leqslant T} d_k(n)n^{-s} \right|^2 dt = T \sum_{n \leqslant T} d_k^2(n)n^{-2\sigma} + O\left(\sum_{n \leqslant T} d_k^2(n)^{1-2\sigma} \right)$$

$$= T \sum_{n=1}^{\infty} d_k^2(n)n^{-2\sigma} + O(T^{2-2\sigma+\varepsilon}),$$

so that the main contribution in (8.103) comes from the first term on the right-hand side of (8.109). Let now

$$F(s) = \zeta^{2k}(s) - \left(\sum_{n \leqslant T} d_k(n)n^{-s} \right)^2,$$

and apply Lemma 8.3, where we shall take $q = 1$, $\alpha = \sigma_k^* + \delta$, $\beta = 1 + \delta$, $\gamma = \sigma$, where $0 < \delta < \frac{1}{2}$ is a fixed constant which may be chosen arbitrarily small. Since k is fixed, we have then

$$\frac{\beta - \sigma}{\beta - \alpha} = \frac{1 + \delta - \sigma}{1 - \sigma_k^*} \leqslant \frac{1 - \sigma}{1 - \sigma_k^*} + \delta^{1/2},$$

and

$$\frac{\sigma - \alpha}{\beta - \alpha} = \frac{\sigma - \sigma_k^* - \delta}{1 - \sigma_k^*} \leqslant \frac{\sigma - \sigma_k^*}{1 - \sigma_k^*}.$$

Using Theorem 5.2 and the definition of σ_k^* we have

$$\int_1^{2T} |F(\alpha + it)| \, dt \leqslant \int_1^{2T} \left| \zeta\left(\sigma_k^* + \delta + it\right) \right|^{2k} dt$$

$$+ \int_1^{2T} \left| \sum_{n \leqslant T} d_k(n)n^{-\sigma_k^* - \delta - it} \right|^2 dt$$

$$\ll T^{1+\delta} + T^{2-2\sigma_k^*+\varepsilon} \ll T^{1+\delta}.$$

To estimate the integral of $|F(\beta + it)|$ in (8.104) recall that

$$\sum_{ab=n} d_k(a)d_k(b) = d_{2k}(n);$$

hence

$$F(\beta + it) = \sum_{n=1}^{\infty} d_{2k}(n)^{-1-\delta-it} - \left(\sum_{n \leqslant T} d_k(n)n^{-1-\delta-it} \right)^2$$

$$= \sum_{n > T} g_k(n)n^{-1-\delta-it},$$

where $|g_k(n)| \leqslant d_{2k}(n) \ll n^\varepsilon$. Again using Theorem 5.2 and the Cauchy–Schwarz inequality we obtain

$$\int_1^{2T} |F(\beta + it)| \, dt \ll T^{1/2} \left(\int_1^{2T} \left| \sum_{n \geqslant T} g_k(n) n^{-1-\delta-it} \right|^2 dt \right)^{1/2} \ll T^{1/2}.$$

Therefore Lemma 8.3 gives

$$\int_1^T |F(\sigma + it)| \, dt \ll T^A,$$

where

$$A = (1 + \delta)\left(\frac{1 - \sigma}{1 - \sigma_k^*} + \delta^{1/2} \right) + \frac{\sigma - \sigma_k^*}{2 - 2\sigma_k^*} \leqslant \frac{2 - \sigma - \sigma_k^*}{2 - 2\sigma_k^*} + \varepsilon$$

for any $\varepsilon > 0$ if $\delta = \delta(\varepsilon)$ is sufficiently small. Noting that for the exponent of the O-term in (8.110) we have

$$2 - 2\sigma < \frac{2 - \sigma - \sigma_k^*}{2 - 2\sigma_k^*} < 1,$$

it follows that we have proved (8.103) with the error term given by (8.105).

Recalling that by the definition of σ_k^* and Theorem 8.4 we have $\sigma_2^* = \frac{1}{2}$, $\sigma_3^* \leqslant \frac{7}{12}$, $\sigma_4^* \leqslant \frac{5}{8}$, $\sigma_5^* \leqslant \frac{41}{60}$, $\sigma_6^* \leqslant \frac{5}{7}$, we obtain immediately from (8.103) and (8.105) asymptotic formulas which we formulate as

THEOREM 8.5. For σ fixed

$$\int_1^T |\zeta(\sigma + it)|^4 \, dt = T \sum_{n=1}^{\infty} d_2^2(n) n^{-2\sigma} + O(T^{3/2-\sigma+\varepsilon}) \qquad \left(\tfrac{1}{2} < \sigma < 1\right),$$

$$\int_1^T |\zeta(\sigma + it)|^6 \, dt = T \sum_{n=1}^{\infty} d_3^2(n) n^{-2\sigma} + O(T^{(17-12\sigma)/10+\varepsilon}) \qquad \left(\tfrac{7}{12} < \sigma < 1\right),$$

$$\int_1^T |\zeta(\sigma + it)|^8 \, dt = T \sum_{n=1}^{\infty} d_4^2(n) n^{-2\sigma} + O(T^{(11-8\sigma)/6+\varepsilon}) \qquad \left(\tfrac{5}{8} < \sigma < 1\right),$$

$$\int_1^T |\zeta(\sigma + it)|^{10} \, dt = T \sum_{n=1}^{\infty} d_5^2(n) n^{-2\sigma} + O(T^{(79-60\sigma)/38+\varepsilon}) \qquad \left(\tfrac{41}{60} < \sigma < 1\right),$$

$$\int_1^T |\zeta(\sigma + it)|^{12} \, dt = T \sum_{n=1}^{\infty} d_6^2(n) n^{-2\sigma} + O(T^{(9-7\sigma)/4+\varepsilon}) \qquad \left(\tfrac{5}{7} < \sigma < 1\right).$$

Naturally explicit bounds for $R(k, \sigma; T)$ in (8.103) when $k > 6$ may be also obtained by this method; but a general formula would be rather complicated in view of Theorem 8.4, and therefore it seemed reasonable to consider (8.103) explicitly for small values of k only.

Finally, it may be mentioned that one can also investigate power moments of $|\zeta(\sigma + it)|^k$ when $k = 1$ or $k = 2$ by a similar method and obtain

(8.111)

$$\int_1^T |\zeta(\sigma + it)| \, dt = T \sum_{n=1}^\infty d_{1/2}^2(n) n^{-2\sigma} + O(T^{5/4 - \sigma/2}) \qquad (\tfrac{1}{2} < \sigma < 1),$$

and

$$(8.112) \quad \int_1^T |\zeta(\sigma + it)|^2 \, dt = \zeta(2\sigma)T + O(T^{2 - 2\sigma}) \qquad (\tfrac{1}{2} < \sigma < 1),$$

where $d_{1/2}(n)$ is the special case of the generalized divisor function $d_z(n)$ (see Section 13.6), which is defined by

$$\zeta^z(s) = \sum_{n=1}^\infty d_z(n) n^{-s} \qquad (\mathrm{Re}\, s > 1).$$

We shall indicate here the proof of (8.111), for which it is sufficient to obtain a corresponding formula for $[T/2, T]$. From Theorem 1.8 we have

$$\zeta(s) = \sum_{n \leqslant T} n^{-s} + O(T^{-\sigma}),$$

where $s = \sigma + it$, $T/2 \leqslant t \leqslant T$, and $\tfrac{1}{2} < \sigma < 1$. We may write

$$\sum_{n \leqslant T} n^{-s} = \left(\sum_{n \leqslant T^{1/2}} d_{1/2}(n) n^{-s} \right)^2 + \sum_{T^{1/2} < n \leqslant T} g(n) n^{-s},$$

where clearly $|g(n)| \leqslant 1$. Then we have

$$(8.113) \quad \int_{T/2}^T |\zeta(s)| \, dt = \int_{T/2}^T \left| \sum_{n \leqslant T^{1/2}} d_{1/2}(n) n^{-s} \right|^2 dt$$

$$+ O\left(\int_{T/2}^T \left| \sum_{T^{1/2} < n \leqslant T} g(n) n^{-s} \right| dt \right) + O(T^{1-\sigma}).$$

The first integral on the right-hand side of (8.113) equals by Theorem 5.2

$$\frac{1}{2} T \sum_{n=1}^\infty d_{1/2}^2(n) n^{-2\sigma} + O(T^{1-\sigma}),$$

while using the Cauchy–Schwarz inequality and Theorem 5.2 the second integral is

$$\ll T^{1/2}\left(\sum_{T^{1/2}<n\leqslant T} g^2(n)n^{-2\sigma}T + \sum_{T^{1/2}<n\leqslant T} g^2(n)n^{1-2\sigma}\right)^{1/2} \ll T^{5/4-\sigma/2},$$

which proves (8.111). The proof of (8.112) is analogous, and the result is by a log factor in the error term sharper than (1.76).

Notes

Power moments for $\zeta(s)$ are discussed in Chapter 7 of E. C. Titchmarsh (1951), but the results of this chapter are sharper—for example, our Theorem 8.4 is stronger than Titchmarsh's Theorem 7.10. Instead of the functions $M(A)$ and $m(\sigma)$, introduced by the author (1980b, 1983b), Titchmarsh defines two other related functions $\mu_k(\sigma)$ and σ_k and formulates his results in terms of these functions. For $k \geqslant 1$ and $\frac{1}{2} < \sigma < 1$ fixed, $\mu_k(\sigma)$ is the infimum of $\xi > 0$ such that

$$\int_1^T |\zeta(\sigma + it)|^{2k}\, dt \ll T^{1+\xi},$$

while σ_k is the infimum of numbers $\sigma > \frac{1}{2}$ for which

(8.114) $$\int_1^T |\zeta(\sigma + it)|^{2k}\, dt \ll T;$$

but we have found it more convenient to work with $M(A)$ and $m(\sigma)$ than with $\mu_k(\sigma)$ and σ_k.

An alternative approach to (8.26) is via the approximate functional equation (4.10), where we let $T/2 \leqslant t \leqslant T$, $s = \frac{1}{2} + it$, and $T/4\pi \leqslant x \leqslant T/2\pi$. We then have

$$\left|\zeta\left(\tfrac{1}{2} + it\right)\right|^4 \ll \left|\sum_{n\leqslant x} d(n)n^{-1/2-it}\right|^2 + \left|\sum_{n\leqslant t^2/4\pi^2 x} d(n)n^{-1/2-it}\right|^2 + \log^2 T;$$

hence multiplying by dx/x and integrating from $T/4\pi$ to $T/2\pi$ we obtain

$$\left|\zeta\left(\tfrac{1}{2} + it\right)\right|^4 \ll \int_{T/4\pi}^{T/2\pi}\left|\sum_{n\leqslant x} d(n)n^{-1/2-it}\right|^2 \frac{dx}{x}$$

$$+ \int_{T/4\pi}^{T/2\pi}\left|\sum_{n\leqslant t^2/4\pi^2 x} d(n)n^{-1/2-it}\right|^2 \frac{dx}{x} + \log^2 T,$$

and the change of variable $y = t^2/(4\pi^2 x)$ in the second integral leads to

$$\left|\zeta\left(\tfrac{1}{2} + it\right)\right|^4 \ll \int_{T/8\pi}^{T/\pi}\left|\sum_{n\leqslant x} d(n)n^{-1/2-it}\right|^2 \frac{dx}{x} + \log^2 T.$$

In the last expression the length of the sum does not depend on t, so that summing over $t = t_r$ and using Theorem 5.3 we obtain (8.26) as before. However, for $k \geqslant 6$ the lengths of the Dirichlet polynomials are too large for any known form of the approximate functional equation to produce a good upper bound for $\int_0^T |\zeta(\frac{1}{2} + it)|^k \, dt$, and here we used a new method, based on the mean value estimate for integrals over short intervals (Theorem 7.2), which was introduced by D. R. Heath-Brown (1978a).

The results of Section 8.3 are the sharpest ones known and may be found in the author's paper (1980b). This work generalizes and sharpens the results of D. R. Heath-Brown (1978a), where he proved the important result (8.35). Heath-Brown's proof of (8.35) uses Atkinson's formula (see Chapter 15) and the averaging integral (A.38) to obtain first what is essentially our Theorem 7.2, and then Lemma 8.1 is derived in the case $(\varkappa, \lambda) = (\frac{1}{2}, \frac{1}{2})$ and $(\varkappa, \lambda) = (\frac{1}{6}, \frac{4}{6})$ with the use of the Halász–Montgomery inequality (see Section A.11). Our approach is simpler, since it avoids Atkinson's formula in the proof of Theorem 7.2, and as shown in connection with (8.55) the use of the Halász–Montgomery inequality is also not necessary. A completely different method for the proof of (8.35) (with T^ε in place of the log factor) is due to H. Iwaniec (1979/1980) (see Section 7.6 of Chapter 7).

The first estimate in (8.56) improves (9) of Heath-Brown (1978a), where he had $R \ll TV^{-6}\log^8 T$ for $V \geqslant T^{2/13}(\log T)^{16/13}$, and $\frac{2}{13} = 0.153846\cdots > 0.152\dot{7} = \frac{11}{72}$. It may be remarked that our range for which the first bound in (8.56) holds is very close to the optimal result that the method of proof allows. Namely, (8.28) yields $R \ll TV^{-6}\log^8 T$ for $V \geqslant T^c$ and any $c > \lambda/(2 - 2\varkappa + 4\lambda)$. As mentioned several times in this text, a general method for finding the minimal value of $f(\varkappa, \lambda)$ [where $(\varkappa, \lambda)$ is an exponent pair and f is a "nice" function, say rational] has not been found yet. Thus each problem has to be tackled separately, and H. -E. Richert has kindly informed me that he has calculated

$$\inf_{(\varkappa, \lambda)} \lambda/(2 - 2\varkappa + 4\lambda) = 0.15274776\ldots,$$

whereas $\frac{11}{72} = 0.152\dot{7}$, so that this value is very close to the optimal one. The function which appears above is of the form $f(\varkappa, \lambda) = (a\varkappa + b\lambda + c)/(d\varkappa + e\lambda + f)$. An algorithm by which one can find the exponent pair that minimizes this type of function has been recently found by S. W. Graham (unpublished).

Bounds for $M(A)$ in Theorem 8.3 may be sharpened by replacing T^ε in (8.1) by a suitable log factor, however, this is of no importance. The idea to use $R = R^\alpha R^{1-\alpha}$ in the proof of Theorem 8.3 corresponds to the use of Hölder's inequality for integrals on the left-hand side of (8.1).

Results of this chapter are concerned primarily with upper bounds of the type (8.1) and (8.2). Some lower bounds will be given in Chapter 9, and it will be seen that the gap between these two types of results is fairly large for $k > 4$. One can reasonably hope that $M(6) = 1$ will be proved in the foreseeable future, and this would represent an important achievement in zeta-function theory.

The material in Section 8.4 is from the author's paper (1980b), only Theorem 8.1 was used to make the lower bounds for $m(\sigma)$ a continuous function in the interval $(\frac{1}{2}, \frac{7}{8}]$. The results are the sharpest ones yet, but the estimate $m(\frac{5}{8}) \geqslant 8$ [follows from the first bound in (8.80)] has been obtained first by D. R. Heath-Brown (1981b) by a somewhat different approach. No effort has been made (except when $\sigma = \frac{2}{3}$) to obtain the best possible estimates for $m(\sigma)$ that our method allows, as this would involve tedious computations with exponent pairs, and the possible improvements would be rather small. It is only when $\sigma \to 1$ that the bounds of Theorem 8.4 are superseded by $m(\sigma) \gg (1 - \sigma)^{-3/2}$, which follows from the estimates of Chapter 6.

The useful technique of dividing T into subintervals of length $\leqslant T_0$ and multiplying by $1 + T/T_0$, used in the proof of Lemma 8.2, seems to have been introduced first by M. N. Huxley (1972a, p. 117), and will be used repeatedly in Chapter 11.

A recent result, obtained independently from the methods of this chapter, is due to S. W. Graham (1982). In the notation of (8.9) it states that $a(\frac{5}{7}) = 14$, $b(\frac{5}{7}) = 196$, and like other estimates of Theorems 8.2 and 8.4 this can be used in proving zero-density estimates by employing the techniques which will be developed in Chapter 11.

Section 8.5 is from the author's monograph (1983b). The convexity estimate of Lemma 8.3 is essentially given on pp. 49–50 of H.-E. Richert (1960). This work contains various types of convexity results which occur in connection with summability properties of the Riesz means $C_\varkappa(x) = \sum_{n \leqslant x} c_n \log^\varkappa x/n$ $(\varkappa > -1)$ associated with the sequence $\{c_n\}_{n=1}^\infty$. Lemma 8.3 provides at once the convexity of Titchmarsh's function σ_k, as defined by (8.114), and is more general than estimates given on pp. 126–127 of E. C. Titchmarsh (1951), which correspond to the case $q = 2$ in (8.104). General convexity estimates for analytic functions may be applied to various problems in zeta-function theory, and for an example see D. R. Heath-Brown (1981a), who applied a form of R. M. Gabriel's convexity theorem (1927) to obtain (9.42). Gabriel's general theorem, whose proof is much more involved than the proof of Lemma 8.3, may be stated as follows: Let $\mathscr{D}$ be a simply connected domain symmetrical about a straight line $\mathscr{L}$ lying in $\mathscr{D}$. Let the boundary of $\mathscr{D}$ be a simple curve $\mathscr{K} = \mathscr{K}_1 \cup \mathscr{K}_2$, where $\mathscr{K}_1$ and $\mathscr{K}_2$ lie on opposite sides of $\mathscr{L}$. If $f(z)$ is regular in $\mathscr{D}$ and continuous on $\mathscr{K}$, then

$$\left(\int_{\mathscr{L}} |f(z)|^{2/(a+b)} |dz| \right)^{(a+b)/2} \leqslant \left(\int_{\mathscr{K}_1} |f(z)|^{1/a} |dz| \right)^{a/2} \left(\int_{\mathscr{K}_2} |f(z)|^{1/b} |dz| \right)^{b/2},$$

where a, $b > 0$ are any two real numbers.

Theorem 8.1 could have been proved by using a two-variable convexity theorem of R. M. Gabriel (1927) [this is also used by E. C. Titchmarsh (1951) on p. 203 in the proof of the zero-density bound (11.26)], but the self-contained approach of Section 8.2 seemed preferable.

The idea to use convexity arguments of Lemma 8.3 may be also found in R. T. Turganaliev (1981), who used them in a somewhat different context. Turganaliev investigated the asymptotic formula

$$(8.115) \qquad \int_0^T |\zeta(\sigma + it)|^{2\lambda} \, dt = T \sum_{n=1}^\infty d_\lambda^2(n) n^{-2\sigma} + O(T^{1-\varkappa+\varepsilon}),$$

where $\varepsilon > 0$ is arbitrary, $0 < \lambda < 2$, $s = \sigma + it$, and $\frac{1}{2} < \sigma < 1$. He proved that (8.115) holds with some $\varkappa = \varkappa(\sigma, \lambda) > 0$, where it is understood that λ does not have to be an integer. Under the simplifying assumption that the Riemann hypothesis holds, it is seen that the function

$$(8.116) \qquad F(s) = \zeta^{2\lambda}(s) - \left(\sum_{n \leqslant T} d_\lambda(n) n^{-s} \right)^2$$

has an analytic continuation for $\operatorname{Re} s > \frac{1}{2}$, so that Lemma 8.3 may be applied with $\alpha = \frac{1}{2} + \delta$, giving (8.115) with $\varkappa = \sigma - \frac{1}{2}$. However, Turganaliev succeeds in proving (8.115) unconditionally, where $\varkappa = \varkappa(\sigma, \lambda) > 0$ is not explicitly determined, but depends among other things on the quality of estimates for the zero-density function $N(\sigma, T)$. The range $0 < \lambda < 2$ to which Turganaliev restricts himself is motivated by the fact that the proof uses $M(4) = 1$, and as we have seen, an analogous estimate of this type for $A > 4$ is still not known to hold. The method of Turganaliev can be presumably adapted to yield analogues of Theorem 8.5 when k is not an integer in (8.103), only the error term will be weaker than the one in (8.105).

/ CHAPTER NINE /
OMEGA RESULTS

9.1 INTRODUCTION

In this chapter we shall turn our attention to Ω-results for the zeta-function. These are statements of the form

$$\zeta(s) = \Omega(u(t)), \qquad 1/\zeta(s) = \Omega(v(t)),$$

where $f(t) = \Omega(g(t))$ means the negation of $f(t) = o(g(t))$, that is, that the inequality $|f(t)| > Cg(t)$ is satisfied for some arbitrarily large values of t with a suitable constant $C > 0$. Similarly, $f(t) = \Omega_+(g(t))$ and $f(t) = \Omega_-(g(t))$ mean that the inequalities $f(t) > Cg(t)$ and $f(t) < -Cg(t)$ hold respectively for some arbitrarily large values of t and a suitable $C > 0$, while $f(t) = \Omega_\pm(g(t))$ means that both $f(t) = \Omega_+(g(t))$ and $f(t) = \Omega_-(g(t))$ hold. Omega results therefore follow from bounds of the form

$$(9.1) \qquad \max_{T \leqslant t \leqslant T+Y} |f(t)| \gg g(Y) \qquad (Y = o(T), T \to \infty),$$

and our main concern will be to obtain some suitable lower bounds of the type (9.1). In previous chapters we have been trying largely to obtain good O-estimates for $\zeta(s)$ or its power moments, that is, upper bounds which hold for all sufficiently large values of t. A comparison of previous O-results and Ω-results of this chapter shows that there is a large gap between them, which reflects the great difficulty in determining the behavior of $\zeta(s)$ in the critical strip. Intuitively, one feels that it is in general the Ω-results which lie closer to the truth than O-results, since the former involve statements about some indefinitely large values of t, while the latter must hold for all sufficiently large values of t.

In Section 9.2 we shall use some simple results from the theory of Diophantine approximation (approximation of reals by integers) to prove

Ω-results for $\zeta(s)$ when $\sigma \geq 1$. However, these methods are insufficient to provide satisfactory Ω-results in the critical strip, and in this case we shall employ a method based on complex integration. In Section 9.3 we shall prove two preparatory lemmas, and in Section 9.4 we shall obtain lower bounds for $|\zeta(\sigma + it)|$ of the type (9.1) when $\frac{1}{2} \leq \sigma \leq 1$. Finally, in Section 9.5 we shall deal with lower bounds for the integrals $\int_{T-Y}^{T+Y} |\zeta(\sigma + it)|^{2k} \, dt$ when $\sigma \geq \frac{1}{2}$.

9.2 OMEGA RESULTS WHEN $\sigma \geq 1$

We shall require two well-known classical results (Lemma 9.1 and Lemma 9.3) from Diophantine approximation, due to P. G. L. Dirichlet and L. Kronecker, respectively.

LEMMA 9.1. Given N real numbers $a_1, a_2, \ldots, a_N$, a positive integer q, and a positive number t_0, we can find a number t in the range

$$(9.2) \qquad t_0 \leq t \leq t_0 q^N$$

and integers $x_1, x_2, \ldots, x_N$ such that

$$(9.3) \qquad |ta_n - x_n| \leq 1/q \qquad (n = 1, 2, \ldots, N).$$

Proof of Lemma 9.1. The proof is based on the use of the "Dirichlet box principle," which asserts that if there are $m + 1$ objects in m boxes, then at least one box contains at least two objects. Consider the N-dimensional unit cube with a vertex at the origin and edges along the coordinate axes. Divide each edge into q equal parts, and thus the cube into q^N equal compartments. Now consider $q^N + 1$ points in the cube, congruent (mod 1) to the points $(ua_1, ua_2, \ldots, ua_N)$, where $u = 0, t_0, 2t_0, \ldots, q^N t_0$. At least two of these points must lie in the same compartment, and if these two points correspond to $u = u_1, u = u_2$ $(u_1 < u_2)$, then $t = u_2 - u_1$ clearly satisfies the requirements of the lemma. If $m \geq 1$, then the same argument shows that the interval $[t_0, mq^N t_0]$ contains at least two solutions of (9.3), and any two solutions differ by at least t_0.

LEMMA 9.2. If $f(x)$ is positive and continuous for $0 \leq a \leq x \leq b$, then

$$(9.4) \qquad \lim_{n \to \infty} \left(\int_a^b f^n(x) \, dx \right)^{1/n} = \max_{a \leq x \leq b} f(x),$$

and a similar result holds for an integral in any number of dimensions.

Proof of Lemma 9.2. This simple result will be needed in the proof of Kronecker's Lemma 9.3. Let $M = \max_{a \leq x \leq b} f(x)$. Then obviously

$$(9.5) \qquad \left(\int_a^b f^n(x) \, dx \right)^{1/n} \leq (b - a)^{1/n} M.$$

Given any $\varepsilon > 0$, there is a subinterval $[a_1, b_1]$ of $[a, b]$ throughout which $f(x) \geqslant M - \varepsilon$, hence

(9.6) $\qquad \left(\int_a^b f^n(x) \, dx \right)^{1/n} \geqslant \left(\int_{a_1}^{b_1} (M - \varepsilon)^n \, dx \right)^{1/n} = (b_1 - a_1)^{1/n} (M - \varepsilon).$

Letting $n \to \infty$ we obtain (9.4) from (9.5) and (9.6), and a similar argument works in the general case too.

LEMMA 9.3. Let $a_1, a_2, \ldots, a_N$ be reals linearly independent over the integers, that is, there is no relation $\lambda_1 a_1 + \cdots + \lambda_N a_N = 0$ in which the coefficients $\lambda_1, \ldots, \lambda_N$ are integers not all zero. Let $b_1, \ldots, b_N$ be any real numbers, and q a given positive number. Then we can find a number t and integers $x_1, \ldots, x_N$ such that

(9.7) $\qquad |ta_n - b_n - x_n| \leqslant 1/q \qquad (n = 1, 2, \ldots, N).$

Proof of Lemma 9.3. If all the numbers b_n are zero, the result follows from Lemma 9.1, which incidentally provides an upper bound for t in (9.2), while Lemma 9.3 does not. Let

$$F(t) = 1 + \sum_{n=1}^N e(ta_n - b_n), \qquad \varphi(t) = |F(t)|.$$

Obviously, $0 \leqslant \varphi(t) \leqslant N + 1$, and we shall prove that

(9.8) $\qquad \limsup_{t \to \infty} \varphi(t) = N + 1,$

which will imply (9.7). To see this assume (9.8) for the moment and let $\varphi(t) \geqslant N + 1 - \eta, 0 < \eta < 1$. Then if $v_n = ta_n - b_n$, $z = e(v_n) = x + iy$, we have

$$N + 1 - \eta \leqslant \varphi(t) \leqslant N - 1 + |1 + e(v_n)|;$$

hence

$$2 - \eta \leqslant |1 + e(v_n)| \leqslant 2 \qquad (n = 1, 2, \ldots, N).$$

Since $x^2 + y^2 = 1$ we have

$$|1 + z|^2 = (1 + x)^2 + y^2 = (1 + x)^2 + 1 - x^2 = 2 + 2x$$

$$\geqslant (2 - \eta)^2 > 4 - 4\eta;$$

hence $1 - 2\eta \leqslant x \leqslant 1$, and $y^2 = 1 - x^2 = (1 - x)(1 + x) \leqslant 4\eta$. Thus

$$|e(ta_n - b_n) - 1| = |z - 1| < 4\eta^{1/2} \qquad (n = 1, 2, \ldots, N),$$

which in view of the continuity and periodicity of $e(x)$ implies (9.7).

To prove (9.8) consider

$$G(\Phi_1, \ldots, \Phi_N) = 1 + \sum_{n=1}^{N} e(\Phi_n),$$

where $\Phi_1 \ldots, \Phi_N$ are independent real variables lying in $[0, 1]$. Then the upper bound for $|G|$ is $N + 1$, which is attained for $\Phi_1 = \cdots = \Phi_N = 0$.

We consider the polynomial expansion of $F^k(t)$ and $G^k(\Phi_1, \ldots, \Phi_N)$, where k is a large integer. Since the numbers $a_1, \ldots, a_N$ are linearly independent over the integers no two terms in the expansion of $F^k(t)$ fall together, so that the expansions of F^k and G^k contain the same number of terms and also the moduli of the corresponding terms are equal. Thus if

$$G^k(\Phi_1, \ldots, \Phi_N) = 1 + \sum_q c_q e(\lambda_{q,1}\Phi_1 + \cdots + \lambda_{q,N}\Phi_N),$$

then

$$F^k(t) = 1 + \sum_q c_q e(\lambda_{q,1}(a_1 t - b_1) + \cdots + \lambda_{q,N}(a_N t - b_N));$$

hence squaring the modulus and integrating term by term we see that

$$F_k = \lim_{T \to \infty} (2T)^{-1} \int_{-T}^{T} |F(t)|^{2k} \, dt = G_k$$

$$= \int_0^1 \cdots \int_0^1 |G(\Phi_1, \ldots, \Phi_N)|^{2k} \, d\Phi_1 \ldots d\Phi_N$$

$$= 1 + \sum_q c_q^2.$$

But the N-dimensional analogue of Lemma 9.2 gives

$$\lim_{k \to \infty} G_k^{1/2k} = N + 1;$$

hence also

$$\limsup_{t \to \infty} \varphi(t) = \lim_{k \to \infty} F_k^{1/2k} = N + 1.$$

Therefore (9.8) holds and the lemma is proved.

We proceed now to apply the lemmas above to $\zeta(s)$. If $\sigma > 1$, then trivially for all t

$$|\zeta(s)| = \left| \sum_{n=1}^{\infty} n^{-s} \right| \leqslant \sum_{n=1}^{\infty} n^{-\sigma} = \zeta(\sigma),$$

and it seems interesting that an analogous lower bound, that is, the Ω-result also holds. This is shown by

THEOREM 9.1. If $0 < \varepsilon < 1$ and $\sigma > 1$, then for some indefinitely large values of t

$$(9.9) \qquad |\zeta(s)| \geqslant (1 - \varepsilon)\zeta(\sigma).$$

Proof of Theorem 9.1. For all values of $N \geqslant 2$

$$\zeta(s) = \sum_{n=1}^{N} n^{-\sigma}e^{-it\log n} + \sum_{n=N+1}^{\infty} n^{-\sigma-it};$$

hence

$$|\zeta(s)| \geqslant \left| \sum_{n=1}^{N} n^{-\sigma}e^{-it\log n} \right| - \left| \sum_{n=N+1}^{\infty} n^{-\sigma-it} \right|$$

$$\geqslant \mathrm{Re}\left(\sum_{n=1}^{N} n^{-\sigma}e^{-it\log n} \right) - \sum_{n=N+1}^{\infty} n^{-\sigma}$$

$$= \sum_{n=1}^{N} n^{-\sigma}\cos(t\log n) - \sum_{n=N+1}^{\infty} n^{-\sigma}.$$

By Lemma 9.1 there is a number t such that $t_0 \leqslant t \leqslant t_0 q^N$ and integers $x_1, \ldots, x_N$ such that, for a given N and $q \geqslant 4$,

$$\left| (2\pi)^{-1}t\log n - x_n \right| \leqslant 1/q \qquad (n = 1, 2, \ldots, N).$$

Hence for these n we have $\cos(t\log n) \geqslant \cos(2\pi/q)$; therefore

$$\sum_{n=1}^{N} n^{-\sigma}\cos(t\log n) \geqslant \cos(2\pi/q)\sum_{n=1}^{N} n^{-\sigma} > \cos(2\pi/q)\zeta(\sigma) - \sum_{n=N+1}^{\infty} n^{-\sigma}.$$

Hence

$$|\zeta(s)| \geqslant \cos(2\pi/q)\zeta(\sigma) - 2\sum_{n=N+1}^{\infty} n^{-\sigma},$$

and since

$$\zeta(\sigma) > \int_{1}^{\infty} u^{-\sigma}\,du = (\sigma - 1)^{-1}, \qquad \sum_{n=N+1}^{\infty} n^{-\sigma} < \int_{N}^{\infty} u^{-\sigma}\,du = \frac{N^{1-\sigma}}{\sigma - 1},$$

we obtain

$$(9.10) \qquad |\zeta(s)| \geqslant \left(\cos(2\pi/q) - 2N^{1-\sigma}\right)\zeta(\sigma).$$

Since $1 - \sigma < 0$ and $\lim\limits_{q \to \infty} \cos(2\pi/q) = 1$, the theorem follows if N and q are sufficiently large.

Recalling that $\zeta(\sigma) \to \infty$ as $\sigma \to 1$ it follows as an immediate corollary of Theorem 9.1 that $\zeta(s)$ is unbounded in the open region

$$(9.11) \qquad \sigma > 1, \qquad t > \delta > 0.$$

Another corollary is that $\zeta(1 + it)$ is unbounded as $t \to \infty$. Indeed, if $\zeta(1 + it)$ were bounded, then since $\zeta(2 + it)$ is bounded too, it would follow by the Phragmén–Lindelöf principle (see Section A.8) that $\zeta(s)$ is bounded in the strip $1 \leqslant \sigma \leqslant 2, t > \delta$, which is false because $\zeta(s)$ is unbounded in the region (9.11). A more precise form of this follows if we take $t_0 = 1$ and $q = 6$ in the proof of Theorem 9.1. Then (9.10) gives

$$|\zeta(s)| \geqslant (\tfrac{1}{2} - 2N^{1-\sigma})/(\sigma - 1) \qquad (1 \leqslant t \leqslant 6^N).$$

Let now $N = [8^{1/(\sigma-1)}]$. Then

$$|\zeta(s)| \geqslant \frac{1}{4(\sigma - 1)} \geqslant \frac{\log(N-1)}{4\log 8} > A\log N > A_1 \log\log t.$$

Therefore this argument gives

THEOREM 9.2. However large t_1 may be, there are values of s in the region $\sigma > 1, t > t_1$, for which

$$(9.12) \qquad |\zeta(s)| > A\log\log t,$$

and also

$$(9.13) \qquad \zeta(1 + it) = \Omega(\log\log t).$$

The second part of the theorem follows from the first by applying the Phragmén–Lindelöf principle to the function

$$f(s) = \frac{\zeta(s)}{\log\log s},$$

where $\log\log s$ is the branch of the logarithm which is real for $s = \sigma > 1$, and is restricted to $|s| > 1, \sigma > 0, t > 0$. Then $f(s)$ is regular for $1 \leqslant \sigma \leqslant 2$ and

$t > \delta$, and uniformly in σ we have $|\log\log s| \sim \log\log t$ as $t \to \infty$. Hence $f(2 + it) \to 0$ as $t \to \infty$, and then if $f(1 + it) \to 0$ as $t \to \infty$, we obtain that $f(s) \to 0$ uniformly in the strip, which contradicts (9.12).

Finally, we shall prove in this section an omega result for $1/\zeta(s)$. Actually, we shall consider $\log\zeta(s)$, which depends on the series $\Sigma_p p^{-s}$, to which Kronecker's Lemma 9.3 may be applied. The result is

THEOREM 9.3. The function $1/\zeta(s)$ is unbounded in the region $\sigma > 1$, $t > \delta > 0$.

Proof of Theorem 9.3. For $\sigma \geqslant 1$ we have

$$(9.14) \quad \left| \log\zeta(s) - \sum_p p^{-s} \right| = \left| \sum_p \sum_{m=2}^{\infty} m^{-1} p^{-ms} \right|$$

$$\leqslant \sum_p \sum_{m=2}^{\infty} p^{-m} = \sum_p p^{-1}(p-1)^{-1} = C,$$

and

$$\mathrm{Re}\left(\sum_p p^{-s} \right) = \sum_{n=1}^{\infty} p_n^{-\sigma}\cos(t\log p_n) \leqslant \sum_{n\leqslant N} p_n^{-\sigma}\cos(t\log p_n) + \sum_{n=N+1}^{\infty} p_n^{-\sigma},$$

where p_n is the nth prime. The numbers $\log p_1, \ldots, \log p_N$ are linearly independent over the integers, since a relation of the form

$$\lambda_1 \log p_1 + \cdots + \lambda_N \log p_N = 0$$

is equivalent to

$$p_1^{\lambda_1} \cdots p_N^{\lambda_N} = 1,$$

and by the fundamental theorem of arithmetic this is possible only if $\lambda_1 = \cdots = \lambda_N = 0$, if the λ's are assumed to be integers. Therefore the numbers $(\log p_n)/2\pi$ are also linearly independent and Lemma 9.3 shows that we can find a number t and integers $x_1, \ldots, x_N$ such that

$$\left| (2\pi)^{-1} t\log p_n - \tfrac{1}{2} - x_n \right| \leqslant \tfrac{1}{6} \quad (n = 1, 2, \ldots, N),$$

or

$$\left| t\log p_n - \pi - 2\pi x_n \right| \leqslant \pi/3 \quad (n = 1, 2, \ldots, N).$$

Hence for these values of n

$$\cos(t\log p_n) = -\cos(t\log p_n - \pi - 2\pi x_n) \leqslant -\cos(\pi/3) = -\tfrac{1}{2},$$

implying

$$\mathrm{Re}\left(\sum_p p^{-s}\right) \leqslant -\frac{1}{2}\sum_{n=1}^N p_n^{-\sigma} + \sum_{n=N+1}^\infty p_n^{-\sigma}.$$

But from elementary prime number theory it is well known that $\sum p_n^{-1}$ is divergent, so that we can make $\sum p_n^{-\sigma} > H$ for any $H > 0$ if σ is sufficiently close to 1. Having fixed σ, we can then choose N so large that

$$\sum_{n=1}^N p_n^{-\sigma} > \tfrac{3}{4}H, \qquad \sum_{n=N+1}^\infty p_n^{-\sigma} < \tfrac{1}{4}H,$$

giving

$$\mathrm{Re}\left(\sum_p p^{-s}\right) < -\tfrac{3}{8}H + \tfrac{1}{4}H = -\tfrac{1}{8}H.$$

Since H may be arbitrary large, it follows from (9.14) that $\log|\zeta(s)|$ takes arbitrary large negative values, which proves the theorem.

9.3 LEMMAS ON CERTAIN ORDER RESULTS

In this section we shall prove two results which will enable us in Section 9.4 to derive satisfactory lower bounds for $\max_{T \leqslant t \leqslant T+Y}|\zeta(\sigma + it)|$ when $\tfrac{1}{2} \leqslant \sigma \leqslant 1$. These are

LEMMA 9.4. Let $\log^c T \leqslant Y \leqslant T$, $T \geqslant T_0$, $\tfrac{1}{2} \leqslant \sigma_0 < 1$, and $c > 0$ fixed, and let further

(9.15) $$M = \max_{T \leqslant t \leqslant T+Y}|\zeta(\sigma_0 + it)|.$$

Then there exists a constant $C > 0$ such that in the region $\sigma_0 \leqslant \sigma \leqslant 2$, $T + Y/3 \leqslant t \leqslant T + 2Y/3$, we have

(9.16) $$|\zeta(s)| \leqslant C(M + 1).$$

LEMMA 9.5. For any integer $k \geqslant 2$ and $\tfrac{1}{2} < \sigma \leqslant 2$ let

(9.17) $$f_k(\sigma) = \sum_{n=1}^\infty d_k^2(n)n^{-2\sigma}.$$

Then for σ fixed and some $0 < C_1 < C_2$

$$(9.18) \qquad \exp\left(\frac{C_1 k^{1/\sigma}}{\log k}\right) < f_k(\sigma) < \exp\left(\frac{C_2 k^{1/\sigma}}{\log k}\right),$$

while for $\delta = \sigma - \frac{1}{2} \leqslant 2/\log k$ and $k \geqslant k_0$

$$(9.19) \quad \exp\left(\frac{C_1 k^2}{\delta \log k}\right) < f_k(\sigma) < \exp(C_2 k^2 L), \qquad L = \log\left(\frac{e}{\delta \log k}\right).$$

Proof of Lemma 9.4. Let M_1 be the maximum of $|\zeta(s)|$ when $s = \sigma + it$ lies in the region described in the lemma, and let $\mathcal{D}$ be the rectangle with vertices $\sigma_0 + i(T + Y/3)$, $\sigma_0 + i(T + 2Y/3)$, $2 + i(T + Y/3)$, and $2 + i(T + 2Y/3)$. Then by the maximum modulus principle

$$M_1 = \max_{s \in \mathcal{D}} |\zeta(s)| = |\zeta(s')|$$

for some point s' on $\mathcal{D}$. But then again by the maximum modulus principle

$$\left|\zeta(s')\right| = \left|\zeta(s')\exp\left((s' - s')^{4N+2}\right)\right| \leqslant \max_{s \in \mathcal{E}} \left|\zeta(s)\exp\left((s - s')^{4N+2}\right)\right|,$$

where N is a sufficiently large integer and $\mathcal{E}$ is the rectangle with vertices $\sigma_0 + iT$, $\sigma_0 + iT + iY$, $2 + iT$, and $2 + iT + iY$. If $s' = \sigma' + it'$, then for some $\theta \geqslant \frac{1}{10}$ and s on the horizontal sides of $\mathcal{E}$ we have

$$\left|\zeta(s)\exp\left((s - s')^{4N+2}\right)\right| \leqslant T^{1/6}\left|\exp\left\{(\sigma - \sigma' + i(t - t'))^{4N+2}\right\}\right|$$

$$\leqslant T^{1/6}\exp(-\theta Y^{4N+2}) \leqslant 1,$$

provided that N is sufficiently large, since $i^{4N+2} = -1$ if N is a natural number and $|t - t'| \geqslant Y/3$. On the side of $\mathcal{E}$ with $\sigma = \sigma_0$ we have $|\zeta(s)| \leqslant M$ and on $\sigma = 2$ we have $|\zeta(s)| \leqslant \zeta(2) = \pi^2/6$ while the exponential factor is bounded; hence (9.16) follows.

Proof of Lemma 9.5. From the multiplicativity of $d_k^2(n)$ we have

$$f_k(\sigma) = \prod_p A_p, \qquad A_p = 1 + \sum_{r=1}^{\infty} d_k^2(p^r) p^{-2r\sigma},$$

$$(9.20) \quad d_k(p^r) = \frac{k(k+1)\cdots(k+r-1)}{r!} \leqslant kd_{k+1}(p^{r-1}) \qquad (k, r \geqslant 1),$$

$$(9.21) \quad 1 + k^2 p^{-2\sigma} < A_p < \left(\sum_{r=0}^{\infty} d_k(p^r) p^{-\sigma}\right)^2 = (1 - p^{-\sigma})^{-2k};$$

hence

$$A_p < 1 + k^2 p^{-2\sigma} \sum_{r=1}^{\infty} d_{k+1}^2(p^{r-1}) p^{-2(r-1)\sigma} < 1 + k^2 p^{-2\sigma} (1 - p^{-\sigma})^{-2(k+1)}$$

because (9.20) and (9.21) hold. For $p > k^{1/\sigma}$

$$(1 - p^{-\sigma})^{-2k-2} < (1 - k^{-1})^{-2k-2} < C$$

for some absolute C, say for $C = 1000$. Thus for $p > k^{1/\sigma}$

$$A_p < 1 + 1000 k^2 p^{-2\sigma}.$$

Therefore we may write

(9.22)

$$\prod_p (1 + k^2 p^{-2\sigma}) < f_k(\sigma) < \prod_{p > Bk^{1/\sigma}} (1 + 1000 k^2 p^{-2\sigma}) \prod_{p \leqslant Bk^{1/\sigma}} (1 - p^{-\sigma})^{-2k},$$

where $B > 0$ is such a constant that $1000 k^2 p^{-2\sigma} < \frac{1}{2}$ for $p > Bk^{1/\sigma}$. Using the prime number theorem (see Chapter 12) in the weak form $\pi(x) \sim \mathrm{li}\, x$ we obtain then

$$\log f_k(\sigma) < \sum_{p > Bk^{1/\sigma}} \log(1 + 1000 k^2 p^{-2\sigma}) + \sum_{p \leqslant Bk^{1/\sigma}} \log(1 - p^{-\sigma})^{-2k}$$

$$\ll k^2 \sum_{p > Bk^{1/\sigma}} p^{-2\sigma} + k \sum_{p \leqslant Bk^{1/\sigma}} p^{-\sigma}$$

$$\ll k^2 \int_{Bk^{1/\sigma}}^{\infty} \frac{t^{-2\sigma}}{\log t} \, dt + \frac{k(k^{1/\sigma})^{1-\sigma}}{\log(k^{1/\sigma})}$$

$$\ll \frac{k^2 (k^{1/\sigma})^{1-2\sigma}}{\log(k^{1/\sigma})} + \frac{k^{1/\sigma}}{\log k} \ll \frac{k^{1/\sigma}}{\log k}.$$

This establishes the upper bound in (9.18), and to obtain the upper bound in (9.19) one proceeds similarly, taking into account that

$$\log \prod_{p \geqslant Bk^{1/\sigma}} (1 + 1000 k^2 p^{-2\sigma}) \leqslant \log\left(\prod_{p \geqslant Bk^{1/\sigma}} (1 - p^{-2\sigma})^{-1} \right)^{1000 k^2}$$

$$= \log\left(\zeta(2\sigma) \prod_{p < Bk^{1/\sigma}} (1 - p^{-2\sigma}) \right)^{1000 k^2},$$

and $\zeta(2\sigma) \ll \delta^{-1}$, while for $0 < \sigma - \frac{1}{2} \leqslant 2/\log k$

$$\prod_{p \leqslant Bk^{1/\sigma}} (1 - p^{-2\sigma}) \ll 1/\log k.$$

The lower bound in (9.18) is obtained analogously as the one in (9.19), which is as follows. For $p > (2k)^{1/\sigma}$ we have

$$1 + k^2 p^{-2\sigma} > \exp\left(\tfrac{1}{2} k^2 p^{-2\sigma}\right),$$

since $e^{x/2} < 1 + x$ for $0 < x \le \frac{1}{2}$. Hence (9.22) yields

(9.23)
$$f_k(\sigma) > \exp\left(\tfrac{1}{2} k^2 \sum_{p > (2k)^{1/\sigma}} p^{-2\sigma}\right).$$

However, by the prime number theorem

$$\sum_{p > (2k)^{1/\sigma}} p^{-2\sigma} \gg \int_{(3k)^{1/\sigma}}^{\infty} \frac{dt}{t^{2\sigma} \log t} \gg (\log k)^{-1} \left.\frac{t^{1-2\sigma}}{1 - 2\sigma}\right|_{(3k)^{1/\sigma}}^{\infty}$$

$$\gg (\log k)^{-1} \frac{(3k)^{(1-2\sigma)/\sigma}}{2\sigma - 1}$$

$$\ge (2\delta \log k)^{-1} \exp\left((1 - 2\sigma)\sigma^{-1} \log(3k)\right) \gg (\delta \log k)^{-1},$$

and (9.23) gives

$$f_k(\sigma) > \exp\left(C_1 \frac{k^2}{\delta \log k}\right),$$

as asserted. This completes the proof of Lemma 9.5.

9.4   OMEGA RESULTS FOR $\frac{1}{2} \le \sigma \le 1$

We proceed now to Ω-results for the order of $\zeta(s)$ in the critical strip $\frac{1}{2} \le \sigma \le 1$, aiming at lower bounds of the type (9.1). The methods of Section 9.2 seem to be of no use here, and we shall employ a complex integration technique and the lemmas of Section 9.3 to prove

THEOREM 9.4.   For any fixed $C > 1, \log^C T \le Y \le T, T \ge T_0$

(9.24)
$$\max_{T \le t \le T + Y} \left|\zeta(\tfrac{1}{2} + it)\right| \ge \exp\left\{A_1 \left(\frac{\log Y}{\log \log Y}\right)^{1/2}\right\},$$

(9.25)
$$\max_{T \le t \le T + Y} \left|\zeta(\sigma + it)\right| \ge \exp\left(A_2 \frac{\log^{1-\sigma} Y}{\log \log Y}\right) \qquad (\tfrac{1}{2} < \sigma < 1),$$

(9.26)
$$\max_{T \le t \le T + Y} \left|\zeta(1 + it)\right| \ge A_3 \log \log Y,$$

where A_1, A_2, A_3 are positive, absolute constants.

Proof of Theorem 9.4. From the Mellin integral (A.7) we have

$$\sum_{n=1}^{\infty} d_k(n)n^{-s_1}e^{-n/Y} = (2\pi i)^{-1}\int_{2-i\infty}^{2+i\infty}\zeta^k(s_1 + w)\Gamma(w)Y^w dw,$$

where $s_1 = \sigma + 2/(\log k) + it$ and k is a large positive integer. Changing the path of integration to the line $\operatorname{Re} w = -1/\log k$, we obtain by the residue theorem

$$\sum_{n=1}^{\infty} d_k(n)n^{-s_1}e^{-n/Y} = \zeta^k(s_1) + (2\pi i)^{-1}\int_{-1/\log k-i\infty}^{-1/\log k+i\infty}\zeta^k(s_1 + w)\Gamma(w)Y^w dw.$$

Now we suppose $\log\log Y \ll k \leqslant \log Y, T + Y/4 \leqslant t \leqslant T + Y/2$. If $C_1, C_2, \ldots$ denote absolute constants, then using Lemma 9.4 and Stirling's formula (A.34) in the weakened form

$$\Gamma(w) \ll e^{-|\operatorname{Im} w|}|w|^{-1}$$

we obtain

(9.27)

$$\left|\sum_{n=1}^{\infty} d_k(n)n^{-s_1}e^{-n/Y}\right| \leqslant (C_1(M + 1))^k$$

$$+ (C_2(M + 1))^k\int_{-1/\log k-i\log^2 T}^{-1/\log k+i\log^2 T}|\Gamma(w)|\,|Y^{-1/\log k}|\,dw|$$

$$+ \int_{\log^2 T}^{\infty}\left|\zeta\left(\sigma + \frac{1}{\log k} \pm it + iv\right)\right|^k e^{-v}\,dv$$

$$\ll (C_3(M + 1))^k\log k + \int_{\log^2 T}^{\infty}(T + v)^{k/6}e^{-v}\,dv$$

$$\leqslant (C_4(M + 1))^k,$$

since we may suppose that $C_3(M + 1) > 1$, and for $v \geqslant T \geqslant T_0$

$$(2v)^{k/6}e^{-v/2} \leqslant e^{-(\log^2 T)/6}.$$

Here $M = \max_{T\leqslant t\leqslant T+Y}|\zeta(\sigma + it)|$, and therefore integration of (9.27) gives

(9.28) $$4Y^{-1}\int_{T+Y/4}^{T+Y/2}\left|\sum_{n=1}^{\infty} d_k(n)n^{-\sigma_1-it}e^{-n/Y}\right|^2 dt \leqslant (C_4(M + 1))^{2k}.$$

By Theorem 5.2 we see that the left-hand side of (9.28) is equal to

$$(9.29) \quad \sum_{n=1}^{\infty} d_k^2(n) n^{-2\sigma_1} e^{-2n/Y} + O\left(Y^{-1} \sum_{n=1}^{\infty} d_k^2(n) n^{1-2\sigma_1} e^{-2n/Y}\right)$$

$$= \sum_{n \leqslant Y \log^2 Y} d_k^2(n) n^{-2\sigma_1} e^{-2n/Y} + O\left(Y^{-1} \sum_{n \leqslant Y \log^2 Y} d_k^2(n) n^{1-2\sigma_1} e^{-2n/Y}\right)$$

$$+ O\left(\sum_{n > Y \log^2 Y} d_k^2(n) n^{-2\sigma_1} e^{-2n/Y}\right) \geqslant \sum_{n \leqslant Y \log^2 Y} d_k^2(n) n^{-2\sigma_1}$$

$$- C_5 Y^{-1} \sum_{n \leqslant Y \log^2 Y} d_k^2(n) n^{1-2\sigma_1} - C_6 e^{-2 \log^2 Y} f_k(\sigma_1),$$

where $f_k(\sigma)$ is given by (9.17) and where we used $e^{-x} \geqslant 1 - x \ (x > 0)$. Next we have

$$\sum_{n \leqslant Y \log^2 Y} d_k^2(n) n^{-2\sigma_1} \geqslant f_k(\sigma_1) - \sum_{n \geqslant Y} d_k^2(n) n^{-2\sigma_1}$$

$$\geqslant f_k(\sigma_1) - Y^{-2/\log k} \sum_{n \geqslant Y} d_k^2(n) n^{-2(\sigma_1 - 1/\log k)}$$

$$\geqslant f_k(\sigma_1) - Y^{-2/\log k} f_k(\sigma_1 - 1/\log k).$$

Similarly, for any $A > 0$ we may write

$$(9.30) \quad Y^{-1} \sum_{n \leqslant Y \log^2 Y} d_k^2(n) n^{1-2\sigma_1} \leqslant A \sum_{n \leqslant AY} d_k^2(n) n^{-2\sigma_1}$$

$$+ (\log^2 Y)(AY)^{-2/\log k} \sum_{AY < n \leqslant Y \log^2 Y} d_k^2(n) n^{-2(\sigma_1 - 1/\log k)}$$

$$\leqslant A f_k(\sigma_1) + C_7 (\log^2 Y) Y^{-2/\log k} f_k(\sigma_1 - 1/\log k).$$

Choosing $A = (2C_5)^{-1}$, we obtain from (9.28)–(9.30)

$$(9.31) \quad (C_4(M+1))^{2k} > C_8 f_k(\sigma_1) - C_9 (\log^2 Y) Y^{-2/\log k} f_k(\sigma_1 - 1/\log k).$$

Here we distinguish the cases $\sigma = \frac{1}{2}$ and $\frac{1}{2} < \sigma < 1$, and use (9.18) and (9.19) to bound $f_k(\sigma_1)$ and $f_k(\sigma_1 - 1/\log k)$, where $\sigma_1 = \sigma + 2/\log k$. In the former case we have $\delta = 2/\log k$ or $\delta = 1/\log k$ and consequently (9.19) gives

$$(C_4(M+1))^{2k} > \exp(C_{10} k^2) - C_{11}(\log^2 Y) \exp\left(C_{12} k^2 - 2\frac{\log Y}{\log k}\right).$$

For $k = [C_{13}(\log Y/\log\log Y)^{1/2}]$ we have $C_{12}k^2 - 2\log Y/\log k \leqslant 0$; hence

$$(C_4(M+1))^{2k} > \exp(C_{10}k^2) - C_{11}\log^2 Y > \tfrac{1}{2}\exp(C_{10}k^2),$$

which gives

$$C_4(M+1) > \exp(C_{12}k) \gg \exp\left\{C_{12}C_{13}\left(\frac{\log Y}{\log\log Y}\right)^{1/2}\right\},$$

thus proving (9.24). In case $\tfrac{1}{2} < \sigma < 1$ we obtain with (9.18)

$$(C_4(M+1))^{2k} > \exp\left(C_{14}\frac{k^{1/\sigma}}{\log k}\right) - C_{15}(\log^2 Y)\exp\left(-2\frac{\log Y}{\log k} + C_{16}\frac{k^{1/\sigma}}{\log k}\right),$$

and (9.25) follows with $k = [C_{17}(\log Y)^\sigma]$. Finally suitably modifying the foregoing argument when $\sigma = 1$ we obtain without difficulty (9.26).

9.5 LOWER BOUNDS FOR POWER MOMENTS WHEN $\sigma = \tfrac{1}{2}$

We proceed now to lower bounds for $\int_{T-Y}^{T+Y}|\zeta(\tfrac{1}{2} + it)|^{2k}\,dt$, $k \geqslant 1$. This will depend on the fact that for $k \geqslant 1$ a fixed integer we have

$$(9.32) \quad \sum_{n \leqslant x} d_k^2(n)n^{-1} = (C_k + o(1))\log^{k^2} x \qquad (C_k > 0, x \to \infty),$$

which follows by partial summation from

$$(9.33) \quad \sum_{n \leqslant x} d_k^2(n) = (D_k + o(1))x\log^{k^2-1}x \qquad (D_k > 0, x \to \infty).$$

A convenient way to obtain (9.33) is to note that $d_k(p) = k$ for all p, so that by multiplicativity of $d_k^2(n)$ we have, for $\sigma > 1$,

$$F_k(s) = \sum_{n=1}^{\infty} d_k^2(n)n^{-s}$$

$$= \prod_p \left(1 + d_k^2(p)p^{-s} + d_k^2(p^2)p^{-2s} + \cdots\right) = \zeta^{k^2}(s)G_k(s),$$

where $G_k(s)$ is absolutely convergent for $\sigma > \tfrac{1}{2}$. Therefore an application of the inversion formula (A.10) gives

$$(9.34) \quad \sum_{n \leqslant x} d_k^2(n) = (2\pi i)^{-1}\int_{1+\varepsilon-iT}^{1+\varepsilon+iT}\zeta^{k^2}(s)G_k(s)s^{-1}x^s\,ds + O(x^{1+\varepsilon}T^{-1}).$$

We replace here the segment of integration $[1 + \varepsilon - iT, 1 + \varepsilon + iT]$ by the segment $[1 - \varepsilon - iT, 1 - \varepsilon + iT]$ using the residue theorem. Noting that for $\sigma \geq 1 - \varepsilon$ the integrand has only a pole of order k^2 at $s = 1$ with residue $\sim D_k x \log^{k^2-1} x$, we obtain (9.33) from (9.34) if $T = x^{2\varepsilon}$ and ε is sufficiently small.

Now we formulate our lower bound result, which is

THEOREM 9.5. If $k \geq 1$ is a fixed integer, $\log^3 T \leq Y \leq T, T \geq T_0$, then uniformly in Y

$$(9.35) \qquad \int_{T-Y}^{T+Y} \left| \zeta(\tfrac{1}{2} + it) \right|^{2k} dt \gg Y (\log Y)^{k^2} (\log \log Y)^{-2}.$$

Proof of Theorem 9.5. Let $s = \tfrac{1}{2} + (\log Y)^{-1} + it, T - Y \leq t \leq T + Y$. Then we have

$$\sum_{n=1}^{\infty} d_k(n) n^{-s} (e^{-n/Y} - e^{-n})$$

$$= (2\pi i)^{-1} \int_{2-i\infty}^{2+i\infty} \zeta^k(s + w) \Gamma(w) (Y^w - 1) \, dw$$

$$= (2\pi i)^{-1} \int_{-1/\log Y - i\infty}^{-1/\log Y + i\infty} \zeta^k(s + w) \Gamma(w) (Y^w - 1) \, dw,$$

since the pole $w = 0$ of $\Gamma(w)$ cancels the zero of $Y^w - 1$. By Stirling's formula (A.34) we see that the portion of the last integral for which $|\operatorname{Im} w| \geq \log^2 T$ is $o(1)$, while for $|\operatorname{Im} w| \leq \log^2 T, \operatorname{Re} w = -1/\log Y$ we use

$$\Gamma(w) \ll e^{-|\operatorname{Im} w|} |w|^{-1}$$

to obtain

$$\sum_{n=1}^{\infty} d_k(n) n^{-s} e^{-n/Y} \ll 1 + \int_{-\log^2 T}^{\log^2 T} \left| \zeta(\tfrac{1}{2} + it + iv) \right|^k e^{-|v|} \frac{dv}{|1/\log Y + iv|}.$$

An application of the Cauchy–Schwarz inequality gives now

$$(9.36) \qquad \left| \sum_{n=1}^{\infty} d_k(n) n^{-s} e^{-n/Y} \right|^2$$

$$\ll 1 + \log \log Y \int_{-\log^2 T}^{\log^2 T} \left| \zeta(\tfrac{1}{2} + it + iv) \right|^{2k} \frac{e^{-|v|} \, dv}{|1/\log Y + iv|},$$

since

$$(9.37)$$

$$\int_{-\log^2 T}^{\log^2 T} \frac{e^{-|v|} \, dv}{|1/\log Y + iv|} \ll \log Y \int_0^{1/\log Y} dv + \int_{1/\log Y}^{1} v^{-1} \, dv + \int_1^{\log^2 T} e^{-v} \, dv$$

$$\ll \log \log Y.$$

Now we integrate (9.36) over t for $T - Y \leqslant t \leqslant T + Y$. The left-hand side of (9.36) gives by using Theorem 5.2

$$\int_{T-Y}^{T+Y} \left| \sum_{n=1}^{\infty} d_k(n) n^{-s} e^{-n/Y} \right|^2 dt$$

$$= 2Y \sum_{n=1}^{\infty} d_k^2(n) n^{-1-2/\log Y} e^{-2n/Y} + O\left(\sum_{n=1}^{\infty} d_k^2(n) n^{-2/\log Y} e^{-n/Y} \right)$$

$$\geqslant 2e^{-2} Y \sum_{n \leqslant Y} d_k^2(n) n^{-1} e^{-2n/Y} - C_1 \sum_{n \leqslant Y \log^2 Y} d_k^2(n) e^{-n/Y} + o(1)$$

$$\geqslant 2e^{-2} Y \sum_{n \leqslant Y} d_k^2(n) n^{-1} - C_2 \sum_{n \leqslant Y} d_k^2(n) - C_1 \sum_{Y < n < Y \log^2 Y} d_k^2(n) e^{-n/Y} + o(1)$$

$$\gg Y \log^{k^2} Y.$$

Here we used $e^{-x} \geqslant 1 - x$ $(x \geqslant 0)$, (9.32), (9.33), and partial summation to obtain

$$\sum_{Y < n \leqslant Y \log^2 Y} d_k^2(n) e^{-n/Y} \ll Y \log^{k^2-1} Y + Y^{-1} \int_Y^{Y \log^2 Y} t \left(\log^{k^2-1} t \right) e^{-t/Y} dt$$

$$\ll Y \log^{k^2-1} Y + Y \int_1^{\log^2 Y} x \log^{k^2-1}(xY) e^{-x} dx$$

$$\ll Y \log^{k^2-1} Y + Y \log^{k^2-1} Y \int_1^{\infty} x e^{-x} dx$$

$$\ll Y \log^{k^2-1} Y.$$

Therefore using (9.37) again we have

$$Y \log^{k^2} Y \ll \log \log Y \int_{-\log^2 T}^{\log^2 T} \frac{e^{-|v|} dv}{|1/\log Y + iv|} \int_{T-Y}^{T+Y} |\zeta(\tfrac{1}{2} + it + iv)|^{2k} dt$$

$$\ll \log \log Y \int_{-\log^2 T}^{\log^2 T} \frac{e^{-|v|} dv}{|1/\log Y + iv|} \int_{T-Y-\log^2 T}^{T+Y+\log^2 T} |\zeta(\tfrac{1}{2} + it)|^{2k} dt$$

$$\ll (\log \log Y)^2 \int_{T-Y-\log^2 T}^{T+Y+\log^2 T} |\zeta(\tfrac{1}{2} + it)|^{2k} dt,$$

and replacing $Y + \log^2 T$ by Y one obtains immediately (9.35).

This completes the proof of Theorem 9.5, but one feels that the factor $(\log \log Y)^{-2}$ in (9.35) is extraneous to the nature of the problem under consideration, since it comes from applying (9.37) twice. This is indeed so, since by employing more refined averaging techniques it is possible to remove this factor and to obtain

$$(9.38) \qquad \int_{T-Y}^{T+Y} \left|\zeta(\tfrac{1}{2} + it)\right|^{2k} dt \gg Y \log^{k^2} Y \qquad (\log^\varepsilon T \leqslant Y \leqslant T),$$

where k may be taken even as $k = N/2$, $N \geqslant 1$ a fixed integer.

Finally, we prove a general result concerning lower bounds for mean values of $|\zeta(\sigma + it)|$, $\sigma \geqslant \frac{1}{2}$. This is

THEOREM 9.6. If $k \geqslant 1$ is a fixed integer, $\sigma \geqslant \frac{1}{2}$ is fixed, $\log^{1+\varepsilon} T \leqslant Y \leqslant T$, $T \geqslant T_0$, then uniformly in σ

$$(9.39) \qquad \int_{T-Y}^{T+Y} |\zeta(\sigma + it)|^k dt \gg Y.$$

Proof of Theorem 9.6. Let $\sigma_1 = \sigma + 2$, $s_1 = \sigma_1 + it$, $T - Y/2 \leqslant t \leqslant T + Y/2$. Then $\zeta(s_1) \gg 1$ and

$$(9.40) \qquad \int_{T-Y/2}^{T+Y/2} \left|\zeta(\sigma_1 + it)\right|^k dt \gg Y.$$

Let now $\mathscr{D}$ be the rectangle with vertices $\sigma + iT + iY$, $\sigma + iT - iY$, $\sigma_2 + iT + iY$, $\sigma_2 + iT - iY$, $\sigma_2 = \sigma + 3$ and let X be a parameter which satisfies $T^{-c} \leqslant X \leqslant T^c$ for some $c > 0$. The residue theorem then gives

$$\zeta^k(s_1) = (2\pi i)^{-1} \int_{\mathscr{D}} \frac{\zeta^k(w)}{w - s_1} \exp\left((w - s_1)^2\right) X^{s_1 - w} dw.$$

On the horizontal sides of $\mathscr{D}$ we have $|\mathrm{Im}(w - s_1)| \geqslant Y/2$; hence

$$\left|\exp\left((w - s_1)^2\right)\right| \leqslant \exp(-C \log^{1+\varepsilon} T).$$

Therefore the condition $T^{-c} \leqslant X \leqslant T^c$ ensures that

$$\zeta^k(\sigma_1 + it) \ll o(1) + X^2 \int_{T-Y}^{T+Y} |\zeta(\sigma + iv)|^k \left|\exp\left((-2 + iv - it)^2\right)\right| dv$$

$$+ X^{-1} \int_{T-Y}^{T+Y} \left|\exp\left((1 + iv - it)^2\right)\right| dv.$$

Integrating this estimate and using (9.40) we obtain

$$(9.41) \qquad Y \ll X^2 \int_{T-Y}^{T+Y} |\zeta(\sigma + iv)|^k \, dv \left(\int_{T-Y/2}^{T+Y/2} e^{-(v-t)^2} \, dt \right)$$

$$+ X^{-1} \int_{T-Y}^{T+Y} dv \left(\int_{T-Y/2}^{T+Y/2} e^{-(v-t)^2} \, dt \right)$$

$$\ll X^2 \int_{T-Y}^{T+Y} |\zeta(\sigma + iv)|^k \, dv + X^{-1} Y.$$

Let now

$$I = \int_{T-Y}^{T+Y} |\zeta(\sigma + iv)|^k \, dv,$$

and choose first $X = Y^\varepsilon$. Then (9.41) gives $I \gg Y^{1-2\varepsilon}$, showing that I cannot be too small. Now choose $X = Y^{1/3} I^{-1/3}$, so that

$$T^{-k/18} \ll X \ll Y.$$

With this choice of X (9.41) reduces to

$$Y \ll Y^{2/3} I^{1/3},$$

and (9.39) follows.

Notes

The content of Section 9.2 represents standard material which may be found, for example, in Chapter 8 of E. C. Titchmarsh (1951). Titchmarsh in fact proves more about bounds for $|\zeta(1 + it)|$ than we do, namely,

$$\limsup_{t \to \infty} \frac{|\zeta(1 + it)|}{\log \log t} \geqslant e^\gamma, \qquad \limsup_{t \to \infty} \frac{1/|\zeta(1 + it)|}{\log \log t} \geqslant 6\pi^{-2} e^\gamma.$$

Later in his book (in Chapter 14) Titchmarsh shows, assuming the Riemann hypothesis to be true,

$$\limsup_{t \to \infty} \frac{|\zeta(1 + it)|}{\log \log t} \leqslant 2 e^\gamma, \qquad \limsup_{t \to \infty} \frac{1/|\zeta(1 + it)|}{\log \log t} \leqslant 12\pi^{-2} e^\gamma.$$

Thus on the Riemann hypothesis it is only a factor of 2 which remains in doubt, for otherwise the order of $|\zeta(1 + it)|^{\pm 1}$ is precisely determined.

Sections 9.3 and 9.4 are based on the work of R. Balasubramanian and K. Ramachandra (1977), which is a continuation of K. Ramachandra (1974a) and (1977). The bound in (9.24) is the best one known, and although one could explicitly evaluate the constant A_1 in (9.24), its value would turn out to be small. However, by a more complicated analysis Balasubramanian and

Ramachandra [see Ramachandra (1981)] have succeeded in showing that (9.24) holds with $A_1 = \frac{3}{4}$. The bound in (9.25) is also the best known result, except when $Y \gg T$, when it is superseded by

$$\log|\zeta(\sigma + iT)| = \Omega_+\left(\log^{1-\sigma}T(\log\log T)^{-\sigma}\right).$$

This result holds for $\frac{1}{2} < \sigma < 1$ fixed and is due to H. L. Montgomery (1977).

Theorem 9.4 is much stronger than the corresponding results of E. C. Titchmarsh (1951), Chapter 8. Titchmarsh proves there the bound (Theorem 7.19)

$$\int_0^\infty \left|\zeta\left(\tfrac{1}{2} + it\right)\right|^{2k} e^{-\delta t}\, dt \geqslant C_k \delta^{-1}\left(\log \delta^{-1}\right)^{k^2} \qquad [0 < \delta \leqslant \delta_0(k)],$$

where $k \geqslant 1$ is a fixed integer. This result seems to imply only

$$\int_0^T \left|\zeta\left(\tfrac{1}{2} + it\right)\right|^{2k} dt = \Omega_+\left(T \log^{k^2} T\right),$$

which is weaker than (9.38).

The lower bound in (9.38) was proved by K. Ramachandra (1978, 1980a), who developed a method capable of dealing with general Dirichlet series. A particularly interesting result of his concerns derivatives of the zeta-function in short intervals. Namely, if $r \geqslant 0$ is a fixed integer, then he proved

$$\int_{T-Y}^{T+Y} \left|\zeta^{(r)}\left(\tfrac{1}{2} + it\right)\right| dt \gg Y(\log Y)^{r+1/4} \qquad \left(\log^C T \leqslant Y \leqslant T\right).$$

Ramachandra also succeeded in proving

$$\int_0^T \left|\zeta\left(\tfrac{1}{2} + it\right)\right| dt \ll T(\log T)^{1/4},$$

but it has still not been proved that

$$\int_0^T \left|\zeta\left(\tfrac{1}{2} + it\right)\right| dt \sim CT(\log T)^{1/4} \qquad (C > 0, T \to \infty),$$

although this seems very probable. D. R. Heath-Brown (1981a) used convexity arguments to show that

(9.42) $$\int_0^T \left|\zeta\left(\tfrac{1}{2} + it\right)\right|^{2k} dt \gg T(\log T)^{k^2}$$

holds for all rational $k \geqslant 0$, and for all real $k \geqslant 0$ if the Riemann hypothesis is true. A slightly weaker result of the type above when $k > 0$ is irrational is proved by K. Ramachandra (1983, 1984a, b) where some related problems are also discussed.

Theorem 9.6 (due to K. Ramachandra) is based on Theorem 3 of R. Balasubramanian (1978), which is much more general than Theorem 9.6. Indeed, the only facts about $\zeta(s)$ that we used in our proof were $\zeta(s) \gg 1$ for $\sigma \geqslant 2$ and $\zeta(s) \ll t^{1/6}$ for $\sigma \geqslant \frac{1}{2}$, so that obviously the proof will work for much more general Dirichlet series than $\zeta(s)$. Although weaker than Theorem 9.5 when $\sigma = \frac{1}{2}$, Theorem 9.6 has the advantage that the exponent of $|\zeta|$ does not have to be even as in Theorem 9.5.

If $Y = T$ and $\sigma < 1$, then the integrand of $\int_{\mathcal{D}}$ in the proof of Theorem 9.6 has the pole $w = 1$ lying on $\mathcal{D}$. In this case one needs to make a small indentation around $w = 1$ for the proof still to be valid.

/ CHAPTER TEN /
ZEROS ON THE CRITICAL LINE

10.1 LEVINSON'S METHOD

The Riemann–von Mangoldt formula (1.44) asserts that

$$(10.1) \qquad N(T) = \frac{T}{2\pi}\log\frac{T}{2\pi} - \frac{T}{2\pi} + O(\log T),$$

where $N(T)$ is the number of zeros of $\zeta(s)$ in the region $0 < \sigma < 1, 0 < t \leqslant T$. The precise location of zeros of $\zeta(s)$ remains to this day the greatest mystery of zeta-function theory, and the famous Riemann hypothesis may be recast in the form $N(T) = N_0(T)$ for all $T > 0$, where $N_0(T)$ is the number of zeros of $\zeta(s)$ of the form $s = \frac{1}{2} + it, 0 < t \leqslant T$. The first nontrivial result, that $N_0(T) \to \infty$ as $T \to \infty$, is due to G. H. Hardy in 1914, and his method actually gives

$$(10.2) \qquad N_0(T) > CT \qquad (C > 0, T \geqslant T_0).$$

This was improved in 1942 by A. Selberg to

$$(10.3) \qquad N_0(T) > CT\log T \qquad (C > 0, T \geqslant T_0),$$

and in view of (10.1) it follows that (up to the value of C) (10.3) is the best possible. Although Selberg's method could be used to yield an effective value of C in (10.3), this value was not made specific. Further significant progress in this field was achieved in 1974 by N. Levinson, who proved

$$(10.4) \qquad N_0(T + U) - N_0(T) > C(N(T + U) - N(T))$$

$$\left[U = TL^{-10}, L = \log(T/2\pi) \right],$$

with $C = \frac{1}{3}$. His paper (1974) was quite appropriately titled: "More than one third of zeros of Riemann's zeta-function are on $\sigma = 1/2$". Although subsequent research has increased somewhat the value $C = \frac{1}{3}$ in (10.4), it does not seem very likely that Levinson's method alone can ever yield $N_0(T) \sim N(T)$ as $T \to \infty$, which would of course still be much weaker than the Riemann hypothesis.

Levinson's method is deep and the technical details complicated, but the basic ideas of the proof of (10.4) can be explained relatively quickly. Therefore we shall not give a complete proof of (10.4), but only a sketch of the fundamental ideas involved. Instead, in Section 10.2 we shall prove Theorem 10.1, which gives a good lower bound for the number of zeros on the critical line in short intervals.

The basic principles of Levinson's method are as follows. The functional equation for $\zeta(s)$ may be written as

(10.5) $h(s)\zeta(s) = h(1-s)\zeta(1-s), \qquad h(s) = \pi^{-s/2}\Gamma(s/2).$

If $|\arg s| \leq \pi - \delta$ and $|\operatorname{Im} \log(s/2\pi)| < \pi$, then by Stirling's formula we have

(10.6) $h(s) = e^{f(s)}, \qquad f(s) = \frac{1}{2}(s-1)\log\frac{s}{2\pi} - \frac{s}{2} + \frac{1}{2}\log 2 + O\left(\frac{1}{|s|}\right),$

which by differentiation gives

(10.7) $$\frac{h'(s)}{h(s)} = f'(s) = \frac{1}{2}\log\frac{s}{2\pi} + O\left(\frac{1}{|s|}\right).$$

Hence for $|\sigma| \leq 3$, $t \geq t_0$

$$f'(s) + f'(1-s) = \log(t/2\pi) + O(1/t).$$

Taking the derivative of (10.5) and eliminating $\zeta(1-s)$ we obtain

(10.8) $h(s)\zeta(s)(f'(s) + f'(1-s)) = -h(s)\zeta'(s) - h(1-s)\zeta'(1-s).$

Now if $s = \frac{1}{2} + it$, then the right-hand side of (10.8) is the sum of two complex conjugates; hence $\zeta(\frac{1}{2} + it) = 0$ whenever

(10.9) $\arg\left\{h(\frac{1}{2} + it)\zeta'(\frac{1}{2} + it)\right\} \equiv \frac{\pi}{2}(\operatorname{mod} \pi).$

It follows from (10.8) that

$$\zeta'(s) = -\chi(s)\{(f'(s) + f'(1-s))\zeta(1-s) + \zeta'(1-s)\},$$

where as usual $\chi(s) = h(1-s)/h(s)$, that is, $\zeta(s) = \chi(s)\zeta(1-s)$. Therefore

$\zeta(\tfrac{1}{2} + it) = 0$ whenever

$$(10.10) \quad \arg\left\{ h\left(\tfrac{1}{2} + it\right)\left(\left(f'\left(\tfrac{1}{2} + it\right) + f'\left(\tfrac{1}{2} - it\right)\right)\zeta\left(\tfrac{1}{2} + it\right) + \zeta'\left(\tfrac{1}{2} + it\right)\right)\right\}$$

$$\equiv \frac{\pi}{2} \,(\text{mod } \pi).$$

Since $\arg h(s)$ is readily found from (10.6), it suffices to determine the change in the argument of the function

$$G(s) = \zeta(s) + \zeta'(s)/(f'(s) + f'(1 - s))$$

on $\sigma = \tfrac{1}{2}$. Indeed, if $\arg G(\tfrac{1}{2} + it)$ did not change, it would follow from (10.6) and (10.10) that $\zeta(\tfrac{1}{2} + it)$ would have essentially its full quota of zeros, namely, $N_0(T) = N(T) + O(\log T)$. By the principle of the argument and Jensen's formula (see Lemma 1.1) it may be found that

$$\arg G\left(\tfrac{1}{2} + it\right)\Big|_T^{T+U} = -2\pi N_G(\mathscr{D}) + O(\log T),$$

where $N_G(\mathscr{D})$ is the number of zeros of $G(s)$ in the rectangle $\mathscr{D}$ with vertices $\tfrac{1}{2} + iT$, $3 + iT$, $\tfrac{1}{2} + iT + iU$, $3 + iT + iU$. If $\Psi(s)$ is an integral function, then the zeros of $G(s)$ are the zeros of $\Psi(s)G(s)$, and for $0 < a < \tfrac{1}{2}$ a classical lemma of Littlewood gives

$$(10.11) \quad 2\pi\left(\tfrac{1}{2} - a\right)N_G(\mathscr{D}) \leq \int_T^{T+U} \log|\Psi(a + it)G(a + it)| \, dt + O(UL^{-1})$$

$$\leq \frac{1}{2} U \log\left(U^{-1} \int_T^{T+U} |\Psi(a + it)G(a + it)|^2 \, dt\right)$$

$$+ O(UL^{-1}).$$

The choice of the "mollifying" factor $\Psi(s)$ and the estimation of the last integral are the most delicate parts of the proof. Levinson chose

$$(10.12) \quad \Psi(s) = \sum_{1 \leq j \leq y} b_j j^{-s}, \qquad y = T^{1/2}L^{-20}, \qquad b_j = \frac{\mu(j)\log(y/j)}{j^{1/2-a}\log y},$$

but this is not the optimal choice, and subsequent research has produced more effective, but more complicated, mollifiers.

The estimation of the last integral in (10.11) is reduced via the Riemann–Siegel formula (4.3) to the estimation of the integral

$$(10.13) \qquad \int_T^{T+U} |\Psi(a + it)H(a + it)|^2 \, dt,$$

where

$$H(s) = g_1(s) - L^{-1}g_2(s) + L^{-1}\chi_1(s)g_3(s),$$

$$\chi_1(s) = (T/2\pi)^{1/2-\sigma}\exp\left(\frac{\pi i}{4} - it\log\left(\frac{t}{2\pi e}\right)\right),$$

$$g_1(s) = \sum_{n \leqslant Q} n^{-s}, \qquad g_2(s) = \sum_{n \leqslant Q}(\log n)n^{-s}, \qquad g_3(s) = \sum_{n \leqslant Q}(\log n)n^{s-1},$$

$$Q = (T/2\pi)^{1/2}.$$

After lengthy calculations the proof is completed by showing that for $a = \frac{1}{2} + O(L^{-1})$ the integral in (10.13) does not exceed DU for some suitable positive constant D.

10.2 ZEROS ON THE CRITICAL LINE IN SHORT INTERVALS

We shall now develop a method whose technical details are much simpler than those of Levinson, in order to obtain a good lower bound for the number of zeros of $\zeta(s)$ on the critical line in short intervals. The basic idea, which goes back to Hardy, is to compare certain averages involving the function

(10.14)

$$Z(u) = \chi^{-1/2}(\tfrac{1}{2} + iu)\zeta(\tfrac{1}{2} + iu), \qquad \chi(s) = 2^s\pi^{s-1}\sin(\pi s/2)\Gamma(1 - s).$$

From the functional equation for $\zeta(s)$ in the form $\zeta(s) = \chi(s)\zeta(1 - s)$ we may deduce that $Z(u)$ is real for u real, while $|Z(u)| = |\zeta(\tfrac{1}{2} + iu)|$. These fundamental properties of the function $Z(u)$ will be used in the proof of

THEOREM 10.1. Let $T^{\alpha+\varepsilon} \leqslant G \leqslant T, T \geqslant T_0$, where $\alpha = 0.329021356\ldots$. Then

(10.15) $$N_0(T + G) - N_0(T - G) > CG/\log T$$

for some absolute $C > 0$.

Actually, the proof will yield $T^c \leqslant G \leqslant T$ as the range for which the conclusion of the theorem holds, where

$$c = \inf(\varkappa + \lambda - \tfrac{1}{2}) + \varepsilon,$$

and $(\varkappa, \lambda)$ is any exponent pair. Choosing the exponent pair

$$(\varkappa, \lambda) = (\alpha/2 + \varepsilon, (\alpha + 1)/2 + \varepsilon), \alpha = 0.329021356\ldots$$

which minimizes $\varkappa + \lambda$ (see Section 2.3 of Chapter 2) we obtain the result of the theorem, and presumably the use of two-dimensional methods would enlarge the range of G slightly.

If the Riemann hypothesis were true, then (10.1) would imply that there are at least $CG \log T$ zeros of the form $\frac{1}{2} + it$, $T - G \leqslant t \leqslant T + G$ for $\log^\varepsilon T \leqslant G \leqslant T$, so that the bound $CG/\log T$ of the theorem is less than the "expected" bound $CG \log T$ by a factor of $\log^2 T$. A factor of $\log T$ in our proof is probably lost in introducing the smoothing exponential factor in the integrals I_1 and I_2 in (10.16) and (10.17). While this loss could be perhaps avoided by using a more sophisticated smoothing function, it seems, however, hardly possible to refine our method and obtain a lower bound of the form $CG \log T$. Nevertheless, the bound $CG/\log T$ is still very close to the expected bound $CG \log T$, and even more important it is attained for a relatively small value of G, while (10.4) gives this type of result for $G = T(\log T/2\pi)^{-10}$, which is only by a log factor smaller than T itself.

Proof of Theorem 10.1. We shall contrast the behavior of the integrals

$$(10.16) \qquad I_1 = \int_{t-U}^{t+U} Z(u)\exp\left(-(t-u)^2 U^{-2} L\right) du$$

and

$$(10.17) \qquad I_2 = \int_{t-U}^{t+U} |Z(u)|\exp\left(-(t-u)^2 U^{-2} L\right) du,$$

where $Z(u)$ is defined by (10.14) and

$$T/2 \leqslant t \leqslant T, \qquad 1 \ll U < T^\varepsilon, \qquad L = 1000 \log T.$$

It is sufficient to prove the theorem for the smallest possible value of G, say G_0, since for $G > G_0$ the interval $[T - G, T + G]$ can be split into $\gg G/G_0$ subintervals of length G_0, each of which contains at least $CG_0/\log T$ zeros of $\zeta(s)$. If $\zeta(\frac{1}{2} + iu)$ has no zero for $t - U \leqslant u \leqslant t + U$, then by the properties of $Z(u)$ we have $|I_1| = I_2$. Let now $\mathscr{A}$ denote the set of points from $[T - G, T + G]$ for which $|I_1| = I_2$. The main step in the proof will be to show that

(10.18)

$$m(\mathscr{A}) \ll GU^{-1}L\left(\log UL^{-1/2}\right)^{-1/2}, \qquad G = T^{\alpha+\varepsilon}, \qquad \alpha = 0.329021356\ldots,$$

where $m(\mathscr{A})$ is the measure of $\mathscr{A}$. To achieve this note that for $U > \log^{1/2+\varepsilon} T$

$$(10.19) \quad I_2 \geqslant e^{-1} \int_{t-UL^{-1/2}}^{t+UL^{-1/2}} |Z(u)| \, du = e^{-1} \int_{t+UL^{-1/2}}^{t+UL^{-1/2}} |\zeta(\tfrac{1}{2} + iu)| \, du$$

$$\gg UL^{-1/2}\log^{1/4}(UL^{-1/2}).$$

Here we used (9.38), while the weaker result (9.39) (which can also be seen to hold for $Y \gg \log^{\varepsilon} T$) would also lead to the same lower bound $CG/\log T$ in the theorem. Now by the Cauchy–Schwarz inequality

(10.20)

$$m(\mathscr{A})UL^{-1/2}\log^{1/4}(UL^{-1/2}) \ll \int_{\mathscr{A}} |I_1|\, dt \leqslant (m(\mathscr{A}))^{1/2} \left(\int_{T-G}^{T+G} I_1^2\, dt \right)^{1/2};$$

hence (10.18) follows if we can show that

(10.21) $\displaystyle \int_{T-G}^{T+G} I_1^2\, dt \ll GU \qquad (G = T^{\alpha+\varepsilon},\ \alpha = 0.329021356\ldots).$

Assuming that (10.21) holds we can finish the proof as follows. Divide the interval $[T-G, T+G]$ into subintervals $J_1, J_2, J_3, J_4, \ldots$ of length $2U$ (except the last, which may be shorter). Then at least one of the intervals $J_{3k-2}, J_{3k-1}, J_{3k}$ $(k = 1, 2, \ldots)$ contains a point t_0 such that $\zeta(\tfrac{1}{2} + it_0) = 0$, or otherwise all points of J_{3k-1} belong to $\mathscr{A}$. Denote by R the number of intervals J_{3k-1} whose points all belong to $\mathscr{A}$. Then using (10.18) we have

$$2RU \leqslant m(\mathscr{A}) \ll GU^{-1}L\big(\log UL^{-1/2}\big)^{-1/2},$$

so that there exist at least

$$\big(\tfrac{1}{6} + o(1)\big)GU^{-1} - R > \tfrac{1}{7}GU^{-1} - CGU^{-2}L\big(\log UL^{-1/2}\big)^{-1/2}$$

zeros of $\zeta(s)$ of the form $s = \tfrac{1}{2} + it,\ T - G \leqslant t \leqslant T + G$. Taking $U = AL$, where $A > 0$ is sufficiently large, it is seen that

(10.22) $\big(\tfrac{1}{6} + o(1)\big)GU^{-1} - R > \tfrac{1}{8}GU^{-1} \gg G/\log T,$

which proves the theorem.

To complete the proof it remains to show that the mean value estimate (10.21) holds. First, we shall evaluate explicitly I_1 by using

$$\zeta(s) = \sum_{n \leqslant x} n^{-s} + \chi(s)\sum_{n \leqslant y} n^{s-1} + O(x^{-\sigma}) + O(t^{1/2-\sigma}y^{\sigma-1}),$$

which is the approximate functional equation (4.1). For $s = \tfrac{1}{2} + it + iu$, $t \asymp T,\ |u| \leqslant U \ll T^{\varepsilon},\ x = y = ((t+u)/2\pi)^{1/2}$, we obtain

(10.23) $\displaystyle I_1 = \int_{-U}^{U} \exp(-u^2 U^{-2}L)x^{-1/2}\big(\tfrac{1}{2} + it + iu\big)\zeta\big(\tfrac{1}{2} + it + iu\big)\, du$

$$= S_1 + \bar{S}_1 + O(UT^{-1/4}),$$

where abbreviating $Q = (t/2\pi)^{1/2}$ we have

(10.24) $S_1 = \sum_{n \leqslant Q} n^{-1/2} \int_{-U}^{U} x^{1/2}(\tfrac{1}{2} + it + iu) n^{it+iu} \exp(-u^2 U^2 L)\, du.$

We recall now (1.25), namely,

$$x(s) = (2\pi/t)^{\sigma+it-1/2} \exp(it + \pi i/4)(1 + O(t^{-1})),$$

and set, for brevity's sake,

(10.25) $f(x) = \tfrac{1}{2}x \log(2\pi/x) + \tfrac{1}{2}x + x \log n + \tfrac{1}{8}\pi.$

Using Taylor's formula and

$$|\exp(ix) - \exp(iy)| \leqslant |x - y| \qquad (x, y \text{ real}),$$

it follows that

(10.26) $S_1 = \sum_{n \leqslant Q} n^{-1/2} \exp(if(t)) \int_{-\infty}^{\infty} \exp\left(iu \log\left(\frac{n}{Q}\right) - \frac{iu^2}{4t} - u^2 U^{-2} L \right) du$

$$+ O(U^4 T^{-7/4}) + O(UT^{-3/4}),$$

since

$$\int_{\pm U}^{\pm \infty} \exp\left(-iu \log\left(\frac{n}{Q}\right) - \frac{iu^2}{4t} - u^2 U^{-2} L \right) du \leqslant U \exp\left(-\frac{L}{2} \right) = UT^{-500}.$$

To evaluate the integral in (10.26) we use (A.38), namely

(10.27) $\int_{-\infty}^{\infty} \exp(Az - Bz^2)\, dz = (\pi/B)^{1/2} \exp(A^2/4B) \qquad (\text{Re } B > 0).$

Therefore if we set $X = (4t/i)^{-1} + U^{-2}L$, then

$$X^{-1/2} \exp\left(-(4X)^{-1} \log^2(n/Q) \right)$$

$$= UL^{-1/2} \exp\left(-U^2(4L)^{-1} \log^2(n/Q) \right) + O(U^3 T^{-1}),$$

so that (10.26) becomes

(10.28) $S_1 = \pi^{1/2} UL^{-1/2} \sum_{n \leqslant Q} n^{-1/2} \exp(if(t)) \exp\left(-U^2(4L)^{-1} \log^2(n/Q) \right)$

$$+ O(U^3 T^{-3/4}).$$

Although the last sum has many terms, the presence of the second exponential factor and our choice $L = 1000 \log T$ will make the contribution of many of these terms negligible. If $U = AL$ and $A > 2$, then using the elementary inequality

$$|\log(1 - x)| \geqslant x/2, \qquad 0 \leqslant x \leqslant \tfrac{1}{2},$$

we obtain, for $1 \leqslant n \leqslant Q - QU^{-1}L$,

$$U^2 L^{-1} \log^2(n/Q) \geqslant U^2 L^{-1} \log^2(1 - U^{-1}L) \geqslant L/4 = 250 \log T;$$

hence

$$\exp\left(-U^2(4L)^{-1} \log^2(n/Q)\right) < T^{-60}$$

and

(10.29)

$$S_1 = \pi^{1/2} U L^{-1/2} \sum_{Q - QU^{-1}L < n \leqslant Q} n^{-1/2} \exp(if(t)) \exp\left(-U^2(4L)^{-1} \log^2(n/Q)\right)$$

$$+ O(U^3 T^{-3/4}).$$

This is the expression for S_1 that we need, and we may now proceed with the proof of (10.21). In (10.29) we may replace $Q = (t/2\pi)^{1/2}$ by $(T/2\pi)^{1/2}$ with an error $\ll U L^{-1/2} T^{-1/4}$, since $|t - T| \leqslant G < T^{1/3}$. Thus we have

$$\int_{T-G}^{T+G} I_1^2 \, dt \ll GU + U^2 L^{-1}$$

$$\times \int_{T-G}^{T+G} \left| \sum_{N_1 < n \leqslant N_2} n^{-1/2 - it} \exp\left(-U^2(4L)^{-1} \log^2(n/Q)\right) \right|^2 dt,$$

where

$$N_1 = (T/2\pi)^{1/2} - (T/2\pi)^{1/2} U^{-1} L, \qquad N_2 = (T/2\pi)^{1/2}.$$

Also in $\exp(-U^2(4L)^{-1} \log^2(n/Q))$ we may replace Q by $(T/2\pi)^{1/2}$ with a small error, thus obtaining

$$\int_{T-G}^{T+G} I_1^2 \, dt \ll GU + U^2 L^{-1} \int_{T-G}^{T+G} \left| \sum_{N_1 < n \leqslant N_2} g(n) n^{-1/2 - it} \right|^2 dt,$$

where

$$g(n) = \exp\left\{-U^2(4L)^{-1} \log^2\left((2\pi n^2/T)^{1/2}\right)\right\} \leqslant 1,$$

and $g(n)$ is a monotonic function for $N_1 \leqslant n \leqslant N_2$ which does not depend on t.

Using again (10.27) we obtain

$$\int_{T-G}^{T+G} \left| \sum_{N_1 < n \leqslant N_2} g(n) n^{-1/2 - it} \right|^2 dt$$

$$\leqslant e \int_{T-G}^{T+G} \exp\left(-(T-t)^2 G^{-2} \right) \left| \sum \right|^2 dt$$

$$\leqslant e \int_{-\infty}^{\infty} e^{-t^2 G^{-2}} \left| \sum_{N_1 < n \leqslant N_2} g(n) n^{-1/2 - it - iT} \right|^2 dt \ll GN_1^{-1}(N_2 - N_1)$$

$$+ G \left| \sum_{N_1 < n < m \leqslant N_2} g(m) g(n)(mn)^{-1/2} \exp\left(iT \log \frac{m}{n} \right) \exp\left(-\frac{G^2}{4} \log^2 \frac{m}{n} \right) \right|.$$

Setting $m = n + r$ it is seen that

$$\exp\left(-\frac{G^2}{4} \left(\log \frac{m}{n} \right)^2 \right) < T^{-c}$$

for any fixed $c > 0$ if $r > T^{1/2} G^{-1} L$, and since $N_1^{-1}(N_2 - N_1) \ll U^{-1} L$ we obtain by partial summation

(10.30) $\quad \int_{T-G}^{T+G} I_1^2 \, dt \ll GU$

$$+ GU^2 L^{-1} \sum_{1 \leqslant r \leqslant T^{1/2} G^{-1} L} T^{-1/2} \max_{N_1 \leqslant M_1 < M_2 \leqslant N_2}$$

$$\times \left| \sum_{M_1 \leqslant n \leqslant M_2} \exp(iT \log(1 + r/n)) \right|.$$

For r fixed and $M_1 \leqslant x \leqslant M_2$ we set $F(x) = T \log(1 + r/x)$. Then $F'(x) \gg 1$ and for $k = 1, 2, \ldots$

$$rT^{(1-k)/2} \ll F^{(k)}(x) \ll rT^{(1-k)/2},$$

where the $\ll$ constants depend only on k. Therefore the last sum in (10.30) is

$\ll r^{\varkappa}T^{\lambda/2}$ if $(\varkappa, \lambda)$ is an exponent pair, giving

(10.31) $$\int_{T-G}^{T+G} I_1^2\, dt \ll GU + GU^2 L^{-1} T^{(\lambda-1)/2} \sum_{r \leqslant T^{1/2}G^{-1}L} r^{\varkappa}$$

$$\ll GU + G^{-\varkappa} U^2 L^{\varkappa} T^{(\varkappa+\lambda)/2} \ll GU$$

for $G \geqslant T^{(\varkappa+\lambda)/(2+2\varkappa)}L$, if $U = AL$. But from Lemma 2.8 we have that $(k, l) = (\varkappa/(2+2\varkappa), \frac{1}{2} + \lambda/(2+2\varkappa))$ is an exponent pair if $(\varkappa, \lambda)$ is an exponent pair and moreover

$$k + l - \tfrac{1}{2} = (\varkappa + \lambda)/(2 + 2\varkappa).$$

Therefore following the iteration of the A- and B-processes (Section 2.3 of Chapter 2) for the construction of the exponent pair $(\alpha/2 + \varepsilon, \alpha/2 + \frac{1}{2} + \varepsilon)$, $\alpha = 0.329021356\ldots$ we find that for any given $\varepsilon > 0$ there is an exponent pair $(\varkappa, \lambda)$ such that

$$\int_{T-G}^{T+G} I_1^2\, dt \ll GU, \qquad G \geqslant T^{(\varkappa+\lambda)/(2+2\varkappa)}L, \qquad (\varkappa + \lambda)/(2 + 2\varkappa) < \alpha + 4\varepsilon.$$

This yields (10.21) and completes the proof of Theorem 10.1.

10.3 CONSECUTIVE ZEROS ON THE CRITICAL LINE

Let γ_n denote the imaginary part of the nth zero of $\zeta(s)$ of the form $s = \frac{1}{2} + it$, $t > 0$. One of the most interesting unsettled problems concerning the sequence $\{\gamma_n\}_{n=1}^{\infty}$ is the estimation of the gap between consecutive zeros on the critical line, that is, inequalities of the form

(10.32) $$\gamma_{n+1} - \gamma_n \ll \gamma_n^c \log^d \gamma_n$$

with some $0 \leqslant c < 1$ and $d \geqslant 0$. The first result in this direction is a classical theorem of G. H. Hardy and J. E. Littlewood from 1918 which states that (10.32) holds with $c = \frac{1}{4} + \varepsilon$. This result of Hardy and Littlewood remained the best one for an exceptionally long time, until, working independently, J. Moser and R. Balasubramanian proved in the 1970s that (10.32) holds with $c = \frac{1}{6}$, $d = 5 + \varepsilon$, and $c = \frac{1}{6} + \varepsilon$, respectively. Their methods both reduce the problem of obtaining (10.32) to the estimation of a certain "short" exponential sum after some averaging process. Following Moser's approach and taking into account the particular structure of the exponential sum in question, A. A. Karacuba in 1981 obtained (10.32) with $c = \frac{5}{32}$, $d = 2$. The exponent $\frac{5}{32}$ is interesting, since it is smaller than the exponent $\frac{35}{216}$ (see Corollary 7.1) for the order of $\zeta(\frac{1}{2} + it)$. In this section we shall use the method of Section 10.2, which is simpler than Moser's or Balasubramanian's method, to prove

THEOREM 10.2. For any $\varepsilon > 0$ and $n \geqslant n_0(\varepsilon)$

$$(10.33) \qquad \gamma_{n+1} - \gamma_n < \gamma_n^{\vartheta + \varepsilon}, \qquad \vartheta = 0.1559458\ldots .$$

Note that here $\vartheta = 0.1559458\ldots < 0.15625 = \frac{5}{32}$, so that (10.33) improves Karacuba's result. Theorem 10.2 seems to be the sharpest known result of its type, although the application of multidimensional techniques for the estimation of exponential sums would lead to some further slight improvements.

Proof of Theorem 10.2. The technique of proof is very similar to the one used for Theorem 10.1. However, in contrast with (10.20) there is no need now for the Cauchy–Schwarz inequality, and also the particular structure of the exponential sum S in (10.38) can be effectively put into use, so that the exponent ϑ in Theorem 10.2 is less than twice the exponent α in Theorem 10.1. We contrast again the behavior of I_1 and I_2, as defined by (10.16) and (10.17), taking this time $t = T$, $T^\varepsilon < U < T^{1/4}$, $L = (\log T)^{1+\varepsilon}$, and $Q = (T/2\pi)^{1/2}$. Then by (9.39) we have

$$(10.34) \qquad \int_{T - UL^{-1/2}}^{T + UL^{-1/2}} \left| \zeta\left(\tfrac{1}{2} + iu\right) \right| du \gg UL^{-1/2},$$

while I_2 is estimated as in Section 10.2. Analogously to (10.23) and (10.28) we find that

$$(10.35)$$

$$I_1 = \int_{T-U}^{T+U} \exp\left(-(T-u)^2 U^{-2}L\right) Z(u)\, du = S_1 + \overline{S}_1 + O(UT^{-1/4}),$$

where

$$(10.36)$$

$$S_1 = \pi^{1/2} UL^{-1/2} \sum_{n \leqslant Q} n^{-1/2} \exp(if(T)) \exp\left(-U^2(4L)^{-1} \log^2(n/Q)\right)$$

$$+ O(U^3 T^{-3/4}),$$

and $f(x)$ is defined by (10.25). Again the presence of the second exponential factor in S_1 will make the contribution of many summands negligible. This time we let

$$P = [Q] = \left[(T/2\pi)^{1/2} \right], \qquad n = P - m,$$

and consider first those n for which

$$m > QU^{-1}L^{1+\varepsilon} = (2\pi)^{-1/2}T^{1/2}U^{-1}(\log T)^{(1+\varepsilon)^2}.$$

But we have

$$U^2L^{-1}\log^2(n/Q) = U^2L^{-1}\{\log(1-(m+o(1))Q^{-1})\}^2 \geqslant \tfrac{1}{2}L^{1+\varepsilon},$$

so that the second exponential factor in (10.36) makes the contribution of those n to S_1 negligible. For the remaining n in (10.36) we obtain by partial summation, (10.35), and (10.36) that

(10.37)

$$I_1 \ll UT^{-1/4} + UT^{-1/4}L^{-1/2}\sum_M \max_{M'} \left| \sum_{M<m\leqslant M'<2M} \exp(iT\log(P-m)) \right|,$$

where $P = [(T/2\pi)^{1/2}]$, the maximum is taken over M' satisfying $M < M' \leqslant 2M$, and $\sum_M$ denotes summation over $O(\log T)$ values $M = 2^{-j}QU^{-1}L^{1+\varepsilon}$, $j = 1, 2, \ldots$, so that the exponential sums in (10.37) are short in the sense that $M = o(P)$. Therefore the problem reduces to the estimation of the exponential sum

(10.38) $\quad S = S(M, M', T) = \displaystyle\sum_{M<m\leqslant M'\leqslant 2M} \exp(iT\log(P-m)),$

$$P = \left[(T/2\pi)^{1/2}\right],$$

where $M \ll T^{1/2}U^{-1}L^{1+\varepsilon}$. From the definition of P we have $T = 2\pi(P+\theta)^2$ for some $0 \leqslant \theta < 1$, and therefore

$$T\log(P-m) - T\log P = -T\sum_{j=1}^{\infty} (m/P)^j j^{-1}$$

$$= -2\pi Pm - 2\pi(2P\theta + \theta^2)mP^{-1} - \pi m^2$$

$$-2\pi(2P\theta + \theta^2)m^2(2P^2)^{-1}$$

$$-T\{m^3/(3P^3) + m^4/(4P^4) + \cdots\}.$$

Taking into account that $\exp(2\pi ir) = 1$ for any integer r and considering separately even and odd m (to get rid of πm^2) we obtain

$$|S| \leqslant |S'| + |S''|,$$

where S' comes from even m and equals

(10.39)

$$S' = \sum_{M_1 < m \leqslant M_1' \leqslant 2M_1} \exp(2\pi i F(m)) \qquad (M \ll M_1 \ll M),$$

(10.40)

$$F(x) = c_1 x + c_2 x^2 + T(2\pi)^{-1}\{(2x)^3/(3P^3) + (2x^4)/(4P^4) + \cdots\},$$

and

$$c_1 = 2(2\theta P + \theta^2)P^{-1} = O(1), \qquad c_2 = c_1 P^{-1} = O(P^{-1}).$$

The expression for S'' (coming from odd m) is similar, and thus it will be sufficient to estimate only S' in (10.39). For $M \ll T^{1/4}$ and $M_1 \leqslant x \leqslant 2M_1$ we have $|F'(x)| \gg 1$ and

$$|F^{(k)}(x)| \asymp M^{3-k}T^{-1/2}, \quad k \leqslant 3;$$

$$|F^{(k)}(x)| \asymp T^{1-k/2} \ll M^{3-k}T^{-1/2}, \quad k > 3,$$

where the $\asymp$ constants depend only on k. This means that we may estimate S' by the theory of exponent pairs of Chapter 2 as

(10.41) $$S' \ll A^\varkappa M^\lambda,$$

where $(\varkappa, \lambda)$ is an exponent pair and

(10.42) $$A = \max_{M_1 \leqslant x \leqslant 2M_1} |F'(x)| \ll M^2 T^{-1/2}.$$

Thus for $M \gg T^{1/4}$ we use (10.41) and (10.42), while for $M \ll T^{1/4}$ we use Lemma 2.6 with $k = 3$ [taking $a = M_1$, $b = M_1'$, $f(x) = T\log(P - x)$, $\lambda_3 = T^{-1/2}$] to estimate S'. Then we obtain

(10.43) $$S \ll M^{2\varkappa+\lambda}T^{-\varkappa/2} + T^{5/24}.$$

Inserting this bound in (10.37) and summing over various M we obtain

(10.44)

$$I_1 \ll UT^{-1/4} + UT^{-1/24}\log T + UT^{(\varkappa+\lambda)/2-1/4}U^{-(2\varkappa+\lambda)}(\log T)^{2\varkappa+\lambda-1/2+\varepsilon}.$$

But if $Z(u)$ does not change sign in $[T - U, T + U]$ then $|I_1| = I_2$; hence using (10.34) we have from (10.44)

$$UL^{-1/2} \ll UT^{-1/24}\log T + UT^{(\varkappa+\lambda)/2-1/4}U^{-(2\varkappa+\lambda)}(\log T)^{2\varkappa+\lambda-1/2+\varepsilon},$$

which is a contradiction if for a suitable $\delta = \delta(\varepsilon)$ (which tends to 0 as $\varepsilon \to 0$)

$$(10.45) \qquad\qquad U = T^{(2\varkappa + 2\lambda - 1)/(8\varkappa + 4\lambda)}(\log T)^{1+\delta}.$$

Now we use Lemma 2.8 and then Lemma 2.9 to deduce that if $(\varkappa, \lambda)$ is an exponent pair, then $(\lambda/(2\varkappa + 2), (2\varkappa + 1)/(2\varkappa + 2))$ is also an exponent pair. Replacing $(\varkappa, \lambda)$ by this last exponent pair in (10.45) we obtain the condition for U in the form

$$(10.46) \qquad\qquad U = T^{(\varkappa + \lambda)/(4\varkappa + 4\lambda + 2)}(\log T)^{1+\delta}.$$

Thus for this choice of U the equality $|I_1| = I_2$ is impossible, and then U trivially satisfies $T^\varepsilon < U < T^{1/4}$. The result of this transformation is that the exponent of T in (10.46) is an increasing function of $\varkappa + \lambda$, which means that the best result will be obtained if we take for $(\varkappa, \lambda)$ the exponent pair for which $\varkappa + \lambda$ is minimal. As seen in Chapter 2 we should then take $(\varkappa, \lambda) = (\alpha/2 + \varepsilon, \frac{1}{2} + \alpha/2 + \varepsilon)$, $\alpha = 0.329021356\ldots$, which gives

$$(\varkappa + \lambda)/(4\varkappa + 4\lambda + 2) = 0.15594583\ldots + \varepsilon = \vartheta + \varepsilon.$$

This implies that for $U = T^{\vartheta + \varepsilon}$ the function $Z(u)$ must change sign for $T - U \leqslant u \leqslant T + U$, and taking $\gamma_n = T - U$ we obtain that $\gamma_{n+1} \in [T - U, T + U]$; hence (10.33) follows. It may be noted that the trivial exponent pair $(\varkappa, \lambda) = (0, 1)$ in (10.46) leads to the exponent $c = \frac{1}{6}$ of Balasubramanian and Moser in (10.32), while Karacuba's value $c = \frac{5}{32}$ follows from the standard exponent pair $(\varkappa, \lambda) = (\frac{1}{6}, \frac{2}{3})$.

Notes

For proofs that $N_0(T) \to \infty$ as $T \to \infty$ see G. H. Hardy (1914) and Chapter 2 of K. Chandrasekharan (1970), and for the proof of (10.3) see A. Selberg (1942). Both of these proofs are extensively discussed in Chapter 10 of E. C. Titchmarsh (1951) and Chapter 11 of H. M. Edwards (1974).

N. Levinson's proof of (10.4) with $C = \frac{1}{3}$ is expounded in his papers (1974, 1975a), while in Levinson (1975b) he proved that in (10.4) one may take $C = 0.3474$. This was improved to $C = 0.35$ by Shi-Tuo Lou (1981) and to $C = 0.3658$ by B. Conrey (1983). Conrey's paper deals with zeros of $\xi(s) = (s/2)(s - 1)\pi^{-s/2}\Gamma(s/2)\zeta(s)$ and its derivatives on the critical line. Let $N_m(T)$ be the number of zeros of $\xi^{(m)}(\frac{1}{2} + it)$ $(m = 0, 1, 2, \ldots)$ with $0 < t \leqslant T$ and let

$$\alpha_m = \liminf_{T \to \infty} \frac{N_m(T + U) - N_m(T)}{UL/2\pi},$$

where $U = TL^{-10}$ and $L = \log(T/2\pi)$. Thus $N_0(T)$ is as before the number of zeros of $\zeta(\frac{1}{2} + it)$ with $0 < t \leqslant T$, and Conrey has developed a complicated variant of Levinson's method to bound α_m. His bounds are: $\alpha_0 > 0.3658$, $\alpha_1 > 0.8137$, $\alpha_2 > 0.9584$, $\alpha_3 > 0.9873$, $\alpha_4 > 0.9948$, $\alpha_5 > 0.9970$, $\alpha_m = 1 + O(m^{-2})$ as $m \to \infty$. By a technique similar to the one in the proof of the

Riemann–von Mangoldt formula (see Theorem 1.7), it may be shown that

$$N^{(m)}(T) = \frac{T}{2\pi}\log\frac{T}{2\pi} - \frac{T}{2\pi} + O_m(\log T),$$

where $N^{(m)}(T)$ denotes the number of zeros of $\xi^{(m)}(s)$ with $0 < \sigma < 1$, $0 < t \leq T$. Therefore Conrey's bound $\alpha_m = 1 + O(m^{-2})$ shows that, in a certain sense, almost all zeros of $\xi^{(m)}(s)$ for m large lie on the critical line. It may be remarked that the zeros of $\zeta^{(m)}(s)$ are not in general the zeros of $\zeta^{(m)}(s)$. For a comparison between the number of zeros of $\zeta(s)$ and $\zeta'(s)$ in the region $0 < \sigma < \frac{1}{2}$, $0 < t < T$, see N. Levinson and H. L. Montgomery (1974), while an asymptotic formula for the number of complex zeros of $\zeta^{(m)}(s)$ in the region $0 < t < T$ is given by B. C. Berndt (1970).

In the text we have presented only an outline of Levinson's method, and for the complete proof of (10.4) the reader is referred to N. Levinson (1974, 1975a). For J. E. Littlewood's lemma on the number of zeros of an analytic function in a rectangle, see for example, the Appendix of K. Prachar (1967).

It was noticed independently by D. R. Heath-Brown (1979e) and A. Selberg (unpublished) that Levinson's method yields the bound (10.4) with $N_0(T)$ replaced by $N_1(T)$, the number of simple zeros of $\zeta(s)$ satisfying $s = \frac{1}{2} + it$, $0 < t \leq T$. Recently R. J. Anderson (1983) improved this result by showing that, for $T \geq T_0$,

$$N_1(T) > 0.3532 N(T).$$

Anderson's proof also uses a modification of Levinson's method.

Theorem 10.1 is a hitherto unpublished result of the author. The first result of this type was obtained by G. H. Hardy and J. E. Littlewood (1921), who proved

$$N_0(T + T^{1/2+\varepsilon}) - N_0(T) > A(\varepsilon)T^{1/2+\varepsilon} \quad [T \geq T_0(\varepsilon)],$$

and this was improved in 1942 by A. Selberg to

$$N_0(T + T^{1/2+\varepsilon}) - N_0(T) > A(\varepsilon)T^{1/2+\varepsilon}\log T \quad [T \geq T_0(\varepsilon)].$$

A. Selberg also conjectured that in the inequality above one may replace $\frac{1}{2} + \varepsilon$ by a number less than $\frac{1}{2}$. Recently, J. Moser (1983a) proved that

$$(10.47) \qquad N_0\left(T + T^{5/12}\Psi(T)\log^3 T\right) - N_0(T) > A(\Psi)T^{5/12}\Psi(T)\log^3 T,$$

where $\Psi(T)$ is any function tending to ∞ as $T \to \infty$, and $A(\Psi) > 0$ is a constant depending on Ψ. Moser's method of proof of (10.47) is entirely different from the method of proof of Theorem 10.1. A. A. Karacuba (1984) has elaborated Selberg's method to reduce $\frac{5}{12}$ to $\frac{27}{82} = 0.32926\ldots$ and eliminate $\Psi(T)$ in (10.47); but the exponent α in Theorem 10.1 is still slightly smaller, since it comes from using the optimal exponent pair which minimizes $\varkappa + \lambda - \frac{1}{2}$, if $(\varkappa, \lambda)$ is an exponent pair.

Theorem 10.2 is due to the author (1983b). The bound $c = \frac{1}{4} + \varepsilon$ in (10.32) is due to Hardy and Littlewood (1918); $c = \frac{1}{6}$, $d = 5 + \varepsilon$ was obtained by J. Moser (1976a, 1976b, 1979); and $c = \frac{1}{6} + \varepsilon$ is to be found in R. Balasubramanian (1978). The methods of the last two authors are different, and Balasubramanian's approach stems from his work on the mean square of $|\zeta(\frac{1}{2} + it)|$ and necessitates a lower bound for $\int_{T-H}^{T+H} |\zeta(\frac{1}{2} + it)|\,dt$ of the type considered in Chapter 9.

Moser's method forms part of his extensive study of discrete properties of $Z(t)$, defined by (10.14), which is carried out in a series of papers (see, e.g., 1976a, 1976b, 1979, 1980, 1982, 1983a, 1983b). Moser (1976b, 1979) obtained the conditional result $c = \frac{1}{8} + \varepsilon$ for (10.32) if the Lindelöf hypothesis is true.

The discrete properties of $Z(t)$ were first studied by E. C. Titchmarsh [see Sections 4.17 and 10.6 of Titchmarsh (1951) for an outline of the method], and the central role is played by the sequence $\{t_\nu\}_{\nu=1}^\infty$, where t_ν is the unique root of the equation

$$\vartheta(t) = \pi\nu \,(\nu \geqslant \nu_0 > 0), \qquad Z(t) = e^{i\vartheta(t)}\zeta\!\left(\tfrac{1}{2} + it\right).$$

Therefore we have

$$\vartheta(t) = -\frac{1}{2}t\log\pi + \operatorname{Im}\log\Gamma\!\left(\frac{1}{4} + \frac{1}{2}it\right) = \frac{t}{2}\log\!\left(\frac{t}{2\pi}\right) - \frac{t}{2} - \frac{\pi}{8} + O\!\left(\frac{1}{t}\right),$$

$$\vartheta'(t) = \frac{1}{2}\log(t/2\pi) + O\!\left(\frac{1}{t}\right).$$

Titchmarsh obtained several interesting results concerning sums with $Z(t_\nu)$, and in Titchmarsh (1934a) he proved

(10.48)
$$\sum_{\nu=M+1}^{N} Z(t_\nu)Z(t_{\nu+1}) = -(2 + o(1))(\gamma + 1)N \qquad (N \to \infty),$$

where $M > 0$ is fixed and sufficiently large. Moreover, he conjectured that for some $A \geqslant 0$

(10.49)
$$\sum_{\nu=M+1}^{N} Z^2(t_\nu)Z^2(t_{\nu+1}) \ll N\log^A N.$$

It may be noted that if for some t_ν^* we have $Z(t_\nu^*)Z(t_{\nu+1}^*) < 0$, then there is a zero $\frac{1}{2} + it$ of $\zeta(s)$ satisfying $t_\nu^* \leqslant t \leqslant t_{\nu+1}^*$, so that from (10.48) one has as a corollary Hardy's classical result that there are infinitely many zeros of $\zeta(s)$ on the critical line. This fact shows the connection between the sequence $\{t_\nu\}_{\nu=1}^\infty$ and zeros on the critical line, and explains why the study of problems involving $Z(t_\nu)$ is worthwhile.

J. Moser's results on problems involving $Z(t_\nu)$ are to be found in his papers (1976a, 1976b, 1979, 1980, 1982, 1983a, 1983b), and in particular Moser (1980) proved Titchmarsh's conjecture (10.49) with $A = 4$. From the Cauchy–Schwarz inequality and $\mu(\frac{1}{2}) \leqslant \frac{1}{6}$, it is seen that (10.49) follows from

(10.50)
$$\sum_{\nu=M+1}^{N} Z^4(t_\nu) \ll N\log^A N.$$

Taking into account that $|\zeta(\frac{1}{2} + it)| = |Z(t)|$, it is clear that (10.50) is a consequence of the discrete fourth power moment estimate

(10.51) $\displaystyle\sum_{r \leqslant R} \left|\zeta\!\left(\tfrac{1}{2} + it_r^*\right)\right|^4 \ll T\log^5 T, \qquad |t_r^*| \leqslant T, \qquad |t_r^* - t_s^*| \geqslant 1 \quad \text{for} \quad r \neq s \leqslant R,$

and (10.51) follows from (8.26). Namely,

$$t_{\nu+1} - t_\nu = (2\pi + o(1))(\log t_\nu)^{-1} \qquad (\nu \to \infty),$$

so that each interval $[t, t+1]$ contains $O(\log t)$ numbers t_ν, and then we may define t_r^* by

$$\left|\zeta\!\left(\tfrac{1}{2} + it_r^*\right)\right| = \max_{r \leqslant t_\nu \leqslant r+1} |Z(t_\nu)| \qquad (r = 1, 2, \ldots).$$

Considering separately t^*_{2m} and t^*_{2m-1}, we may obtain the spacing condition $|t^*_r - t^*_s| \geqslant 1$ $(r \neq s)$ in (10.51), and furthermore $N \sim (2\pi)^{-1} T \log T$ if $\nu \leqslant N$ and $t_\nu \leqslant T$. Therefore collecting $O(\log T)$ estimates of the type (10.51) we obtain (10.50) with $A = 5$, and consequently (10.49) too. This is poorer than Moser's result by a log factor, but the derivation sketched here is much simpler. Higher power moments that were discussed in Chapter 8 allow one to estimate in a similar way sums of the type

$$\sum_{\nu = M+1}^{N} Z^{2k}(t_\nu) Z^{2k}(t_{\nu+1}),$$

where $k \geqslant 1$ is a fixed integer.

A. A. Karacuba (1981) proved that (10.32) holds with $c = \frac{5}{32}$, $d = 2$. His method is based on Moser's variant of Titchmarsh's discrete method and leads essentially also to the estimation of the sum S in (10.38); but this sum is obtained here by a different averaging process and is estimated more carefully. Karacuba's paper also contains a result on zeros of $Z^{(k)}(t)$ in short intervals. If $T \geqslant T_0$, $k \geqslant 1$ is a fixed integer, and

$$H \gg T^{1/(2k+6)} (\log T)^{2/(k+1)},$$

then Karacuba proves that every interval of the form $(T, T + H]$ contains a zero of $Z^{(k)}(t)$ of odd order. The main tool in the proof of this result is an approximate functional equation for $Z^{(k)}(t)$. This is similar to the approximate functional equation (4.1) for $\zeta(s)$ when $\sigma = \frac{1}{2}$, $x = y = (t/2\pi)^{1/2}$, and the length of the Dirichlet polynomials approximating $Z^{(k)}(t)$ is $(t/2\pi)^{1/2}$, while the error term is $O(t^{-1/4} \log^k t)$.

/ *CHAPTER ELEVEN* /
ZERO-DENSITY ESTIMATES

11.1 INTRODUCTION

Zero-density estimates involve upper bounds for the function $N(\sigma, T)$, which represents the number of zeros $\rho = \beta + i\gamma$ (β, γ real) of the zeta-function for which $\beta \geqslant \sigma \geqslant 0$, where σ is fixed and $|\gamma| \leqslant T$. Estimates for $N(\sigma, T)$ may be written in the form

$$(11.1) \qquad N(\sigma, T) \ll T^{A(\sigma)(1-\sigma)} \log^C T,$$

where $C \geqslant 0$, or

$$(11.2) \qquad N(\sigma, T) \ll T^{A(\sigma)(1-\sigma)+\varepsilon},$$

where we shall always suppose that the $\ll$ constant is uniform in σ and T, but depends only on ε. In view of the Riemann–von Mangoldt formula (1.44) one has trivially $A(\sigma)(1 - \sigma) = 1$, $C = 1$ in (11.1) for $0 \leqslant \sigma \leqslant \frac{1}{2}$, while for $\sigma > \frac{1}{2}$ obviously $A(\sigma)(1 - \sigma) \leqslant 1$ and $A(\sigma)(1 - \sigma)$ is nonincreasing. Zero-density estimates have a large number of applications in many branches of analytic number theory, and some of these applications to prime number theory will be presented in Chapter 12. It turns out that in some problems (like the estimation of the difference between consecutive primes) results obtainable from the Lindelöf (or even Riemann) hypothesis follow in almost the same degree of sharpness from a much weaker conjecture, namely,

$$(11.3) \qquad A(\sigma) \leqslant 2 \qquad (\tfrac{1}{2} \leqslant \sigma \leqslant 1),$$

which [both in (11.1) and in (11.2)] is known as "the density hypothesis." Since in many applications (11.1) does not have much advantage over the somewhat weaker (11.2), we shall be concerned primarily with estimates of the type

(11.2), formulating our results for convenience as upper bounds for $A(\sigma)$ in (11.2) rather than for $N(\sigma, T)$ itself. In view of the preceding discussion and the well-known fact that there are no zeros on the line $\sigma = 1$, it is sufficient to consider the range $\frac{1}{2} < \sigma < 1$ in (11.1) or (11.2). Except when σ is very close to $\frac{1}{2}$ we shall prove the sharpest known bounds for $A(\sigma)$ in this chapter. To accomplish this we shall use a zero-detection method, which will be fully explained in Section 11.2, and which offers great flexibility in the estimation of $A(\sigma)$. Among other tools we shall use the higher power moment estimates of Chapter 8, and certain double zeta sums, which will be considered in Sections 11.6 and 11.8.

11.2 THE ZERO-DETECTION METHOD

We start from (A.7) with $x = n/Y$, namely,

$$(11.4) \qquad e^{-n/Y} = (2\pi i)^{-1} \int_{2-i\infty}^{2+i\infty} \Gamma(w) Y^w n^{-w} \, dw,$$

and let $M_X(s) = \sum_{n \leqslant X} \mu(n) n^{-s}$, where $s = \sigma + it$, $\log^2 T \leqslant |t| \leqslant T$, $1 \ll X \leqslant Y \ll T^C$, $\mu(n)$ is the Möbius function, and $X = X(T)$, $Y = Y(T)$ are parameters which will be suitably chosen. In view of the elementary relation

$$\sum_{d|n} \mu(d) = \begin{cases} 1 & n = 1, \\ 0 & n > 1, \end{cases}$$

it is seen that each zero $\rho = \beta + i\gamma$ of $\zeta(s)$ counted by $N(\sigma, T)$ satisfies

$$(11.5) \quad e^{-1/Y} + \sum_{n > X} a(n) n^{-\rho} e^{-n/Y}$$

$$= (2\pi i)^{-1} \int_{2-i\infty}^{2+i\infty} \zeta(\rho + w) M_X(\rho + w) Y^w \Gamma(w) \, dw,$$

where

$$(11.6) \qquad a(n) = \sum_{d|n, \, d \leqslant X} \mu(d), \qquad |a(n)| \leqslant d(n) < n^\varepsilon,$$

since the absolutely convergent series $\zeta(\rho + w) = \sum_{n=1}^{\infty} n^{-\rho - w}$ and $M_X(\rho + w)$ may be multiplied and then integrated termwise using (11.4).

Now the line of integration in (11.5) is moved to $\operatorname{Re} w = \frac{1}{2} - \beta$. For $|\gamma| \geqslant \log^2 T$ the residue at the pole $w = 1 - \rho$ of the integrand is $o(1)$ by (A.34), and the pole $w = 0$ of $\Gamma(w)$ is canceled by the zero $w = 0$ of $\zeta(\rho + w)$. Also using (A.34) one has

$$(11.7) \quad \int_{\operatorname{Re} w = 1/2 - \beta} = o(1) + (2\pi i)^{-1} \int_{-\log^2 T}^{\log^2 T} \zeta(\tfrac{1}{2} + i\gamma + iv)$$

$$\times M_X(\tfrac{1}{2} + i\gamma + iv) \Gamma(\tfrac{1}{2} - \beta + iv) Y^{1/2 - \beta + iv} \, dv,$$

while trivially

(11.8)
$$\sum_{n > Y \log^2 Y} a(n) n^{-\rho} e^{-n/Y} = o(1)$$

as $Y \to \infty$. But then in (11.5) $\exp(-1/Y) \to 1$, so that each $\rho = \beta + i\gamma$ counted by $N(\sigma, T)$ satisfies at least one of the following conditions:

(11.9)
$$\sum_{X < n \leqslant Y \log^2 Y} a(n) n^{-\rho} e^{-n/Y} \gg 1,$$

(11.10)

$$\int_{-\log^2 T}^{\log^2 T} \zeta(\tfrac{1}{2} + i\gamma + iv) M_X(\tfrac{1}{2} + i\gamma + iv) \Gamma(\tfrac{1}{2} - \beta + iv) Y^{1/2 - \beta + iv} dv \gg 1,$$

(11.11)
$$|\gamma| \leqslant \log^2 T.$$

The number of zeros ρ satisfying (11.11) is trivially $O(\log^3 T)$, since each strip $T \leqslant t \leqslant T + 1$ contains $O(\log T)$ zeros by the Riemann–von Mangoldt formula (1.44). By the same argument we may choose R_1 zeros satisfying (11.9) and R_2 zeros satisfying (11.10) so that the imaginary parts of these zeros differ from each other by at least $2 \log^4 T$ and therefore

(11.12)
$$N(\sigma, T) \ll (R_1 + R_2 + 1) \log^5 T.$$

A suitable choice of X in most of our zero-density results will be

(11.13)
$$X = T^\varepsilon,$$

so that trivially

(11.14)
$$M_X(\tfrac{1}{2} + i\gamma + iv) \ll T^\varepsilon \quad \text{for} \quad |v| \leqslant \log^2 T.$$

With this choice of X we shall regulate the length of the Dirichlet polynomial appearing in (11.9). Note that each ρ counted by R_1 satisfies

(11.15)
$$\sum_{N < n \leqslant 2N} a(n) n^{-\rho} e^{-n/Y} \gg 1/\log Y$$

for at least one of $O(\log Y)$ values $T^\varepsilon \leqslant N = 2^{-j} Y \log^2 Y$, $j = 1, 2, \ldots$, and we may consider representative zeros of those counted by R_1 which are $\gg R_1/\log Y$ in number and which satisfy (11.15) with a particular N. The exact size of N is not important, since we are going to raise (11.15) to the power k, where k is a natural number depending on N such that $N^k = M$, $(2N)^k = P \leqslant T^C$; whence $k \ll 1$ and

$$\sum_{M < n \leqslant P} b(n) n^{-\rho} \gg 1/\log^k Y$$

with $b(n) \ll d_{2k}(n) \ll n^{\varepsilon}$ and $P \ll M$. We split this last sum into subsums of length not exceeding M and choose k so that $N^k \leqslant Y^r \log^{2r} Y < N^{k+1}$, $k \geqslant r \geqslant 2$ is satisfied, where r is a fixed integer. Then we have

$$(11.16) \qquad Y^{r^2/(r+1)} \log^{2r^2/(r+1)} Y \ll M \ll Y^r \log^{2r} Y,$$

and in view of existing power moments for the zeta-function it turns out that the practical choice for r in (11.16) is $r = 2$, which gives

$$(11.17) \qquad Y^{4/3} \log^{8/3} Y \ll M \ll Y^2 \log^4 Y.$$

Using the estimate preceding (11.16) and partial summation we obtain

$$R_1 \ll \log^D T \left\{ \sum_{\rho} \left(\left| \sum_{M < n \leqslant 2M} b(n) n^{-\sigma - i\gamma} \right| M^{\sigma - \beta} \right. \right.$$

$$\left. \left. + \int_M^{2M} \left| \sum_{M < n \leqslant u} b(n) n^{-\sigma - i\gamma} \right| M^{\sigma - \beta - 1} du \right) \right\}$$

for some D satisfying $D \asymp 1$. Therefore relabeling the $b(n)$'s [i.e., letting $b(n) = 0$ for $u < n \leqslant 2M$] we see that R_1 is estimated by

$$(11.18) \quad R_1 \ll \sum_{\rho} \left| \sum_{M < n \leqslant 2M} b(n) n^{-\sigma - i\gamma} \right|^{\alpha} \log^D T \qquad (\alpha = 1 \text{ or } \alpha = 2).$$

Here $D \asymp 1$, $b(n) \ll n^{\varepsilon}$, M satisfies (11.17) and $\sum_{\rho}$ denotes summation over representative zeros $\rho = \beta + i\gamma$ $(\beta \geqslant \sigma)$ of R_1. The choice $\alpha = 1$ or $\alpha = 2$, which corresponds to the use of (A.39) or (A.40), is merely for technical reasons since (11.18) obviously holds for any fixed $\alpha > 0$.

To estimate R_2 we set for $r = 1, 2, \ldots, R_2$

$$\left| \zeta\left(\tfrac{1}{2} + i\gamma_r + iv'\right) \right| = \max_{-\log^2 T \leqslant v \leqslant \log^2 T} \left| \zeta\left(\tfrac{1}{2} + i\gamma_r + iv\right) \right|$$

and

$$t_r = \gamma_r + v',$$

where $\gamma_1, \gamma_2, \ldots, \gamma_{R_2}$ are imaginary parts of zeros satisfying (11.10). Then from (11.10) we infer that

$$(11.19) \qquad 1 \ll T^{\varepsilon} Y^{1/2 - \sigma} \left| \zeta\left(\tfrac{1}{2} + it_r\right) \right| \qquad (r = 1, 2, \ldots, R_2).$$

For $r \neq s$ obviously $|t_r - t_s| \geqslant \log^4 T$, and so raising (11.19) to the power $A \geqslant 4$ we have

$$(11.20) \quad R_2 \ll T^\varepsilon \sum_{r \leqslant R_2} \left| \zeta\left(\tfrac{1}{2} + it_r\right) \right|^A Y^{A(1/2-\sigma)} \ll T^{M(A)+\varepsilon} Y^{A(1/2-\sigma)},$$

where $M(A)$ is defined by (8.1). We may also utilize directly Theorem 8.2 to estimate R_2, since with $V = T^{-\varepsilon} Y^{\sigma-1/2}$ in (11.19) we obtain from (8.28)

$$(11.21) \quad R_2 \ll T^{1+\varepsilon} Y^{3-6\sigma} + T^{(\varkappa+\lambda+\varepsilon)/\varkappa} Y^{(1-2\sigma)(1+2\varkappa+2\lambda)/\varkappa},$$

where $(\varkappa, \lambda)$ is any exponent pair for which $\varkappa > 0$.

Having thus prepared the groundwork for zero-density estimates we shall proceed to specific results, with the remark that the estimation of R_1 is in general more difficult than the estimation of R_2, for which good bounds (11.20) and (11.21) exist. Several techniques for bounding R_1 will be presented, but it turns out that for $\tfrac{1}{2} < \sigma \leqslant \tfrac{3}{4}$ the mean value theorem for Dirichlet polynomials (Theorem 5.3) is the best available tool, while for $\sigma \geqslant \tfrac{3}{4}$ the best results are obtained by applications of the Halász–Montgomery inequality (A.39) or (A.40), which offers a considerable flexibility of approach.

11.3 THE INGHAM–HUXLEY ESTIMATES

The estimates in question provide upper bounds for $A(\sigma)$ in the whole range $\tfrac{1}{2} \leqslant \sigma \leqslant 1$, and they are contained in

THEOREM 11.1.

$$(11.22) \qquad A(\sigma) \leqslant 3/(2 - \sigma) \qquad \left(\tfrac{1}{2} \leqslant \sigma \leqslant \tfrac{3}{4}\right),$$

$$(11.23) \qquad A(\sigma) \leqslant 3/(3\sigma - 1) \qquad \left(\tfrac{3}{4} \leqslant \sigma \leqslant 1\right),$$

$$(11.24) \qquad A(\sigma) \leqslant \tfrac{12}{5} \qquad \left(\tfrac{1}{2} \leqslant \sigma \leqslant 1\right).$$

The estimate (11.24) is a simple consequence of (11.22) (proved by A. E. Ingham in 1940) and (11.23) (proved by M. N. Huxley in 1972), since for $\tfrac{1}{2} \leqslant \sigma \leqslant \tfrac{3}{4}$ the function $3/(2 - \sigma)$ is increasing, while for $\tfrac{3}{4} \leqslant \sigma \leqslant 1$ the function $3/(3\sigma - 1)$ is decreasing and their common value at $\sigma = \tfrac{3}{4}$ is $\tfrac{12}{5}$. The significance of (11.24) is that it is the best-known estimate of the type $A(\sigma) \leqslant C$ ($C \geqslant 2$ an absolute constant) valid in the whole range $\tfrac{1}{2} \leqslant \sigma \leqslant 1$, and estimates of this sort are often needed in applications.

Proof of Theorem 11.1. To obtain (11.22) we use (11.17), (11.18) with $\alpha = 2$, and the mean value estimate for Dirichlet polynomials in the form (5.14). This gives

$$R_1 \ll T^\varepsilon \sum_r \left| \sum_{M < n \leqslant 2M} b(n) n^{-\sigma - i\gamma_r} \right|^2 \ll T^\varepsilon (T + M) M^{1-2\sigma}$$

$$\ll T^\varepsilon (Y^{4-4\sigma} + TY^{4(1-2\sigma)/3}),$$

where Σ_r denotes summation over representative zeros $\rho = \beta + i\gamma_r$ of R_1 zeros which satisfy (11.18). Using (11.20) with $M(4) = 1$, it follows from (11.12) that

$$N(\sigma, T) \ll T^\varepsilon (Y^{4-4\sigma} + TY^{4(1-2\sigma)/3} + 1) \ll T^{3(1-\sigma)/(2-\sigma)+\varepsilon}$$

for $Y = T^{3/(8-4\sigma)}$, and this gives (11.22). It is perhaps surprising that Ingham's bound (11.22) withstands improvement for almost half a century (except when σ is very close to $\frac{1}{2}$), and the main reason for this seems to be that $M(A) = 1$ is still known to hold only for $A \leqslant 4$.

The main difficulty in estimating R_1 in general is the presence of the coefficients $b(n)$, which are nonmonotonic and therefore cannot be removed by partial summation techniques such as (A.19) or (A.20). An obvious way to remove the $b(n)$'s is the use of the Halász–Montgomery inequality, and for the proof of (11.23) we shall use (A.39) with $\xi = \{\xi_n\}_{n=1}^\infty$ and $\xi_n = b(n)n^{-\sigma}$ for $M < n \leqslant 2M$ and zero otherwise, and $\varphi_r = \{\varphi_{r,n}\}_{n=1}^\infty$ and $\varphi_{r,n} = n^{-it_r}$ for $M < n \leqslant 2M$ and zero otherwise, where we have denoted the ordinates of representative zeros of R_1 by t_r. Then from (A.43) $\|\xi\|^2 \ll T^\varepsilon M^{1-2\sigma}$; hence (11.18) with $\alpha = 1$ and (A.39) give

$$R_1^2 \ll T^\varepsilon R_1 M^{2-2\sigma} + T^\varepsilon M^{1-2\sigma} \sum_{r \neq s} \left| \sum_{M < n \leqslant 2M} n^{-it_r + it_s} \right|.$$

The effect of this procedure is that the inner sum above is an exponential sum to which the techniques of Chapter 2 are applicable. Indeed we estimate this sum as

$$\ll M|t_r - t_s|^{-1} + T^{1/2}$$

by the exponent pair $(\varkappa, \lambda) = (\frac{1}{2}, \frac{1}{2})$ if $|t_r - t_s| \gg M$, and if this is not satisfied then by Lemma 1.2 and Lemma 2.1. Therefore we have

$$R_1^2 \ll T^\varepsilon R_1 M^{2-2\sigma} + T^\varepsilon M^{2-2\sigma} \sum_{r \neq s} |t_r - t_s|^{-1} + T^\varepsilon R_1^2 T^{1/2} M^{1-2\sigma}.$$

But in view of $|t_r - t_s| \geqslant \log^4 T$ it follows that

$$\sum_{r \neq s} |t_r - t_s|^{-1} \ll \sum_{r \leqslant R_1} \log^{-4} T \sum_{n \leqslant 3T} n^{-1} \ll R_1 \log^{-3} T,$$

and therefore we obtain

$$R_1 \ll T^\varepsilon M^{2-2\sigma}$$

if $T \ll M^{4\sigma-2-\varepsilon}$. Thus we divide T into subintervals of length at most $T_0 = M^{4\sigma-2-\varepsilon}$ so that if R_0 denotes the number of representative zeros of R_1 in each of these intervals, then the above estimate holds with R_0 in place of R_1, and (11.17) gives then

$$(11.25) \qquad R_1 \ll R_0(1 + T/T_0) \ll T^\varepsilon(M^{2-2\sigma} + TM^{4-6\sigma})$$

$$\ll T^\varepsilon(Y^{4-4\sigma} + TY^{(16-24\sigma)/3}).$$

From (11.20) with $M(12) \leqslant 2$ we have finally

$$N(\sigma, T) \ll T^\varepsilon(Y^{4-4\sigma} + TY^{(16-24\sigma)/3} + T^2Y^{6-12\sigma} + 1) \ll T^{3(1-\sigma)/(3\sigma-1)+\varepsilon}$$

with $Y = T^{3/(12\sigma-4)}$, which completes the proof of Theorem 11.1.

The estimates furnished by Theorem 11.1 may be replaced by slightly sharper bounds of the form (11.1), and in particular it can be proved that

$$(11.26) \quad N(\sigma, T) \ll T^{3(1-\sigma)/(2-\sigma)}\log^5 T \qquad (\tfrac{1}{2} \leqslant \sigma \leqslant \tfrac{3}{4}),$$

$$(11.27) \quad N(\sigma, T) \ll T^{3(1-\sigma)/(3\sigma-1)}\log^{44} T \qquad (\tfrac{3}{4} \leqslant \sigma \leqslant 1),$$

and

$$(11.28) \quad N(\sigma, T) \ll T^{(5\sigma-3)(1-\sigma)/(\sigma^2+\sigma-1)}\log^9 T \qquad (\tfrac{3}{4} \leqslant \sigma \leqslant 1)$$

hold. Since for $\tfrac{3}{4} \leqslant \sigma \leqslant 1$ we have $(5\sigma - 3)/(\sigma^2 + \sigma - 1) \geqslant 3/(3\sigma - 1)$, it follows that (11.27) is sharper than (11.28), but the log factor in (11.28) is better. Moreover, $(5\sigma - 3)/(\sigma^2 + \sigma - 1) \leqslant \tfrac{12}{5}$ for $\tfrac{3}{4} \leqslant \sigma \leqslant 1$, so that from (11.26) and (11.28) we obtain

$$(11.29) \qquad N(\sigma, T) \ll T^{12(1-\sigma)/5}\log^9 T \qquad (\tfrac{1}{2} \leqslant \sigma \leqslant 1).$$

The proofs of (11.26)–(11.28) may be carried out similarly to the corresponding proofs of Theorem 11.1, only the choice of the parameter X in (11.9) and (11.10) will not be $X = T^\varepsilon$ as in (11.13), and instead of the crude estimate $a(n) \ll n^\varepsilon$ [or $b(n) \ll n^\varepsilon$ for $b(n)$ in (11.18)] we shall use $|a(n)| \leqslant d(n)$ in estimating sums of the form $\Sigma a^2(n)n^{-2\sigma}$, which arise after either the mean value theorem for Dirichlet polynomials or the Halász–Montgomery inequality is applied. We present now the proof of (11.26) with a somewhat poorer log factor, by supposing that the imaginary parts of representative zeros differ by $2B \log T$, where $B > 0$ is a large, fixed constant.

To bound R_1 zeros for which (11.9) holds we observe that there exist $\gg R_1/\log T$ of these zeros for which

$$\sum_{M<n\leqslant 2M} a(n)n^{-\sigma-i\gamma_r}e^{-n/Y} \gg 1/\log Y \qquad [|a(n)| \leqslant d(n)]$$

holds for some M satisfying $X \leqslant M \leqslant Y\log^2 Y$; hence by the mean value estimate (5.14) we obtain

$$R_1 \ll \log^4 T\left(T \sum_{M<n\leqslant 2M} d^2(n)n^{-2\sigma} + \sum_{M\leqslant n\leqslant 2M} d^2(n)n^{1-2\sigma}\right)e^{-M/Y}$$

$$\ll (T+M)M^{1-2\sigma}e^{-M/Y}\log^7 T \ll (TX^{1-2\sigma} + Y^{2-2\sigma})\log^7 T.$$

To bound R_2 we may suppose that $\beta \geqslant \sigma \geqslant \frac{1}{2} + 1/\log T$ in view of the Riemann–von Mangoldt formula (1.44). Hence raising (11.10) to the power $\frac{4}{3}$ we obtain for R_2 numbers t_r such that $|t_r| \leqslant 2T$, $|t_r - t_s| \geqslant B\log T$ $(r \neq s)$,

$$R_2 Y^{(4\sigma-2)/3}\log^{-4/3}T \ll \sum_{r\leqslant R_2} \left|\zeta(\tfrac{1}{2} + it_r)\right|^{4/3}\left|M_X(\tfrac{1}{2} + it_r)\right|^{4/3}$$

$$\leqslant \left(\sum_{r\leqslant R_2} \left|\zeta(\tfrac{1}{2} + it_r)\right|^4\right)^{1/3}\left(\sum_{r\leqslant R_2} \left|M_X(\tfrac{1}{2} + it_r)\right|^2\right)^{2/3}$$

$$\ll (T\log^5 T)^{1/3}\left(T\sum_{n\leqslant X}n^{-1} + \sum_{n\leqslant X}1\right)^{2/3} \ll T\log^{7/3}T,$$

provided that $X \ll T\log T$, where we used Hölder's inequality, $M(4) = 1$ and (5.14). Therefore

$$N(\sigma,T) \ll (R_1 + R_2 + 1)\log^2 T \ll (Y^{2-2\sigma} + TX^{1-2\sigma} + TY^{(2-4\sigma)/3})\log^9 T$$

$$\ll T^{3(1-\sigma)/(2-\sigma)}\log^9 T$$

with $X = T$, $Y = T^{3/(4-2\sigma)}$, and similarly with some modifications in the proof of (11.23) one can obtain (11.27) and (11.28).

11.4 ESTIMATES FOR σ NEAR UNITY

In this section we shall prove two results, Theorems 11.2 and 11.3, which provide the sharpest known bounds of $N(\sigma,T)$ when σ is close to 1. For the estimation of R_1 we shall again use the Halász–Montgomery inequality, and combine it with order results for $\zeta(s)$ when σ is near 1 and power moment estimates of Chapter 8. We shall prove

THEOREM 11.2.

(11.30) $\quad A(\sigma) \leqslant 4/(2\sigma + 1) \qquad \left(\frac{17}{18} \leqslant \sigma \leqslant 1\right)$,

(11.31) $\quad A(\sigma) \leqslant 24/(30\sigma - 11) \qquad \left(\frac{155}{174} = 0.8908\ldots \leqslant \sigma \leqslant \frac{17}{18}\right)$.

THEOREM 11.3. Let $M(\alpha, T) = \max_{1 \leqslant t \leqslant T} |\zeta(\alpha + it)|$ for $\frac{1}{2} \leqslant \alpha \leqslant 1$. Then for $\frac{9}{10} \leqslant \sigma \leqslant 1$

(11.32) $\qquad N(\sigma, T) \ll (M(5\sigma - 4, 3T))^{7/6} \log^{169/12} T$,

and, in particular, for $\frac{152}{155} \leqslant \sigma \leqslant 1$,

(11.33) $\qquad N(\sigma, T) \ll T^{1600(1-\sigma)^{3/2}} \log^{15} T$

and

(11.34) $\qquad N(\sigma, T) \ll T^{35(1-\sigma)/36} \log^{16} T$.

Proof of Theorem 11.2. As in the proof of (11.23) we shall utilize (A.39), but the choice of ξ and φ_r will be different. We shall take $\xi = \{\xi_n\}_{n=1}^{\infty}$, $\xi_n = b(n)(e^{-n/2M} - e^{-n/M})^{-1/2} n^{-\sigma}$ for $M < n \leqslant 2M$ and zero otherwise, and $\varphi_r = \{\varphi_{r,n}\}_{n=1}^{\infty}$ with $\varphi_{r,n} = (e^{-n/2M} - e^{-n/M})^{1/2} n^{-it_r}$ $(n = 1, 2, \ldots)$. Writing R for the number of representative zeros of R_1 and $t_1, \ldots, t_R$ for their imaginary parts, we have then from (A.39) and (11.18)

(11.35) $\quad R_1^2 \ll T^{\varepsilon}\left(M^{2-2\sigma} R_1 + M^{1-2\sigma} \sum_{r \neq s \leqslant R} |H(it_r - it_s)|\right)$,

where for t real we have from (4.60), with $h = k = 1$,

(11.36) $\quad H(it) = \sum_{n=1}^{\infty} (e^{-n/2M} - e^{-n/M}) n^{-it}$

$$= (2\pi i)^{-1} \int_{2-i\infty}^{2+i\infty} \zeta(w + it)((2M)^w - M^w) \Gamma(w)\, dw,$$

since $1 \ll e^{-n/2M} - e^{-n/M} \ll 1$ for $M < n \leqslant 2M$, $\|\xi\|^2 \ll T^{\varepsilon} M^{1-2\sigma}$, and $H(0) \ll M$. Moving the line of integration in (11.36) to $\mathrm{Re}\, w = \frac{1}{2}$ we encounter a simple pole at $w = 1 - it$ with residue $\ll M e^{-|t|}$ by (A.34), so that

(11.37)

$$H(it) = (2\pi i)^{-1} \int_{1/2-i\infty}^{1/2+i\infty} \zeta(w + it)((2M)^w - M^w) \Gamma(w)\, dw + O(M e^{-|t|}).$$

Also in view of (A.34) the integral in (11.37) is $o(1)$ for $M \ll T^c$ if $|\operatorname{Im} w| \geqslant \log^2 T$, which gives

(11.38)

$$\sum_{r \neq s \leqslant R} |H(it_r - it_s)| \ll M \sum_{r \neq s \leqslant R} e^{-|t_r - t_s|} + o(R^2)$$

$$+ M^{1/2} \int_{-\log^2 T}^{\log^2 T} \sum_{r \neq s \leqslant R} \left| \zeta\left(\tfrac{1}{2} + it_r - it_s + iv \right) \right| dv.$$

The first sum on the right-hand side of (11.38) is $O(R)$ since the t_r's are at least $\log^4 T$ apart, and to bound the second sum we fix each s and set $\tau_r = t_r - t_s + v$. Then $|\tau_r| \leqslant 3T$ for $r = 1, \ldots, R$ and $|\tau_{r_1} - \tau_{r_2}| \geqslant \log^4 T$ for $r_1 \neq r_2$. We use Hölder's inequality and (8.31) to obtain

$$(11.39) \quad \sum_{r \leqslant R} \left| \zeta\left(\tfrac{1}{2} + i\tau_r \right) \right| \leqslant R^{5/6} \left(\sum_{r \leqslant R, \, |\zeta| \geqslant T^{2/13}} \left| \zeta\left(\tfrac{1}{2} + i\tau_r \right) \right|^6 \right)^{1/6}$$

$$+ R^{18/19} \left(\sum_{r \leqslant R, \, |\zeta| \leqslant T^{2/13}} \left| \zeta\left(\tfrac{1}{2} + i\tau_r \right) \right|^{19} \right)^{1/19}$$

$$\ll T^\varepsilon \left(R^{5/6} T^{1/6} + R^{18/19} T^{3/19} \right).$$

Inserting (11.39) into (11.38) and using then (11.35) we obtain

$$(11.40) \quad R_1 \ll T^\varepsilon \left(M^{2-2\sigma} + TM^{9-12\sigma} + T^3 M^{19(3-4\sigma)/2} \right),$$

and to bound R_2 we use (11.21) with $(\varkappa, \lambda) = \left(\tfrac{2}{7}, \tfrac{4}{7} \right)$, so that

$$(11.41) \quad R_2 \ll T^\varepsilon \left(TY^{3-6\sigma} + T^3 Y^{19(1/2-\sigma)} \right).$$

We now first use (11.40) to estimate the number of points R_0 lying in an interval of length not exceeding $T_0 = M^{(72\sigma - 53)/6}$. Then $R_0 \ll T^\varepsilon M^{2-2\sigma}$ for $\sigma \leqslant \tfrac{11}{12}$ and

$$(11.42) \quad R_1 \ll R_0(1 + T/T_0) \ll T^\varepsilon \left(M^{2-2\sigma} + TM^{(65-84\sigma)/6} \right)$$

$$\ll T^\varepsilon \left(Y^{4-4\sigma} + TY^{2(65-84\sigma)/9} \right)$$

for $\sigma \geqslant \tfrac{65}{84}$. With $Y = T^{6/(30\sigma - 11)}$ it follows from (11.41) and (11.42) that

$$N(\sigma, T) \ll T^\varepsilon \left(Y^{4-4\sigma} + TY^{2(65-84\sigma)/9} + TY^{3-6\sigma} + T^3 Y^{19(1/2-\sigma)} \right)$$

$$\ll T^{24(1-\sigma)/(30\sigma - 11) + \varepsilon}$$

for $\frac{155}{174} \leqslant \sigma \leqslant \frac{11}{12}$. For $\sigma \geqslant \frac{11}{12}$ we repeat the foregoing procedure choosing this time $T_0 = M^{10\sigma-7}$ in (11.40) to obtain $R_0 \ll T^\varepsilon M^{2-2\sigma}$ and

$$(11.43) \quad R_1 \ll R_0(1 + T/T_0) \ll T^\varepsilon(M^{2-2\sigma} + TM^{9-12\sigma})$$

$$\ll T^\varepsilon(Y^{4-4\sigma} + TY^{12-16\sigma}) \ll T^\varepsilon(Y^{4-4\sigma} + TY^{3-6\sigma}).$$

Choosing $Y = T^{1/(2\sigma+1)}$ for $\sigma \geqslant \frac{17}{18}$ and $Y = T^{6/(30\sigma-11)}$ for $\frac{11}{12} \leqslant \sigma \leqslant \frac{17}{18}$, respectively, we complete the proof on comparing (11.41) and (11.43) and using

$$N(\sigma, T) \ll T^\varepsilon(R_1 + R_2 + 1).$$

Proof of Theorem 11.3. We shall use the zero-detection method of Section 11.2, only now in (11.5) we move the line of integration to $\mathrm{Re}\, w = \alpha - \beta < 0$ for some suitable $\frac{1}{2} \leqslant \alpha \leqslant 1$. Hence reasoning as in Section 11.2 we obtain that (11.12) holds for R_1 zeros $\rho_r = \beta_r + i\gamma_r$, $\beta_r \geqslant \sigma$ for which

$$(11.44) \quad \sum_{X < n \leqslant Y\log^2 Y} a(n)n^{-\sigma-i\gamma_r} \gg 1 \quad [|a(n)| \leqslant d(n)],$$

and for R_2 zeros ρ_r for which

$$(11.45)$$

$$\int_{-\log^2 T}^{\log^2 T} \zeta(\alpha + i\gamma_r + iv)M_X(\alpha + i\gamma_r + iv)\Gamma(\alpha - \beta + iv)Y^{\alpha-\beta+iv}\,dv \gg 1.$$

There exists a number M ($X \leqslant M \leqslant Y\log^2 Y$) such that

$$(11.46) \quad \sum_{M < n \leqslant 2M} a(n)n^{-\sigma-i\gamma_r} \gg 1/\log Y$$

for R ($\gg R_1/\log Y$) zeros, where the imaginary parts of these zeros are denoted by $t_1, \ldots, t_R$. Applying the Halász–Montgomery inequality (A.39) similarly as in (11.35) we obtain

$$R^2 \ll \log^2 Y\left(\sum_{M < n \leqslant 2M} d^2(n)e^{-2n/Y}n^{-2\sigma}\right)\left(RM + \sum_{r \neq s \leqslant R} |H(it_r - it_s)|\right),$$

where $H(it)$ is defined by (11.36). Moving the line of integration in the expression for $H(it)$ to $\mathrm{Re}\, w = \alpha$ it is found that

$$R^2 \ll \log^5 T\left(RM^{2-2\sigma} + R^2M^{1+\alpha-2\sigma}M(\alpha, 3T)\right)e^{-M/Y};$$

hence we have for $\sigma \geqslant (\alpha + 1)/2$

$$(11.47)$$

$$R_1 \ll R\log T \ll \max_{X \leqslant M = 2^k \leqslant Y\log^2 Y} M^{2-2\sigma}e^{-M/Y}\log^6 T \ll Y^{2-2\sigma}\log^6 T$$

if

(11.48) $$X^{2\sigma-1-\alpha} \gg M(\alpha, 3T)\log^5 T.$$

To estimate R_2 we may suppose first that $\sigma < 1 - C/\log T$ in view of the zero-free region (1.55); hence from (11.45) it follows that

$$M_X(\alpha + it_r) \gg Y^{\sigma-\alpha}\big(M(\alpha, 2T)\log T\big)^{-1}$$

for R_2 points t_r such that $|t_r| \leqslant 2T$, $|t_r - t_s| > \log^4 T$ for $r \neq s \leqslant R_2$. Then there is a number N $(1 \ll N \leqslant X)$ such that

(11.49) $$\sum_{N < n \leqslant 2N} \mu(n)n^{-\alpha-it_r} \gg Y^{\sigma-\alpha}\big(M(\alpha, 2T)\log^2 T\big)^{-1}$$

for R_0 $(\gg R_2/\log T)$ numbers t_r. Applying the Halász–Montgomery inequality as in the previous case we obtain

$$R_0^2 \ll M^2(\alpha, 2T)Y^{2\alpha-2\sigma}\log^4 T\big(R_0 N^{2-2\alpha} + N^{1-\alpha}M(\alpha, 3T)R_0^2\big);$$

hence

(11.50) $$R_2 \ll X^{2-2\alpha}Y^{2\alpha-2\sigma}M^2(\alpha, 2T)\log^5 T$$

if

(11.51) $$Y^{2\sigma-2\alpha} \gg M^3(\alpha, 3T)X^{1-\alpha}\log^4 T.$$

Therefore a suitable choice of X and Y will be

$$X = \big\{C_1 M(\alpha, 3T)\log^5 T\big\}^{1/(2\sigma-1-\alpha)},$$

$$Y = \big\{C_2 M(\alpha, 3T)\big\}^{(3\sigma-2\alpha-1)/(\sigma-\alpha)(2\sigma-1-\alpha)}(\log T)^{(8\sigma-9\alpha+1)/2(\sigma-\alpha)(2\sigma-1-\alpha)},$$

and we see that the bound for R_2 is smaller than the bound for R_1. Hence (11.12) gives

(11.52) $$N(\sigma, T) \ll Y^{2-2\sigma}\log^{11} T$$

$$\ll \big\{C_2 M(\alpha, 3T)\big\}^{2(1-\sigma)(3\sigma-2\alpha-1)/(\sigma-\alpha)(2\sigma-1-\alpha)}\log^B T,$$

where

$$B = 11 + \frac{(8\sigma - 9\alpha + 1)(1 - \sigma)}{(\sigma - \alpha)(2\sigma - 1 - \alpha)}$$

$$\big[\tfrac{1}{2} \leqslant \alpha \leqslant 1, (\alpha + 1)/2 \leqslant \sigma \leqslant 1 - C/\log T\big].$$

Now we aim to choose α in such a way that using

$$(11.53) \qquad M(\alpha, T) \ll T^{D(1-\alpha)^{3/2}} \log^{2/3} T \qquad (\alpha_0 \leqslant \alpha \leqslant 1),$$

with $D = 122$ (Theorem 6.3) we obtain a (relative) minimum in (11.52). If $\alpha = k\sigma - (k-1) \ (k > 2)$, then in view of (11.52) we see that it is appropriate to take $k = 5$; hence $\alpha = 5\sigma - 4$, $\sigma \geqslant (\alpha + 1)/2$ is satisfied, and $\alpha \geqslant \frac{1}{2}$ for $\sigma \geqslant \frac{9}{10}$. Thus (11.32) follows from (11.52). With (11.53) we obtain (11.33) if σ is sufficiently close to unity, say $\sigma \geqslant 1 - 10^{-4}$. But if $\sigma < 1 - 10^{-4}$ then (11.34) improves (11.33); hence we may confine our attention to (11.34) now. Using convexity and (7.59) with $l = 6$ [or (7.57)] we have

$$(11.54) \qquad \zeta(\sigma + it) \ll t^{(1-\sigma)/6} \log t \qquad \left(\tfrac{28}{31} \leqslant \sigma \leqslant 1\right);$$

hence

$$(11.55) \qquad M(5\sigma - 4, T) \ll T^{5(1-\sigma)/6} \log T$$

for $5\sigma - 4 \geqslant \frac{28}{31}$, that is, for $\sigma \geqslant \frac{152}{155} = 0.98064\ldots$, and inserting (11.55) in (11.32) we obtain (11.34). The exponent $\frac{35}{36}$ in (11.34) could be still somewhat decreased by shortening the range for σ; however, our purpose in presenting (11.34) was to obtain $A(\sigma) < 1$ for a reasonably large range of σ.

11.5 REFLECTION PRINCIPLE ESTIMATES

For the remainder of the chapter we shall be dealing with zero-density estimates for $\sigma \geqslant \frac{3}{4}$ and X and M will satisfy (11.13) and (11.17), respectively. In this section we shall present a method due to M. Jutila, which consists of employing the reflection principle inequality (4.67) with $s = it_r - it_s$ [where t_r denotes ordinates of representative zeros of R_1 satisfying (11.18)] and $Y = M$. Similarly to (11.35) we obtain this time

$$(11.56) \qquad R_1^2 \ll T^\varepsilon \left(R_1 M^{2-2\sigma} + M^{1-2\sigma} \sum_{r \neq s \leqslant R} |K(it_r - it_s)| \right),$$

where R is the number of representative zeros of R_1, and where with $h = \log^2 T$ and $t \neq 0$ real we have

$$(11.57) \qquad K(it) = \sum_{n=1}^{\infty} \left(e^{-(n/2M)^h} - e^{-(n/M)^h} \right) n^{-it}$$

$$\ll 1 + M^{1/2} \int_{-h^2}^{h^2} \left| \sum_{n \leqslant 4T/M} n^{-1/2 + it + iv} \right| dv$$

by (4.67). Therefore from (11.56) and (11.57) we infer that

$$(11.58) \qquad R_1^2 \ll T^\varepsilon \Big(R_1 M^{2-2\sigma} + M^{1-2\sigma} R_1^2 + M^{(3-4\sigma)/2} $$

$$\times \int_{-h^2}^{h^2} \sum_{r \neq s \leqslant R} \Big| \sum_{n \leqslant 4T/M} n^{-1/2 + it_r - it_s + iv} \Big| \, dv.$$

As we are interested in the range $\sigma \geqslant \frac{3}{4}$, the term $M^{1-2\sigma} R_1^2$ may be discarded in (11.58), and on applying Hölder's inequality with $k \geqslant 1$ an integer we have

(11.59)

$$R_1 \ll T^\varepsilon M^{2-2\sigma} + T^\varepsilon M^{k(3-4\sigma)/2} \max_{|v| \leqslant h^2} \left(\sum_{r \neq s \leqslant R} \Big| \sum_{n \leqslant 4T/M} n^{-1/2 + it_r - it_s + iv} \Big|^{2k} \right)^{1/2}$$

$$\ll T^\varepsilon \Bigg\{ M^{2-2\sigma} + M^{k(3-4\sigma)/2}$$

$$\times \max_{|v| \leqslant h^2} \max_{N \leqslant (4T/M)^k} \left(\sum_{r \neq s \leqslant R} \Big| \sum_{N < n \leqslant 2N} f(n) n^{-1/2 + it_r - it_s + iv} \Big|^2 \right)^{1/2} \Bigg\},$$

where $f(n) \ll n^\varepsilon$, and f is independent of t_r and t_s. The point of this approach is that now the coefficients may be removed from the last double sum by appealing to the following simple

LEMMA 11.1. Let $a_1, \ldots, a_N$ be complex numbers such that $|a_1| \leqslant A$, $\ldots, |a_N| \leqslant A$ and let $M \geqslant N$. Then for any fixed $0 \leqslant \sigma \leqslant 1$

$$(11.60) \qquad \sum_{r, s \leqslant R} \Big| \sum_{n \leqslant N} a_n n^{-\sigma - it_r + it_s} \Big|^2 \leqslant A^2 \sum_{r, s \leqslant R} \Big| \sum_{n \leqslant M} n^{-\sigma - it_r + it_s} \Big|^2.$$

Proof of Lemma 11.1. The left-hand side of (11.60) is

$$\sum_{m, n \leqslant N} a_m \bar{a}_n (mn)^{-\sigma} \sum_{r, s \leqslant R} (m/n)^{-it_r + it_s} = \sum_{m, n \leqslant N} a_m \bar{a}_n (mn)^{-\sigma} \Big| \sum_{r \leqslant R} (m/n)^{it_r} \Big|^2$$

$$\leqslant A^2 \sum_{m, n \leqslant M} (mn)^{-\sigma} \Big| \sum_{r \leqslant R} (m/n)^{it_r} \Big|^2 = A^2 \sum_{r, s \leqslant R} \Big| \sum_{n \leqslant M} n^{-\sigma - it_r + it_s} \Big|^2.$$

Therefore applying Lemma 11.1 to the last sum in (11.59) with $a_n = f(n) n^{iv}$, $\sigma = \frac{1}{2}$, $A = T^\varepsilon$, we obtain

$$(11.61) \qquad R_1 \ll T^\varepsilon M^{2-2\sigma} + T^\varepsilon M^{k(3-4\sigma)/2} \max_{N \leqslant (4T/M)^k} (S(N))^{1/2},$$

where we define

$$(11.62) \qquad S(N) = \sum_{r,\,s \leqslant R} \left| \sum_{N < n \leqslant 2N} n^{-1/2 - it_r + it_s} \right|^2 ,$$

$$|t_r| \leqslant T, \qquad |t_r - t_s| \geqslant \log^4 T \quad \text{for} \quad r \neq s \leqslant R,$$

and $R \leqslant R_1$ is the representative set of zeros of R_1. Therefore $S(N)$ may be called a "double zeta sum," since it is similar to a sum of Dirichlet polynomials approximating $|\zeta(\tfrac{1}{2} + it_r - it_s)|$ by the approximate functional equation (4.1). The estimation of $S(N)$ represents the main step in obtaining density estimates from (11.61). The results that will eventually follow then from (11.61) will provide good bounds for $A(\sigma)$ in the range $\tfrac{3}{4} < \sigma < 1$, when σ is not close to $\tfrac{3}{4}$ or 1, and in particular we shall show that $A(\sigma) \leqslant 2$ ("density hypothesis") holds for $\sigma \geqslant \tfrac{11}{14} = 0.78571\ldots$. In the next section we shall deal with the sums $S(N)$, while in Section 11.7 we shall obtain zero-density results from (11.61) and estimates of $S(N)$.

11.6 DOUBLE ZETA SUMS

There are several ways to treat the double zeta sums $S(N)$ defined by (11.62). First, note that the terms with $r = s$ in (11.62) contribute $\ll RN$, and if the terms with $r \neq s$ were small one would expect

$$(11.63) \qquad S(N) \ll T^\varepsilon (RN + R^2)$$

to hold. This bound is very strong and is, however, certainly well beyond reach at present for all N. Although no restriction on N (with respect to T) has been made in the definition of $S(N)$, one may safely suppose that $N \leqslant T$, since for $N > T$ the sharp bound (11.63) does hold. This is not hard to see, since using Lemma 11.1, for $N > T$ we have

$$S(N) \ll RN + N^{-1} \sum_{r \neq s \leqslant R} \left| \sum_{N < n \leqslant 2N} n^{-it_r + it_s} \right|^2$$

$$\ll RN + R^2 + N^{-1} \sum_{r \neq s \leqslant R} N^2 |t_r - t_s|^{-2} \ll RN + R^2,$$

where in view of $N > T \geqslant \tfrac{1}{2}|t_r - t_s|$ we were able to use Lemma 1.2 and Lemma 2.1. Thus in what follows we may always assume that $N \leqslant T$.

Next, with $c_n = e^{-n/2N} - e^{-n/N}$ we have $1 \ll c_n \ll 1$ for $N < n \leqslant 2N$ and $c_n > 0$ for all $n \geqslant 1$. Observe that (11.60) remains true if $|a_n| \leqslant A b_n$ and the inner sum on the right-hand side of (11.60) is replaced by

$\sum_{n \leqslant M} b_n n^{-\sigma - it_r + it_s}$, so that with $M = \infty$ we obtain

(11.64) $\qquad S(N) \ll \sum_{r, s \leqslant R} \left| \sum_{n=1}^{\infty} c_n n^{-1/2 - it_r + it_s} \right|^2 = S^*(N),$

say. The contribution of the terms $r = s$ to $S^*(N)$ is $\ll RN$. To estimate the contribution of the remaining terms note that for $|t| \geqslant \log^2 T$ we have

(11.65)

$$\sum_{n=1}^{\infty} c_n n^{-1/2 - it} = (2\pi i)^{-1} \int_{1-i\infty}^{1+i\infty} \zeta(w + \tfrac{1}{2} + it) \Gamma(w)((2N)^w - N^w) \, dw$$

$$= (2\pi i)^{-1} \int_{-i\infty}^{i\infty} \zeta(w + \tfrac{1}{2} + it) \Gamma(w)((2N)^w - N^w) \, dw + o(1),$$

since the integrand is regular at $w = 0$ [because of the zero of $(2N)^w - N^w$], and the residue at the pole $w = \tfrac{1}{2} - it$ is $o(1)$ by (A.34). Likewise the last integral in (11.65) may be broken at $|\text{Im} \, w| = \log^2 T$ with an error which is $o(1)$, and we obtain

$$S^*(N) \ll RN + R^2 + \max_{|v| \leqslant \log^2 T} \sum_{r \neq s \leqslant R} \left| \zeta(\tfrac{1}{2} + it_r - it_s + iv) \right|^2,$$

so that (11.64) gives

(11.66) $\quad S(N) \ll RN + R^2 + \max_{|v| \leqslant \log^2 T} \sum_{r \neq s \leqslant R} \left| \zeta(\tfrac{1}{2} + it_r - it_s + iv) \right|^2.$

Here we shall use (8.28) and Hölder's inequality similarly as in the proof of (11.39), by fixing each s and summing over r. Hence if $(\varkappa, \lambda)$ is an exponent pair $(\varkappa > 0)$ and $H(T) = T^{\lambda/(1 - 2\varkappa + 4\lambda)}$, then the last expression in (11.66) is

$$\ll \left(\sum_{|\zeta| \geqslant H(3T)} |\zeta|^6 \right)^{1/3} R^{4/3} + \left(\sum_{|\zeta| < H(3T)} |\zeta|^{(2 + 4\varkappa + 4\lambda)/\varkappa} \right)^{\varkappa/(1 + 2\varkappa + 2\lambda)}$$

$$\times R^{(2 + 2\varkappa + 4\lambda)/(1 + 2\varkappa + 2\lambda)}$$

$$\ll T^{\varepsilon} \big(R^{5/3} T^{1/3} + R^{(2 + 3\varkappa + 4\lambda)/(1 + 2\varkappa + 2\lambda)} T^{(\varkappa + \lambda)/(1 + 2\varkappa + 2\lambda)} \big).$$

Inserting this bound in (11.66) we obtain

LEMMA 11.2. For any exponent pair $(\varkappa, \lambda)$ with $\varkappa > 0$

(11.67)

$$S(N) \ll RN + R^{5/3} T^{1/3 + \varepsilon} + R^{(2 + 3\varkappa + 4\lambda)/(1 + 2\varkappa + 2\lambda)} T^{(\varkappa + \lambda + \varepsilon)/(1 + 2\varkappa + 2\lambda)}.$$

The particular choice $(\varkappa, \lambda) = (\frac{1}{2}, \frac{1}{2})$ gives immediately

$$(11.68) \qquad\qquad S(N) \ll RN + R^{11/6}T^{1/3+\varepsilon}.$$

As was remarked already at the beginning of this section, the terms $r = s$ in $S(N)$ make a contribution of order RN, and it seems natural to expect that $S(N)$ is in some sense an increasing function of N (for T fixed). A useful result in this direction will be proved now, which is

LEMMA 11.3. For $U \geqslant N \log T$

$$(11.69) \qquad\qquad S(N) \ll T^\varepsilon S(U).$$

Proof of Lemma 11.3. Let us define, for a fixed $K \geqslant 1$,

$$g_{r,s}(n) = n^{-1/2 - it_r + it_s}, \qquad S(N, K) = \sum_{r,\,s \leqslant R} \left| \sum_{N < n \leqslant KN} g_{r,s}(n) \right|^2$$

with $|t_r| \leqslant T$, $|t_r - t_s| \geqslant \log^4 T$ for $r \neq s \leqslant R$, so that in this notation $S(N) = S(N, 2)$. Consider

$$H(\mathbf{e}) = \sum_{r,\,s \leqslant R} \left| \sum_{N < n \leqslant KN} g_{r,s}(n) \right|^2 \left| \sum_{M < m \leqslant 3M/2} e_m g_{r,s}(m) \right|^2$$

$$= \sum_{r,\,s \leqslant R} \left| \sum_{MN < k \leqslant 3KMN/2} a_k g_{r,s}(k) \right|^2,$$

where

$$a_k = \sum_{k = mn,\; M < m \leqslant 3M/2,\; N < n \leqslant KN} e_m \ll d(k) \ll T^\varepsilon,$$

if we suppose that $M \ll T^c$, and the components e_m of the vector $\mathbf{e}$ are each ± 1. By summing $H(\mathbf{e})$ over $2^{[M/2]}$ possible vectors $\mathbf{e}$ and using Lemma 11.1 we obtain

$$\sum_{\mathbf{e}} H(\mathbf{e}) \leqslant 2^{[M/2]} T^\varepsilon S(MN, 3K/2).$$

On the other hand

$$\sum_{\mathbf{e}} H(\mathbf{e}) = \sum_{r,\,s \leqslant R} \left| \sum_n \right|^2 \sum_{M < m_1,\, m_2 \leqslant 3M/2} g_{r,s}(m_1) \overline{g_{r,s}(m_2)} \sum_{\mathbf{e}} e_{m_1} e_{m_2}$$

$$= 2^{[M/2]} \sum_{r,\,s \leqslant R} \left| \sum_n \right|^2 \sum_{M < m \leqslant 3M/2} m^{-1} \gg 2^{[M/2]} S(N, K),$$

so that

$$(11.70) \qquad S(N, K) \ll T^\varepsilon S(MN, 3K/2).$$

Here we used the relation

$$\sum_e e_{m_1} e_{m_2} = \begin{cases} 2^{[M/2]} & \text{if } m_1 = m_2, \\ 0 & \text{if } m_1 \neq m_2, \end{cases}$$

since if $m_1 \neq m_2$, then $2^{[M/2]-1}$ summands are $+1$ and the other $2^{[M/2]-1}$ are -1, canceling each other. To obtain (11.69) from (11.70) we first use the Cauchy–Schwarz inequality and write

$$(11.71) \qquad S(N) \ll \sum_{j=0}^{2} S(N_j, K), \qquad K = 2^{1/3}, \qquad N_j = NK^j,$$

and then apply (11.70) with $M = M_j = 1 + [U/N_j]$, so that $U \leqslant M_j N_j$, and for U/N sufficiently large we have $3KM_j N_j/2 \leqslant 2U$. Since $U \geqslant N \log T$ by hypothesis we have that U/N is large and therefore (11.70) and (11.71) give, by the use of Lemma 11.1,

$$S(N) \ll T^\varepsilon \sum_{j=0}^{2} S(M_j N_j, 3K/2) \ll T^\varepsilon S(U).$$

LEMMA 11.4.

$$(11.72) \qquad S(N) \ll T^\varepsilon\big(RN + R^2 + S(T \log^2 T/N)\big).$$

Proof of Lemma 11.4. As already noted, the result is nontrivial only for $N \leqslant T$. Letting $h = \log^2 T$ we have by the proof of Lemma 11.1 (with $M = \infty$)

$$S(N) \ll N^{-1} \sum_{r,\,s \leqslant R} \left| \sum_{n=1}^{\infty} \left(e^{-(n/2N)^h} - e^{-(n/N)^h}\right) n^{-it_r+it_s} \right|^2,$$

since the exponential term is positive for all $n \geqslant 1$ and it is $\asymp 1$ for $N < n \leqslant 2N$. Here we once again use the reflection principle estimate (4.67) and Lemma 11.1 to obtain

$$S(N) \ll RN + R^2 + T^\varepsilon \sum_{r,\,s \leqslant R} \left| \sum_{n \leqslant 4T/N} n^{-1/2 - it_r + it_s} \right|^2$$

$$\ll RN + R^2 + T^\varepsilon \sum_{j=0}^{O(\log T)} S(2^{1-j}T/N)$$

$$\ll RN + R^2 + T^\varepsilon S(T \log^2 T/N),$$

where in the last step Lemma 11.3 was used. This proves Lemma 11.4.

The preceding lemmas enable us to deduce another explicit estimate for $S(N)$, which is for some ranges of R and N sharper than the bound provided by Lemma 11.2. This is

LEMMA 11.5.

(11.73) $$S(N) \ll T^\varepsilon(RN + R^2 + R^{5/4}T^{1/2}),$$

and for $N \geq T^{2/3}\log^4 T$ the term $R^{5/4}T^{1/2}$ may be omitted in (11.73).

Proof of Lemma 11.5. By using Lemmas 11.1 and 11.3 we have with the aid of the Cauchy–Schwarz inequality

$$S(N) \leq R\left(\sum_{r,\,s \leq R}\left|\sum_{N < n \leq 2N} n^{-1/2 - it_r + it_s}\right|^4\right)^{1/2} \ll R\big(S(2N^2\log T)\big)^{1/2}T^\varepsilon.$$

Then using Lemmas 11.4 and 11.3

$$S(N) \ll T^\varepsilon RN + T^\varepsilon R^2 + T^\varepsilon R\big(S(2T^2N^{-2}\log^5 T)\big)^{1/2}$$

$$\ll T^\varepsilon\big(RN + R^2 + RS^{1/2}(N)\big)$$

for $N \geq T^{2/3}\log^4 T$, and simplifying we obtain at once (11.73) without $R^{5/4}T^{1/2}$. Therefore the range $N \geq T$ for which the bound (11.63) holds may be extended to $N \geq T^{2/3}\log^4 T$, and any further improvement would be very interesting.

To prove (11.73) we proceed similarly using (11.72) and setting

$$U = \max\big(N\log T,\ R^{1/4}T^{1/2}\log T\big).$$

Then

$$S(N) \ll T^\varepsilon S(U) \ll T^\varepsilon\big(RU + R^2 + S(T\log^2 T/U)\big)$$

$$\ll T^\varepsilon\big(RU + R^2 + R\big(S(T^2U^{-2}\log^5 T)\big)^{1/2}\big),$$

and using (11.68) we have

$$S(N) \ll T^\varepsilon\big(RU + R^2 + R^{3/2}TU^{-1} + R^{23/12}T^{1/6}\big)$$

$$\ll T^\varepsilon\big(RN + R^{5/4}T^{1/2} + R^{23/12}T^{1/6}\big).$$

Repeating the same procedure but using the above estimate in place of (11.68) we obtain

$$S(N) \ll T^\varepsilon\big(RN + R^{47/24}T^{1/12} + R^{5/4}T^{1/2} + R^{13/8}T^{1/4}\big).$$

But it is easily seen that $R^{13/8}T^{1/4} \ll R^{5/4}T^{1/2} + R^2$, so that the last term in the above estimate may be discarded, and repeating the procedure k times we have

$$S(N) \ll T^\varepsilon \big\{ RN + R^{5/4}T^{1/2} + R^{2-(6(2^k))^{-1}}T^{(3(2^k))^{-1}} \big\},$$

so that (11.73) follows on taking k sufficiently large.

11.7 ZERO-DENSITY ESTIMATES FOR $\frac{3}{4} < \sigma < 1$

We now have at our disposal good estimates for $S(N)$ furnished by the lemmas of Section 11.6, and we shall use (11.61), (11.67), and (11.73) to obtain, for $k \geqslant 2$ an integer,

$$R_1 \ll T^\varepsilon \big\{ M^{2-2\sigma} + M^{k(3-4\sigma)/2}R_1 + M^{k(1-2\sigma)}T^{k/2}R_1^{1/2}$$

$$+ M^{k(3-4\sigma)/2}\min\big(R_1^{5/8}T^{1/4}, \ R_1^{5/6}T^{1/6}$$

$$+ R_1^{(2+3\varkappa+4\lambda)/(2+4\varkappa+4\lambda)}T^{(\varkappa+\lambda)/(2+4\varkappa+4\lambda)}\big)\big\}.$$

In view of $\sigma > \frac{3}{4}$ the term $M^{k(3-4\sigma)/2}R_1$ may be omitted and after simplifying the estimate above we obtain

(11.74)

$$R_1 \ll T^\varepsilon \big\{ M^{2-2\sigma} + M^{k(2-4\sigma)}T^k$$

$$+ \min\big(T^{2/3}M^{4k(3-4\sigma)/3}, \ TM^{3k(3-4\sigma)} + T^{(\varkappa+\lambda)/\varkappa}M^{k(1+2\varkappa+2\lambda)(3-4\sigma)/\varkappa}\big)\big\}.$$

To make the first two terms on the right-hand side of (11.74) equal we choose

(11.75) $$T = T_0 = M^{((4k-2)\sigma+2-2k)/k}.$$

With this choice of T_0 the remaining terms in (11.74) do not exceed $T^\varepsilon M^{2-2\sigma}$ for

(11.76) $$\sigma \geqslant \min\left(\frac{6k^2 - 5k + 2}{8k^2 - 7k + 2}, \right.$$

$$\left. \max\left(\frac{9k^2 - 4k + 2}{12k^2 - 6k + 2}, \ \frac{3k^2(1 + 2\varkappa + 2\lambda) - (4\varkappa + 2\lambda)k + 2\varkappa + 2\lambda}{4k^2(1 + 2\varkappa + 2\lambda) - (6\varkappa + 4\lambda)k + 2\varkappa + 2\lambda} \right) \right)$$

Thus we obtain, provided that (11.76) holds,

$$R_1 \ll T^\varepsilon M^{2-2\sigma}(1 + T/T_0) \ll T^\varepsilon\left(M^{2-2\sigma} + TM^{2-2\sigma}M^{\{(2k-2+(2-4k)\sigma)/k\}}\right)$$

$$\ll T^\varepsilon\left(Y^{4-4\sigma} + TY^{4(4k-2-(6k-2)\sigma)/3k}\right),$$

since $4k - 2 - (6k - 2)\sigma \leqslant 0$ for $\sigma \geqslant \frac{3}{4}$ and $k \geqslant 2$. Using (11.20) with $M(12) \leqslant 2$ we have

$$(11.77) \quad N(\sigma, T) \ll T^\varepsilon\left(Y^{4-4\sigma} + TY^{4(4k-2-(6k-2)\sigma)/3k} + T^2Y^{6-12\sigma}\right)$$

$$\ll T^{3k(1-\sigma)/((3k-2)\sigma+2-k)+\varepsilon}$$

for $Y = T^{3k/((12k-8)\sigma+8-4k)}$.

Before we proceed to estimates arising from specific values of k, it may be noted that by letting $k \to \infty$ all the expressions in (11.76) tend to $\frac{3}{4}$ and (11.77) gives then $A(\sigma) \leqslant 3/(3\sigma - 1)$ when k is very large, which is just M. N. Huxley's estimate (11.23). The functions appearing in (11.76) are decreasing functions of k, while the function that appears in (11.77) in the estimate for $N(\sigma, T)$ is an increasing function of k, so that there is no simple k which will furnish the sharpest bound obtainable by this method in the whole range $\sigma \geqslant \sigma_0 > \frac{3}{4}$. Taking first $k = 2$ we see that $A(\sigma) \leqslant 3/(2\sigma)$ holds for

$$(11.78) \qquad \sigma \geqslant \min\left(\frac{4}{5}, \max\left(\frac{15}{9}, \frac{6 + 9\varkappa + 11\lambda}{8 + 11\varkappa + 13\lambda}\right)\right),$$

and the choice $(\varkappa, \lambda) = \left(\frac{97}{251}, \frac{132}{251}\right)$ gives $A(\sigma) \leqslant 3/(2\sigma)$ for $\sigma \geqslant \frac{3831}{4791} = 0.799624\ldots$. For $k = 3$ we have $A(\sigma) \leqslant 9/(7\sigma - 1)$ for

$$(11.79) \qquad \sigma \geqslant \min\left(\frac{41}{53}, \max\left(\frac{71}{92}, \frac{27 + 55\varkappa + 50\lambda}{36 + 56\varkappa + 62\lambda}\right)\right),$$

and therefore $A(\sigma) \leqslant 9/(7\sigma - 1)$ holds for $\sigma \geqslant \frac{41}{53} = 0.773584\ldots$. Since $9/(7\sigma - 1) \leqslant 2$ for $\sigma \geqslant \frac{11}{14}$ we obtain also

$$(11.80) \qquad A(\sigma) \leqslant 2 \qquad \left(\sigma \geqslant \frac{11}{14} = 0.785714\ldots\right),$$

which is the best-known range for which the density hypothesis holds. Finally, we shall also consider the case $k = 4$, when from the first expression on the right-hand side of (11.76) we see that $A(\sigma) \leqslant 6/(5\sigma - 1)$ holds for $\sigma \geqslant \frac{13}{17} = 0.764705\ldots$. The estimates that were just discussed may be collected together to give

THEOREM 11.4.

(11.81) $\quad A(\sigma) \leqslant 3/(2\sigma) \qquad \left(\tfrac{3831}{4791} = 0.799624 \ldots \leqslant \sigma \leqslant 1\right),$

(11.82) $\quad A(\sigma) \leqslant 2 \qquad \left(\tfrac{11}{14} = 0.785714 \ldots \leqslant \sigma \leqslant 1\right),$

(11.83) $\quad A(\sigma) \leqslant 9/(7\sigma - 1) \qquad \left(\tfrac{41}{53} = 0.773584 \ldots \leqslant \sigma \leqslant 1\right),$

(11.84) $\quad A(\sigma) \leqslant 6/(5\sigma - 1) \qquad \left(\tfrac{13}{17} = 0.764705 \ldots \leqslant \sigma \leqslant 1\right).$

11.8 ZERO-DENSITY ESTIMATES FOR σ CLOSE TO $\tfrac{3}{4}$

Although all estimates of the type (11.77) improve (11.23) for a fixed k, none can be made yet to hold in the whole range $\sigma \geqslant \tfrac{3}{4}$. Therefore it is of interest to find an estimate which will improve (11.23) for $\sigma \geqslant \tfrac{3}{4}$. This may be done, and the result is

THEOREM 11.5.

(11.85) $\qquad A(\sigma) \leqslant 3/(7\sigma - 4) \qquad \left(\tfrac{3}{4} \leqslant \sigma \leqslant \tfrac{10}{13}\right),$

(11.86) $\qquad A(\sigma) \leqslant 9/(8\sigma - 2) \qquad \left(\tfrac{10}{13} \leqslant \sigma \leqslant 1\right).$

Note that both (11.85) and (11.23) give $A(\tfrac{3}{4}) \leqslant \tfrac{12}{5}$, but Theorem 11.5 improves (11.23) in the whole range $\sigma > \tfrac{3}{4}$. Nevertheless, (11.84) is still sharper, so that the importance of (11.85) lies in the range $\tfrac{3}{4} < \sigma \leqslant \tfrac{13}{17}$. The main idea of proof is the use of a double zeta-function sum [different from (11.62)] at the line $\sigma = \tfrac{3}{4}$. For a fixed θ satisfying $\tfrac{1}{2} < \theta < 1$ let us define

(11.87) $\qquad S_1(\theta) = \sum_{r,\,s \leqslant R} \left| \zeta(\theta + it_r - it_s + iv') \right|^2,$

where the real numbers $t_1, \ldots, t_R$ satisfy $|t_r| \leqslant T$, $|t_r - t_s| \geqslant 2\log^4 T$ for $r \neq s \leqslant R$, and v' is defined by

(11.88) $\quad \left| \zeta(\theta + it_r - it_s + iv') \right| = \max_{-\log^2 T \leqslant v \leqslant \log^2 T} \left| \zeta(\theta + it_r - it_s + iv) \right|.$

Furthermore we define $S_1(\tfrac{1}{2})$ analogously as $S_1(\theta)$, only for technical reasons in the definition of v' the maximum of v is to be taken over the interval $[-2\log^2 T, 2\log^2 T]$. The proof of Theorem 11.5 will require a good bound for $S_1(\frac{3}{4})$, which is furnished by

LEMMA 11.6.

(11.89) $$S_1(\tfrac{3}{4}) \ll T^\varepsilon (R^2 + R^{11/8} T^{1/4}).$$

Proof of Lemma 11.6. The most important step in the proof of (11.89) consists of showing that

(11.90) $$S_1(\tfrac{3}{4}) \ll T^\varepsilon R^2 + T^\varepsilon R^{3/4} \big(S_1(\tfrac{1}{2})\big)^{1/2}.$$

To obtain (11.90) we may start from (11.4) with $s = \frac{3}{4} + it_r - it_s + iv'$ $(r \neq s)$, $1 \ll Y \ll T^C$, and move the line of integration to $\operatorname{Re} w = -\frac{1}{4}$. We encounter a pole at $w = 1 - s$ with residue $o(1)$ in view of (A.34) and a pole at $w = 0$ with residue $\zeta(s)$. Therefore

(11.91) $$\zeta^2(s) \ll 1 + \left| \sum_{n \leqslant Y} e^{-n/Y} n^{-s} \right|^2$$

$$+ Y^{-1/2} \int_{-\log^2 T}^{\log^2 T} e^{-|y|} \left| \zeta \big(\tfrac{1}{2} + it_r - it_s + iv' + iy \big) \right|^2 dy,$$

and summing over r, $s \leqslant R$ it follows, by the Cauchy–Schwarz inequality,

(11.92)

$$S_1(\tfrac{3}{4}) \ll T^\varepsilon R^2 + \sum_{r, s \leqslant R} \left| \sum_{n \leqslant Y} e^{-n/Y} n^{-3/4 - it_r + it_s - iv'} \right|^2 + Y^{-1/2} S_1(\tfrac{1}{2})$$

$$\ll T^\varepsilon R^2 + R \left(\sum_{r, s \leqslant R} \left| \sum_{n \leqslant Y^2} c(n) n^{-3/4 - it_r + it_s - iv'} \right|^2 \right)^{1/2} + Y^{-1/2} S_1(\tfrac{1}{2}),$$

where $c(n) \ll n^\varepsilon$. To estimate the sum under the square root we use Lemma 11.1 and obtain, similarly as in (11.64),

$$\sum_{r, s \leqslant R} \left| \sum_{n \leqslant Y^2} \right|^2 \ll \log T \left(\max_{M \leqslant Y^2} \sum_{r, s \leqslant R} \left| \sum_{M < n \leqslant 2M} \right|^2 \right)$$

$$\ll T^\varepsilon \max_{M \leqslant Y^2} M^{-3/2} \sum_{r, s \leqslant R} \left| H(it_r - it_s) \right|^2,$$

where for t real we have, as in (11.36),

$$H(it) = \sum_{n=1}^{\infty} (e^{-n/2M} - e^{-n/M})n^{-it}$$

$$= (2\pi i)^{-1} \int_{2-i\infty}^{2+i\infty} \zeta(w + it)((2M)^w - M^w)\Gamma(w)\, dw,$$

and therefore trivially $H(0) \ll M$. For $t = t_r - t_s \neq 0$ we move the line of integration in the above integral to $\mathrm{Re}\, w = \frac{3}{4}$, encountering a pole at $w = 1 - it_r + it_s$ with residue $o(1)$ by (A.34), and we obtain

$$\sum_{r,\, s \leqslant R} \left| \sum_{n \leqslant Y^2} c(n)n^{-3/4 - it_r + it_s} \right|^2 \ll T^\varepsilon \max_{M \leqslant Y^2} \left\{ M^{-3/2}\left(RM^2 + R^2 + M^{3/2}S_1\left(\tfrac{3}{4}\right)\right) \right\}$$

$$\ll T^\varepsilon \left(RY + R^2 + S_1\left(\tfrac{3}{4}\right) \right).$$

Inserting this estimate in (11.92) and simplifying, we obtain with $Y = R^{-3/2}S_1(\tfrac{1}{2})$

$$S_1\left(\tfrac{3}{4}\right) \ll T^\varepsilon R^2 + T^\varepsilon R^{3/2}Y^{1/2} + Y^{-1/2}S_1\left(\tfrac{1}{2}\right) \ll T^\varepsilon \left\{ R^2 + R^{3/4}\left(S_1\left(\tfrac{1}{2}\right)\right)^{1/2} \right\},$$

which is precisely (11.90). To obtain finally (11.89) from (11.90) we need an adequate bound for $S_1(\tfrac{1}{2})$. From the reflection principle equation (4.66) we have with $s = \tfrac{1}{2} + it_r - it_s + iv'$ $(r \neq s)$, $h = \log^2 T$, $\alpha = \tfrac{1}{2} - \varepsilon$, $M = 4T/Y$, $k = 1$,

$$(11.93) \quad \zeta(s) \ll 1 + \left| \sum_{n \leqslant Y} e^{-(n/Y)^h}n^{-s} \right| + T^\varepsilon \int_{-h^2}^{h^2} \left| \sum_{n \leqslant 4T/Y} n^{s-1-\varepsilon+iy} \right| dy.$$

To make the lengths of the sums in (11.93) equal we choose $Y = 2T^{1/2}$. Squaring, summing over r, $s \leqslant R$, using Lemma 11.1 and Lemma 11.3 it follows that

$$(11.94) \qquad S_1\left(\tfrac{1}{2}\right) \ll T^\varepsilon \left(R^2 + S\left(2T^{1/2}\log T\right) \right),$$

where $S(N)$ is defined by (11.62). If we use (11.73) to bound $S(2T^{1/2}\log T)$, then (11.89) follows at once from (11.90) and (11.94).

Proof of Theorem 11.5. It remains yet to prove Theorem 11.5 with the aid of Lemma 11.6. We use (11.35), but now in (11.36) we move the line of integration to $\mathrm{Re}\, w = \tfrac{1}{4}$ and employ the functional equation (1.23) and the asymptotic formula (1.25). Instead of (11.37) we obtain now, with the aid of the Cauchy–Schwarz inequality,

$$\sum_{r\neq s\leqslant R} \left|H(it_r - it_s)\right| \ll M \sum_{r\neq s\leqslant R} e^{-|t_r - t_s|} + R^2 + (MT)^{1/4}$$

$$\times \int_{-\log^2 T}^{\log^2 T} \sum_{r\neq s\leqslant R} \left|\zeta\left(\tfrac{3}{4} + it_r - it_s + iv\right)\right| dv$$

$$\ll RM + R^2 + T^{\varepsilon+1/4}M^{1/4}R\left(S_1\left(\tfrac{3}{4}\right)\right)^{1/2}.$$

Therefore if we use this estimate and Lemma 11.6 in (11.35), it follows after some simplification that

$$(11.95) \quad R_1 \ll T^\varepsilon M^{2-2\sigma} + R_1 T^{\varepsilon+1/4}M^{(5-8\sigma)/4} + T^{\varepsilon+6/5}M^{(20-32\sigma)/5}.$$

For R_0 points lying in an interval of length $T = T_0 = M^{8\sigma-5-\varepsilon}$ we have

$$R_0 \ll T^\varepsilon\left(M^{2-2\sigma} + T_0^{6/5}M^{(20-32\sigma)/5}\right) \ll T^\varepsilon M^{2-2\sigma}$$

for $\sigma \leqslant \tfrac{10}{13}$, and therefore for $\tfrac{3}{4} \leqslant \sigma \leqslant \tfrac{10}{13}$ with $Y = T^{3/(28\sigma-16)}$ we obtain

$$R_1 \ll R_0(1 + T/T_0) \ll T^\varepsilon M^{2-2\sigma}(1 + T/T_0)$$

$$\ll T^\varepsilon\left(Y^{4-4\sigma} + TY^{(28-40\sigma)/3}\right) \ll T^{3(1-\sigma)/(7\sigma-4)+\varepsilon}.$$

Using (11.20) with $M(4) = 1$ it is seen that for $\tfrac{3}{4} \leqslant \sigma \leqslant \tfrac{11}{14}$ one has $TY^{2-4\sigma} \leqslant TY^{(28-40\sigma)/3}$, hence (11.85) follows. Analogously we obtain (11.86) if in (11.95) we choose $T_0 = M^{(11\sigma-5)/3}$.

Notes

The contents of this chapter are based on Chapter 9 of the author's monograph (1983b). The zero-detection method of Section 11.2 is to be found in general form (to include zero-density estimates for L-functions) in Chapter 12 of H. L. Montgomery (1971) (see also 1969), where a comprehensive survey of results up to 1971 is given.

The results of this chapter are the sharpest ones hitherto, except when σ is very close to $\tfrac{1}{2}$, in which case the estimate

$$(11.96) \qquad\qquad N(\sigma,T) \ll T^{1-(\sigma-1/2)/4}\log T \qquad \left(\tfrac{1}{2} \leqslant \sigma \leqslant 1\right)$$

of A. Selberg (1946) is sharper. Recently, M. Jutila (1982) improved (11.96) by replacing $\tfrac{1}{4}$ by $1 - \delta$ for any fixed $0 < \delta < 1$.

The useful procedure of estimating R_1 by its representatives for which (11.16) and (11.18) hold has been introduced by M. Jutila (1972), while (as mentioned in Chapter 8) the technique of dividing T into subintervals of length T_0 and then multiplying the resulting estimate by $1 + T/T_0$ is due to M. N. Huxley (1972a).

A. E. Ingham (1940) proved $N(\sigma,T) \ll T^{3(1-\sigma)/(2-\sigma)}\log^5 T$ by a method different from the one presented in Section 11.2, and which seems to be more complicated. Ingham's result is given as Theorem 9.19 (B) by E. C. Titchmarsh (1951).

For details of (11.27) and (11.28) see M. N. Huxley (1972a, Chapter 23; 1972b).

As stated in the proof of Ingham's estimate (11.22), the main reason why this bound withstands improvement for such a long time is the lack of $M(A) = 1$ for $A > 4$. This also explains the seemingly mysterious choice $r = 2$ in (11.16). Namely, if in (11.16) we choose $r = 3$ ($r \geq 4$ can be analyzed analogously), then following the proof of Theorem 11.1 we have

$$R_1 \ll T^\varepsilon M^{1-2\sigma}(T + M) \ll T^\varepsilon\big(Y^{6-6\sigma} + TY^{9(1-2\sigma)/4}\big) \qquad \big(\tfrac{1}{2} \leq \sigma \leq \tfrac{3}{4}\big),$$

$$R_1 \ll T^\varepsilon\big(M^{2-2\sigma} + TM^{4-6\sigma}\big) \ll T^\varepsilon\big(Y^{6-6\sigma} + TY^{9(2-3\sigma)/2}\big) \qquad \big(\tfrac{3}{4} \leq \sigma \leq 1\big).$$

If we now use $R_1 \ll T^{1+\varepsilon}Y^{2-4\sigma}$ or $R_2 \ll T^{2+\varepsilon}Y^{6-12\sigma}$ [coming from $M(4) = 1$ and $M(12) \leq 2$, respectively] we see that we get poorer estimates for $N(\sigma, T)$ than the ones furnished by Theorem 11.1. However, if $M(6) = 1$ were known to hold, then from (11.20) and the estimates above we would obtain

$$N(\sigma, T) \ll T^\varepsilon\big(Y^{6-6\sigma} + TY^{9(1-2\sigma)/4} + TY^{3-6\sigma}\big) \ll T^{8(1-\sigma)/(5-2\sigma)+\varepsilon} \qquad \big(\tfrac{1}{2} \leq \sigma \leq \tfrac{3}{4}\big),$$

$$N(\sigma, T) \ll T^\varepsilon\big(Y^{6-6\sigma} + TY^{9(2-3\sigma)/2} + TY^{3-6\sigma}\big) \ll T^{4(1-\sigma)/(5\sigma-2)+\varepsilon} \qquad \big(\tfrac{3}{4} \leq \sigma \leq 1\big)$$

with $Y = T^{4/(15-6\sigma)}$ and $Y = T^{2/(15\sigma-6)}$, respectively. This would improve Theorem 11.1 and give $A(\sigma) \leq \tfrac{16}{7}$ for the whole range $\tfrac{1}{2} \leq \sigma \leq 1$. Thus we see one more reason for the importance of the sixth power moment estimate $M(6) = 1$.

Theorem 11.2 is due to the author (1980b), and improves on $A(\sigma) \leq 4/(4\sigma - 1)$ $(\tfrac{25}{28} \leq \sigma \leq 1)$, which was obtained by D. R. Heath-Brown (1979b).

If the Lindelöf hypothesis that $\zeta(\tfrac{1}{2} + it) \ll t^\varepsilon$ is true, then trivially for $Y > T^{\varepsilon_1}$ one has $R_2 \ll T^\varepsilon$ for $\sigma > \tfrac{1}{2}$, and by (11.36)

$$\sum_{r,\,s \leq R} |H(it_r - it_s)| \ll RM + R^2 M^{1/2}T^\varepsilon;$$

hence by (11.35) one has for $\sigma > \tfrac{3}{4}$

$$R_1 \ll T^\varepsilon M^{2-2\sigma} \ll T^\varepsilon Y^{4-4\sigma}.$$

Choosing $Y = T^\varepsilon$ one obtains

$$N(\sigma, T) \ll T^\varepsilon \qquad \big(\tfrac{3}{4} + \delta \leq \sigma \leq 1\big),$$

where $\varepsilon = \varepsilon(\delta)$ may be made arbitrarily small for any $\delta > 0$, which is a result of G. Halász and P. Turán (1969). For a nice survey of Turán's conditional and unconditional results concerning zero-density estimates, the reader is referred to P. Turán (1971).

Theorem 11.3 [in the form (11.52)] is due to H. L. Montgomery (1971) (Theorem 12.3). Montgomery chooses $\alpha = 4\sigma - 3$ and claims then in Corollary 12.5 that (11.53) with $D = 100$ [due to H. -E. Richert (1967)] leads to

(11.97) $$N(\sigma, T) \ll T^{167(1-\sigma)^{3/2}} \log^{17}T \qquad \big(\tfrac{1}{2} \leq \sigma \leq 1\big).$$

However there is an error in his calculation, since (11.53) with $D = 100$ and $\alpha = 4\sigma - 3$ implies (11.97) with $1333.3\ldots$ in place of 167, while (11.32) gives with $D = 100$ the constant $1304.37\ldots$ in (11.33).

Section 11.5 is based on M. Jutila's paper (1977), which among other things contains the bound $\sigma \geq \tfrac{11}{14}$ for which the density hypothesis $A(\sigma) \leq 2$ holds, and this is still the best-known result of this type. The history of bounds for density is also mentioned in Jutila (1977) and is as follows. Let a be such a constant for which the density hypothesis $A(\sigma) \leqslant 2$ holds for $\sigma \geqslant a$. Then H. L. Montgomery (1971) proved $a \leqslant \frac{9}{10}$; M. N. Huxley (1972b), $a \leqslant \frac{5}{6}$; K. Ramachandra (1975a), $a \leqslant \frac{21}{26}$; F. Forti and C. Viola (1973), $a \leqslant 0.8059\ldots$; M. N. Huxley (1973b, 1975a, 1975b), $a \leqslant \frac{4}{5}$; and some intermediate results, M. Jutila (1972), $a \leqslant \frac{43}{54}$.

Lemma 11.2 is due to the author (1983b), while the remaining lemmas of Section 11.6 are due to D. R. Heath-Brown (1979f). An alternative proof of Lemma 11.5 may be given with the aid of (11.66). If we use (11.93) and Lemmas 11.3 and 11.4 we obtain

$$S_1\left(\tfrac{1}{2}\right) \ll T^\varepsilon \left(S(2Y \log T) + S(2TY^{-1} \log T) \right) \ll T^\varepsilon \left(RY + R^2 + S(2TY^{-1} \log^3 T) \right).$$

But now by the Cauchy–Schwarz inequality we have

$$S(2TY^{-1} \log^3 T) \ll T^\varepsilon R \left(S(T^2 Y^{-2} \log^8 T) \right)^{1/2}$$

$$\ll T^\varepsilon \left(R^{3/2} TY^{-1} + R^2 + S_1^{1/2}\left(\tfrac{1}{2}\right) R \right),$$

where we again used Lemmas 11.1 and 11.3 and (11.68), and $S_1\left(\tfrac{1}{2}\right)$ is the double zeta-sum that appears in (11.68). From the last two bounds we have

$$S_1\left(\tfrac{1}{2}\right) \ll T^\varepsilon \left(RY + R^{3/2} TY^{-1} + R^2 \right) \ll T^\varepsilon \left(R^{5/4} T^{1/2} + R^2 \right)$$

with $Y = R^{1/4} T^{1/2}$, and Lemma 11.5 follows then at once from (11.66).

Ranges given for various estimates by Theorem 11.4 are the best ones known. (11.81) was proved by M. N. Huxley (1973b, 1975a, 1975b) to hold for $\sigma \geqslant \frac{37}{42}$, and by the author (1979b) for $\sigma \geqslant \frac{4}{5}$, while the present range was given by the author (1980b). (11.83) was proved by D. R. Heath-Brown (1979b) for $\sigma \geqslant \frac{11}{14}$ and by the author (1979b) for $\sigma \geqslant \frac{74}{95}$. A proof of (11.84) for $\sigma \geqslant \frac{67}{87}$ is given by the author's paper (1979b), which also contains zero-density estimates coming from bounds for $\mu(\sigma)$.

Lemma 11.6 and Theorem 11.5 are due to the author (1984b). An argument similar to the one used in the proof of Lemma 11.6 gives

$$S(\theta) \ll T^\varepsilon R^2 + T^\varepsilon R^{3\theta - 3/2} \left(S(1/2) \right)^{2 - 2\theta},$$

where $S(\theta)$ is defined by (11.87), and the above estimate for $\theta = \frac{3}{4}$ reduces to (11.90).

/ CHAPTER TWELVE /
THE DISTRIBUTION OF PRIMES

12.1 GENERAL REMARKS

The zeta-function was introduced in mathematics as an analytic tool for studying prime numbers. Therefore it is only natural that some of the most important applications of the zeta-function belong to prime number theory. Here we shall be concerned with some of the most important of these applications, but the possibilities of zeta-function theory in the theory of primes are by no means exhausted by the subjects treated in this chapter. As usual, p_n will denote the nth prime number,

$$(12.1) \qquad \pi(x) = \sum_{p \leqslant x} 1$$

will be the number of primes not exceeding x,

$$(12.2) \qquad \theta(x) = \sum_{p \leqslant x} \log p$$

is the Chebyshev function, while in this chapter only

$$(12.3) \qquad \psi(x) = \sum_{n \leqslant x} \Lambda(n) = \sum_{p^m \leqslant x} \log p,$$

since no confusion will arise with the otherwise used notation $\psi(x) = x - [x] - \frac{1}{2}$, and both of these notations are standard.

Many problems in prime number theory may be formulated in terms of the functions π, θ, and ψ, and it turns out that the weighted function ψ is in a certain sense the most "natural" of all three, since it possesses a (relatively simple) explicit formula [see (12.6)], which relates the order of $\psi(x) - x$ to a certain sum over nontrivial zeros of the zeta-function. Moreover, we have

$$(12.4) \quad \theta(x) = \psi(x) + O\left(\sum_{p^m \leqslant x, \, m \geqslant 2} \log p\right) = \psi(x) + O(x^{1/2}\log x),$$

$$(12.5) \quad \pi(x) = \int_{3/2}^{x} \frac{d\theta(t)}{\log t} = \frac{\theta(x)}{\log x} + \int_{3/2}^{x} \frac{\theta(t)\,dt}{t\log^2 t},$$

and in most problems it is enough to work with $\psi(x)$ and if necessary "translate" the results from $\psi(x)$ to $\theta(x)$ and $\pi(x)$ via (12.4) and (12.5).

Already Riemann, whose work was in many aspects decades beyond that of his contemporaries, stated the elegant formula

$$(12.6) \quad \psi(x) = x - \sum_{\rho} \frac{x^\rho}{\rho} - \frac{\zeta'(0)}{\zeta(0)} - \frac{1}{2}\log(1 - x^{-2}) \qquad (x > 1, \, x \neq p^m),$$

where the sum is over nontrivial zeros $\rho = \beta + i\gamma$ of $\zeta(s)$, and means

$$\sum_{\rho} \frac{x^\rho}{\rho} = \lim_{T \to \infty} \sum_{|\gamma| \leqslant T} \frac{x^\rho}{\rho}.$$

This formula was proved by H. von Mangoldt in 1895, and despite its beauty it is somewhat impractical because of the difficulties in estimating the "tails" of the series in (12.6). A truncated form of (12.6), due to E. Landau, will be proved in the next section and will serve as the basis for most of our investigations concerning $\psi(x)$.

Broadly speaking problems in prime number theory fall into two main categories: *global problems* and *local problems*, and we shall be concerned here with both types of problems. Global problems involve asymptotic formulas, sums, estimations, and so on of $\pi(x)$ and $\psi(x)$, while local problems involve questions dealing with individual p_n, such as problems with $p_{n+1} - p_n$, the difference between consecutive primes.

The most important global problem in prime number theory is the so-called prime number theorem, which consists of the estimation of $\psi(x) - x$ [or equivalently $\pi(x) - \mathrm{li}\,x$]. Already from Theorem 1.5 (not to mention the much stronger Theorem 6.1) we know that $\beta < 1$ if $\rho = \beta + i\gamma$ is a nontrivial zero of $\zeta(s)$, so that one expects the series in (12.6) to be of a lower order of magnitude than x, that is, one expects $\psi(x) = (1 + o(1))x$ as $x \to \infty$. This assertion is indeed true, and represents the simplest form of the prime number theorem. The prime number theorem will be discussed in Section 12.3, where we shall use the strongest known zero-free region (6.1) to deduce

$$(12.7) \quad \psi(x) = x + O\left\{x\exp\left(-C\log^{3/5}x(\log\log x)^{-1/5}\right)\right\} \qquad (C > 0),$$

which represents the strongest known form of the prime number theorem. In Section 12.4 we shall discuss the analogous formulas for $M(x) = \sum_{n \leqslant x} \mu(n)$ and $\psi_k(x) = \sum_{n \leqslant x} \Lambda_k(n)$, where $\Lambda_k(n)$ is the generalized von Mangoldt function (see Section 1.7 of Chapter 1). There are no elegant formulas for $\psi(x)$ and $\pi(x)$ because of the influence of complex zeros of $\zeta(s)$, as evident in (12.6). This influence is indeed a profound one, and in Section 12.3 it will be pointed out that (12.7) is essentially equivalent to the zero-free region (6.1).

A local problem in prime number theory which has always attracted much attention is the determination of θ and θ' in the inequality

$$(12.8) \qquad p_{n+1} - p_n \ll p_n^{\theta} \log^{\theta'} p_n.$$

A well-known unproved conjecture related to (12.8) states that between any two squares there is a prime, and H. Cramér conjectured even

$$(12.9) \qquad p_{n+1} - p_n \ll \log^2 p_n.$$

The latter conjecture is hopeless by today's methods, since even the Riemann hypothesis does not seem to imply anything more than

$$(12.10) \qquad p_{n+1} - p_n \ll p_n^{1/2} \log p_n,$$

which will be proved in Theorem 12.10.

In Sections 12.5 and 12.6 we shall consider some questions involving primes in short intervals, and in particular inequalities of the type (12.8) will be discussed. These inequalities may be derived from an asymptotic formula of the form

$$(12.11) \quad \psi(x+h) - \psi(x) = (1 + o(1))h \qquad [h = o(x), \ x \to \infty]$$

by taking $x = p_n$, and we shall prove the strongest known result of the type (12.11) in Theorem 12.8 by employing zero-density estimates of Chapter 11. However, in recent years it has been discovered that the linear sieve may be successfully combined with analytic methods to provide bounds of the type

$$(12.12) \qquad \psi(x+h) - \psi(x) \gg h$$

with a smaller value of h than the one needed for the asymptotic formula (12.11) to hold. Following the method of H. Iwaniec and M. Jutila we shall prove (12.12) with $h = x^{13/23}$ in Theorem 12.11, [and consequently (12.8) too with $\theta = \frac{13}{23}$] and then indicate how further sharpenings may be obtained.

In general, instead of primes one may consider all the corresponding problems where primes are replaced by "almost primes," that is, integers having a restricted number of prime factors. In this case one usually gets sharper results than in the case when only primes are being considered, and in

Section 12.7 we shall investigate the distribution of almost primes P_2 in short intervals. By P_2 we mean a number with at most two prime factors (equal or distinct), and likewise by P_r we mean a number with at most r prime factors.

An unconditional result concerning the existence of P_2's in short intervals will follow from the asymptotic formula given by Theorem 12.12, while a sharp result for P_2's in almost all short intervals is furnished by Theorem 12.13.

Finally, in Section 12.8 we shall obtain some results involving sums of $p_{n+1} - p_n$. Following the method developed by D. R. Heath-Brown we shall introduce $N^*(\sigma, T)$, a sort of a new zero-density function, and upper bounds for this function will lead to good upper bounds for the sums

$$(12.13) \qquad \sum_{p_n \leqslant x} (p_{n+1} - p_n)^2, \qquad \sum_{p_n \leqslant x,\; p_{n+1}-p_n > p_n^{1/2}} (p_{n+1} - p_n).$$

Sharp conditional results (e.g., assuming the Riemann hypothesis) for the sums in (12.13) were obtained long ago by H. Cramér and A. Selberg, but the results which will be proved in Section 12.8 are unconditional and hitherto the sharpest ones.

12.2 THE EXPLICIT FORMULA FOR $\psi(x)$

In this section we shall prove E. Landau's classical formula for $\psi(x)$, which will represent our main tool in the investigation of the distribution of primes. The method of proof may be adapted without difficulties to yield also a proof of Riemann's formula (12.6). The result is

THEOREM 12.1. Let $\rho = \beta + i\gamma$ denote complex zeros of $\zeta(s)$. Then for $T \geqslant T_0$ uniformly in T we have

$$(12.14) \qquad \psi(x) = x - \sum_{|\gamma| \leqslant T} \frac{x^\rho}{\rho} + O\left(xT^{-1}(\log xT)^2\right) + O(\log x).$$

The basic idea of the proof is to use an inversion formula for Dirichlet series. As the first step we shall prove the following lemma, which may be used to yield a proof of the general inversion formula (A.10).

LEMMA 12.1. Let $\delta(y) = 0$ if $0 < y < 1$, $\delta(1) = \frac{1}{2}$, $\delta(y) = 1$ if $y > 1$, and

$$I(y, T) = (2\pi i)^{-1} \int_{c-iT}^{c+iT} y^s s^{-1} \, ds.$$

Then for $y, c, T > 0$ we have

$$(12.15) \qquad |I(y,T) - \delta(y)| < \begin{cases} y^c \min(1, T^{-1}|\log y|^{-1}) & \text{if } y \neq 1, \\ cT^{-1} & \text{if } y = 1. \end{cases}$$

Proof of Lemma 12.1. Suppose first that $0 < y < 1$. The function $y^s s^{-1}$ tends to zero as $\sigma \to \infty$ uniformly in t. Therefore by Cauchy's theorem

$$I(y,T) = -(2\pi i)^{-1} \int_{c+iT}^{\infty+iT} y^s s^{-1} \, ds + (2\pi i)^{-1} \int_{c-iT}^{\infty-iT} y^s s^{-1} \, ds.$$

Now we have

$$\left| \int_{c\pm iT}^{\infty \pm iT} y^s s^{-1} \, ds \right| \leqslant T^{-1} \int_c^{\infty} y^\sigma \, d\sigma = y^c T^{-1} |\log y|^{-1},$$

while for the other inequality we replace the vertical segment of integration by a circular arc with center 0, radius $R = (c^2 + T^2)^{1/2}$, which lies to the right of the segment. Thus on the arc $|y^s| \leqslant y^c$ and $|s| \leqslant R$; hence

$$|I(y,T)| \leqslant \frac{1}{2\pi} \left(\pi R \frac{y^c}{R} \right) < y^c.$$

When $y > 1$ we replace the vertical segment of integration by the same circular arc, only now we take that part of the arc which lies to the left of the segment. The contour then includes the pole of the integrand at $s = 0$, which gives the residue $1 = \delta(y)$, while the integrand over the arc is bounded as in the previous case. Finally by direct computation we find that

$$I(1,T) = (2\pi)^{-1} \int_{-T}^{T} \frac{dt}{c+it} = \pi^{-1} \int_0^T \frac{c \, dt}{c^2 + t^2} = \pi^{-1} \int_0^{T/c} \frac{du}{1+u^2}$$

$$= 1/2 - \pi^{-1} \int_{T/c}^{\infty} \frac{du}{1+u^2},$$

and the last integral is less than c/T.

Proof of Theorem 12.1. Let $c = 1 + (\log x)^{-1}$ and consider the integral

(12.16) $$J(x,T) = (2\pi i)^{-1} \int_{c-iT}^{c+iT} -\frac{\zeta'(s) x^s \, ds}{\zeta(s) s}$$

For $\sigma > 1$ we have (1.5), where the series converges absolutely. Therefore Lemma 12.1 gives

(12.17)

$$\sum_{n \leqslant x}' \Lambda(n) = J(x,T) + O\left(\sum_{n=1,\, n \neq x}^{\infty} \Lambda(n)(x/n)^c \min(1, T^{-1}|\log x/n|^{-1}) \right)$$

$$+ O(T^{-1} \log x),$$

where Σ' in (12.17) denotes that the last term in the sum is $\frac{1}{2}\Lambda(x)$ if x is an integer.

To estimate the series in (12.17) consider first the terms with $n \leqslant \frac{3}{4}x$ or $n \geqslant \frac{5}{4}x$. Then $|\log x/n| \gg 1$, and since $x^c = ex$ the contribution of these terms is

$$\ll xT^{-1} \sum_{n=1}^{\infty} \Lambda(n)n^{-c} = \left(xT^{-1}\zeta'(c)\right)/\zeta(c)$$

$$\ll xT^{-1}(c-1)^{-1} = T^{-1}x\log x.$$

It remains to estimate the terms with $\frac{3}{4} < n \leqslant \frac{5}{4}x$. Consider here first the terms for which $\frac{3}{4}x < n \leqslant x - 2$ and set $[x] - n = r$. Then $1 \leqslant r \leqslant x/4$ and

$$\log\frac{x}{n} = \log\frac{x}{[x]-r} \geqslant \left|\log\left(1 - \frac{r}{[x]}\right)\right| \geqslant \frac{r}{2x},$$

so that the contribution of the terms with $\frac{3}{4}x < n \leqslant x - 2$ is

$$\ll xT^{-1} \sum_{1 \leqslant r \leqslant x/4} \Lambda(x-r)r^{-1} \ll T^{-1}x\log x \sum_{1 \leqslant r \leqslant x/4} r^{-1} \ll T^{-1}x\log^2 x.$$

Analogously the terms with $x + 2 \leqslant n \leqslant \frac{5}{4}x$ contribute also $\ll T^{-1}x\log^2 x$, while at most five terms of the interval $x - 2 \leqslant n \leqslant x + 2$ contribute a total which is $\ll \log x$. Therefore

$$(12.18) \qquad \sum_{n \leqslant x} \Lambda(n) = J(x,T) + O\left(T^{-1}x\log^2 x\right) + O(\log x),$$

and the next step in the proof is to replace the vertical segment of integration $[c - iT, c + iT]$ in $J(x,T)$ by the other three sides of the rectangle with vertices $c - iT$, $c + iT$, $-U + iT$, $-U - iT$, where U is a large odd integer. If $T \neq \gamma$ for any zero $\rho = \beta + i\gamma$ of $\zeta(s)$, then the residue theorem gives

$$(12.19) \quad \psi(x) = x - \sum_{|\gamma| < T} \frac{x^\rho}{\rho} - \frac{\zeta'(0)}{\zeta(0)} - \sum_{0 < 2m < U} \frac{x^{-2m}}{-2m}$$

$$+ O\left(T^{-1}x\log^2 x\right) + O(\log x) + (2\pi i)^{-1}$$

$$\times \left\{\left(\int_{c-iT}^{-U-iT} + \int_{-U-iT}^{-U+iT} + \int_{-U+iT}^{c+iT}\right) - \frac{\zeta'(s)x^s\,ds}{\zeta(s)s}\right\},$$

since the integrand in $J(x,T)$ has poles as $s = 1$, $s = \rho$, $s = 0$, and $s = -2m$ $(m \geqslant 1)$.

From (1.49) we see that the number of zeros ρ for which $|\gamma - T| < 1$ is $O(\log T)$, thus the differences of ordinates of these zeros cannot be all $o(1/\log T)$. Hence we can choose T (varying it by a bounded amount, if necessary) so that $|\gamma - T| \gg (\log T)^{-1}$ for all zeros ρ.

The integrals in (12.19) will all contribute to the error terms in (12.14) and to estimate them we need a suitable bound for $\zeta'(s)/\zeta(s)$. For $s = \sigma + iT$, $-1 \leqslant \sigma \leqslant 2$ Eq. (1.52) gives

$$(12.20) \qquad \frac{\zeta'(s)}{\zeta(s)} = \sum_{\rho,\, |\gamma-T|<1} \frac{1}{s - \rho} + O(\log T) \ll \log^2 T \qquad (-1 \leqslant \sigma \leqslant 2)$$

since by our choice of T we have $|\gamma - T| \gg (\log T)^{-1}$ for all ρ, and the number of summands is here $\ll \log T$. To obtain bounds for $\sigma \leqslant -1$, we write the functional equation (1.23) as

$$\zeta(1 - s) = 2^{1-s} \pi^{-s} \cos(\pi s/2) \Gamma(s) \zeta(s),$$

and take the logarithmic derivative to obtain

$$(12.21) \qquad \frac{-\zeta'(1 - s)}{\zeta(1 - s)} = -\log 2\pi - \frac{\pi}{2} \tan \frac{\pi s}{2} + \frac{\Gamma'(s)}{\Gamma(s)} + \frac{\zeta'(s)}{\zeta(s)}.$$

The second term on the right-hand side of (12.21) is bounded if $|s - (2m + 1)| \geqslant \frac{1}{2}$, that is, if

$$|(1 - s) + 2m| \geqslant \tfrac{1}{2}.$$

If $1 - \sigma \leqslant -1$, then $\Gamma'(s)/\Gamma(s) = O(\log 2|1 - s|)$ by (A.35), while the last term in (12.21) is bounded. Thus

$$(12.22) \qquad \zeta'(s)/\zeta(s) \ll \log(2|s|) \qquad (\sigma \leqslant -1),$$

provided that circles of radii $\frac{1}{2}$ around the trivial zeros $s = -2m$ of $\zeta(s)$ are excluded. Using (12.20) and (12.22) we have

$$\int_{-U\pm iT}^{c\pm iT} \ll T^{-1}\log^2 T \int_{-\infty}^{c} x^\sigma \, d\sigma \ll x \log^2 T/(T \log x),$$

while (12.22) gives

$$\int_{-U-iT}^{-U+iT} \ll U^{-1}\log U \int_{-T}^{T} x^{-U} \, dt \ll T \log U / U x^U = o(1) \qquad (U \to \infty).$$

Inserting these estimates in (12.19) we obtain (12.14) when we note that for any fixed real A the Riemann–von Mangoldt formula (1.44) gives

$$\sum_{|\gamma|<T} \frac{x^\rho}{\rho} = \sum_{|\gamma|\leqslant T+A} \frac{x^\rho}{\rho} + O(xT^{-1}\log T).$$

12.3 THE PRIME NUMBER THEOREM

After having established the explicit formula (12.14) it will be a simple matter to deduce the (strongest known) prime number theorem from the (strongest known) zero-free region (6.1). Indeed, it will be equally easy to consider the more general zero-free region given by (6.53), namely,

$$(12.23) \quad \zeta(s) \neq 0 \quad \text{for} \quad \sigma \geq 1 - C_1(\log t)^{-a/(a+1)}(\log \log t)^{-1/(a+1)},$$

where $a > 0$, $C_1 > 0$, and $t \geq t_0$. Hence for any zero $\rho = \beta + i\gamma$ of $\zeta(s)$ with $|\gamma| \leq T$ we have

$$\beta \leq 1 - C_1(\log T)^{-a/(a+1)}(\log \log T)^{-1(a+1)}.$$

Supposing that $x^\varepsilon \leq T \leq x$ we obtain then from (12.14)

$$\psi(x) = x + O\left(x^{1 - C_1(\log T)^{-a/(a+1)}}(\log \log T)^{-1/(a+1)} \sum_{|\gamma| \leq T} 1/|\gamma|\right)$$

$$+ O(T^{-1}x \log^2 x)$$

$$= x + O\left\{x \exp\left(-C_2 \log x (\log T)^{-a/(a+1)}(\log \log T)^{-1/(a+1)}\right)\right\}$$

$$+ O(T^{-1}x \log^2 x)$$

for some $C_2 > 0$, where we used (1.53). To make the error terms approximately equal we choose

$$(12.24) \qquad T = \exp\left\{(\log x)^{(a+1)/(2a+1)}(\log \log x)^{-1/(2a+1)}\right\},$$

so that we obtain then with a suitable $C_3 > 0$

$$(12.25) \quad \psi(x) = x + O\left\{x \exp\left(-C_3(\log x)^{(a+1)/(2a+1)}(\log \log x)^{-1/(2a+1)}\right)\right\}.$$

Theorem 6.1 shows that we have (12.23) with $a = 2$, which means that we have proved the prime number theorem in the form given by

THEOREM 12.2.   There is an absolute constant $C > 0$ such that

$$(12.26) \qquad \psi(x) = x + O\left\{x \exp\left(-C(\log x)^{3/5}(\log \log x)^{-1/5}\right)\right\}.$$

By using (12.4) and (12.5) we obtain easily from (12.26) a corresponding result for $\pi(x)$ itself, namely,

(12.27)

$$\pi(x) = \operatorname{li} x + O\left\{ x\exp\left(-C(\log x)^{3/5}(\log\log x)^{-1/5}\right)\right\}. \qquad (C > 0)$$

It may be remarked that the use of a weaker zero-free region than (6.1) would correspondingly lead to a weaker form of the error term in (12.26) and (12.27). Unfortunately, the error term in (12.26) is not $O(x^{1-\delta})$ for any $\delta > 0$, and such an error term cannot be obtained unless $\zeta(s)$ does not vanish in the region $\sigma > 1 - \delta_1$ for some $0 < \delta_1 < \frac{1}{2}$. In fact more can be said about the equivalence between a zero-free region and the order of the error term in the prime number theorem, and we shall prove here

THEOREM 12.3.   Let $\frac{1}{2} \le \theta < 1$ be fixed. Then

(12.28) $\psi(x) = x + O\left(x^{\theta}\log^2 x\right)$

if and only if

(12.29) $\zeta(s) \ne 0 \quad \text{for} \quad \sigma > \theta.$

Proof of Theorem 12.3.   Suppose that (12.28) holds. For $\sigma > 1$ we have

$$-\frac{\zeta'(s)}{\zeta(s)} = \sum_{n=1}^{\infty} \Lambda(n)n^{-s} = \int_{1-0}^{\infty} y^{-s}\,d\psi(y) = s\int_{1}^{\infty}\psi(y)y^{-s-1}\,dy.$$

If we set $\psi(x) = x + R(x)$, then we obtain

(12.30) $-\dfrac{\zeta'(s)}{\zeta(s)} = \dfrac{s}{s-1} + s\displaystyle\int_{1}^{\infty} R(y)y^{-s-1}\,dy \qquad (\sigma > 1).$

Now if $R(x) \ll x^{\theta}\log^2 x$, then the integral in (12.30) is clearly a regular function of s for $\sigma > \theta$, and consequently $\zeta(s)$ cannot vanish in this region.

To deduce that (12.29) implies (12.28) we use (12.14). Then for $\rho = \beta + i\gamma$, $|\gamma| \le T$ we have $\beta \le \theta$ and choosing $T = x^{1-\theta}$ we obtain with (1.53)

$$\psi(x) = x + O\left(x^{\theta}\sum_{|\gamma|\le T} 1/|\gamma|\right) + O\left(xT^{-1}\log^2 x\right)$$

$$= x + O\left(x^{\theta}\log^2 x\right) + O\left(xT^{-1}\log^2 x\right) = x + O\left(x^{\theta}\log^2 x\right).$$

In particular this discussion shows that the Riemann hypothesis is equivalent to the statement that

(12.31) $\psi(x) = x + O\left(x^{1/2}\log^2 x\right),$

but (12.31) does not seem to be any less difficult to prove than the original form of the Riemann hypothesis.

In connection with the equivalence of a zero-free region for $\zeta(s)$ and the order of the error term in the prime number theorem it may be also remarked that P. Turán proved the equivalence of

$$(12.32) \quad \psi(x) = x + O\big(x\exp\big(-C_1(\log x)^{\varkappa}\big)\big) \qquad (0 < \varkappa < 1, C_1 > 0),$$

and

$$(12.33) \quad \zeta(s) \neq 0 \quad \text{for} \quad \sigma \geqslant 1 - C_2(\log t)^{(\varkappa-1)/\varkappa} \qquad [C_2 > 0, t \geqslant t_0(\varkappa)].$$

From Theorem 12.1 it is easily seen that (12.33) implies (12.32), so it is the other implication which is considerably harder to prove. Thus Theorem 12.2 turns out to be essentially an equivalent of the zero-free region (6.1), but it seems less difficult to obtain forms of the prime number theorem from the zero-free region than conversely.

A natural question that may be asked in connection with Theorem 12.3 is what can be said about omega results concerning $\psi(x)$? We shall prove here a simple omega result for $\psi(x)$, namely,

THEOREM 12.4.

$$(12.34) \qquad\qquad \psi(x) = x + \Omega(x^{1/2}).$$

Proof of Theorem 12.4. For $\sigma > 1$ we obtain from (12.30)

$$(12.35) \qquad\qquad \int_i^{\infty} c(y)y^{-s}\,dy = f(s),$$

where for a suitable $c > 0$

(12.36)

$$c(x) = (\psi(x) - x + cx^{1/2})/x, \qquad f(s) = -\frac{\zeta'(s)}{s\zeta(s)} - \frac{1}{s-1} + \frac{c}{s-\frac{1}{2}}.$$

Suppose that $c(x) \geqslant 0$ for $x \geqslant x_0 = x_0(c)$. If (12.34) is not true, then the integral in (12.35) converges for $\sigma > \frac{1}{2}$. Hence for $\sigma > \frac{1}{2}$

$$(12.37) \quad |f(\sigma + it)| \leqslant \int_1^{x_0} |c(y)|y^{-\sigma}\,dy + \int_{x_0}^{\infty} c(y)y^{-\sigma}\,dy$$

$$= \int_1^{x_0} (|c(y)| - c(y))y^{-\sigma}\,dy + f(\sigma)$$

$$\leqslant 2\int_1^{x_0} |c(y)|y^{-1/2}\,dy + f(\sigma) = K + f(\sigma),$$

where K is an absolute constant. Now we take $t = \gamma_1$, where $\rho_1 = \frac{1}{2} + i\gamma_1$ is the first zero on the critical line with $\gamma_1 > 0$ (a numerical calculation shows that ρ_1 is simple and approximately $\gamma_1 = 14.13$). If we multiply (12.37) by $\sigma - \frac{1}{2}$ and let $\sigma \to \frac{1}{2} + 0$, then we obtain

$$\lim_{\sigma \to 1/2 + 0} \frac{|\zeta'(\sigma + i\gamma_1)|(\sigma - \frac{1}{2})}{|\sigma + i\gamma_1||\zeta(\sigma + i\gamma_1)|} = \frac{1}{|\frac{1}{2} + i\gamma_1|} \leqslant c.$$

This is a contradiction if $c = (1 - \varepsilon)|\frac{1}{2} + i\gamma_1|^{-1}$, so that we must have $c(x) < 0$ for some arbitrarily large values of x. A similar result also holds for the function $c(x) = (-\psi(x) + x + cx^{1/2})/x$. In fact, the proof shows that

$$\limsup_{x \to \infty} \frac{\psi(x) - x}{x^{1/2}} \geqslant \left|\tfrac{1}{2} + i\gamma_1\right|^{-1} \geqslant 0.07, \qquad \liminf_{x \to \infty} \frac{\psi(x) - x}{x^{1/2}} \leqslant -0.07,$$

and with some more effort we could also show that

$$(12.38) \qquad \pi(x) = \operatorname{li} x + \Omega\left(\frac{x^{1/2}}{\log x}\right).$$

However, note that in omega results of this type partial summation is not sufficient to deduce (12.38) from (12.34), that is, $\psi(x)$ and $\pi(x)$ have to be treated separately. The best-known omega results in this problem, due to J. E. Littlewood, are only by a $\log\log\log x$ factor better than (12.34) and (12.38), and they are

$$(12.39) \qquad \psi(x) = x + \Omega_\pm(x^{1/2}\log\log\log x)$$

and

$$(12.40) \qquad \pi(x) = \operatorname{li} x + \Omega_\pm\left(\frac{x^{1/2}}{\log x}\log\log\log x\right).$$

Another interesting question is to consider the mean value of $\psi(x) - x$. Here we have

$$(12.41) \qquad x^2 \ll \int_1^x (\psi(y) - y)^2 \, dy,$$

which will follow easily from the Cauchy–Schwarz inequality and

THEOREM 12.5.

$$(12.42) \qquad x^{3/2} \ll \int_1^x |\psi(y) - y| \, dy.$$

Proof of Theorem 12.5. We shall show that

$$(12.43) \quad \int_1^x |\psi(y) - y| \, dy > \frac{C_1 |\zeta'(\rho_0)|}{|\rho_0|^4} x^{1+\beta_0} - C_2 x^{5/4} \quad (C_1, C_2 > 0)$$

if $\rho_0 = \beta_0 + i\gamma_0$, $\beta_0 \geq \frac{1}{2}$ is any simple nontrivial zero of $\zeta(s)$, and taking for ρ_0 any known zero on the critical line we obtain (12.42) from (12.43). As in the proof of Theorem 12.3 we shall write $R(x) = \psi(x) - x$, and using (12.30) we obtain

$$H(s) = \frac{-(s-2)\zeta'(s-1) - (s-1)\zeta(s-1)}{(s-1)(s-2)\zeta(s-1)}$$

$$= \int_1^\infty R(y) y^{-s} \, dy \quad (\sigma > 2).$$

We introduce also the functions

$$h(s) = \frac{(s-2)\zeta(s-1)}{(s-1-\rho_0)(s+1)^4},$$

$$w(u) = (2\pi i)^{-1} \int_{3-i\infty}^{3+i\infty} e^{us} h(s) \, ds \quad (u \text{ real}),$$

and start from the formula

$$(12.44) \quad (2\pi i)^{-1} \int_{3-i\infty}^{3+i\infty} h(s) H(s) e^{s \log x} \, ds = \int_1^\infty R(y) w(\log x - \log y) \, dy,$$

which results from interchange of integration.

We estimate the integral on the left-hand side of (12.44) by moving the line of integration to $\sigma = \frac{5}{4}$. The only pole of the integrand that is passed is $s = 1 + \rho_0$, so that we easily obtain that the left-hand side of (12.44) equals

$$(12.45) \quad -\zeta'(\rho_0)(1 - \rho_0^{-1})(\rho_0 + 2)^{-4} x^{1+\rho_0} + O(x^{5/4}).$$

On the other hand, if $u \leq 0$ we replace the line of integration $\sigma = 3$ in the integral for $w(u)$ by the circular arc $|s| = R$, $\sigma \geq 3$ and corresponding parts of the line $\sigma = 3$. If $R \to \infty$ we obtain then $w(u) = 0$ for $u \leq 0$, while if $u \geq 0$ we replace the path of integration by parts of the line $\sigma = 0$ and the circular arc $|s| = R$, $\sigma \leq 0$. Here we encounter a pole at $s = -1$ with residue $O(1)$ if $R > 1$, and letting $R \to \infty$ we obtain $w(u) = O(1)$ if $u \geq 0$. Therefore

$$(12.46) \quad \int_1^\infty R(y) w(\log x - \log y) \, dy \ll \int_1^x |R(y)| \, dy = \int_1^x |\psi(y) - y| \, dy,$$

and (12.43) follows on comparing (12.44)–(12.46). The $\ll$ constant in (12.42) could be explicitly evaluated at the cost of some calculations.

12.4 THE GENERALIZED VON MANGOLDT FUNCTION AND THE MÖBIUS FUNCTION

As a further application of complex integration and the zero-free region (6.1) we shall obtain now asymptotic formulas for the functions

$$\psi_k(x) = \sum_{n \leqslant x} \Lambda_k(n) = \sum_{n \leqslant x} \sum_{d|n} \mu(d)(\log n/d)^k,$$

where $k \geqslant 1$ is a fixed integer and

$$M(x) = \sum_{n \leqslant x} \mu(n).$$

The results, which are the sharpest ones known, are contained in

THEOREM 12.6. For $k \geqslant 1$ a fixed integer there exists an absolute constant $C_k > 0$ and computable constants $a_{0,k} = k$, $a_{1,k}, \ldots, a_{k-1,k}$ such that

(12.47)

$$\psi_k(x) = \sum_{n \leqslant x} \Lambda_k(n) = \left(a_{0,k}\log^{k-1}x + a_{1,k}\log^{k-2}x + \cdots + a_{k-1,k}\right)x$$

$$+ O\left(x\exp\left(- C_k\log^{3/5}x(\log\log x)^{-1/5}\right)\right).$$

THEOREM 12.7. There is an absolute constant $C > 0$ such that

(12.48) $$M(x) = \sum_{n \leqslant x} \mu(n) \ll x\exp\left(-C\log^{3/5}x(\log\log x)^{-1/5}\right).$$

Note that in (12.48) there is no main term. This reflects the oscillating nature of $\mu(n)$ and $M(x)$ and the fact that the generating series $1/\zeta(s)$ of $\mu(n)$ does not have a pole at $s = 1$, while $-\zeta^{(k)}(s)/\zeta(s)$, the generating series of $\Lambda_k(n)$, does. To prove these theorems we shall need the following

LEMMA 12.2. Suppose that $f(s)$ is regular in the circle $|s - s_0| \leqslant r$, where $f(s_0) \neq 0$, $|f(s)/f(s_0)| \leqslant M$, $|f'(s_0)/f(s_0)| < Ar^{-1}\log M$ for some $A > 0$. If $f(s) \neq 0$ in the part $\sigma \geqslant \sigma_0 - 2r_1$ of the circle $|s - s_0| \leqslant r$, where $0 < r_1 < r/4$, then for $|s - s_0| \leqslant r_1$

(12.49) $$|f'(s)/f(s)| < Cr^{-1}\log M$$

for some suitable $C > 0$, which depends only on A.

LEMMA 12.3. There is an absolute constant $C > 0$ such that

$$(12.50) \qquad 1/\zeta(s) = O\left(\log^{2/3}T(\log\log T)^{1/3}\right)$$

in the region

$$(12.51) \qquad \sigma \geqslant 1 - C(\log t)^{-2/3}(\log\log t)^{-1/3}, \qquad T_0 < t \leqslant T.$$

LEMMA 12.4. Let $k \geqslant 1$ be a fixed integer. Then in the region (12.51) we have

$$(12.52) \qquad \zeta^{(k)}(s) = O\left((\log T)^{(2/3)(k+1)+\varepsilon}\right).$$

Proof of Lemma 12.2. From the reasoning preceding (6.61) of Lemma 6.7 we have

$$(12.53) \qquad \left|\frac{f'(s)}{f(s)} - \sum_\rho \frac{1}{s-\rho}\right| < C_1 r^{-1}\log M$$

for some $C_1 > 0$, where $|s - s_0| \leqslant r/4$, and ρ denotes zeros of $f(s)$ in $|s - s_0| \leqslant r/4$. By hypothesis $|f'(s_0)/f(s_0)| < Ar^{-1}\log M$, and (12.53) holds for $s = s_0$, so that for some $B > 0$ we have

$$\left|\sum_\rho \frac{1}{s_0 - \rho}\right| < Br^{-1}\log M.$$

But for all ρ we have $|s_0 - \rho| \geqslant 2r_1$, so that for s in the circle $|s - s_0| \leqslant r_1$ we have

$$|s - \rho| \geqslant \tfrac{1}{2}|s_0 - \rho|.$$

Since $\mathrm{Re}(s_0 - \rho) \geqslant 2r_1$ for $|s - s_0| \leqslant r_1$ and all ρ, we infer that

$$\left|\frac{1}{s-\rho} - \frac{1}{s_0-\rho}\right| = \frac{|s-s_0|}{|s-\rho||s_0-\rho|} \leqslant \frac{r_1}{\tfrac{1}{2}|s_0-\rho|^2} \leqslant \frac{\mathrm{Re}(s_0-\rho)}{|s_0-\rho|^2}$$

$$= \mathrm{Re}\frac{1}{s_0-\rho}.$$

Therefore for $|s - s_0| \leqslant r_1$

$$\left|\sum_\rho \left(\frac{1}{s-\rho} - \frac{1}{s_0-\rho}\right)\right| \leqslant \mathrm{Re}\sum_\rho \frac{1}{s_0-\rho} \leqslant \left|\sum_\rho \frac{1}{s_0-\rho}\right| \leqslant Br^{-1}\log M;$$

hence

$$(12.54) \qquad \left|\sum_\rho \frac{1}{s-\rho}\right| \leqslant 2Br^{-1}\log M,$$

and (12.49) follows from (12.53) and (12.54).

Proof of Lemma 12.3. We shall apply Lemma 12.2 with

$$f(s) = \zeta(s), \quad s_0 = 1 + A_1 h(t_0) + it_0, \quad h(t_0) = (\log t_0)^{-2/3}(\log\log t_0)^{-1/3},$$

$$T_0 < t \leqslant t_0 + 1 \leqslant T + 1, \quad r = A_2(\log t_0 / \log\log t_0)^{-2/3},$$

where A_1, $A_2 > 0$ are suitable absolute constants. In the circle $|s - s_0| \leqslant r$ we have

$$\frac{\zeta(s)}{\zeta(s_0)} \ll \frac{(\log T)^{2/3+\varepsilon}}{\sigma_0 - 1} < \log^2 T$$

for $0 < \varepsilon < \frac{1}{3}$ and suitable A_2 by (6.52), and

$$\frac{\zeta'(s_0)}{\zeta(s_0)} \ll \frac{1}{\sigma_0 - 1} \ll (\log t_0)^{2/3}(\log\log t_0)^{1/3} \ll r^{-1}\log\log T;$$

hence we may take $M = \log^2 T$, and by (6.1) $\zeta(s) \neq 0$ for $\sigma \geqslant 1 - A_1 h(t_0)$ for suitable $A_1 > 0$. Finally if we choose $2r_1 = (3/2)A_1 h(t_0)$, then all conditions of Lemma 12.2 hold and (12.49) gives

$$(12.55) \quad \zeta'(s)/\zeta(s) \ll (\log T)^{2/3}(\log\log T)^{1/3}, \quad |s - s_0| \leqslant \tfrac{3}{4}A_1 h(t_0),$$

and in particular (12.55) holds for

$$t = t_0, \quad \sigma \geqslant 1 - \tfrac{1}{4}A_1(\log t_0)^{-2/3}(\log\log t_0)^{-1/3}.$$

Now if

$$(12.56) \quad 1 - \tfrac{1}{4}A_1 h(t) \leqslant \sigma \leqslant 1 + h(t),$$

then

$$\log\frac{1}{|\zeta(s)|} = -\operatorname{Re}\log\zeta(s)$$

$$= -\operatorname{Re}\log\zeta(1 + h(t) + it) + \int_\sigma^{1+h(t)}\operatorname{Re}\frac{\zeta'(u + it)}{\zeta(u + it)}\, du$$

$$\leqslant \log\zeta(1 + h(t)) + \int_\sigma^{1+h(t)}Ch^{-1}(t)\, du \leqslant \log Ch^{-1}(t) + O(1).$$

Hence (12.50) follows in the range (12.56), and for larger σ it follows trivially from

$$|\zeta'(s)/\zeta(s)| \leqslant -\zeta'(\sigma_1)/\zeta(\sigma_1) \ll (\sigma_1 - 1)^{-1} \quad (\sigma \geqslant \sigma_1 > 1).$$

Note that there would be no extra difficulty in formulating Lemmas 12.3 and 12.4 to follow from a more general zero-free region than (12.51), for example, like in Lemma 6.8. However, for our purposes the present formulation is sufficient, and besides (12.50) is the best-known result of its type.

Proof of Lemma 12.4. As in the proof of Lemma 12.3, let $h(x) = (\log x)^{-2/3}(\log \log x)^{-1/3}$. By (6.51) with $a = 2$ we have $\zeta(s) \ll \log^{2/3+\varepsilon}T$ if for suitable $C > 0$

$$(12.57) \qquad \sigma \geqslant 1 - Ch(t)\log \log t, \qquad T_0 < t \leqslant T.$$

Now let $s_0 = \sigma_0 + it_0$ be a point in the region

$$(12.58) \qquad \sigma \geqslant 1 - Ah(t), \qquad T_1 < t \leqslant T - 1,$$

where $A > 0$ is arbitrary but fixed, and $T_1 > 0$ is sufficiently large. If $r = (\log t_0)^{-2/3}$, then all points of the circle $|s - s_0| \leqslant r$ lie in the region (12.57). To prove this we have to show that

$$(12.59) \qquad \sigma_1 \geqslant 1 - Ch(t_1)\log \log t_1$$

if $s_1 = \sigma_1 + it_1$ is a point of the circle $|s - s_0| \leqslant r$. Now

$$\sigma_1 \geqslant \sigma_0 - r \geqslant 1 - Ah(t_0) - r,$$

and $h(t_0 + r)\log \log(t_0 + r) \leqslant h(t_1)\log \log t_1$, since $t_0 + r \geqslant t_1$ and clearly $h(x)\log \log x$ is decreasing for $x \geqslant x_0$. Therefore (12.59) follows from

$$Ch(t_0 + r)\log \log(t_0 + r) \geqslant (A + 1)\log^{-2/3}t_0 \geqslant Ah(t_0) + r.$$

The first inequality above reduces to

$$\left(\frac{C}{A + 1}\right)^{3/2} \log \log(t_0 + r) \geqslant \log\left(1 + \frac{r}{t_0}\right),$$

which is obvious, since the left-hand side exceeds 1, while $\log(1 + r/t_0) \leqslant 2r/t_0 \leqslant \frac{1}{2}$ for $t_0 > T_1$.

To finish the proof, note that by the formula for the kth derivative of regular functions we obtain

$$\zeta^{(k)}(s_0) = (2\pi i)^{-1}k! \int_{|s-s_0|=r} \frac{\zeta(s)\,ds}{(s - s_0)^{k+1}} \ll \int_0^{2\pi} |\zeta(s_0 + re^{it})| |r^{-k}\,dt$$

$$\ll r^{-k}\log^{(2/3)+\varepsilon}(t_0 + r\sin t) \ll (\log T)^{(2/3)(k+1)+\varepsilon},$$

since for $|s - s_0| = r$ (12.57) holds. This proves the lemma for $T_1 \leqslant t \leqslant T - 1$, and replacing T by $T + 1$ we just change the constant C in (12.51).

Proof of Theorems 12.6 and 12.7. For $\sigma > 1$ let

$$F(s) = (-1)^k \zeta^{(k)}(s)/\zeta(s) = \sum_{n=1}^{\infty} \Lambda_k(n) n^{-s}.$$

From $\Lambda_1(n) = \Lambda(n) \ll \log n$ and (1.102) we have

$$\Lambda_k(n) \ll \log^k n,$$

so that from the inversion formula (A.10) with $b = 1 + (\log x)^{-1}$, $x_0 \leqslant T \leqslant x^{1-\varepsilon}$, it follows that

$$\sum_{n \leqslant x} \Lambda_k(n) = (2\pi i)^{-1} \int_{b-iT}^{b+iT} F(s) s^{-1} x^s \, ds + O(T^{-1} x \log^{k+1} x).$$

If $a = 1 - C(\log T)^{-2/3}(\log \log T)^{-1/3}$ and $C > 0$ is a suitable constant, then by (6.1) and the residue theorem we obtain

$$(12.60) \quad (2\pi i)^{-1} \int_{b-iT}^{b+iT} F(s) s^{-1} x^s \, ds = \operatorname*{Res}_{s=1} F(s) s^{-1} x^s + I_1 + I_2 + I_3,$$

where

$$I_1 = (2\pi i)^{-1} \int_{b-iT}^{a-iT} F(s) s^{-1} x^s \, ds, \qquad I_3 = (2\pi i)^{-1} \int_{a+iT}^{b+iT} F(s) s^{-1} x^s \, ds,$$

$$I_2 = (2\pi i)^{-1} \int_{a-iT}^{a+iT} F(s) s^{-1} x^s \, ds,$$

since $s = 1$ is the only pole of $F(s) s^{-1} x^s$ in the region $\sigma \geqslant a$, $|t| \leqslant T$. Using Lemma 12.4 we obtain

$$|I_1| \leqslant (2\pi)^{-1} \int_a^b |F(\sigma - iT)| x^\sigma |\sigma - iT|^{-1} \, d\sigma$$

$$\ll (\log T)^{(2/3)(k+1)+\varepsilon} T^{-1} \int_a^b x^\sigma \, d\sigma \ll x T^{-1} \log^{k+2} T,$$

and the same estimate holds also for I_3, while

$$|I_2| \leqslant (2\pi)^{-1} \int_{-T}^{T} |F(a + it)| |a + it|^{-1} x^a \, dt$$

$$\ll (1 - a)^k x^a \int_0^{T_0} (a^2 + t^2)^{-1/2} \, dt + x^a \int_{T_0}^{T} (\log T)^{(2/3)(k+1)+\varepsilon} \frac{dt}{t}$$

$$\ll x^a \log^{k+2} x.$$

Next, we have

$$\operatorname*{Res}_{s=1} F(s)s^{-1}x^s = \lim_{s \to 1} \left(\frac{(-1)^k(s-1)^k \zeta^{(k)}(s)x^s}{(k-1)!s\zeta(s)} \right)^{(k-1)}$$

$$= x \left(\sum_{j=0}^{k-1} a_{j,k} \log^{k-1-j} x \right),$$

where

$$a_{0,k} = \lim_{s \to 1} \frac{(-1)^k(s-1)^k \zeta^{(k)}(s)}{(k-1)!s\zeta(s)} = \frac{k!}{(k-1)!} = k,$$

and the other constants $a_{j,k}$ $(j = 1, \ldots, k-1)$ are also computable. Hence (12.60) and the estimates for I_1, I_2, and I_3 give

$$(12.61) \quad \sum_{n \leqslant x} \Lambda_k(n) = \left(k \log^{k-1}x + a_{1,k} \log^{k-2}x + \cdots + a_{k-1,k} \right) x$$

$$+ O(xT^{-1}\log^{k+2}x) + O(x^a \log^{k+2}x).$$

To make the error terms of the same order of magnitude we shall choose

$$(12.62) \qquad\qquad T = \exp\left(\log^{3/5} x (\log\log x)^{-1/5} \right).$$

Then for $x \geqslant x_0$ and a suitable $C > 0$ we have

$$xT^{-1}\log^{k+2}x \leqslant x \exp\left(-\tfrac{1}{2}\log^{3/5}x(\log\log x)^{-1/5} \right),$$

$$x^a \log^{k+2}x \leqslant x \exp\left(-C \log^{3/5}x(\log\log x)^{-1/5} \right);$$

hence (12.47) follows from (12.61). In case $k = 1$ (12.47) reduces to Theorem 12.2, which means that we have obtained another proof of the prime number theorem, only this time without the explicit formula for $\psi(x)$.

It remains to settle the proof of (12.48). Using again the inversion formula (A.10) with $b = 1 + (\log x)^{-1}$, $T_0 < T \leqslant x^{1-\varepsilon}$, we find that

$$M(x) = \sum_{n \leqslant x} \mu(n) = (2\pi i)^{-1} \int_{b-iT}^{b+iT} \frac{x^s \, ds}{s\zeta(s)} + O(xT^{-1}\log x)$$

$$= (2\pi i)^{-1} \left(\int_{b-iT}^{a-iT} + \int_{a-iT}^{a+iT} + \int_{a+iT}^{b+iT} \right) \frac{x^s \, ds}{s\zeta(s)} + O(xT^{-1}\log x),$$

since now there is no pole at $s = 1$. Applying Lemma 12.3 and taking T as in (12.62) we complete the proof of (12.48) similarly as the proof of (12.47).

With the aid of (12.48) (or directly by complex integration and Lemma 12.3) one can also prove

$$\sum_{n \leqslant x} \lambda(n) \ll x \exp\left(-C \log^{3/5}x (\log\log x)^{-1/5}\right) \qquad (C > 0),$$

where $\lambda(n)$ is the so-called Liouville function. It is defined by the convolution relation

$$\lambda(n) = \sum_{d^2 | n} \mu(n/d^2),$$

or by the Dirichlet-series representation

$$\sum_{n=1}^{\infty} \lambda(n)n^{-s} = \zeta(2s)/\zeta(s) \qquad (\sigma > 1).$$

12.5 VON MANGOLDT'S FUNCTION IN SHORT INTERVALS

We shall now use the explicit formula (12.14) for $\psi(x)$ and zero-density estimates of Chapter 11 to investigate the asymptotic behavior of $\psi(x + h) - \psi(x)$. From (12.14) one might argue heuristically that

$$(12.63) \quad \psi(x + h) - \psi(x) = (1 + o(1))h \qquad [h = o(x) \text{ as } x \to \infty],$$

and Theorem 12.2 shows that this is certainly true if $h \geqslant x \exp(-C \log^{3/5}x(\log\log x)^{-1/5})$ for some suitable $C > 0$. However, it turns out that this can be much improved, and the range for h for which (12.63) holds will be our main topic in this section. The importance of the asymptotic formula (12.63) is that it gives at once

$$(12.64) \qquad\qquad p_{n+1} - p_n \ll p_n^{\theta}(\log p_n)^{\theta'}$$

for $0 < \theta < 1$, $\theta' \geqslant 0$ of (12.63) holds for $h \gg x^{\theta}(\log x)^{\theta'}$. Namely if the interval $(x, x + h]$ contains no primes, then

$$\psi(x + h) - \psi(x) = \sum_{x < p^m \leqslant x+h, \, m \geqslant 2} \log p \ll \log^2 x \sum_{x < n^2 \leqslant x+h} 1$$

$$\ll \log^2 x \left(1 + (x + h)^{1/2} - x^{1/2}\right)$$

$$\ll \log^2 x \left(1 + hx^{-1/2}\right) = o(h)$$

if $h \gg x^{\theta}(\log x)^{\theta'}$, while for $x \geq x_0$ (12.63) gives

$$\psi(x + h) - \psi(x) > h/2.$$

Therefore $(x, x + h]$ contains at least one prime, and taking $x = p_n$ this implies $p_{n+1} \in (p_n, p_n + Cp_n^{\theta}(\log p_n)^{\theta'}]$, hence (12.64) follows. We shall prove now

THEOREM 12.8. Suppose that uniformly for $\frac{1}{2} \leq \sigma \leq 1$

$$(12.65) \qquad\qquad N(\sigma, T) \ll T^{A(\sigma)(1 - \sigma)} \log^D T$$

with some $D \geq 1$ and $A(\sigma)(1 - \sigma) \leq 1$, where for some $\frac{1}{2} < u < 1$ we have $A(\sigma) \leq 2$ for $\sigma \geq u$, and for $\frac{1}{2} \leq \sigma \leq u$ we have $A(\sigma) \leq C$ with some $C > 2$. Then (12.63) holds for $h \geq x^{1-C^{-1}} \log^M x$, where M is any number satisfying

$$(12.66) \qquad\qquad M > \frac{D + 2}{C(1 - u)}.$$

From zero-density estimates (11.29) and (11.82) we see that we may take $C = \frac{12}{5}$, $D = 9$, and $u = \frac{11}{14} + \varepsilon$ to obtain as corollaries of Theorem 12.8

$$(12.67) \quad \psi(x + h) - \psi(x) = (1 + o(1))h \qquad (x^{7/12} \log^{22} x \leq h \leq x),$$

and

$$(12.68) \qquad\qquad p_{n+1} - p_n \ll p_n^{7/12} \log^{22} p_n.$$

Proof of Theorem 12.8. From (12.14) with $x_0 < T \leq x$ we have

$$(12.69) \quad \psi(x + h) - \psi(x) = h - \sum_{|\gamma| \leq T} x^{\rho} C(x, \rho) + O(xT^{-1} \log^2 x),$$

where trivially

$$(12.70) \qquad\qquad C(x, \rho) = \frac{(1 + h/x)^{\rho} - 1}{\rho} \ll 1/|\gamma|,$$

and also

$$(12.71) \qquad |C(x, \rho)| = \left| \int_1^{1+h/x} t^{\rho - 1} \, dt \right| \leq \int_1^{1+h/x} t^{\beta - 1} \, dt \leq hx^{-1}.$$

For convenience of writing we now set $U = x/h$ and suppose $U < T$. Then we may write

$$\sum_{|\gamma| \leq T} x^{\rho} C(x, \rho) = S_1 + S_2 + S_3,$$

say, where

$$S_1 = \sum_{u\leqslant\beta<1,\,|\gamma|\leqslant U} x^\rho C(x,\rho) \ll U^{-1} \sum_{u\leqslant\beta<1,\,|\gamma|\leqslant U} x^\beta,$$

$$S_2 = \sum_{u\leqslant\beta<1,\,U<|\gamma|\leqslant T} x^\rho C(x,\rho) \ll \sum_{u\leqslant\beta<1,\,U<|\gamma|\leqslant T} x^\beta/|\gamma|,$$

$$S_3 = \sum_{0<\beta<u,\,|\gamma|\leqslant T} x^\rho C(x,\rho).$$

To estimate S_1 and S_2 by means of the zero-density function $N(\sigma, T)$ note that

(12.72)

$$\sum_{u\leqslant\beta<1,\,|\gamma|\leqslant U} (x^\beta - x^u) = \log x \sum_{u\leqslant\beta<1,\,|\gamma|\leqslant U} \int_u^\beta x^\sigma\,d\sigma = \log x \int_u^1 N(\sigma,U)x^\sigma\,d\sigma,$$

which implies

(12.73)
$$S_1 \ll \log x \max_{u\leqslant\sigma\leqslant 1} x^\sigma N(\sigma,U)U^{-1}.$$

Likewise, it follows on using (12.72)

$$\sum_{\beta\geqslant u,\,U<|\gamma|\leqslant T} x^\beta(1/|\gamma| - 1/T) = \sum_{\beta\geqslant u,\,U<|\gamma|\leqslant T} x^\beta \int_{|\gamma|}^T t^{-2}\,dt$$

$$= \int_U^T t^{-2} \sum_{u\leqslant\beta\leqslant 1,\,U<|\gamma|\leqslant T} x^\beta\,dt$$

$$= \int_U^T t^{-2}x^u\{N(u,T) - N(u,U)\}\,dt$$

$$+ \int_U^T t^{-2}\log x \int_u^1 x^\sigma\{N(\sigma,t) - N(\sigma,U)\}\,d\sigma\,dt,$$

and from this estimate and (12.72) we obtain

(12.74)
$$S_1 + S_2 \ll \log^2 x \max_{u\leqslant\sigma\leqslant 1} x^\sigma \max_{U\leqslant t\leqslant T} N(\sigma,t)t^{-1}.$$

Now we take $h = x^{1-C^{-1}}\log^M x$, $T = x^{C^{-1}}\log^{3-M}x$, $U = x/h = x^{C^{-1}}\log^{-M}x$, where $C > 2$, $A(\sigma) \leqslant C$, and M satisfies (12.66). By (6.1) we have $N(\sigma, t) = 0$ for

(12.75) $\quad \sigma \geqslant 1 - Kh(x)$, $h(x) = \log^{-2/3}x(\log\log x)^{-1/3}$ $\quad (K > 0)$, and since $A(\sigma)(1 - \sigma) \leqslant 1$ we obtain

$$S_1 + S_2 \ll xU^{-1}\log^{D+2}x \max_{u \leqslant \sigma \leqslant 1 - Kh(x)} \left(xU^{-A(\sigma)}\right)^{(\sigma-1)}$$

$$\ll xU^{-1}\log^{D+2}x \max_{u \leqslant \sigma \leqslant 1 - Kh(x)} x^{(1-2C^{-1})(\sigma-1)}$$

$$\ll h\log^{D+2}x \exp\left(-K(1 - 2C^{-1})\log^{1/3}x(\log\log x)^{-1/3}\right) = o(h),$$

where we used $xU^{-A(\sigma)} > x^{1-2C^{-1}}$, $1 - 2C^{-1} > 0$ [instead of (12.75) any zero-free region $\sigma \geqslant 1 - \log^{\varepsilon-1}t$ would suffice for this proof].

Using the trivial estimate $N(\sigma, T) \ll T\log T$ for $0 < \sigma \leqslant \frac{1}{2}$ and proceeding as in the proof of (12.73) we obtain

$$S_3 = \sum_{0 < \beta < 1/2, |\gamma| \leqslant T} x^\rho C(x, \rho) + \sum_{1/2 \leqslant \beta < u, |\gamma| \leqslant T} x^\rho C(x, \rho)$$

$$\ll x^{1/2}U^{-1}T\log T + \log^2 x \max_{1/2 \leqslant \sigma \leqslant u} x^\sigma \max_{U \leqslant t \leqslant T} N(\sigma, t)t^{-1}$$

$$\ll hx^{-1/2}T\log x + \log^{D+2}x \max_{1/2 \leqslant \sigma \leqslant u} x^\sigma U^{A(\sigma)(1-\sigma)-1}$$

$$\ll hx^{-1/2}T\log x + x^{-C^{-1}}\log^{D+M+2}x \max_{1/2 \leqslant \sigma \leqslant u} x^\sigma \left(x^{C^{-1}}\log^{-M}x\right)^{C(1-\sigma)}$$

$$= hx^{-1/2}T\log x + h(\log x)^{D+2-MC(1-u)} = o(h),$$

provided that (12.66) holds.

This completes the proof of Theorem 12.8, but if instead of requiring (12.63) to hold for all $x \geqslant x_0$ one asks just for "almost all" x, then the range for h given by Theorem 12.8 can be considerably improved. Problems involving "almost all" x occur frequently in analytic number theory, and the meaning of this expression in our case is that the measure of $x \in [X, 2X]$ for which (12.63) does not hold is $o(X)$ as $X \to \infty$. With this in mind we formulate

THEOREM 12.9. Suppose that $N(\sigma, T) \ll T^{C(1-\sigma)}\log^D T$ holds uniformly for $\frac{1}{2} \leqslant \sigma \leqslant 1$ with some $C \geqslant 2$ and $D \geqslant 1$. Then for almost all x we have $\psi(x + h) - \psi(x) = (1 + o(1))h$, provided that

(12.76) $h \geqslant x^{1-2C^{-1}}\log^B x$ $[h = o(x), B \geqslant 0]$,

where $B = B(C, D)$ is a constant that may be explicitly evaluated.

From (11.29) we know that we may take $C = \frac{12}{5}$, $D = 9$, so that as a corollary of Theorem 12.9 we obtain that for almost all x there is a prime in the interval

$$(12.77) \qquad \left(x, \ x + x^{1/6} \log^{B} x \right],$$

which may be compared to the unconditional estimate (12.68), and following the proof given below it is seen that one may take $B = 4 \times 10^{7}$ in (12.77).

Proof of Theorem 12.9. Let for $\frac{1}{2} < u < 1$ fixed, $x \leqslant y \leqslant 2x$, $U = x^{2/C} \log^{-B} x$, $T = U \log^{3} x$, and $e^{\delta} = 1 + U^{-1}$

$$(12.78)$$

$$\Delta(y) = \psi(y + y/U) - \psi(y) - y/U + \sum_{u \leqslant \beta \leqslant 1, |\gamma| \leqslant T} y^{\rho}(e^{\delta\rho} - 1)/\rho.$$

If we can prove that for suitably chosen u and U

$$(12.79) \qquad \sum_{u \leqslant \beta \leqslant 1, |\gamma| \leqslant T} y^{\rho}(e^{\delta\rho} - 1)/\rho = o(x/U) \qquad (x \to \infty)$$

holds, then the theorem will follow from

$$(12.80) \qquad \int_{x}^{2x} |\Delta(y)|^{2} \, dy = o(x^{3} U^{-2}) \qquad (x \to \infty).$$

To obtain (12.79) we shall choose u to satisfy

$$1 - (C/4A)^{2} < u < 1,$$

so that for $\beta \geqslant u$ we have

$$A(1 - \beta)^{3/2} < C(1 - \beta)/4,$$

where $A = 1600$ is the constant appearing in (11.33). Thus using $(e^{\delta\rho} - 1)/\rho \ll 1/U$, (11.33), (12.73), (12.74), and (12.75) we obtain

$$\sum_{u \leqslant \beta \leqslant 1, |\gamma| \leqslant T} y^{\rho}(e^{\delta\rho} - 1)/\rho \ll U^{-1} \sum_{|\gamma| \leqslant x^{2/C}, u \leqslant \beta \leqslant 1 - Kh(x)} y^{\beta}$$

$$\ll U^{-1} \log x \max_{u \leqslant \sigma \leqslant 1 - Kh(x)} x^{\sigma} N(\sigma, x^{2/C})$$

$$\ll xU^{-1} \log^{16} x \max_{u \leqslant \sigma \leqslant 1 - Kh(x)} x^{2AC^{-1}(1-\sigma)^{3/2} + \sigma - 1}$$

$$\ll xU^{-1} (\log^{16} x) x^{-Kh(x)/2} = o(x/U).$$

This proves (12.79), and to show that (12.80) holds we use the explicit formula (12.14) for $\psi(x)$ to obtain

$$|\Delta(y)|^2 \ll \left| \sum_{\beta < u, |\gamma| \leqslant T} y^\rho (e^{\delta\rho} - 1)/\rho \right|^2 + o(x^2 U^{-2}).$$

Squaring out and integrating we have

$$\int_x^{2x} \left| \sum_{\beta < u, |\gamma| \leqslant U} y^\rho (e^{\delta\rho} - 1)/\rho \right|^2 dy$$

$$= \sum_{\beta < u, |\gamma| \leqslant U} \sum_{\beta' < u, |\gamma'| \leqslant U} \frac{(2x)^{\bar\rho + \rho' + 1} - x^{\bar\rho + \rho' + 1}}{\bar\rho\rho'(\bar\rho + \rho' + 1)} (e^{\delta\bar\rho} - 1)(e^{\delta\rho'} - 1)$$

$$\ll U^{-2} \cdot \sum_{\beta \leqslant u, |\gamma| \leqslant U} x^{2\beta+1} \sum_{\beta' \leqslant \beta, |\gamma'| \leqslant U} \frac{1}{1 + |\gamma - \gamma'|}$$

$$\ll U^{-2} \log^2 x \sum_{\beta \leqslant u, |\gamma| \leqslant U} x^{2\beta+1} \ll U^{-2} \log^3 x \max_{\sigma \leqslant u} x^{2\sigma+1} N(\sigma, U).$$

Here we used $(e^{\delta\rho} - 1)/\rho \ll 1/U$, the argument that leads to (12.72) and the estimate

$$(12.81) \qquad \sum_{\beta' \leqslant u, |\gamma'| \leqslant U} \frac{1}{1 + |\gamma - \gamma'|} \ll \log^2 x,$$

which follows from (1.53). Proceeding analogously, only using $(e^{\delta\rho} - 1)/\rho \ll 1/|\gamma|$, we also obtain

$$\int_x^{2x} \left| \sum_{\beta < u, U < |\gamma| \leqslant T} y^\rho (e^{\delta\rho} - 1)/\rho \right|^2 dy \ll U^{-1} \log^4 x \max_{\sigma \leqslant u} x^{2\sigma+1} \max_{U \leqslant t \leqslant T} N(\sigma, t) t^{-1}.$$

Therefore we obtain

$$\int_x^{2x} |\Delta(y)|^2 dy \ll U^{-1} \log^4 x \max_{1/2 \leqslant \sigma \leqslant u} x^{2\sigma+1} \max_{U \leqslant t \leqslant T} N(\sigma, t) t^{-1} + o(x^3 U^{-2})$$

$$\ll x U^{-2} \log^4 x \max_{1/2 \leqslant \sigma \leqslant u} x^{2\sigma} U^{A(\sigma)(1-\sigma)} \log^D x + o(x^3 U^{-2})$$

$$\ll x U^{-2} \log^{D+4} x \max_{1/2 \leqslant \sigma \leqslant u} x^{2\sigma} (x^{2/C} \log^{-B} x)^{C(1-\sigma)} + o(x^3 U^{-2})$$

$$\ll x^3 U^{-2} (\log x)^{D+4+BC(u-1)} + o(x^3 U^{-2}) = o(x^3 U^{-2}),$$

provided that $B > (D + 4)/(C - Cu)$, hence Theorem 12.9 follows. The smallest value for B which will be attained by this method is if we choose $u = 1 - (C/4A)^2 + \varepsilon$.

12.6 THE DIFFERENCE BETWEEN CONSECUTIVE PRIMES

We have seen in Theorem 12.8 how zero-density estimates lead to the asymptotic formula (12.63). In particular the method of proof shows that the density hypothesis in the form

$$N(\sigma, T) \ll T^{2-2\sigma+\varepsilon} \qquad \left(\tfrac{1}{2} \leqslant \sigma \leqslant 1\right)$$

gives

$$(12.82) \quad \psi(x + h) - \psi(x) = (1 + o(1))h \qquad (x^{1/2+\varepsilon} \leqslant h \leqslant x)$$

and consequently also

$$(12.83) \qquad\qquad p_{n+1} - p_n \ll p_n^{1/2+\varepsilon}.$$

It is interesting to note that in principle zero-density estimates cannot give much more than (12.83) for the difference between consecutive primes. This will be seen from the following

THEOREM 12.10. If the Riemann hypothesis is true, then

$$(12.84) \qquad\qquad p_{n+1} - p_n \ll p_n^{1/2}\log p_n.$$

Proof of Theorem 12.10. Suppose that the Riemann hypothesis is true and that the interval $[x, x + 2h]$ contains no primes for some $x^\varepsilon < h < x^{1-\varepsilon}$ [at the end of the proof we shall take $x = p_n$ and deduce (12.84)]. Then using (12.14) with $T = x$ we have

(12.85)

$$o(h^2) = \int_x^{x+h}(\psi(y + h) - \psi(y))\, dy = h^2 - \sum_{|\gamma|\leqslant x} G(\rho) + O(h \log^2 x),$$

where

$$(12.86) \qquad\qquad G(\rho) = \int_x^{x+h} \frac{(y + h)^\rho - y^\rho}{\rho}\, dy,$$

and $\rho = \tfrac{1}{2} + i\gamma$ if the Riemann hypothesis is true. For $|\gamma| \leqslant U \ (< x)$ we use the trivial $N(T) \ll T \log T$ to obtain

$$\sum_{|\gamma| \leqslant U} G(\rho) = \int_x^{x+h} \int_y^{y+h} z^{-1/2} \left(\sum_{|\gamma| \leqslant U} z^{i\gamma} \right) dz \, dy \ll h^2 x^{-1/2} U \log U.$$

For $|\gamma| > U$ we evaluate the integral in (12.86) to obtain

$$G(\rho) = \frac{(x+2h)^{\rho+1} - 2(x+h)^{\rho+1} + x^{\rho+1}}{\rho(\rho+1)} \ll x^{3/2} |\gamma|^{-2},$$

which gives by using (1.53)

$$\sum_{U < |\gamma| \leqslant x} G(\rho) \ll x^{3/2} \sum_{|\gamma| \gtrsim U} |\gamma|^{-2} \ll x^{3/2} U^{-1} \log U.$$

From the estimates for $\Sigma G(\rho)$ and (12.85) we then obtain

$$(12.87) \quad h^2 + o(h^2) = O(h^2 x^{-1/2} U \log U) + O(x^{3/2} U^{-1} \log U),$$

and choosing $U = xh^{-1}$ (12.87) reduces to

$$h^2 + o(h^2) = O(hx^{1/2} \log x),$$

which is impossible for $h = Cx^{1/2} \log x$ and $C > 0$ sufficiently large. Thus the interval $[x, x + 2Cx^{1/2} \log x]$ must contain a prime for $x \geqslant x_0$, which means that for $x = p_n$

$$p_{n+1} \in [x, x + 2Cx^{1/2} \log x],$$

and this obviously implies (12.84).

Theorem 12.10 is the best result that one can obtain at present by assuming the Riemann hypothesis. By assuming some additional conjectures (e.g., the pair correlation hypothesis mentioned in Section 1.9 of Chapter 1) (12.84) can be slightly improved, but not to $p_{n+1} - p_n = o(p_n^{1/2})$ at present.

We shall now present another method, which also gives unconditional estimates of the type (12.64), and supersedes (12.68). Namely, for the derivation of (12.64) from (12.63) one really does not need the asymptotic formula (12.63), and a lower bound of the type $\psi(x+h) - \psi(x) \gg h$ suffices, or equivalently

$$(12.88) \qquad \pi(x) - \pi(x-y) > \frac{Ay}{\log x} \qquad (A > 0, \, x \geqslant x_0)$$

with a suitable y. Lower bounds of the correct order of magnitude occur frequently in sieve theory, and the method of obtaining a good bound of the type (12.88) (which with $y \geqslant x^{13/23}$ is due to H. Iwaniec and M. Jutila) is a combination of sieve and analytic results, the latter being to a large extent dependent on estimates from zeta-function theory. From the theory of the linear sieve we shall borrow a suitable lower-bound estimate, which we formulate as

LEMMA 12.5. Let $\frac{1}{2} < \theta < 1$, $y = x^{\theta}$, $2 < v < 3$, $z = x^{1/v}$, $P(z) = \prod_{p < z} p$,

$$\mathscr{A} = \{n : x - y < n \leqslant x\}, \quad S(\mathscr{A}, z) = |\{a \in \mathscr{A}; (a, P(z)) = 1\}|.$$

If $M, N \geqslant 2$, $z \geqslant 2$, and $\varepsilon > 0$ is arbitrary, then

$$(12.89) \quad S(\mathscr{A}, z) > \prod_{p < z} (1 - 1/p) y \{ f(s) + E \} - 2^{\varepsilon^{-10}} R(\mathscr{A}, M, N),$$

where $s = \log MN / \log z$, $E = E(\varepsilon, s, M, N) \ll \varepsilon s + \varepsilon^{-8} (\log MN)^{-1/3}$,

$$f(s) = 2e^{\gamma} s^{-1} \log(s - 1) \qquad (2 \leqslant s \leqslant 4),$$

and for $|a_m| \leqslant 1$, $|b_n| \leqslant 1$ which depend at most on M, N, z and ε we have

$$(12.90) \quad R(\mathscr{A}, M, N) = \sum_{m < M} \sum_{n < N} a_m b_n \left(\left[\frac{x}{mn} \right] - \left[\frac{x - y}{mn} \right] - \frac{y}{mn} \right).$$

For the product appearing in (12.89) we have

$$(12.91) \qquad \prod_{p \leqslant x} (1 - 1/p) = \frac{e^{-\gamma} + o(1)}{\log x} \qquad (x \to \infty),$$

which by taking logarithms follows from the asymptotic formula

$$(12.92) \quad A(x) = \sum_{p \leqslant x} 1/p = \log \log x + \gamma - \sum_{p} \sum_{m \geqslant 2} m^{-1} p^{-m} + O(1/\log x).$$

The last formula is a well-known result from elementary number theory, and follows from the prime number theorem by partial summation without the explicit determination of the constant $\gamma - \sum_p \sum_m m^{-1} p^{-m}$. A simple way to evaluate the constant in question is to note that, for $\sigma > 1$,

$$(12.93) \qquad f(\sigma) = \sum_{p} p^{-\sigma} = (\sigma - 1) \int_{2}^{\infty} A(t) t^{-\sigma} dt.$$

Then supposing that

$$A(t) = \log \log t + B + O(1/\log t),$$

using the fact that

$$\gamma = -\int_0^\infty e^{-u} \log u \, du = -\Gamma'(1)$$

and

$$f(\sigma) = \log \zeta(\sigma) - \sum_p \sum_{m \geqslant 2} m^{-1} p^{-m\sigma} = -\log(\sigma - 1) + o(1) - \sum_p \sum_{m \geqslant 2} m^{-1} p^{-m},$$

we find, on comparing the expressions for $f(\sigma)$ and letting $\sigma \to 1 + 0$,

$$B = \gamma - \sum_p \sum_{m \geqslant 2} m^{-1} p^{-m}.$$

By a more refined sieve technique than the one necessary for Lemma 12.5 and elaboration of various analytic estimations the constant θ in (12.64) was reduced from the value $\theta = \frac{13}{23} = 0.56521\ldots$ of H. Iwaniec and M. Jutila to $\theta = \frac{11}{20} + \varepsilon = 0.55 + \varepsilon$ (due to D. R. Heath-Brown and H. Iwaniec), and the latest record is $\theta = \frac{1}{2} + \frac{1}{21} = 0.54761\ldots$ (obtained recently by H. Iwaniec and J. Pintz). This is not far from $\theta = \frac{1}{2} + \varepsilon$, which is essentially the present limit of both sieve methods and zero-density methods [such as those used for the proof of (12.63)]. The latest sieve and analytic techniques used to obtain values of θ less than $\frac{11}{20}$ are complicated, and besides the method does not seem to have given yet the optimal value that it possibly can. It is for these reasons that we shall sketch here only the proof of the following

THEOREM 12.11. For $y = x^\theta$, $\theta \geqslant \frac{13}{23} = 0.56521\ldots$, and $x \geqslant x_0$

$$(12.94) \qquad \pi(x) - \pi(x - y) > \frac{y}{177 \log x}.$$

Proof of Theorem 12.11. By taking $x - y = p_n$ (12.94) clearly implies (12.64) with $\theta = \frac{13}{23}$, $\theta' = 0$. Let $q(n) = \min\{p : p | n\}$ and $S(\mathcal{A}, z)$ as in Lemma 12.5. Then

$$(12.95) \quad S(\mathcal{A}, z) = \sum_{x - y < n \leqslant x, \, q(n) \geqslant x^{1/v}} 1$$

$$= \pi(x) - \pi(x - y) + \sum_{x - y < n < x; \, n = p_1 p_2; \, p_1, \, p_2 \geqslant x^{1/v}} 1.$$

Thus in order to obtain the desired lower bound for $\pi(x) - \pi(x - y)$ we shall use the lower bound (12.89), but we need an upper bound for $R(\mathcal{A}, M, N)$ and the last sum in (12.95). We shall now sketch the proof that, if $\frac{5}{9} < \theta < 1$, $0 < \alpha < \min(\frac{1}{2}, (5\theta - 1)/4)$, then there exists a number $\eta = \eta(\theta, \alpha) > 0$ such that

$$(12.96) \qquad R(\mathscr{A}, x^{\alpha}, x^{\alpha}) \ll yx^{-\eta},$$

where the $\ll$ constant depends at most on θ and α.

To obtain (12.96) consider, for $2 \leqslant M, N \leqslant x$, the Dirichlet polynomials

$$A(s) = \sum_{M \leqslant m < 2M} a_m m^{-s}, \qquad B(s) = \sum_{N \leqslant n < 2N} b_n n^{-s}.$$

Then by the inversion formula (A.10) we may write

$$\sum_{M \leqslant m < 2M} \sum_{N \leqslant n < 2N} a_m b_n \left(\left[\frac{x}{mn} \right] - \left[\frac{x-y}{mn} \right] \right)$$

$$= \sum_{x-y < kmn \leqslant x, \ M \leqslant m < 2M, \ N \leqslant n < 2N} a_m b_n$$

$$= (2\pi i)^{-1} \int_{2-iT}^{2+iT} \zeta(s) A(s) B(s) \frac{x^s - (x-y)^s}{s} ds + O(yx^{-\eta})$$

$$= \sum_{M \leqslant m < 2M} \sum_{N \leqslant n < 2N} a_m b_n (mn)^{-1} y$$

$$+ (2\pi i)^{-1} \int_{\sigma-iT}^{\sigma+iT} \zeta(s) A(s) B(s) \frac{x^s - (x-y)^s}{s} ds + O(yx^{-\eta})$$

$$= \sum_{M \leqslant m < 2M} \sum_{N \leqslant n < 2N} a_m b_n (mn)^{-1} y$$

$$+ O \left(yx^{\sigma-1} \int_{-T}^{T} |\zeta(\sigma + it) A(\sigma + it) B(\sigma + it)| \, dt \right) + O(yx^{-\eta}).$$

Here $\frac{1}{2} \leqslant \sigma < 1$ is arbitrary but fixed, and we use the residue theorem since $\zeta(s)$ has a simple pole at $s = 1$ with residue 1. Moreover, T is a value from the interval $[\frac{1}{2}x^{1+2\eta}y^{-1}, x^{1+2\eta}y^{-1}]$, since clearly in any interval $[T_1, 2T_1]$, $T_1 \geqslant 2$ there is a T such that

$$\int_{1/2}^{2} |\zeta(\sigma \pm iT)| \, d\sigma \ll \log T.$$

The last integral containing $|\zeta AB|$ is estimated by using the Cauchy–Schwarz inequality, the mean value theorem for Dirichlet polynomials, the Halász–Montgomery inequality (A.40), and power moments for $\zeta(s)$ of Chapter 8. Thus from Lemma 12.5 and (12.96) we obtain

$$(12.97) \qquad S(\mathscr{A}, x^{1/v}) > \frac{y}{\log x} \{ \lambda(\theta, v) - \varepsilon \},$$

where $x \geqslant x_0(\varepsilon, \theta)$ and

$$\lambda(\theta, v) = \frac{4}{5\theta - 1} \log\left(\frac{5\theta - 1}{2} v - 1\right),$$

$$\frac{5}{9} < \theta < \frac{3}{5}, \qquad \frac{4}{5\theta - 1} < v < \frac{8}{5\theta - 1}.$$

Let us for $2 < u < (1 - \theta)^{-1}$ define

$$(12.98) \quad T(\mathscr{A}; x^{1/v}, x^{1/u}) = \sum_{x^{1/u} \leqslant p < x^{1/2}} \sum_{x-y < n \leqslant x;\ p|n,\ q(n) \geqslant x^{1/v}} 1$$

$$= \sum_{x^{1/u} \leqslant p < x^{1/2}} \sum_{x-y < pp_1 \leqslant x,\ p_1 \geqslant x^{1/v}} 1$$

$$= \sum_{x-y < n \leqslant x,\ n=pp_1,\ x^{1/u} \leqslant p < x^{1/2},\ p_1 \geqslant x^{1/v}} 1.$$

Therefore

$$S(\mathscr{A}, x^{1/v}) - T(\mathscr{A}; x^{1/v}, x^{1/u})$$

$$= \pi(x) - \pi(x - y) + \sum_{x-y < n \leqslant x,\ n=pp_1,\ x^{1/v} \leqslant p < x^{1/u},\ p_1 \geqslant x^{1/v}} 1$$

$$= \pi(x) - \pi(x - y) + \sum_{x^{1/v} \leqslant p < x^{1/u}} \left\{ \pi\left(\frac{x}{p}\right) - \pi\left(\frac{x - y}{p}\right) \right\},$$

since all primes $p_1 \geqslant (x - y)/p$ must exceed $x^{1/v}$. Hence

$$(12.99) \quad \pi(x) - \pi(x - y) = S(\mathscr{A}, x^{1/v}) - T(\mathscr{A}; x^{1/v}, x^{1/u})$$

$$- \sum_{x^{1/v} \leqslant p < x^{1/u}} \left\{ \pi\left(\frac{x}{p}\right) - \pi\left(\frac{x - y}{p}\right) \right\}.$$

We already have a suitable lower bound for $S(\mathscr{A}, x^{1/v})$, namely, (12.97), and we need an upper bound for $T(\mathscr{A}; x^{1/v}, x^{1/u})$ and the last sum in (12.99). By the upper-bound sieve analogue of (12.89) and an analytic argument similar to the one used for the proof of (12.96) we find that

$$(12.100) \qquad T(\mathscr{A}; x^{1/v}, x^{1/u}) < \frac{y}{\log x} \{ \Lambda(\theta, u) + \varepsilon \},$$

where for $x \geq x_0(\varepsilon, \theta)$, $(5\theta + 7)/12 < 1/u + 3/v$ we have

$$\Lambda(\theta, u) = \frac{24}{7 + 5\theta} \log \frac{(7 + 5\theta)u - 12}{2(1 + 5\theta)}.$$

To estimate the last sum in (12.99) we consider sums of the type

(12.101) $\quad \psi(x, y; P) = \sum_{P \leq p < 2P} \left\{ \psi\left(\frac{x}{p}\right) - \psi\left(\frac{x - y}{p}\right) \right\}$

with $x^{1/v} \leq P < x^{1/u}$, since it is easier to work with $\psi(x)$ than with $\pi(x)$. The estimation of the sum in (12.101) is carried out by the inversion formula (A.10), where we take $b = 1 + (\log x)^{-1}$, $2 \leq T \leq x$. If

$$K(s) = \sum_{P \leq p < 2P} p^{-s},$$

then

(12.102) $\quad \psi(x, y; P) = (2\pi i)^{-1} \int_{b-iT}^{b+iT} \frac{-\zeta'(s)}{s\zeta(s)} K(s)\{x^s - (x - y)^s\}\, ds$

$$+ O(x^{1+\varepsilon}T^{-1})$$

$$= y \sum_{P \leq p < 2P} 1/p - \sum_{|\gamma| \leq T} \rho^{-1} K(\rho)\{x^\rho - (x - y)^\rho\}$$

$$+ O(x^{1+\varepsilon}T^{-1}).$$

Here we used the fact that $s = 1$ is a pole of the integrand, which gives rise to the term

$$yK(1) = y \sum_{P \leq p < 2P} 1/p.$$

Also arguments analogous to the ones needed in the proof of the explicit formula (12.14) were used, in particular the fact that (12.20) holds uniformly for $-1 < \sigma < 2$ with a suitable choice of T.

To deal with the sum over nontrivial zeros $\rho = \beta + i\gamma$ of $\zeta(s)$ in (12.102) note that

$$\sum_{|\gamma| \leq T} \rho^{-1} K(\rho)\{x^\rho - (x - y)^\rho\}$$

$$= \sum_{|\gamma| \leq T} K(\rho) \int_{x-y}^{x} t^{\rho-1}\, dt$$

$$= \int_{x-y}^{x} \left(\sum_{|\gamma| \leq T} K(\rho) t^{\rho-1}\, dt \right) \ll yx^{-1} \sum_{|\gamma| \leq T} |K(\rho)| x^\beta,$$

while

$$\sum_{|\gamma|\leqslant T} |K(\rho)|(x^\beta - 1) = \log x \sum_{|\gamma|\leqslant T} |K(\rho)| \int_0^\beta x^\sigma \, d\sigma$$

$$= \log x \int_0^1 \sum_{|\gamma|\leqslant T, \, \beta \geqslant \sigma} |K(\rho)| x^\sigma \, d\sigma$$

$$\ll \log x \max_{0 \leqslant \sigma \leqslant 1} x^\sigma \sum_{|\gamma|\leqslant T, \, \beta \geqslant \sigma} |K(\rho)|.$$

Thus the problem of the estimation of $\psi(x, y; P)$ reduces to the estimation of the "weighted" zero-density function

(12.103) $$K(\sigma, T) = \sum_{|\gamma|\leqslant T, \, \beta \geqslant \sigma} |K(\rho)|,$$

which differs from the ordinary zero-density function $N(\sigma, T)$ by the factor $|K(\rho)|$. The appearance of the zero-density function $K(\sigma, T)$ is probably the most salient analytic feature of this approach to the estimation of $\pi(x) - \pi(x - y)$. It is interesting that, for the range relevant in our application, the function $K(\sigma, T)$ can be estimated even better than the seemingly simpler function $N(\sigma, T)$. The bounds for $K(\sigma, T)$ are contained in

LEMMA 12.6. Let $K(s) = \sum_{Q \leqslant q < 2Q} a_q q^{-s}$, where $|a_q| \leqslant 1$, and let $K(\sigma, T)$ be defined by (12.103). Then uniformly for $0 \leqslant \sigma \leqslant 1$ we have

(12.104) $$K(\sigma, T) \ll \begin{cases} (T^{16/5}/Q)^{1-\sigma} \log^C T & \text{if } T^{4/5} < Q \leqslant T, \\ (T^{6/5} Q)^{1-\sigma} \log^C T & \text{if } T \leqslant Q < T^{6/5}. \end{cases}$$

Proof of Lemma 12.6. First by an application of the Cauchy–Schwarz inequality we obtain

(12.105) $$K^2(\sigma, T) \leqslant N(\sigma, T) \sum_{|\gamma|\leqslant T, \, \beta \geqslant \sigma} |K(\rho)|^2.$$

For the range $T \leqslant Q < T^{6/5}$ we use $N(\sigma, T) \ll T^{12(1-\sigma)/5} \log^C T$ [see (11.29)], and the mean value for Dirichlet polynomials (Theorem 5.3), by choosing a representative set of γ_r's which satisfy $|\gamma_r| \leqslant T$, $|\gamma_{r+1} - \gamma_r| \geqslant 1$ and $|K(\rho_r)|$ is maximal in suitable vertical strips of unit width. Thus we obtain

$$K^2(\sigma, T) \ll T^{12(1-\sigma)/5}(T + Q)Q^{1-2\sigma} \log^C T \ll T^{12(1-\sigma)/5}Q^{2-2\sigma} \log^C T,$$

and this gives the second bound in (12.104). For the range $T^{4/5} < Q \leqslant T$ we consider separately various ranges of σ. For $\sigma \leqslant \frac{3}{4}$ we use again (12.105), Theorem 5.3, and (11.26) to obtain

$$K^2(\sigma, T) \ll T^{(3-3\sigma)/(2-\sigma)}TQ^{1-2\sigma} \log^C T \ll T^{16(2-2\sigma)/5}Q^{2\sigma-2} \log^C T,$$

since $Q^{3-4\sigma} \leqslant T^{3-4\sigma}$ and $3/(4 - 2\sigma) \leqslant \frac{6}{5}$ for $\sigma \leqslant \frac{3}{4}$. The same type of argument works for $\frac{3}{4} \leqslant \sigma \leqslant \frac{7}{9}$ with the estimate (11.27). For $\frac{7}{9} \leqslant \sigma \leqslant 1$ we use the same zero-density estimate, but instead of Theorem 5.3 we use the Halász–Montgomery inequality (A.40) with

$$\varphi_r = \{\varphi_{r,n}\}_{n=1}^{\infty}, \varphi_{r,n} = \left(e^{-n/(2Q)} - e^{-n/Q}\right)^{1/2} n^{-i\gamma_r},$$

so that for $r \neq s$

$$(\varphi_r, \varphi_s) = (2\pi i)^{-1} \int_{2-i\infty}^{2+i\infty} \zeta(w + i\gamma_r - i\gamma_s)\Gamma(w)((2Q)^w - Q^w)\, dw,$$

similarly as in zero-density estimates in Chapter 11. Moving the line of integration in the integrand to $\text{Re}\, w = 0$ we obtain

$$\sum_{|\gamma| \leqslant T, \beta \geqslant \sigma} |K(\rho)|^2 \ll (Q + N(\sigma, T)T^{1/2})Q^{1-2\sigma}\log{}^C T,$$

and the first estimate in (12.104) follows.

With Lemma 12.6 at our disposal the proof of Theorem 12.11 is almost complete. Taking $T = x^{1+2\eta}y^{-1}$ in (12.102) we find, using Lemma 12.6, that for

(12.106) $\frac{16}{5}(1 - \theta) < 1 + 1/v, \qquad \frac{6}{5}(1 - \theta) < 1 - 1/u$

we have

$$\psi(x, y; P) = y \sum_{P \leqslant p < 2P} 1/p + O(y \log^{-4}x),$$

which gives

$$\sum_{P \leqslant p < 2P} \left\{ \pi\left(\frac{x}{p}\right) - \pi\left(\frac{x-y}{p}\right) \right\} \sim \sum_{P \leqslant p < 2P} \frac{1}{\log x/p} \left\{ \psi\left(\frac{x}{p}\right) - \psi\left(\frac{x-y}{p}\right) \right\}$$

$$\sim y \sum_{P \leqslant p < 2P} \frac{1}{p \log x/p} \qquad (x \to \infty).$$

But if $y = x^c, 0 < c < 1$, then by the prime number theorem

$$\sum_{p \leqslant y} \frac{1}{p \log x/p} = \int_{3/2}^{y} \frac{d\pi(t)}{t \log x/t} \sim \int_{3/2}^{y} \frac{dt}{t \log t \log x/t}$$

$$= (\log x)^{-1} \int_{3/2}^{y} \left(\sum_{k=0}^{\infty} (\log t/\log x)^k \right) \frac{dt}{t \log t}$$

$$\sim \frac{\log \log y}{\log x} + \sum_{k=0}^{\infty} \frac{\log^{k+1} y}{(k + 1)\log^{k+2} x}$$

$$= \frac{1}{\log x}\left(\log \log x + \log \frac{c}{1 - c} \right) \qquad (x \to \infty).$$

Taking in this formula $c = 1/u$ and $c = 1/v$ and subtracting we obtain then for $x \to \infty$

(12.107)

$$\sum_{x^{1/v} \leqslant p < x^{1/u}} \left\{ \pi\left(\frac{x}{p}\right) - \pi\left(\frac{x - y}{p}\right) \right\} \sim \frac{y}{\log x} \log \frac{v - 1}{u - 1} = \frac{y}{\log x} \Omega(u, v)$$

say, which is asymptotic formula for the crucial sum appearing in (12.99). Substituting all the preceding estimates in (12.99) we finally obtain

$$\pi(x) - \pi(x - x^{\theta}) > \frac{x}{177 \log x} \qquad (x \geqslant x_0)$$

if $\frac{5}{9} < \theta \leqslant \frac{7}{12}$, and there are parameters u and v which satisfy (12.106) and

(12.108) $$\lambda(\theta, v) - \Lambda(\theta, u) - \Omega(u, v) > \frac{1}{177}.$$

The optimal values for u and v are the ones for which the interval $[1/u, 1/v]$ is as long as possible. We take $u = \frac{23}{9}$, $v = \frac{23}{11}$ [so that (12.106) is equalized], and then a calculation shows that (12.108) holds for all θ which satisfy $\frac{13}{23} \leqslant \theta \leqslant \frac{7}{12}$, which completes the proof of Theorem 12.11, since the values of θ larger than $\frac{7}{12}$ are being taken care of by the asymptotic formula (12.67).

12.7 ALMOST PRIMES IN SHORT INTERVALS

The zero-density estimate (12.104) can be used also in a simple way to obtain results about "almost primes" P_2 in short intervals. We shall prove now

THEOREM 12.12.   Let $0 < \delta < \frac{1}{4}$ be a fixed number and let $Q = x^{5/11 - \delta}$, $h = x^{6/11 + \delta}$. Then

(12.109) $$\sum_{Q < p \leqslant 2Q} \left(\psi\left(\frac{x + h}{p}\right) - \psi\left(\frac{x}{p}\right) \right) = h \sum_{Q < p \leqslant 2Q} 1/p + o(h/\log x).$$

Proof of Theorem 12.12.   From the explicit formula (12.14) for $\psi(x)$ we see that the left-hand side of (12.109) is equal to

(12.110) $$\sum_{Q < p \leqslant 2Q} h/p - \sum_{|\gamma| \leqslant T} \rho^{-1}\{(x + h)^{\rho} - x^{\rho}\}$$

$$\times \sum_{Q < p \leqslant 2Q} p^{-\rho} + O(xT^{-1}\log^2 x) + O(Q),$$

where we shall choose $T = Q \log^4 x = x^{5/11-\delta} \log^4 x$. As in the discussion preceding (12.103) we obtain

$$\sum_{|\gamma| \leqslant T} \rho^{-1} \{ (x+h)^{\rho} - x^{\rho} \} \sum_{Q < p \leqslant 2Q} p^{-\rho} \ll \sum_{|\gamma| \leqslant T} h x^{\beta-1} |K(\rho)|$$

$$\ll \log x \max_{0 \leqslant \sigma \leqslant 1} h x^{\sigma-1} K(\sigma, T),$$

where

$$K(s) = \sum_{Q < p \leqslant 2Q} p^{-s},$$

and $K(\sigma, T)$ is the weighted zero-density function introduced in (12.103). If we use the zero-free region in the form given by (12.75), then an application of Lemma 12.6 gives

$$\sum_{|\gamma| \leqslant T} \rho^{-1} \{ (x+h)^{\rho} - x^{\rho} \} K(\rho) \ll h \log^C x \max_{0 \leqslant \sigma \leqslant 1 - Kh(x)} x^{\sigma-1} T^{11(1-\sigma)/5}$$

$$\ll h (\log^C x) \exp(-\varepsilon K (\log x) h(x))$$

$$= o(h/\log x) \qquad [\varepsilon = \varepsilon(\delta)].$$

Inserting this bound in (12.110) we obtain

$$\sum_{Q < p \leqslant 2Q} \left\{ \psi\left(\frac{x+h}{p}\right) - \psi\left(\frac{x}{p}\right) \right\} = h \sum_{Q < p \leqslant 2Q} 1/p + o(h/\log x)$$

$$+ O\left(x^{6/11+\delta} \log^{-2} x \right) + O\left(x^{5/11} \right)$$

$$= h \sum_{Q < p \leqslant 2Q} 1/p + o(h/\log x),$$

which is the assertion of the theorem.

In (12.109) $\psi(x)$ can be replaced by $\theta(x)$, and moreover since

$$\sum_{Q < p \leqslant 2Q} 1/p \gg \{ \pi(2Q) - \pi(Q) \}/Q \gg 1/\log Q \gg 1/\log x,$$

we obtain from (12.109)

(12.111) $$\sum_{Q < p \leqslant 2Q} \left\{ \pi\left(\frac{x+h}{p}\right) - \pi\left(\frac{x}{p}\right) \right\} \gg h/\log^2 x.$$

Therefore (12.111) implies that at least one of the differences $\pi((x + h)/p)$ $- \pi(x/p)$ must be positive for some p, and consequently the interval $[x/p, (x + h)/p]$ contains a prime p_1. This means that we have proved that the interval $[x, x + h]$ contains a number $P_2 = pp_1$ if $h = x^{6/11+\delta}$, and by a more careful analysis, analogous to the one carried out in the proof of Theorem 12.8, we could improve this to $h = x^{6/11}\log^{19}x$. However, as in the case of the difference between consecutive primes one can take advantage of the recent sieve techniques and combine them with analytic methods to prove a result about P_2's in short intervals, which substantially improves on the exponent $\frac{6}{11} = 0.545454 \cdots$. The latest published result in this direction (due to H. Iwaniec and M. Laborde) is that there is a P_2 in $(x, x + h]$ if $h = x^{0.45}$.

If one asks for P_2's in almost all intervals $(x, x + h]$, then a very satisfactory result can be obtained. This is given by

THEOREM 12.13.   For almost all x the interval $(x, x + x^\varepsilon]$ contains a P_2.

Proof of Theorem 12.13.   Let $\varepsilon > 0$ be given and consider the integral

$$(12.112) \qquad I(x) = x^{-1}\int_x^{2x}\left|\Phi(y) - yV^{-1}P(1)\right|^2 dy,$$

where

$$(12.113) \quad P(s) = \sum_{U<p\leqslant 2U} p^{-s}, \qquad \Phi(y) = \sum_{y<pp'\leqslant y(1+1/V), U<p\leqslant 2U} \log p',$$

p, p' are primes, and $x^{1/3} < U, V < x\log^{-2}x$. If we can show that

$$(12.114) \qquad I(x) = o\left\{\left(xV^{-1}P(1)\right)^2\right\} \qquad (x \to \infty),$$

then for almost all y from $[x, 2x]$

$$\Phi(y) \sim yV^{-1}P(1) \qquad (x \to \infty),$$

and there must be a $P_2 = pp'$ in the interval $(y, y + yV^{-1}]$, and the theorem follows with

$$(12.115) \qquad\qquad V = x^{1-\varepsilon}.$$

Now for $\sigma > 1$ we have

$$\sum_p (\log p)p^{-s} = -\frac{\zeta'(s)}{\zeta(s)} + G(s), \qquad G(s) = \sum_p \frac{\log p}{p^s(1 - p^s)}.$$

Choosing $\eta = 1 - (\log x)^{-3/4}$ we have by the inversion formula (A.10) and the residue theorem

$$\Phi(y) - yV^{-1}P(1)$$

$$= (2\pi i)^{-1} \int_{\eta - iU}^{\eta + iU} \left(G(s) - \frac{\zeta'(s)}{\zeta(s)} \right) P(s)\{(1 + 1/V)^s - 1\} s^{-1} y^s \, ds$$

$$+ O(U^{-1} x \log^3 x),$$

where we used (12.50) and (12.52) to estimate $\int_{\eta \pm iU}^{2 \pm iU}$. Inserting this expression in (12.112) and integrating over y we obtain, since $((1 + 1/V)^s - 1)/s \ll 1/V$,

$$I(x) \ll x^{2\eta} V^{-2} \log^2 x \int_{-U}^{U} \int_{-U}^{U} \frac{|P(\eta + iu) P(\eta + iv)|}{1 + |u - v|} \, du \, dv + x^2 U^{-2} \log^6 x.$$

Using $|ab| \leq \frac{1}{2}|a|^2 + \frac{1}{2}|b|^2$ it is seen that the double integral is

$$\ll \int_{-U}^{U} |P(\eta + iu)|^2 \, du \int_{-U}^{U} \frac{dv}{1 + |u - v|} \ll U^{2(1-\eta)} \log U,$$

hence

$$I(x) \ll \left(xV^{-1}P(1)\right)^2 \left\{ \log^5 x \left(\frac{U}{x}\right)^{2(1-\eta)} + \left(\frac{V}{U}\right)^2 \log^8 x \right\}.$$

If as in (12.115) we choose $V = x^{1-\varepsilon}$, $U = V\exp(\log^{1/4} x)$, then we have

$$I(x) \ll \left(xV^{-1}P(1)\right)^2 \exp(-\log^{1/5} x) = o\left\{\left(xV^{-1}P(1)\right)^2\right\},$$

which ends the proof of Theorem 12.13.

12.8 SUMS OF DIFFERENCES BETWEEN CONSECUTIVE PRIMES

We have seen in Theorem 12.10 that even the Riemann hypothesis does not seem to imply anything sharper than $p_{n+1} - p_n \ll p_n^{1/2} \log p_n$. Therefore it seems natural to ask about the frequency of p_n's for which $p_{n+1} - p_n > p_n^{1/2}$, and about estimates for sums of $(p_{n+1} - p_n)^2$. Analytic methods involving the zeta-function and capable of dealing with this type of problem evolved in recent years, and the sharpest known results are furnished by

THEOREM 12.14. For any $\varepsilon > 0$

(12.116)
$$\sum_{p_n \leqslant x,\, p_{n+1} - p_n > p_n^{1/2}} (p_{n+1} - p_n) \ll x^{3/4+\varepsilon},$$

(12.117)
$$\sum_{p_n \leqslant x} (p_{n+1} - p_n)^2 \ll x^{23/18+\varepsilon}.$$

The proofs of these results are due to D. R. Heath-Brown and depend on estimates for the zero-density function $N^*(\sigma, T)$, contained in

LEMMA 12.7. Let $N^*(\sigma, T)$ denote the number of ordered sets of zeros $\rho_j = \beta_j + i\gamma_j$ $(j = 1, 2, 3, 4)$, each counted by $N(\sigma, T)$, which satisfy $|\gamma_1 + \gamma_2 - \gamma_3 - \gamma_4| \leqslant 1$. Then uniformly in σ

(12.118)
$$N^*(\sigma, T) \ll \begin{cases} T^{(10-11\sigma)/(2-\sigma)+\varepsilon} & \left(\tfrac{1}{2} \leqslant \sigma \leqslant \tfrac{2}{3}\right), \\ T^{(18-19\sigma)/(4-2\sigma)+\varepsilon} & \left(\tfrac{2}{3} \leqslant \sigma \leqslant \tfrac{3}{4}\right), \\ T^{(12-12\sigma)/(4\sigma-1)+\varepsilon} & \left(\tfrac{3}{4} \leqslant \sigma \leqslant 1\right). \end{cases}$$

Proof of Theorem 12.14. Let $x \geqslant x_0$, $x^\varepsilon \leqslant U \leqslant x^{1-\varepsilon}$, $T = U\log^3 x$, $e^\delta = 1 + U^{-1}$, $\tfrac{1}{2} < u < 1$,

(12.119)
$$x < p_n \leqslant y \leqslant (p_n + p_{n+1})/2 < p_{n+1} \leqslant 2x,$$

and

(12.120)
$$\Delta(y) = \psi(y + y/U) - \psi(y) - y/U + \sum_{u \leqslant \beta \leqslant 1,\, |\gamma| \leqslant T} y^\rho(e^{\delta\rho} - 1)/\rho.$$

If we suppose that

(12.121)
$$p_{n+1} - p_n \geqslant 4xU^{-1},$$

then

$$y + y/U < p_n + (p_{n+1} - p_n)/2 + (p_{n+1} - p_n)2x/4x = p_{n+1}.$$

In that case the interval $(y, y + y/U]$ contains no primes, and if u and U are chosen in such a way that

(12.122)
$$\sum_{u \leqslant \beta \leqslant 1,\, |\gamma| \leqslant T} y^\rho(e^{\delta\rho} - 1)/\rho = o(y/U) \qquad (x \to \infty)$$

holds, then from the explicit formula (12.14) for $\psi(x)$ we obtain

$$\Delta(y) = yU^{-1} + O\!\left(\sum_{y < p^m \leqslant y + y/U,\, m \geqslant 2} \log p\right) + o(y/U),$$

which implies

$$|\Delta(y)| \gg xU^{-1}$$

if y satisfies (12.119). Thus writing $I_n = [p_n, (p_n + p_{n+1})/2]$ we obtain

$$\int_{I_n} |\Delta(y)|^2 \, dy \ll x^2 U^{-2}(p_{n+1} - p_n),$$

and

$$\int_{I_n} |\Delta(y)|^4 \, dy \gg x^4 U^{-4}(p_{n+1} - p_n).$$

If U is a function of x alone, then by summing over disjoint intervals I_n we have

(12.123)
$$\sum_{x < p_n < p_{n+1} \leqslant 2x, \, p_{n+1} - p_n \geqslant 4x/U} (p_{n+1} - p_n)$$

$$\ll \min\left(x^{-2}U^2 \int_x^{2x} |\Delta(y)|^2 \, dy, \; x^{-4}U^4 \int_x^{2x} |\Delta(y)|^4 \, dy\right).$$

Using

(12.124)
$$|(e^{\delta \rho} - 1)/\rho| = \left|\int_0^\delta e^{t\rho} \, dt\right| \leqslant \int_0^\delta e^{t\beta} \, dt \leqslant e\delta \ll 1/U,$$

and proceeding as in the proof of Theorem 12.9 we find that

$$\int_x^{2x} |\Delta(y)|^2 \, dy \ll U^{-2} \log^3 x \max_{\sigma \leqslant u} x^{2\sigma + 1} N(\sigma, T).$$

To estimate the integral of the fourth power of $|\Delta(y)|$ in (12.123) divide the interval $[0, u]$ into $O(\log x)$ subintervals J_k of the form

$$J_k = \left[\frac{k}{\log x}, \frac{k+1}{\log x}\right], \qquad k = 0, 1, 2, \ldots .$$

Using Hölder's inequality we have

$$|\Delta(y)|^4 \ll \log^3 x \sum_k \left| \sum_{\beta \in J_k, |\gamma| \leqslant T} y^\rho (e^{\delta \rho} - 1)/\rho \right|^4;$$

whence using (12.124) and integrating we obtain

$$\int_{x}^{2x} |\Delta(y)|^4 \, dy$$

$$\ll \log^3 x \sum_{k} \sum_{\substack{\beta_1,\ldots,\beta_4 \in J_k \\ |\gamma_1| \leqslant T, \ldots, |\gamma_4| \leqslant T}} \frac{(2x)^{\rho_1 + \rho_2 + \bar{\rho}_3 + \bar{\rho}_4 + 1} - x^{\rho_1 + \rho_2 + \bar{\rho}_3 + \bar{\rho}_4 + 1}}{\rho_1 \rho_2 \rho_3 \rho_4 (\rho_1 + \rho_2 + \bar{\rho}_3 + \bar{\rho}_4 + 1)}$$

$$\times (e^{\delta \rho_1} - 1) \cdots (e^{\delta \bar{\rho}_4} - 1)$$

$$\ll U^{-4} \log^3 x \sum_{k} x^{4k/(\log x) + 1} \sum_{\substack{\beta_1,\ldots,\beta_4 \geqslant k/\log x; \, |\gamma_1|, \ldots, |\gamma_4| \leqslant T}}$$

$$\times |\rho_1 + \rho_2 + \bar{\rho}_3 + \bar{\rho}_4 + 1|^{-1}$$

$$\ll U^{-4} \log^3 x \max_{\sigma \leqslant u} x^{4\sigma + 1} M(\sigma, T),$$

where

$$M(\sigma, T) = \sum_{\substack{\beta_1,\ldots,\beta_4 \geqslant \sigma; \, |\gamma_1| \leqslant T, \ldots, |\gamma_4| \leqslant T}} (1 + |\gamma_1 + \gamma_2 - \gamma_3 - \gamma_4|)^{-1}.$$

We shall show now that

(12.125) $$M(\sigma, T) \ll N^*(\sigma, T) \log T,$$

where $N^*(\sigma, T)$ is defined in Lemma 12.7. To see this, write $g(\nu)$ for the number of solutions of

$$|\gamma_1 + \gamma_2 - \gamma_3 - \gamma_4 - \nu| \leqslant \tfrac{1}{2}; \qquad \sigma \leqslant \beta_j \leqslant 1, \qquad |\gamma_j| \leqslant T \qquad (j = 1, 2, 3, 4)$$

where ν is an integer. Then

$$|\nu| - \tfrac{1}{2} \leqslant |\gamma_1 + \gamma_2 - \gamma_3 - \gamma_4| \leqslant |\nu| + \tfrac{1}{2}$$

and

$$M(\sigma, T) \ll \sum_{|\nu| \leqslant 4T} g(\nu)/(1 + |\nu|).$$

An elementary calculation shows that

$$\int_{-1}^{1} (1 - |y|) e^{-2\pi i x y} \, dy = \left(\frac{\sin \pi x}{\pi x} \right)^2 \qquad (x \text{ real}),$$

and therefore by the inverse Fourier transform we obtain

$$(12.126) \quad \int_{-\infty}^{\infty} e(xy) \left(\frac{\sin \pi x}{\pi x} \right)^2 dx = \int_{-\infty}^{\infty} e^{2\pi i xy} \left(\frac{\sin \pi x}{\pi x} \right)^2 dx$$

$$= \begin{cases} 1 - |y| & (|y| \leqslant 1), \\ 0 & (|y| > 1). \end{cases}$$

If we define now

$$F(x) = \sum_{\sigma \leqslant \beta \leqslant 1, \, |\gamma| \leqslant T} e(x\gamma),$$

then using (12.126) we obtain

$$g(\nu) \leqslant 2 \int_{-\infty}^{\infty} |F(x)|^4 e(-\nu x) \left(\frac{\sin \pi x}{\pi x} \right)^2 dx \leqslant 2 \int_{-\infty}^{\infty} |F(x)|^4 \left(\frac{\sin \pi x}{\pi x} \right)^2 dx$$

$$\leqslant 2 N^*(\sigma, T);$$

whence

$$M(\sigma, T) \ll N^*(\sigma, T) \sum_{|\nu| \leqslant 4T} 1/(1 + |\nu|) \ll N^*(\sigma, T) \log T.$$

Therefore (12.123) gives the fundamental formula

$$(12.127) \quad \sum_{x < p_n < p_{n+1} \leqslant 2x, \, p_{n+1} - p_n \geqslant 4x/U} (p_{n+1} - p_n)$$

$$\ll \log^4 x \min \Big\{ \max_{\sigma \leqslant u} x^{2\sigma - 1} N(\sigma, U \log^3 x),$$

$$\times \max_{\sigma \leqslant u} x^{4\sigma - 3} N^*(\sigma, U \log^3 x) \Big\}.$$

We choose now $U = 8x^{1/2}$, and then (12.121) is implied by $p_{n+1} - p_n > p_n^{1/2}$. Writing estimates for $N^*(\sigma, T)$ in the form

$$(12.128) \quad N^*(\sigma, T) \ll T^{B(\sigma)(1 - \sigma) + \varepsilon},$$

we obtain from (12.127)

(12.129)

$$\sum_{x < p_n \leqslant 2x, \, p_{n+1} - p_n > p_n^{1/2}} (p_{n+1} - p_n) \ll x^\varepsilon \Big(x^{7/12} + \max_{\sigma \leqslant u} x^{4\sigma - 3 + B(\sigma)(1 - \sigma)/2} \Big),$$

where the term $x^{7/12}$ comes from estimating $p_{n+1} - p_n$ by (12.68) for at most one prime p_n ($\leqslant 2x$) for which $2x < p_{n+1}$.

Now we use the zero-free region $\zeta(s)$ in the form (12.75) and the zero-density bound $A(\sigma) \leqslant 3/(2\sigma)$, $\sigma \geqslant \frac{4}{5}$ [see (11.81)], to obtain

$$\sum_{4/5 \leqslant \sigma \leqslant 1, |\gamma| \leqslant T} y^\rho (e^{\delta\rho} - 1)/\rho \ll U^{-1}\log x \max_{4/5 \leqslant \sigma \leqslant 1 - Kh(x)} x^\sigma N(\sigma, T)$$

$$\ll xU^{-1}\log^3 x \max_{4/5 \leqslant \sigma \leqslant 1 - Kh(x)} \left(xU^{-(15/8+\varepsilon)}\right)^{\sigma-1}$$

$$\ll xU^{-1}x^{-K(1/20-\varepsilon)h(x)} \ll xU^{-1}\log^{-A}x$$

for any fixed $A > 0$ if $\varepsilon = \frac{1}{40}$, say. Therefore (12.122) holds with $u = \frac{4}{5}$, and using bounds for $B(\sigma)$ in (12.128) given by Lemma 12.7 we find that

$$\max_{1/2 \leqslant \sigma \leqslant 4/5} (4\sigma - 3 + B(\sigma)(1 - \sigma)/2) = 4(\tfrac{3}{4}) - 3 + B(\tfrac{3}{4})(1 - \tfrac{3}{4})/2 = \tfrac{3}{4},$$

which gives then (12.116), since for $0 \leqslant \sigma \leqslant \frac{1}{2}$ the trivial estimate $B(\sigma)(1 - \sigma) = 3$ [follows, e.g., from (12.135)–(12.137)] suffices.

It may be remarked that the foregoing method may be employed to yield bounds for the more general sum

$$(12.130) \quad S(x; c) = \sum_{p_n \leqslant x, \, p_{n+1} - p_n > p_n^c} (p_{n+1} - p_n) \qquad (0 < c < \tfrac{13}{23}).$$

Also supposing that $4x/U \leqslant p_{n+1} - p_n \leqslant 8x/U$ we obtain from (12.127)

$$(12.131) \quad \sum_{x < p_n \leqslant 2x, \, 4x/U \leqslant p_{n+1} - p_n \leqslant 8x/U} (p_{n+1} - p_n)^2$$

$$\ll x^2 U^{-2} + xU^{-1}\log^4 x \min\Big\{ \max_{\sigma \leqslant u} x^{2\sigma-1}N(\sigma, T),$$

$$\times \max_{\sigma \leqslant u} x^{4\sigma-3}N^*(\sigma, T)\Big\}.$$

This is the crucial relation in the proof of (12.117). Estimating small differences of $p_{n+1} - p_n$ trivially we may restrict U to the range $x^\eta \leqslant U \leqslant x^{1-\eta}$ for some small $\eta > 0$; hence we may choose $u = 1 - \delta$ in (12.120). Next, we divide $[0, u]$ into subintervals J_k as in the estimation of the integral of the fourth power of $|\Delta(y)|$, and consider $O(\log x)$ subsums

$$(12.132) \quad S_k = \sum_{\beta \in J_k, \, |\gamma| \leqslant T} y^\rho (e^{\delta\rho} - 1)/\rho$$

of which $\Delta(y)$ consists. Those values of k for which $S_k = O(xU^{-1}\log^{-2}x)$ contribute a total of $o(xU^{-1})$ to $\Delta(y)$, while if $S_k \gg xU^{-1}\log^{-2}x$, then

$$xU^{-1}\log^{-2}x \ll U^{-1}x^{k/\log x}N\big(k/\log x, U\log^3 x\big);$$

whence

$$\log^{-c}x \ll \big(U^{A(k/\log x)}x^{-1}\big)^{(1-k/\log x)},$$

if we suppose that $N(\sigma, T) \ll T^{A(\sigma)(1-\sigma)}\log^C T$. Raising the last inequality to the power $(1 - k/\log x)^{-1}$ we obtain

(12.133) $$\qquad\qquad U^{A(\sigma)} \gg x\log^{-c}x$$

for $\sigma = k/\log x \le 1 - 10^{-7}$, while using the zero-free region (6.1) and the bound $A(\sigma) \le 1600(1 - \sigma)^{3/2}$ [see (11.33)] we obtain that the contribution of S_k's with $\sigma = k/\log x > 1 - 10^{-7}$ to $\Delta(y)$ is $o(x/U)$, so that we may consider only those $\sigma = k/\log x$ for which (12.133) holds. Further note that the right-hand side of (12.131) is

$$\ll x^2U^{-2} + x^{2+\varepsilon}U^{-1}\min\bigg\{\max_{1/2\le\sigma\le u} \big(U^{A(\sigma)}x^{-2}\big)^{1-\sigma}, \max_{1/2\le\sigma\le u} \big(U^{B(\sigma)}x^{-4}\big)^{1-\sigma}\bigg\}.$$

Therefore if

(12.134) $$\qquad\qquad U^{B(\sigma)-A(\sigma)} \ge x^2,$$

then using $A(\sigma)(1 - \sigma) \le 1$ we find that

$$x^2U^{-1}\big(U^{A(\sigma)}x^{-2}\big)^{(1-\sigma)} \le x^{\alpha(\sigma)},$$

where

$$\alpha(\sigma) = 4\sigma - 2 + 2\{B(\sigma)(1 - \sigma) - 1\}/\{B(\sigma) - A(\sigma)\}.$$

If (12.134) does not hold, then using alternatively (12.133) we obtain

$$x^2U^{-1}\big(U^{B(\sigma)}x^{-4}\big)^{(1-\sigma)} \ll x^{\alpha(\sigma)} + x^{\beta(\sigma)+\varepsilon},$$

where

$$\beta(\sigma) = 4\sigma - 2\{B(\sigma)(1 - \sigma) - 1\}/A(\sigma).$$

Therefore by summing over various $U = 2^m$ we obtain from (12.131)

$$\sum_{p_n\le x}(p_{n+1} - p_n)^2 \ll x^\varepsilon\bigg\{x^{7/6} + \max_{1/2\le\sigma\le 1}\big(x^{\alpha(\sigma)} + x^{\beta(\sigma)}\big)\bigg\},$$

where for $\alpha(\sigma)$ and $\beta(\sigma)$ we use the bounds of Chapter 11 and Lemma 12.7. In the range $\frac{1}{2} \leqslant \sigma \leqslant \frac{3}{4}$ we use $A(\sigma) \leqslant (3 - 3\sigma)/(2 - \sigma) \leqslant \frac{12}{5}$ and $B(\sigma) \leqslant (36 - 8\sigma)/5$ to obtain

$$\beta(\sigma) \leqslant \alpha(\sigma) \leqslant (7 + 12\sigma - 8\sigma^2)/(12 - 4\sigma) \leqslant \tfrac{23}{18},$$

where the maximum occurs at $\sigma = \frac{3}{4}$. For $\sigma \geqslant \frac{3}{4}$ we use $B(\sigma) \leqslant 12/(4\sigma - 1)$ and bounds for $A(\sigma)$ of Chapter 11 to see that neither $\alpha(\sigma)$ nor $\beta(\sigma)$ exceeds $\frac{23}{18}$, so that (12.117) follows.

Proof of Lemma 12.7. To complete the proof of Theorem 12.14 it remains to prove (12.118). Each vertical strip $T \leqslant t \leqslant T + 1$ contains $O(\log T)$ zeros ρ of $\zeta(s)$ and

$$|\gamma_1 + \gamma_2 - \gamma_3 - \gamma_4| \leqslant 1$$

implies

$$[\gamma_1] + [\gamma_2] = [\gamma_3] + [\gamma_4] + j \qquad (j = 0, \pm 1, \pm 2).$$

Therefore we have

$$(12.135) \quad N^*(\sigma, T) \ll \log^4 T \int_0^1 \sum_{|j| \leqslant 2} \left| \sum_{|[\gamma]| \leqslant T, \beta \geqslant \sigma} e([\gamma]x) \right|^4 e(-jx) \, dx$$

$$\ll \log^4 T \int_0^1 \left| \sum_{|[\gamma]| \leqslant T, \beta \geqslant \sigma} e([\gamma]x) \right|^4 dx = R_0 \log^4 T,$$

where

$$(12.136) \qquad R_0 = \sum_{[\gamma_1]+[\gamma_2]=[\gamma_3]+[\gamma_4]}^* 1,$$

and Σ^* denotes summation over all quadruples of zeros $\rho_j = \beta_j + i\gamma_j$ ($j = 1, 2, 3, 4$) for which $\beta_j \geqslant \sigma, |[\gamma_j]| \leqslant T$. If $\gamma_1, \gamma_2, \gamma_3$ are arbitrary, then $[\gamma_4]$ is uniquely determined from the equation $[\gamma_4] = [\gamma_1] + [\gamma_2] - [\gamma_3]$; hence trivially

$$(12.137) \qquad R_0 \ll N^3(\sigma, T).$$

In the inner sum in the first bound in (12.135) we may further distinguish between zeros ρ for which either (11.9) or (11.10) or (11.11) holds, and consequently the foregoing discussion gives in fact

$$(12.138) \quad N^*(\sigma, T) \ll T^\varepsilon (1 + \min(T^3 Y^{6-12\sigma}, T^6 Y^{18-36\sigma}) + R^*),$$

where

$$(12.139) \qquad R^* = \underset{[\gamma_1]+[\gamma_2]=[\gamma_3]+[\gamma_4]}{{\sum}'} 1,$$

and Σ' denotes summation over zeros $\rho_j = \beta_j + i\gamma_j$ ($j = 1, 2, 3, 4$) counted by $N(\sigma, T)$ for which (11.9) holds. We first raise (11.15) to a power k ($\geqslant 1$) such that $Y^{1/2} < M \leqslant Y$, and then raise to higher powers to obtain $M_1 = M^{k_1}$ which we use to bound $R(= R_1$ in the notation of Chapter 11) and $M_2 = M^{k_2}$ which we use to bound R^*. We take k_2 such that

$$M^{k_2} \leqslant Y^2 < M^{1+k_2};$$

whence $k_2 = 2$ or $k_2 = 3$ and $Y^{4/3} < M_2 \leqslant Y^2$. We shall chose $k_1 = k_2$ or $k_1 = 1 + k_2$ using the following procedure.

For $\sigma \leqslant \frac{3}{4}$ we take $Y = T^{1/2}$, whence $T^3 Y^{6-12\sigma} = T^{6-6\sigma}$, which is sufficient for Lemma 12.7 in the range $\frac{1}{2} \leqslant \sigma \leqslant \frac{3}{4}$, and by the mean value for Dirichlet polynomials (Theorem 5.3) we have

$$(12.140) \qquad R \ll T^{\varepsilon}(T + M_1) M_1^{1-2\sigma}.$$

For

$$(12.141) \qquad T^{1/(2-\sigma)} \leqslant M_2 \leqslant T = Y^2$$

we take $k_1 = k_2$, so that (12.140) gives

$$(12.142) \qquad R \ll T^{1+\varepsilon} M_1^{1-2\sigma},$$

while for

$$(12.143) \qquad Y^{4/3} = T^{2/3} \leqslant M_2 \leqslant T^{1/(2-\sigma)}$$

we pick $k_1 = 1 + k_2$. Then $M_1 = M^{1+k_2} \geqslant Y^2 = T$; whence

$$(12.144) \qquad R \ll T^{\varepsilon} M_1^{2-2\sigma} \ll T^{\varepsilon} M_2^{3-3\sigma}.$$

In case $\sigma \geqslant \frac{3}{4}$ we proceed similarly, only we choose $Y = T^{1/(4\sigma-1)}$ and instead of (12.140) we use (11.25), namely,

$$R \ll T^{\varepsilon}\left(M_1^{2-2\sigma} + TM_1^{4-6\sigma}\right).$$

For $T^{1/(3\sigma-1)} \leqslant M_2 \leqslant Y^2$ we take $k_1 = k_2$, whence

$$(12.145) \qquad R \ll T^{\varepsilon} M_2^{2-2\sigma}$$

for

(12.146)
$$T^{1/(4\sigma-2)} \leqslant M_2 \leqslant Y^2,$$

and

$$R \ll T^{1+\varepsilon}M_2^{4-6\sigma}$$

for

(12.147)
$$T^{1/(3\sigma-1)} \leqslant M_2 \leqslant Y^{1/(4\sigma-2)}.$$

If

(12.148)
$$Y^{4/3} \leqslant M_2 \leqslant T^{1/(3\sigma-1)}$$

we pick $k_1 = 1 + k_2$. For $k_2 = 2$ we have $M_1 = M_2^{3/2}$, and since $M_2 \geqslant Y^{4/3}$ we obtain

$$TM_1^{4-6\sigma} = TM_2^{6-9\sigma} \leqslant M_2^{3-3\sigma},$$

while for $k_2 = 3$ we have

$$TM_1^{4-6\sigma} = TM_2^{(16-24\sigma)/3} \leqslant M_2^{3-3\sigma},$$

since $M_1 = M_2^{4/3}$ and $M_2 \geqslant Y^{3/2}$. Thus for either value of k_2

(12.149)
$$R \ll T^\varepsilon M_2^{3-3\sigma}.$$

We may now estimate the sum R^* in (12.139). We write $[\gamma] = t_u = t$, $\rho = \beta + i\gamma = \sigma + it + \delta$, $\beta \geqslant \sigma$, $\delta = i(\gamma - [\gamma]) + \beta - \sigma$. Then with

$$S(y, t) = S(y) = \sum_{M < n \leqslant y} b(n)n^{-\sigma-it}$$

we have by partial summation

$$T^{-\varepsilon} \ll \sum_{M < n \leqslant P} b(n)n^{-\rho} \ll |S(P)| + \int_M^P |S(y)|y^{-1}\,dy \qquad (P = 2^k M),$$

by the equation preceding (11.16). Thus for each $[\gamma] = t_u$ appearing in R^*

(12.150)
$$T^{-\varepsilon} \ll |S(P, t_u)| + \int_M^P |S(y, t_u)|y^{-1}\,dy.$$

We may write

$$R^* = \sum_{t_u + t_v = t_w + t_x} 1,$$

and with the notation

$$m(t) = \sum_{t=t_u+t_v-t_w} 1, \qquad n(t) = \sum_{t=t_u+t_v} 1$$

(12.150) gives

$$R^* = \sum_{t_x} m(t_x) \ll T^\varepsilon \sum_{t_x} m(t_x) \left\{ |S(P, t_x)|^2 + \int_M^P |S(y, t_x)|^2 y^{-1} dy \right\}.$$

Set now $t_x = t - t_w$. Then

$$\sum_{t_x} m(t_x) |S(y, t_x)|^2 \leqslant \sum_{t, t_w} n(t) \left| \sum_{M < n \leqslant y} b(n) n^{-\sigma - it + it_w} \right|^2,$$

since the right-hand side counts all solutions $t = t_u + t_v$, including those t for which $t - t_w = t_x$. But the right-hand side of the last inequality equals

$$\sum_{M < n_1, n_2 \leqslant y} b(n_1) \overline{b(n_2)} (n_1 n_2)^{-\sigma} \sum_t n(t)(n_2/n_1)^{it} \sum_{t_w} (n_1/n_2)^{it_w}$$

$$\ll T^\varepsilon M^{1-2\sigma} \sum_{M < n_1, n_2 \leqslant P} (n_1 n_2)^{-1/2} \left| \sum_t n(t)(n_2/n_1)^{it} \right| \left| \sum_{t_w} (n_1/n_2)^{it_w} \right|$$

$$\ll T^\varepsilon M^{1-2\sigma} \left(\sum_{M < n_1, n_2 \leqslant P} (n_1 n_2)^{-1/2} \left| \sum_t \right|^2 \right)^{1/2}$$

$$\times \left(\sum_{M < n_1, n_2 \leqslant P} (n_1 n_2)^{-1/2} \left| \sum_{t_w} \right|^2 \right)^{1/2}$$

$$\ll T^\varepsilon M^{1-2\sigma} \left(\sum_{t, t'} n(t) n(t') \sum_{M < n_1, n_2 \leqslant P} (n_1 n_2)^{-1/2} (n_2/n_1)^{it-it'} \right)^{1/2}$$

$$\times S^{1/2}(T^\varepsilon M)$$

$$\ll T^\varepsilon M^{1-2\sigma} \left(\sum_{t_u, t_v, t_w, t_x} \left| \sum_{M < n \leqslant P} n^{-1/2 + it_u + it_v - it_w - it_x} \right|^2 \right)^{1/2} S^{1/2}(T^\varepsilon M),$$

where $S(N)$ is the double zeta-sum defined by (11.62) and the t_r's are now integers (the spacing condition $|t_r - t_s| \geqslant \log^4 T$ $(r \neq s)$ may be imposed by multiplying the bound for R^* by an unimportant log factor). The sum

$$S_1(N) = \sum_{t_u,t_v,t_w,t_x \leqslant R} \left| \sum_{N<n\leqslant 2N} n^{-1/2+it_u+it_v-it_w-it_x} \right|^2$$

may be estimated by the technique analogous to the one used in Chapter 11 in bounding $S(N)$. Therefore

$$S_1(N) \ll R^*N + R^4$$

$$+ \max_{|v| \leqslant \log^2 T} \sum_{t_u+t_v \neq t_w+t_x} \left| \zeta\left(\tfrac{1}{2} + it_u + it_v - it_w - it_x + iv\right) \right|^2$$

$$\ll R^*N + R^4 + T^\varepsilon\left(S_1(T^\varepsilon Y) + S_1(T^{1+\varepsilon}Y^{-1})\right)$$

$$\ll R^*N + T^\varepsilon R^4 + T^\varepsilon\left(R^*Y + S_1(T^{1+\varepsilon}Y^{-1})\right),$$

if we use the analogue of Lemma 11.4 which is established similarly. By the Cauchy–Schwarz inequality

$$S_1(N) \ll R^2\left\{S_1(T^\varepsilon N^2)\right\}^{1/2};$$

hence

$$\sum = \sum_{t_u+t_v \neq t_w+t_x} |\zeta|^2 \ll T^\varepsilon\left\{R^*Y + R^4 + R^2\left(R^*T^2Y^{-2} + \sum\right)^{1/2}\right\},$$

which implies

$$\sum \ll T^\varepsilon\left(R^*Y + R^4 + R^2(R^*)^{1/2}TY^{-1}\right) \ll T^\varepsilon\left(R(R^*)^{3/4}T^{1/2} + R^4\right)$$

for $Y = T^{1/2}R(R^*)^{-1/4}$. This means that we have

$$S_1(N) \ll T^\varepsilon\left(R^*N + R^4 + R(R^*)^{3/4}T^{1/2}\right),$$

and, similarly as in Lemma 11.5 for $S(N)$, the term $R(R^*)^{3/4}T^{1/2}$ may be discarded if $N \geqslant T^{2/3}$.

Therefore for $M = M_2$, by the foregoing discussion, we have

$$(12.151) \qquad R^* \ll T^\varepsilon M_2^{1-2\sigma}\left\{S_1(T^\varepsilon M_2)S(T^\varepsilon M_2)\right\}^{1/2}$$

$$\ll T^\varepsilon M_2^{1-2\sigma}\left(R^*M_2 + R^4 + R(R^*)^{3/4}T^{1/2}\right)^{1/2}$$

$$\times\left(RM_2 + R^2 + R^{5/4}T^{1/2}\right)^{1/2},$$

and for $M_2 \geqslant T^{2/3}$ (12.151) reduces to

$$R^* \ll T^\varepsilon M_2^{1-2\sigma} \left(RM_2 + R^2 \right)^{1/2} \left(R^* M_2 + R^4 \right)^{1/2};$$

hence for $M_2 \geqslant T^{2/3}$

(12.152) $\qquad R^* \ll T^\varepsilon \left(RM_2^{4-4\sigma} + R^3 M_2^{1-2\sigma} + R^{5/2} M_2^{(3-4\sigma)/2} \right).$

To finish the proof of the lemma suppose first $\sigma \leqslant \frac{3}{4}$. Then $M_2 \geqslant T^{2/3}$, and we estimate R^* by (12.142), (12.144), and (12.152). If (12.142) holds, then

$$R^* \ll T^\varepsilon \left(M_2^{10-11\sigma} + M_2^{7-7\sigma} + M_2^{(18-19\sigma)/2} \right)$$

$$\ll T^\varepsilon \left(T^{(10-11\sigma)/(2-\sigma)} + T^{6-6\sigma} + T^{(18-19\sigma)/(4-2\sigma)} \right),$$

while if (12.144) holds

$$R^* \ll T^\varepsilon \left(M_2^{10-11\sigma} + M_2^{7-7\sigma} + M_2^{(18-19\sigma)/2} \right)$$

$$\ll T^\varepsilon \left(T^{(10-11\sigma)/(2-\sigma)} + T^{(7-7\sigma)/(2-\sigma)} + T^{(18-19\sigma)/(4-2\sigma)} \right),$$

and we obtain the first two estimates of Lemma 12.7.

For $\sigma \geqslant \frac{3}{4}$ we have $R \ll M_2$. If we suppose additionally that

(12.153) $\qquad\qquad\qquad RM_2 \geqslant R^{5/4} T^{1/2},$

then (12.151) reduces to

$$R^* \ll T^\varepsilon \left((R^*)^{1/2} R^{1/2} M_2^{2-2\sigma} + R^{5/2} M_2^{(3-4\sigma)/2} + (R^*)^{3/8} R T^{1/4} M_2^{(3-4\sigma)/2} \right);$$

whence

$$R^* \ll T^\varepsilon \left(RM_2^{4-4\sigma} + R^{5/2} M_2^{(3-4\sigma)/2} + R^{8/5} T^{2/5} M_2^{(12-26\sigma)/5} \right).$$

In case (12.145) and (12.146) hold we obtain

$$R^* \ll T^\varepsilon \left(M_2^{6-6\sigma} + M_2^{(13-14\sigma)/2} + T^{2/5} M_2^{(28-32\sigma)/5} \right)$$

$$\ll T^\varepsilon \left(T^{(12-12\sigma)/(4\sigma-1)} + T^{(13-14\sigma)/(4\sigma-1)} + T^{(54-56\sigma)/(20\sigma-5)} \right.$$

$$\left. + T^{(12-12\sigma)/(12\sigma-5)} \right) \ll T^{(12-12\sigma)/(4\sigma-1)+\varepsilon},$$

as required. An analogous argument works if (12.148) holds, because we may restrict ourselves to the range $\sigma \leqslant \frac{25}{28}$, since $A(\sigma) \leqslant 4/(4\sigma - 1)$ follows from Theorem 11.2 for $\sigma \geqslant \frac{25}{28}$, and then the trivial bound (12.137) suffices.

Finally suppose that $\frac{3}{4} \leqslant \sigma \leqslant \frac{25}{28}$ and that (12.153) fails to hold. The bound $R \ll T^{\varepsilon} M_2^{3-3\sigma}$ shows that we must have then (12.148), and (12.151) reduces to

$$R^* \ll T^{\varepsilon} \left(R^{5/4} T^{1/2} M_2^{3-4\sigma} + R^{21/8} T^{1/4} M_2^{1-2\sigma} + R^{9/5} T^{4/5} M_2^{(8-16\sigma)/5} \right).$$

Using $R \ll T^{\varepsilon} M_2^{3-3\sigma}$ the last bound is easily seen to be $\ll T^{(12-12\sigma)/(4\sigma-1)+\varepsilon}$ for $\sigma \leqslant \frac{21}{22}$, completing the proof of Lemma 12.7.

Notes

For (12.4) one needs the well-known elementary inequality $\pi(x) \ll x/\log x$. M. Nair (1982) found recently a simple elementary method which gives

$$(\log 2 - \varepsilon) \frac{N}{\log N} \leqslant \pi(N) \leqslant (\log 4 + \varepsilon) \frac{N}{\log N} \qquad (N \geqslant N_0).$$

For elements of prime number theory the reader is referred to K. Prachar (1967).

The explicit formula (12.6) was stated without proof by B. Riemann (1859), who conjectured also a more complicated formula for $\pi(x) - \mathrm{li}\, x$ in terms of nontrivial zeros ρ of $\zeta(s)$. Both formulas were proved by H. von Mangoldt (1895).

The prime number theorem (12.7) represents the strongest known form of this result. The formula is due to H.-E. Richert and A. Walfisz, and was proved in Walfisz's book (1963). The history of the prime number theorem is a fascinating subject which goes back to the 18th century to A.-M. Legendre and C. F. Gauss. In modern terminology they conjectured

(12.154) $$\pi(x) = \frac{x}{\log x - A + o(1)} \qquad (A = 1.08366\ldots),$$

and

$$\pi(x) = (1 + o(1)) \int_2^x \frac{dt}{\log t} \qquad (x \to \infty),$$

respectively. In view of (12.27) the conjecture of Gauss was certainly true, while that of Legendre was not. Namely, an integration by parts shows that

$$\int \frac{dx}{\log x} = \frac{x}{\log x} + \frac{1!\,x}{\log^2 x} + \frac{2!\,x}{\log^3 x} + \cdots + \frac{(N-1)!\,x}{\log^N x} + O_N\left(\frac{x}{\log^{N+1} x} \right)$$

for any integer $N \geqslant 1$, so that (12.154) is true with $A = 1$. For a more detailed discussion of Legendre's formula see J. Pintz (1980f).

The prime number theorem with an error term was obtained first in the form

(12.155) $$\psi(x) = x + O\left(x \exp\left(-C \log^{1/2} x \right) \right) \qquad (C > 0).$$

This was achieved independently at the end of the 19th century by J. Hadamard (1893, 1896) and C. J. de la Vallée-Poussin (1896, 1899). Their proofs were both analytic and essentially depended on the zero-free region (1.55). A full account may be found in K. Prachar (1967) and A. Walfisz (1963), where a general discussion concerning the influence of the zero-free region on the error term in the prime number theorem is given. It was mentioned in Notes of Chapter 6 that both

N. M. Korobov (1958a) and I. M. Vinogradov (1958) claimed the zero-free region $\sigma \geqslant 1 - C \log^{-2/3} t$, which their method still does not seem to yield. Therefore it is no surprise that they also claimed the prime number theorem in the form

$$(12.156) \qquad \psi(x) = x + O\left(x \exp\left(-C \log^{3/5} x\right)\right) \qquad (C > 0),$$

which is still sometimes quoted in the literature, but the best that one can prove at present is nevertheless (12.7), which is slightly weaker than (12.156).

A veritable sensation among number theorists was caused by A. Selberg (1949) and P. Erdős (1949), who obtained an elementary proof of the prime number theorem in the form $\psi(x) = (1 + o(1)) x$. By "elementary" one means here a proof which does not use complex function theory, but only elementary real analysis. Although "elementary," the proofs of Selberg and Erdős were by no means easy. The starting point is Selberg's formula

$$(12.157) \qquad \sum_{p \leqslant x} \log^2 p + \sum_{pq \leqslant x} \log p \log q = 2x \log x + O(x) \qquad (p, q \text{ primes}),$$

but it would be too great a digression to explain here how (12.157) leads to the proof of $\psi(x) \sim x$. The original arguments of Selberg and Erdős were later simplified, and there exist at present many variants of the elementary proof of the prime number theorem [e.g., in K. Chandrasekharan (1970) or K. Prachar (1967)]. It should be mentioned that in 1962 E. Wirsing (1962, 1964) and E. Bombieri (1962) obtained independently an elementary proof of the prime number theorem in the form

$$\psi(x) = x + O\left(x \log^{-A} x\right),$$

where $A > 0$ is arbitrary, but fixed. This was superseded in 1970 by H. Diamond and J. Steinig (1970), who obtained an elementary proof of

$$(12.158) \qquad \psi(x) = x + O\left(x \exp\left(-\log^{1/7-\varepsilon} x\right)\right).$$

By refining their method A. F. Lavrik and A. Š. Sobirov (1973) proved (12.158) with $1/6 - \varepsilon$. The error term here is poorer than the error term in (12.155), which was the first result obtained by methods of complex analysis, but (12.158) is nevertheless very important since in view of P. Turán's result (1950) [see also J. Pintz (1980c, 1980d)] that (12.32) implies (12.33), we obtain the zero-free region

$$\sigma \geqslant 1 - \left(\log t\right)^{-\varepsilon - 6}, \qquad t \geqslant t_0.$$

This shows that a nontrivial zero-free region for $\zeta(s)$ can be obtained by elementary methods, and for an expert account of elementary methods in prime number theory the reader is referred to the article of H. G. Diamond (1982).

Proofs of E. Landau's classical formula (12.14) may be found in standard works, such as H. Davenport (1980) or K. Chandrasekharan (1970). For a slight improvement of the error terms, see D. A. Goldston (1981) and D. Wolke (1980, 1983).

Theorem 12.4 is due to E. Schmidt (1903). A conditional omega result for $\psi(x)$ was proved by E. Grosswald (1965). Namely, if θ denotes the supremum of real parts of zeros of $\zeta(s)$ and $\theta > \frac{1}{2}$, then Grosswald proves

$$\psi(x) = x + O(x^\theta), \qquad \psi(x) = x + \Omega_\pm(x^\theta),$$

so that in this case the exact order of $\psi(x) - x$ is determined up to the value of the constants involved.

The proof of Theorem 12.5 is due to J. Pintz (1982c), and the same idea readily leads also to

$$(12.159) \qquad \int_1^x |M(x)| \, dx \gg x^{3/2},$$

which implies $M(x) = \Omega(x^{1/2})$, and as mentioned in Notes of Chapter 1, J. Pintz (1982a) contains a sharper result than (12.159). By using the exponential integral (A.38) one has

$$(12.160) \quad (2\pi i)^{-1} \int_{2-i\infty}^{2+i\infty} \frac{e^{ks^2+\mu s}}{s\zeta(s)} \, ds = \frac{1}{2}(\pi k)^{-1/2} \int_1^\infty M(x) \exp\left(\frac{(\log x - \mu)^2}{4k} \right) \frac{dx}{x},$$

where $k, \mu > 0$ may be suitably chosen. This relation is to be found in J. Pintz (1981a), who chooses

$$k = \mu \lambda^{-2}, \qquad \lambda = \sqrt{2}\,(T - 1/2), \qquad \mu \in \left[\log Y, \log Y + \tfrac{1}{2} \right]$$

and estimates both integrals in (12.160) to show that if all zeros of $\zeta(s)$ with $|\gamma| \leqslant T$ lie on $\sigma = \tfrac{1}{2}$ and are simple, then $M(x)$ changes sign in every interval

$$[Y^{1-1/(T-2)}, Y^{1+1/(T-2)}]$$

if $Y > c(T)$, an explicitly calculable constant. Since by the work of van de Lune and te Riele (1983) we may take $T > 3 \times 10^8$, it follows that $M(x)$ changes sign in every interval of the form $[Y^{1-2 \times 10^{-8}}, Y]$ if $Y > C_1$.

In a series of papers, J. Pintz (1977, 1978, 1980a, b, c, d, 1982b, c, 1984) obtains a number of interesting results concerning oscillations of the functions

$$\Delta_1(x) = \pi(x) - \mathrm{li}\,x, \qquad \Delta_2(x) = \Pi(x) - \mathrm{li}\,x = \sum_{p^m \leqslant x} 1/m - \mathrm{li}\,x,$$

$$\Delta_3(x) = \theta(x) - x = \sum_{p \leqslant x} \log p - x, \qquad \Delta_4(x) = \psi(x) - x,$$

which may be considered as error terms in various forms of the prime number theorem, and his papers contain many references to earlier work on the same subject.

Besides using the exponential averaging technique [as in (12.160)], which is also used on several occasions in this text, Pintz makes ample use in his work of P. Turán's well-known power sum method [see P. Turán (1984), where comparative number theory and oscillation theorems are also discussed] to obtain nontrivial lower bounds for exponential sums which arise after the application of exponential averaging techniques. Pintz (1980b) obtained a result stronger than Theorem 12.5, namely, he proved for $Y > Y_0$

$$\int_{Y\exp\left(-5\sqrt{\log Y}\right)}^Y |\psi(x) - x| \, dx > 10^{-4} Y^{3/2},$$

while he announced that $\pi(x) - \mathrm{li}\,x$ changes sign in every interval $[Y\exp(-500(\log\log Y)^3), Y]$ for $Y > Y_1$.

Theorem 12.6 was proved by the author (1977a) and Theorem 12.7 by A. Walfisz (1963). Lemma 12.2 is standard, and proofs may be found in K. Prachar (1967) and E. C. Titchmarsh (1951). Lemma 12.3 is proved by the method of Theorem 3.11 of Titchmarsh (1951), and (12.50) holds if $1/\zeta(s)$ is replaced by $\zeta'(s)/\zeta(s)$.

Theorem 12.8 was proved by the author (1979a) with (12.68) in mind, but (12.68) was superseded at once by the work of H. Iwaniec and M. Jutila (1979) on primes in short intervals, which is explained in Section 12.6. Y. Motohashi (1976) outlines the proof of

$$\sum_{x < n \leqslant x+h} \mu(n) = o(h), \qquad x^{7/12+\varepsilon} \leqslant h \leqslant x,$$

and remarks without proof that his method leads to

$$p_{n+1} - p_n < p_n^{7/12} (\log p_n)^{25}$$

for sufficiently large n.

Theorem 12.9 is on the lines of author's paper (1979a). However, its corollary (12.77) can be improved by a combination of sieve and analytic techniques, much in the way that (12.68) [a corollary of (12.67)] was improved in Section 12.6. Thus G. Harman (1982) proved that for almost all n the interval $(n, n + n^{1/10+\varepsilon}]$ contains a prime, but he did not improve on Theorem 12.9, that is, his method does not show that (12.63) holds for almost all intervals of the type mentioned above.

The proof of Theorem 12.10 that is given in the text is much simpler than the first proof of this result, due to H. Cramér (1937). Assuming the Riemann hypothesis, A. Selberg (1943) proved a stronger result than Theorem 12.10, namely,

$$\sum_{p_n \leqslant x,\, p_{n+1}-p_n \geqslant H} (p_{n+1} - p_n) \ll xH^{-1}\log^2 x \qquad (H \geqslant 1)$$

and he also proved under the Riemann hypothesis

$$\sum_{p_n \leqslant x} (p_{n+1} - p_n)^2 \ll x\log^3 x.$$

As already mentioned in the Notes of Chapter 1, for recent conditional results on $p_{n+1} - p_n$ see the articles of D. R. Heath-Brown (1982), J. Mueller (1981), and the thesis of D. A. Goldston (1981).

The history of the constants θ and θ' such that (12.64) holds is as follows:

$\theta = 1 - \frac{1}{33,000} + \varepsilon$	G. Hoheisel (1930)
$\theta = 1 - \frac{1}{250} + \varepsilon$	H. Heilbronn (1933)
$\theta = \frac{5}{8} + \varepsilon$	A. E. Ingham (1937)
$\theta = \frac{3}{5} + \varepsilon$	H. L. Montgomery (1969)
$\theta = \frac{7}{12} + \varepsilon$	M. N. Huxley (1972a)
$\theta = \frac{7}{12}, \theta' = 22$	A. Ivić (1979a)
$\theta = \frac{13}{23}, \theta' = 0$	H. Iwaniec and M. Jutila (1979)
$\theta = \frac{11}{20} + \varepsilon$	D. R. Heath-Brown and H. Iwaniec (1979)
$\theta = \frac{17}{31} + \varepsilon$	J. Pintz (1981b, in press)
$\theta = \frac{23}{42}, \theta' = 0$	H. Iwaniec and J. Pintz (1984)

For a general result on the linear sieve, which implies Lemma 12.5, see H. Iwaniec (1980b).

Results of section 12.6 deal with upper bounds for $p_{n+1} - p_n$. Concerning lower bounds it should be remarked that P. Erdős (1935) found an ingenious method which gives that

$$(12.161) \qquad p_{n+1} - p_n > C \log p_n \frac{\log_2 p_n \log_4 p_n}{(\log_3 p_n)^2}$$

holds for infinitely many n and some absolute constant $C > 0$, where $\log_k x$ denotes $\log(\log_{k-1} x)$. Erdős's construction for large differences between consecutive primes is still the best one, although the method was refined by R. A. Rankin (1962/1963) to yield (12.161) with $C = e^\gamma - \varepsilon$. (For many unconventional problems and the results of P. Erdős on primes, see his 1977 and 1981 articles.) Erdős (1940) contains the first unconditional proof of the existence of an absolute constant $A > 1$ such that

$$\liminf_{n \to \infty} (p_{n+1} - p_n)/\log p_n \leqslant A.$$

Various explicit values of A were calculated by different authors, and the best one seems to be $A \leqslant 0.4425 \ldots$, due to M. N. Huxley (1973a, 1977); a forthcoming paper of his has $A \leqslant 0.4394$.

Theorem 12.12 is from an unpublished manuscript of D. Wolke. A general result of this type, with x^8 replaced by a suitable log power, was proved by the author (1980a).

For the latest published result that there is a P_2 in $(x, x + h]$, $h = x^{0.45}$, see H. Iwaniec and M. Laborde (1981). Their proof is based on a subtle combination of sieve and analytic methods, which include the theory of exponent pairs and the circle method. A forthcoming article by H. Halberstam and H.-E. Richert will improve the exponent 0.45 to 0.4476.

Theorem 12.13 is from Y. Motohashi (1979), but this is not the strongest known result of this type. D. Wolke (1979) proved that almost all intervals $(n, n + \log^C n]$ contain a P_2 if $C = 5 \times 10^6$, while G. Harman (1981) showed that one can take $C = 7 + \varepsilon$.

Theorem 12.14 is based on D. R. Heath-Brown (1979b, 1979c, 1979d), and represents the best unconditional results of its kind. Heath-Brown's method is partly based on the method of D. Wolke (1975), which contains a proof of

$$\sum_{p_n \leqslant x, \, p_{n+1} - p_n > p_n^{1/2}} (p_{n+1} - p_n) \ll x^b$$

with $b = \frac{29}{30}$. Wolke's paper sparked interest for the estimation of the sum above, and the following values of b were obtained:

$b = \frac{29}{30}$	D. Wolke (1975)
$b = \frac{85}{98} + \varepsilon = 0.86734 \ldots + \varepsilon$	R. J. Cook (1979)
$b = 0.8242 \ldots$	M. N. Huxley (1980)
$b = \frac{215}{266} + \varepsilon = 0.80827 \ldots + \varepsilon$	A. Ivić (1979a)
$b = \frac{29}{36} + \varepsilon = 0.805 + \varepsilon$	R. J. Cook (1981)
$b = \frac{3}{4} + \varepsilon$	D. R. Heath-Brown (1979d).

Cook (1981) contains general estimates for the sum $S(x; c)$ in (12.130), but his results can be improved by the methods of Heath-Brown used for the proof of Theorem 12.14.

If we call $N(\sigma, T) \ll T^{2-2\sigma+\varepsilon}$ the "ordinary" density hypothesis, then in view of the trivial $N^*(\sigma, T) \ll N^3(\sigma, T)\log T$ we may call

$$N^*(\sigma, T) \ll T^{6-6\sigma+\varepsilon}$$

the density hypothesis for $N^*(\sigma, T)$. The most important feature of Heath-Brown's method is Lemma 12.7, which gives a sharp estimate for $N^*(\sigma, T)$. In fact the density hypothesis for $N^*(\sigma, T)$ is seen to hold for $\sigma = \frac{1}{2}, \frac{2}{3}$ and $\sigma \geqslant \frac{3}{4}$, and it would be interesting to extend this range further.

CHAPTER THIRTEEN

THE DIRICHLET DIVISOR PROBLEM

13.1 INTRODUCTION

In this chapter we shall investigate various problems involving $\Delta_k(x)$, the error term in the asymptotic formula for $\sum_{n \leqslant x} d_k(n)$, where for $k \geqslant 2$ fixed $d_k(n)$ is the number of ways n can be written as a product of k factors. The estimation of $\Delta_k(x)$ is known as the Dirichlet divisor problem in honor of P. G. L. Dirichlet, who showed in the middle of the 19th century by elementary arguments that $\Delta_2(x) \ll x^{1/2}$. An extension of his method gives $\Delta_k(x) \ll x^{(k-1)/k} \log^{k-2}x$ in the general case, but our objective is to employ the theory of the zeta-function to obtain much sharper results. In fact zeta-function theory may be successfully utilized in many other problems which may in general be called divisor problems, and some of these will be treated in Chapter 14.

The connection between $d_k(n)$ and $\zeta(s)$ is a natural one, since for $\operatorname{Re} s > 1$

$$\zeta^k(s) = \sum_{n=1}^{\infty} d_k(n)n^{-s},$$

so that one expects that the properties of $\Delta_k(x)$ and $\zeta^k(s)$ are closely connected, and in fact from the inversion formula (A.8) one has

(13.1) $$\sum_{n \leqslant x} {}' d_k(n) = (2\pi i)^{-1} \int_{2-i\infty}^{2+i\infty} \zeta^k(s)x^s s^{-1}\, ds.$$

Moving the line of integration in (13.1) to some $\frac{1}{2} < c < 1$ (but sufficiently close to 1) and noting that the integrand in (13.1) has only a pole of order k at

$s = 1$, one has by the residue theorem

(13.2)

$$\sum_{n \leqslant x}{}' d_k(n) = x P_{k-1}(\log x) + (2\pi i)^{-1} \int_{c-i\infty}^{c+i\infty} \zeta^k(s) x^s s^{-1}\, ds \qquad (\tfrac{1}{2} < c < 1),$$

where $P_{k-1}(t)$ is a polynomial of degree $k - 1$ in t. We write

(13.3)
$$\Delta_k(x) = \sum_{n \leqslant x} d_k(n) - x P_{k-1}(\log x),$$

noting that the coefficients of P_{k-1} may be evaluated by using

(13.4)
$$P_{k-1}(\log x) = \operatorname*{Res}_{s=1} x^{s-1} \zeta^k(s) s^{-1}.$$

In fact from the Laurent expansion [see (1.11) and (1.12)]

(13.5)
$$\zeta(s) = (s-1)^{-1} + \gamma + \sum_{k=1}^{\infty} \gamma_k (s-1)^k$$

and (13.4) one may calculate explicitly the coefficients of P_{k-1} as functions of the γ_k's ($\gamma = \gamma_0$ is Euler's constant). For instance we have

(13.6) $P_1(t) = t + (2\gamma - 1)$,

(13.7) $P_2(t) = \tfrac{1}{2}t^2 + (3\gamma - 1)t + (3\gamma^2 - 3\gamma + 3\gamma_1 + 1)$,

(13.8) $P_3(t) = \tfrac{1}{6}t^3 + (2\gamma - \tfrac{1}{2})t^2 + (6\gamma^2 - 4\gamma + 4\gamma_1 + 1)t$

$$+ \{ -1 + 4(\gamma - \gamma_1 + \gamma_2) - 6\gamma^2 + 4\gamma^3 + 12\gamma\gamma_1 \},$$

and there exist general formulas for the coefficients of P_{k-1}.

Besides investigating order results and mean values of $\Delta_k(x)$ we shall also consider some other related problems involving $\Delta_k(x)$. Following standard notation, we define α_k and β_k as the infima of numbers a_k and b_k, respectively, for which

(13.9)
$$\Delta_k(x) \ll x^{a_k + \varepsilon}, \qquad \int_1^x \Delta_k^2(y)\, dy \ll x^{1 + 2b_k + \varepsilon},$$

so that many problems involving $\Delta_k(x)$ may be formulated in terms of α_k or β_k. The chapter ends with a discussion of the circle problem, whose close connection with the divisor problem for $k = 2$ is exhibited.

13.2 ESTIMATES FOR $\Delta_2(x)$ AND $\Delta_3(x)$

The most convenient way of obtaining estimates for $\Delta_2(x)$ and $\Delta_3(x)$ seems to be the use of

$$(13.10) \quad \Delta_k(x) \ll x^{(k-1)/2k} \left| \sum_{n \leqslant N} d_k(n) n^{-(k+1)/2k} e\left(k(nx)^{1/k} \right) \right|$$

$$+ x^{(k-1+\varepsilon)/k} N^{-1/k} + x^\varepsilon,$$

as given by (3.23). By writing

$$d_k(n) = \sum_{m_1 m_2 \cdots m_k = n} 1$$

the sum over $n \leqslant N$ in (13.10) is transformed then into a multiple exponential sum, and the best results hitherto seem to be those obtainable by the methods of G. Kolesnik. Following more closely the proof of (3.17) it is seen that one can obtain

(13.11)

$$\Delta_2(x) \ll x^{1/4} \left| \sum_{mn \leqslant N} (mn)^{-3/4} e\left(2(mn)^{1/2} \right) \right| + x^{1/2} N^{-1/2} \log^2 x + x^\varepsilon;$$

hence an application of Lemma 7.3 gives, similarly as in the proof of Theorem 7.3,

(13.12)

$$\Delta_2(x) \ll x^\varepsilon + \log^2 x \left\{ \max_{M \leqslant N} \left(x^{3/16} M^{173/152 - 3/4} \right. \right.$$

$$\left. \left. + x^{5/16} M^{119/152 - 3/4} \right) + x^{1/2} N^{-1/2} \right\}$$

$$\ll x^\varepsilon + x^{5/16} N^{5/152} \log^2 x + x^{3/16} N^{59/152} \log^2 x + x^{1/2} N^{-1/2} \log^2 x$$

$$\ll x^{35/108} \log^2 x,$$

if we choose $N = x^{19/54}$. Therefore we obtain

THEOREM 13.1.

$$(13.13) \qquad\qquad \Delta_2(x) \ll x^{35/108} \log^2 x.$$

Here the exponent $35/108$ is exactly twice the exponent for the order of $\zeta(\tfrac{1}{2} + iT)$ in (7.63). This is no coincidence, since the exponential sums to which both problems reduce are of a very similar nature. More light on the intrinsic connection between $\Delta_2(x) = \Delta(x)$ and $\zeta(\frac{1}{2} + iT)$ will be shed in Chapter 15, where we shall discuss Atkinson's formula for the mean square of $|\zeta(\frac{1}{2} + it)|$.

The estimation of $\Delta_3(x)$ is naturally more complicated than the estimation of $\Delta_2(x)$, and is carried out via (13.10) with $k = 3$. The best result yet is

$$(13.14) \qquad \Delta_3(x) \ll x^{43/96+\varepsilon}.$$

This is due to G. Kolesnik (1981a), whose proof is long and complicated and will not be presented here.

13.3 ESTIMATES OF $\Delta_k(x)$ BY POWER MOMENTS OF THE ZETA-FUNCTION

We start from the inversion formula (A.10) which gives

$$(13.15) \qquad \sum_{n \leqslant x} d_k(n) = (2\pi i)^{-1} \int_{1+\varepsilon-iT}^{1+\varepsilon+iT} \zeta^k(w) x^w w^{-1} \, dw$$

$$+ O(x^{1+\varepsilon}T^{-1}) \qquad (T \leqslant x).$$

For σ fixed satisfying $\frac{1}{2} \leqslant \sigma < 1$ we deform the path of integration in (13.15) to obtain by the residue theorem

$$(13.16) \qquad \Delta_k(x) = \sum_{n \leqslant x} d_k(n) - \operatorname*{Res}_{s=1} \zeta^k(s) x^s s^{-1}$$

$$= I_1 + I_2 + I_3 + O(x^{1+\varepsilon}T^{-1}),$$

say, where

(13.17)

$$I_1 = (2\pi i)^{-1} \int_{\sigma-iT}^{\sigma+iT} \zeta^k(w) x^w w^{-1} \, dw \ll x^\sigma + x^\sigma \int_1^T |\zeta(\sigma + iv)|^k v^{-1} \, dv$$

and

$$(13.18) \quad I_2 + I_3 \ll \int_\sigma^{1+\varepsilon} x^\theta |\zeta(\theta + iT)|^k T^{-1} \, d\theta \ll \max_{\sigma \leqslant \theta \leqslant 1+\varepsilon} x^\theta T^{k\mu(\theta)-1+\varepsilon},$$

where $\mu(\sigma)$ is defined by (1.65). From (13.17) it is immediately seen that estimates for power moments of the zeta-function lead to estimates of $\Delta_k(x)$. Our result will be the following

THEOREM 13.2. Let α_k be the infimum of numbers a_k such that $\Delta_k(x) \ll x^{a_k+\varepsilon}$ for any $\varepsilon > 0$. Then

$$\alpha_k \leqslant (3k-4)/4k \qquad\qquad (4 \leqslant k \leqslant 8),$$

$$\alpha_9 \leqslant \tfrac{35}{54}, \; \alpha_{10} \leqslant \tfrac{41}{60}, \; \alpha_{11} \leqslant \tfrac{7}{10},$$

$$\alpha_k \leqslant (k-2)/(k+2) \qquad\quad (12 \leqslant k \leqslant 25),$$

$$\alpha_k \leqslant (k-1)/(k+4) \qquad\quad (26 \leqslant k \leqslant 50),$$

$$\alpha_k \leqslant (31k-98)/32k \qquad\quad (51 \leqslant k \leqslant 57),$$

$$\alpha_k \leqslant (7k-34)/7k \qquad\qquad (k \geqslant 58).$$

Proof of Theorem 13.2. The proof is based on estimates for $m(\sigma)$, as furnished by Theorem 8.4. For a fixed k we choose σ in such a way that $m(\sigma) = k$, where for $m(\sigma)$ we take the estimates given by Theorem 8.4. For $\mu(\sigma)$ in (13.18) we use the bound $\mu(\sigma) \leqslant c(\sigma)$, where $c(\sigma)$ is the piecewise linear function given by (8.73). Note that by Theorem 8.4 and (8.73) we have $m(\sigma) \leqslant 1/c(\sigma)$; hence with $T = x^{1-\sigma}$ we obtain

$$I_2 + I_3 \ll x^{1+\varepsilon}T^{-1} + x^\sigma T^{kc(\sigma)-1+\varepsilon} \ll x^{\sigma+\varepsilon},$$

and therefore (13.17) gives

$$\Delta_k(x) \ll x^{\sigma+\varepsilon}.$$

In this fashion estimates for $9 \leqslant k \leqslant 11$ given by Theorem 13.2 follow at once, and for $4 \leqslant k \leqslant 8$ we use $m(\sigma) \geqslant 4/(3-4\sigma)$ ($\tfrac{1}{2} \leqslant \sigma \leqslant \tfrac{5}{8}$), so that $k = 4/(3-4\sigma)$ gives $\sigma = (3k-4)/4k$. For $4 \leqslant k \leqslant 8$ this value of σ satisfies $\tfrac{1}{2} \leqslant \sigma \leqslant \tfrac{5}{8}$, and $\alpha_k \leqslant (3k-4)/4k$ follows for $4 \leqslant k \leqslant 8$. Next, we take $\sigma = \tfrac{5}{7}$ in (13.17) and (13.18). With $m(\tfrac{5}{7}) \geqslant 12$, $c(\tfrac{5}{7}) = \tfrac{1}{14}$ we have

$$I_1 \ll x^{5/7} + x^{5/7} \int_1^T \left|\zeta\left(\tfrac{5}{7} + it\right)\right|^{12} t^{-1} \left|\zeta\left(\tfrac{5}{7} + it\right)\right|^{k-12} dt$$

$$\ll x^{5/7}T^{(k-12+\varepsilon)/14}$$

for $k \geqslant 12$, and therefore

$$\Delta_k(x) \ll x^{1+\varepsilon}T^{-1} + x^{5/7}T^{(k-12+\varepsilon)/14} \ll x^{(k-2)/(k+2)+\varepsilon}$$

for $12 \leqslant k \leqslant 25$ if $T = x^{4/(k+2)}$.

A similar argument gives $\alpha_k \leqslant (k-1)/(k+4)$ for $k \geqslant 26$ by using $m(\tfrac{5}{6}) \geqslant 26$, $c(\tfrac{5}{6}) = \tfrac{1}{30}$. Also by Theorem 8.4 we have $m(\sigma) \geqslant 98/(31-32\sigma) = k$ for $\tfrac{13}{15} \leqslant \sigma = (31k-98)/32k \leqslant 0.91591\ldots$, which is satisfied for $30 \leqslant k \leqslant 57$. From the last estimate in (8.80) we have $m(\sigma) \geqslant 34/(7-7\sigma) = k$ for $\sigma =$

$(7k - 34)/7k \leqslant 0.91591\ldots$ for $k \geqslant 57$. On comparing then $(k - 1)/(k + 4)$ with $(31k - 98)/32k$ we obtain the full assertion of Theorem 13.2.

For each particular $k \geqslant 13$ the bounds of Theorem 13.2 can be slightly improved by a more careful choice of exponent pairs in the proof of Theorem 8.4, and taking more care one could also derive bounds of the type $\Delta_k(x) \ll x^{\alpha_k} \log^{d_k} x$ for some $d_k \geqslant 0$. The bounds of Theorem 13.2 are the sharpest one known, except when k is large, when a better bound may be obtained by using the order result (6.6).

13.4   ESTIMATES OF $\Delta_k(x)$ WHEN k IS VERY LARGE

We proceed now to estimate $\Delta_k(x)$ when k is very large, using

$$(13.19) \quad \zeta(\sigma + it) \ll t^{D(1-\sigma)^{3/2}}(\log t)^{2/3} \qquad (D = 122,\ t \geqslant t_0,\ \sigma_0 \leqslant \sigma \leqslant 1),$$

which is the estimate given by Theorem 6.3. In (13.17) and (13.18) we choose $\sigma = 1 - 2Ak^{-2/3}$, $T = x^{Ak^{-2/3}}$, where $A > 0$ is an absolute constant which will be determined in a moment. Using (13.19) we obtain

$$I_1 \ll x^{1-2Ak^{-2/3}} T^{Dk(1-\sigma)^{3/2}} \log^{(2k+3)/3} x$$

$$= x^{1-(2A-\sqrt{8}\,DA^{5/2})k^{-2/3}} \log^{(2k+3)/3} x,$$

$$I_2 + I_3 \ll x^{1+\varepsilon} T^{-1} \log^{2k/3} T + \max_{\sigma \leqslant \theta \leqslant 1} x^{\theta} T^{Dk(1-\theta)^{3/2}-1} \log^{2k/3} T$$

$$\ll \left(x^{1-(A-\varepsilon)k^{-2/3}} + x^{1-(2A-\sqrt{8}\,DA^{5/2})k^{-2/3}} \right) \log^{(2k+3)/3} x,$$

since the function

$$f(\theta) = \theta + ADk^{1/3}(1 - \theta)^{3/2} - Ak^{-2/3}$$

attains its maximal value in the interval $[\sigma, 1]$ at one of the endpoints of the interval. Now we choose A to satisfy

$$A = 2A - \sqrt{8}\,DA^{5/2},$$

which with $D = 122$ gives

$$A = \left(\frac{1}{122\sqrt{8}} \right)^{2/3} = 0.02032\ldots,$$

and finally from (13.16) we obtain

THEOREM 13.3. For $k \geqslant 100$ fixed we have

(13.20) $$\Delta_k(x) \ll x^{1-k^{-2/3}/50} \log^{(2k+3)/3} x.$$

Here the relevant range for k is $k \geqslant 15 \times 10^6$, when (13.20) improves $\alpha_k \leqslant (7k - 34)/7k$ $(k \geqslant 58)$, which is the last estimate of Theorem 13.2. Thus Theorem 13.3 is significant only when k is "very large," and with a little more care one could obtain a result similar to (13.20) which is uniform in k.

13.5 ESTIMATES OF β_k

In Section 13.1 we defined β_k as the infimum of b_k for which

$$\int_0^x \Delta_k^2(y)\, dy \ll x^{1+2b_k+\varepsilon}$$

holds for every $\varepsilon > 0$, so that β_k may be thought of as the exponent of the average order of $|\Delta_k(y)|$. The classical elementary results concerning the estimation of β_k are embodied in the following two lemmas, while some specific estimates are given by Theorem 13.4.

LEMMA 13.1. Let γ_k be the infimum of $\sigma > 0$ for which

$$\int_{-\infty}^{\infty} |\zeta(\sigma + it)|^{2k} |\sigma + it|^{-2}\, dt \ll 1.$$

Then $\beta_k = \gamma_k$ and for $\sigma > \beta_k$

(13.21) $$(2\pi)^{-1} \int_{-\infty}^{\infty} |\zeta(\sigma + it)|^{2k} |\sigma + it|^{-2}\, dt = \int_0^{\infty} \Delta_k^2(x) x^{-2\sigma-1}\, dx.$$

Proof of Lemma 13.1. From (13.2) we have

(13.22) $$\Delta_k(x) = (2\pi i)^{-1} \lim_{T \to \infty} \int_{c-iT}^{c+iT} \zeta^k(s) x^s s^{-1}\, ds$$

for some $c < 1$ and close to 1. Since $\zeta^k(s)s^{-1} \to 0$ uniformly in the strip as $t \to \pm\infty$, it is seen on integrating over the rectangle $c' \pm iT,\ c \pm iT,\ \gamma_k < c'$ $< c < 1$, that (13.22) holds for any $c > \gamma_k$. Replacing in (13.22) x by $1/x$, taking $c > \gamma_k$, and using Parseval's identity (A.5) we have

(13.23) $$(2\pi)^{-1} \int_{-\infty}^{\infty} |\zeta(c + it)|^{2k} |c + it|^{-2}\, dt = \int_0^{\infty} \Delta_k^2(1/x) x^{2c-1}\, dx$$

$$= \int_0^{\infty} \Delta_k^2(x) x^{-2c-1}\, dx.$$

This gives, for $\gamma_k < c < 1$,

$$\int_N^{2N} \Delta_k^2(x) x^{-2c-1} \, dx \ll 1, \qquad \int_N^{2N} \Delta_k^2(y) \, dy \ll N^{2c+1},$$

and therefore

$$\int_0^x \Delta_k^2(y) \, dy = \sum_{N=x2^{-j}, \, j \geqslant 1} \int_N^{2N} \Delta_k^2(y) \, dy \ll x^{2c+1};$$

hence $\beta_k \leqslant c, \beta_k \leqslant \gamma_k$.

The other inequality, namely $\beta_k \geqslant \gamma_k$, may be obtained by observing that from (13.22) and (A.1) one has

$$(13.24) \qquad \zeta^k(s) s^{-1} = \int_0^\infty \Delta_k(1/x) x^{s-1} \, dx = \int_0^\infty \Delta_k(x) x^{-s-1} \, dx.$$

The integral in (13.24) is absolutely and uniformly convergent for $\beta_k < \sigma < 1$, since by the Cauchy–Schwarz inequality

$$\int_N^{2N} |\Delta_k(x)| x^{-\sigma-1} \, dx \leqslant \left(\int_N^{2N} \Delta_k^2(x) \, dx \right)^{1/2} \left(\int_N^{2N} x^{-2\sigma-2} \, dx \right)^{1/2}$$

$$\ll N^{\beta_k - \sigma + \varepsilon},$$

and by adding integrals over various intervals $[N, 2N]$ it is seen that the right-hand side of (13.24) is regular for $\beta_k < \sigma < 1$, so that (13.24) holds by analytic continuation in the strip $\beta_k < \sigma < 1$. By the same argument the right-hand side of (13.23) is bounded for $\beta_k < \sigma < 1$, hence (13.23) holds in the same strip, giving $\beta_k \geqslant \gamma_k$, which combined with $\beta_k \leqslant \gamma_k$ yields finally $\beta_k = \gamma_k$.

LEMMA 13.2. For $k = 2, 3, \ldots$

$$\alpha_k \geqslant \beta_k \geqslant (k - 1)/(2k).$$

Proof of Lemma 13.2. The inequality $\alpha_k \geqslant \beta_k$ is obvious, and for the other inequality we have, for $0 < \sigma < \frac{1}{2}$,

$$\int_{-\infty}^\infty |\zeta(\sigma + it)|^{2k} |\sigma + it|^{-2} \, dt \geqslant \int_{T/2}^T |\zeta(\sigma + it)|^{2k} |\sigma + it|^{-2} \, dt$$

$$\gg T^{-2} \int_{T/2}^T |\zeta(\sigma + it)|^{2k} \, dt$$

$$\gg T^{k(1-2\sigma)-2} \int_{T/2}^T |\zeta(1 - \sigma - it)|^{2k} \, dt$$

$$\gg T^{k(1-2\sigma)-1},$$

where we used the functional equation (1.23), the asymptotic formula (1.25), and the lower bound of Theorem 9.6. For $\sigma < (k - 1)/2k$ the last expression above remains unbounded when $T \to \infty$, giving $\gamma_k \geq (k - 1)/2k$, and the result follows from $\beta_k = \gamma_k$.

Lemma 13.2 provides an omega result in the Dirichlet divisor problem, namely it shows that α_k and β_k cannot be smaller than $(k - 1)/2k$. A classical conjecture is that $\alpha_k = \beta_k = (k - 1)/2k$ holds for all $k \geq 2$. This conjecture is very strong, since $\beta_k = (k - 1)/2k$ $(k \geq 2)$ is an equivalent of the Lindelöf hypothesis, as mentioned in Section 1.9 of Chapter 1. To see this equivalence, note that $\beta_k = (k - 1)/2k$ $(k \geq 2)$ implies, by Lemma 13.1, that $\gamma_k = (k - 1)/2k \leq \frac{1}{2}$ for all $k \geq 2$; hence by the definition of γ_k we have

$$\int_{T/2}^{T} \left| \zeta(\tfrac{1}{2} + it) \right|^{2k} dt \ll T^2 \int_{-\infty}^{\infty} \left| \zeta(\tfrac{1}{2} + it) \right|^{2k} \left| \tfrac{1}{2} + it \right|^{-2} dt \ll T^2,$$

and $\zeta(\tfrac{1}{2} + iT) \ll T^\varepsilon$ then follows from Lemma 7.1 on taking k sufficiently large. In the other direction, the Lindelöf hypothesis clearly implies that $m((k + 1)/2k) \geq 2k$ for $k \geq 2$; hence $\beta_k = (k - 1)/2k$ for $k \geq 2$ follows from

LEMMA 13.3. For each integer $k \geq 2$ a necessary and sufficient condition that $\beta_k = (k - 1)/2k$ is that $m((k + 1)/2k) \geq 2k$, where $m(\sigma)$ is defined by (8.2).

Proof of Lemma 13.3. Suppose first that $m((k + 1)/2k) \geq 2k$. Then for $\sigma < (k - 1)/2k$ we have by the functional equation (1.23) and (1.25)

$$\int_{1}^{T} \left| \zeta(\sigma + it) \right|^{2k} dt \ll T^{k(1-2\sigma)} \int_{1}^{T} \left| \zeta(1 - \sigma - it) \right|^{2k} dt$$

$$\ll T^{k(1-2\sigma)+1+\varepsilon}.$$

Therefore for $(k - 1 - \varepsilon)/2k < \sigma < (k + 1 + \varepsilon)/2k$ by convexity of mean values (Lemma 8.3) we have

$$\int_{1}^{T} \left| \zeta(\sigma + it) \right|^{2k} dt \ll T^{1+\varepsilon+(1/2+1/2k-\sigma)k},$$

and the exponent of T is here less than 2 for $\sigma > (k - 1 + \varepsilon)/2k$, giving

(13.25) $$\int_{T/2}^{T} \left| \zeta(\sigma + it) \right|^{2k} \left| \sigma + it \right|^{-2} dt \ll T^{-\delta}$$

for some $\delta = \delta(\varepsilon) > 0$. Replacing in (13.25) T by $T2^{-j}$ and summing over $j \geq 1$ it follows that $\gamma_k \leq (k - 1)/2k$; therefore by Lemma 13.1 we have also $\beta_k \leq (k - 1)/2k$, and Lemma 13.2 finally gives $\beta_k = (k - 1)/2k$.

In the other direction, if $\beta_k = (k-1)/2k$, then by (13.21) we have

$$\int_{T/2}^{T} |\zeta(\sigma + it)|^{2k} dt \ll T^{2+\varepsilon}$$

for $\sigma > (k-1)/2k$, and using again the functional equation for $\zeta(s)$ and convexity of mean values we obtain $m((k+1)/2k) \geqslant 2k$ by following the foregoing argument.

Finally, we prove some explicit estimates for β_k, contained in

THEOREM 13.4. $\beta_k = (k-1)/2k$ for $k = 2, 3, 4,$ and $\beta_5 \leqslant \frac{119}{260} = 0.45769\ldots, \beta_6 \leqslant \frac{1}{2}, \beta_7 \leqslant \frac{39}{70} = 0.55714\ldots.$

Proof of Theorem 13.4. By Theorem 8.4 we have $m(\sigma) \geqslant 4/(3-4\sigma)$ for $\frac{1}{2} \leqslant \sigma \leqslant \frac{5}{8}$, hence $m(\frac{5}{8}) \geqslant 8$. By Lemma 13.3 we obtain at once that $\beta_k = (k-1)/2k$ for $k = 2, 3, 4$, which in view of Lemma 13.2 shows that this is best possible, while for other values of k the estimate $\beta_k = (k-1)/2k$ seems to be beyond reach at present.

Consider now the case $k = 5$. By Lemma 13.1 it will suffice to show

$$\int_{T}^{2T} |\zeta(\sigma + it)|^{10} dt \ll T^{2-\delta}$$

for $\sigma > \frac{119}{260}$ and any fixed $\delta > 0$. From the estimate $m(\frac{41}{60}) \geqslant 10$, furnished by Theorem 8.4, the functional equation (1.23), and (1.25) we have, for $\frac{19}{60} \leqslant \sigma \leqslant \frac{1}{2}$,

$$\int_{T}^{2T} |\zeta(\sigma + it)|^{10} dt \ll T^{(207 - 260\sigma)/44 + \varepsilon},$$

where we also used convexity and the estimate $M(10) \leqslant \frac{7}{4}$ of Theorem 8.3. Since $(207 - 260\sigma)/44 < 2$ for $\sigma > \frac{119}{260}$, we obtain $\beta_5 \leqslant \frac{119}{260}$ as asserted. Similarly, from $M(12) \leqslant 2$ it follows at once that $\beta_6 \leqslant \frac{1}{2}$, while for β_7 we use $M(14) \leqslant \frac{62}{27}$ (Theorem 8.3) and $m(\frac{3}{4}) \geqslant 14$ (Theorem 8.4). Thus by convexity

$$\int_{T}^{2T} |\zeta(\sigma + it)|^{14} dt \ll T^{(132 - 140\sigma)/27 + \varepsilon}$$

for $\frac{1}{2} \leqslant \sigma \leqslant \frac{3}{4}$, and $(132 - 140\sigma)/27 < 2$ for $\sigma > \frac{39}{70}$, proving the last part of the theorem. Other values of β_k for $k \geqslant 8$ may be calculated analogously, but the present form of estimates for $m(\sigma)$ and $M(A)$ would render a general formula for β_k ($k \geqslant 8$) too complicated, and for this reason only estimates for small values of k are explicitly stated here.

13.6   MEAN-SQUARE ESTIMATES OF $\Delta_k(x)$

By definition estimates of β_k are in fact mean-square estimates of $\Delta_k(x)$. However, owing to the importance of the integrals in question it seems appropriate to investigate them more closely. In particular, it would be highly desirable to obtain asymptotic formulas for $\int_1^T \Delta_k^2(x)\,dx$, and we begin the discussion of this problem by proving

THEOREM 13.5.

$$(13.26) \quad \int_1^T \Delta_2^2(x)\,dx = (6\pi^2)^{-1} \sum_{n=1}^{\infty} d^2(n)n^{-3/2}T^{3/2} + O(T^{5/4+\varepsilon}).$$

Proof of Theorem 13.5.   The value of the constant $\sum_{n=1}^{\infty} d^2(n)n^{-3/2}$ is equal to $\zeta^4(\tfrac{3}{2})/\zeta(3) = 14.8316\ldots$. It will be sufficient to prove the corresponding formula for the integral over $[T, 2T]$ and then to replace T by $T/2$, $T/2^2$, and so on, and to add up all the results. We start from the truncated Voronoi formula (3.17) where we take $N = T$. Integrating term by term we obtain

$$(13.27) \quad \int_T^{2T} \Delta_2^2(x)\,dx$$

$$= (2\pi^2)^{-1} \int_T^{2T} x^{1/2} \sum_{m,n \leqslant T} d(m)d(n)(mn)^{-3/4}$$

$$\times \cos(4\pi\sqrt{mx} - \pi/4)\cos(4\pi\sqrt{nx} - \pi/4)\,dx$$

$$+ O\left(T^{1/4+\varepsilon} \int_T^{2T} \left| \sum_{n \leqslant T} d(n)n^{-3/4}\cos(4\pi\sqrt{nx} - \pi/4) \right| dx \right)$$

$$+ O(T^{1+\varepsilon}).$$

In the first sum in (13.27) we distinguish the cases $m = n$ and $m \neq n$. The terms with $m = n$ contribute

$$(13.28) \quad (2\pi^2)^{-1} \sum_{n \leqslant T} \int_T^{2T} d^2(n)n^{-3/2}x^{1/2}\cos^2(4\pi\sqrt{nx} - \pi/4)\,dx$$

$$= (4\pi^2)^{-1} \sum_{n \leqslant T} d^2(n)n^{-3/2} \int_T^{2T} x^{1/2}\left(1 + \cos(8\pi\sqrt{nx} - \pi/2)\right)dx$$

$$= (6\pi^2)^{-1}\left((2T)^{3/2} - T^{3/2}\right) \sum_{n=1}^{\infty} d^2(n)n^{-3/2} + O(T\log^3 T).$$

In (13.28) we have used partial summation and (5.24) to obtain

$$\sum_{n \geq T} d^2(n)n^{-3/2} \ll T^{-1/2}\log^3 T,$$

and we also used (2.3) to estimate

$$\sum_{n \leq T} d^2(n)n^{-3/2} \int_T^{2T} x^{1/2}\cos(8\pi\sqrt{nx} - \pi/4)\,dx \ll T \sum_{n \leq T} d^2(n)n^{-2} \ll T.$$

In view of $2 \cos X \cos Y = \cos(X + Y) + \cos(X - Y)$, it is seen that the terms in (13.27) for which $m \neq n$ are a multiple of

$$(13.29) \quad \sum_{m \neq n \leq T} d(m)d(n)(mn)^{-3/4} \int_T^{2T} x^{1/2}\cos(4\pi\sqrt{mx} - 4\pi\sqrt{nx})\,dx$$

$$+ \sum_{m \neq n \leq T} d(m)d(n)(mn)^{-3/4} \int_T^{2T} x^{1/2}\sin(4\pi\sqrt{mx} + 4\pi\sqrt{nx})\,dx$$

$$= S_1 + S_2,$$

say. Estimating the integrals in S_2 by (2.3) we have

$$(13.30) \quad S_2 \ll T \sum_{m < n \leq T} d(m)d(n)(mn)^{-3/4}(m^{1/2} + n^{1/2})^{-1} \ll T \log^3 T.$$

Analogously, we obtain

$$(13.31) \quad S_1 \ll T \sum_{n < m \leq T} d(m)d(n)(mn)^{-3/4}(m^{1/2} - n^{1/2})^{-1}$$

$$= T\left(\sum_{n \leq m/2} + \sum_{n > m/2}\right)$$

$$= T(S_1' + S_1''),$$

say. We have

$$S_1' \ll \sum_{m \leq T} d(m)m^{-1/4} \sum_{n \leq m/2} d(n)n^{-3/4}(m - n)^{-1}$$

$$\ll \sum_{m \leq T} d(m)m^{-5/4}m^{1/4}\log T \ll \log^3 T,$$

$$S_1'' \ll \sum_{m \leq T} d(m)m^{-1} \sum_{m/2 < n < m} d(n)(m - n)^{-1}$$

$$\ll T^\varepsilon \sum_{m \leq T} d(m)m^{-1} \ll T^\varepsilon.$$

Therefore the first sum in (13.27) is by the preceding estimates equal to

$$(6\pi^2)^{-1}\big((2T)^{3/2} - T^{3/2}\big) \sum_{n=1}^{\infty} d^2(n)n^{-3/2} + O(T^{1+\varepsilon}).$$

The first O-term in (13.27) is estimated by the Cauchy–Schwarz inequality as

$$\ll T^{3/4+\varepsilon}\left(\int_T^{2T} |\sum_{n\leqslant T} d(n)n^{-3/4}\cos(4\pi\sqrt{nx} - \pi/4)|^2 \, dx\right)^{1/2} \ll T^{5/4+\varepsilon}$$

when we square out the modulus under the integral sign and treat the terms $m = n$ and $m \neq n$ similarly as before.

This remark ends the proof of Theorem 13.5, but it should be mentioned that the error term given in Theorem 13.5 is by no means the best possible one. Analyzing more carefully the proof, it may be seen that T^ε in the error term in (13.27) may be replaced by a suitable log power, but this would still be much weaker than the following result of K.-C. Tong (1956):

$$(13.32) \quad \int_1^T \Delta_2^2(x)\, dx = (6\pi^2)^{-1} \sum_{n=1}^{\infty} d^2(n)n^{-3/2}T^{3/2} + O(T \log^5 T).$$

The proof of this formula is beyond the scope of the method used for Theorem 13.5, and requires subtle averaging techniques involving certain exponential integrals. It also seems natural to ask what is the best possible O-result that may be obtained in (13.32). In this direction we shall prove the following

THEOREM 13.6. The asymptotic formula

$$(13.33) \quad \int_1^T \Delta_2^2(x)\, dx = (6\pi^2)^{-1} \sum_{n=1}^{\infty} d^2(n)n^{-3/2}T^{3/2} + O(T^{3/4-\delta})$$

cannot hold for any $\delta > 0$.

Proof of Theorem 13.6. From Theorem 13.5 it follows that there exist arbitrarily large x such that

$$(13.34) \qquad\qquad |\Delta_2(x)| = |\Delta(x)| > Cx^{1/4} = G$$

for some suitable $C > 0$, and from the classical work of G. H. Hardy it follows that C may be taken arbitrary. Suppose now that $|t - x| \leqslant Gx^{-\varepsilon}$. Since $d(n) < n^{\varepsilon/3}$ for $n \geqslant n_0(\varepsilon)$, we have in view of (13.3)

(13.35)

$$|\Delta(t) - \Delta(x)| \leqslant \left| \sum_{t\leqslant n\leqslant x} d(n) \right| + |xP_1(\log x) - tP_1(\log t)| \leqslant Gx^{-\varepsilon/2},$$

and then also

(13.36) $|\Delta(t)| \geq \|\Delta(x)| - |\Delta(t) - \Delta(x)\| \geq G - Gx^{-\varepsilon/2} \geq G/2.$

Next, by the Cauchy–Schwarz inequality and (13.36)

(13.37) $G^2 x^{-\varepsilon} \leq \int_{x-Gx^{-\varepsilon}}^{x+Gx^{-\varepsilon}} |\Delta(t)| \, dt \leq (2Gx^{-\varepsilon})^{1/2} \left(\int_{x-Gx^{-\varepsilon}}^{x+Gx^{-\varepsilon}} \Delta^2(t) \, dt \right)^{1/2}$

$$\leq (2Gx^{-\varepsilon})^{1/2} \left(\tfrac{3}{2} D (2Gx^{-\varepsilon}) 2x^{1/2} + R(x + Gx^{-\varepsilon}) \right.$$

$$\left. - R(x - Gx^{-\varepsilon}) \right)^{1/2},$$

where we have set

(13.38) $R(x) = \int_1^x \Delta^2(y) \, dy - Dx^{3/2}, \qquad D = (6\pi^2)^{-1} \sum_{n=1}^{\infty} d^2(n) n^{-3/2}.$

If $R(x) \ll x^{3/4-\delta}$ for some $\delta > 0$, then (13.37) yields

(13.39) $G^2 x^{-\varepsilon} \leq (12 D)^{1/2} G x^{1/4-\varepsilon} + O(G^{1/2} x^{-\varepsilon/2} x^{3/8-\delta/2}).$

Now if in (13.34) we choose $C > (12 D)^{1/2}$ and $\varepsilon = \delta/2$, then for x sufficiently large (13.39) gives a contradiction which proves the theorem, and small improvements may be obtained by using sharper omega estimates for $\Delta(x)$ than (13.34).

It seems also natural to ask whether the analogues of Theorems 13.5 and 13.6 hold for $\Delta_k(x)$ when $k \geq 3$. In considering this problem it ought to be mentioned that K.-C. Tong (1956) proved a general result concerning asymptotic formulas for $\int_1^x \Delta_k^2(y) \, dy$, which seems to be hitherto the sharpest one. If in analogy with (13.38) we define

(13.40)

$$R_k(x) = \int_1^x \Delta_k^2(y) \, dy - ((4k - 2)\pi^2)^{-1} \sum_{n=1}^{\infty} d_k^2(n) n^{-(k+1)/k} x^{(2k-1)/k},$$

then Tong's result may be formulated as

(13.41) $R_k(x) \ll \begin{cases} x \log^5 x & (k = 2), \\ x^{c_k + \varepsilon}, & c_k = 2 - \dfrac{3 - 4\sigma_k}{2k(1 - \sigma_k) - 1} \quad (k \geq 3), \end{cases}$

where σ_k is the infimum of σ such that for every $\varepsilon > 0$

$$(13.42) \qquad \int_1^T |\mathcal{S}(\sigma + it)|^{2k}\, dt \ll T^{1+\varepsilon},$$

and for (13.41) to hold one should have $\sigma_k \leqslant (k+1)/2k$.

Suppose now that $k = 3$. Then by Theorem 8.4 we have $m(\frac{7}{12}) \geqslant 6$, which in the notation of (13.42) implies $\sigma_3 \leqslant \frac{7}{12}$; hence from (13.41) we infer

$$(13.43) \qquad \int_1^x \Delta_3^2(y)\, dy = (10\pi^2)^{-1} \sum_{n=1}^{\infty} d_3^2(n) n^{-4/3} x^{5/3} + O(x^{14/9+\varepsilon}).$$

This formula is substantially stronger than $\beta_3 = \frac{1}{3}$ only, as given by Theorem 13.4, but with $k = 3$ (13.41) at present exhausts itself in the sense that for $k > 4$ the best estimates for σ_k obtainable from Theorem 8.4 are not sufficiently sharp to ensure that the condition $\sigma_k \leqslant (k+1)/2k$ is satisfied. In the case $k = 4$ we have $\sigma_4 \leqslant \frac{5}{8}$, which gives only $R_4(x) \ll x^{7/4+\varepsilon}$, but this is equivalent to $\beta_4 = \frac{3}{8}$, and was already established by Theorem 13.4. A result analogous to Theorem 13.6 may be obtained in the general case by defining θ_k as the infimum of θ such that

$$(13.44) \qquad R_k(x) \ll x^{\theta}, \qquad \theta < 2 - 1/k.$$

Arguing as in the case $k = 2$, it follows that $\theta_k < (3k-3)/2k$ cannot hold, and it might be conjectured that $\theta_k = (3k-3)/2k$ for all $k \geqslant 2$. If true this conjecture is very strong, since it implies at once the classical conjecture $\alpha_k = (k-1)/2k$ by the following

THEOREM 13.7. Let θ_k be the infimum of numbers θ $(< 2 - 1/k)$ such that (13.44) holds. Then $\alpha_k \leqslant \frac{1}{3}\theta_k$.

Proof of Theorem 13.7. The proof is analogous to the proof of Theorem 13.6. Suppose that $R_k(x) \ll x^{\theta_k+\varepsilon/2}$ for some $\varepsilon > 0$, and suppose further that for some sufficiently large x and a suitable $C > 0$

$$(13.45) \qquad |\Delta_k(x)| > Cx^{\theta_k/3+\varepsilon} = G.$$

Then for $|t - x| \leqslant Gx^{-\varepsilon}$ we have as in (13.35) and (13.36) that

$$|\Delta_k(t)| \geqslant G/2,$$

and therefore analogously to (13.37) we obtain

$$(13.46) \qquad G^2 x^{-\varepsilon} \leqslant (2Gx^{-\varepsilon})^{1/2} \left(\int_{x-Gx^{-\varepsilon}}^{x+Gx^{-\varepsilon}} \Delta_k^2(t)\, dt \right)^{1/2}$$

$$\leqslant (2Gx^{-\varepsilon})^{1/2} \left(E_k Gx^{-\varepsilon} x^{(k-1)/k} + O(x^{\theta_k+\varepsilon/2}) \right)^{1/2}$$

for some $E_k > 0$, and (13.46) implies

$$(13.47) \qquad G^2 x^{-\varepsilon} \leqslant (2E_k)^{1/2} G x^{-\varepsilon} x^{(k-1)/2k} + O(G^{1/2} x^{\theta_k/2 - \varepsilon/4}).$$

However, (13.47) is seen to be impossible for $C > (2E_k)^{1/2}$ in (13.45) because $\theta_k \geqslant (3k - 3)/2k$. Therefore (13.45) cannot hold and we obtain $\alpha_k \leqslant \frac{1}{3}\theta_k$, as asserted.

From (13.32) and (13.43) we have $\theta_2 \leqslant 1, \theta_3 \leqslant \frac{14}{9}$, so that from Theorem 13.7 we deduce $\alpha_2 \leqslant \frac{1}{3}, \alpha_3 \leqslant \frac{14}{27}$, which is superseded by (13.13) and (13.14). In fact the estimates for α_2 and α_3 can be deduced directly from estimates of $\int_{T-G}^{T+G} \Delta_k^2(y)\, dy$. Using the method of proof of Theorem 13.5, it is readily seen that for $T^\varepsilon \leqslant G \leqslant T$ we have

$$(13.48) \qquad \int_{T-G}^{T+G} \Delta_2^2(t)\, dt \ll T^\varepsilon (GT^{1/2} + T),$$

$$(13.49) \qquad \int_{T-G}^{T+G} \Delta_3^2(t)\, dt \ll T^\varepsilon (GT^{2/3} + T^{3/2}),$$

which gives $\alpha_2 \leqslant \frac{1}{3}, \alpha_3 \leqslant \frac{1}{2}$, following the method of proof of Theorem 13.7.

13.7 LARGE VALUES AND POWER MOMENTS OF $\Delta_k(x)$

In view of Theorem 13.5 and (13.43) it is seen that in mean square $\Delta_2(x)$ and $\Delta_3(x)$ are of order $x^{1/4}$ and $x^{1/3}$, respectively, which supports the conjecture $\alpha_2 = \frac{1}{4}, \alpha_3 = \frac{1}{3}$. An interesting problem is to generalize mean-square estimates to higher powers, and to consider integrals of the type

$$(13.50) \qquad \int_1^T |\Delta_k(x)|^A\, dx, \qquad A \geqslant 2.$$

As a starting point in these investigations one may take (3.23), namely

$$(13.51) \qquad \Delta_k(x) \ll x^{(k-1)/2k} \left| \sum_{n \leqslant N} d_k(n) n^{-(k+1)/2k} e\big(k(nx)^{1/k}\big) \right|$$

$$+ x^{(k-1+\varepsilon)/k} N^{-1/k} + x^\varepsilon.$$

The sum over $n \leqslant N$ for $k = 2$ is similar in nature to the sum (7.21) occurring in the investigation of large values of the zeta-function on the critical line in Chapter 8, and the methods that we shall apply to deal with the large values of $\Delta_k(x)$ will be similar. The main obstacle is the presence of the divisor function $d_k(n)$ in (13.51), which will be eliminated by the use of the Halász–Montgomery inequality (A.40). This will lead to a large-values estimate for

$\Delta_k(x)$, and then (similarly as was done for higher power moments of the zeta-function in Chapter 8) we shall estimate (13.50) by majorizing the integral by discrete sums to which our large-values estimate may be applied to bound the number of summands. Although our method will work for general $\Delta_k(x)$, the results are sharp only when $k = 2$ and $k = 3$, and therefore we shall consider only these cases. The basic estimate is the following

THEOREM 13.8. Let $1 \leqslant t_1 < t_2 \cdots < t_R \leqslant T$ and $|t_r - t_s| \geqslant V$ for $r \neq s \leqslant R$. If $\Delta_2(t_r) \gg V > T^{7/32+\varepsilon}$ for $r \leqslant R$, then

$$(13.52) \qquad R \ll T^\varepsilon (TV^{-3} + T^{15/4}V^{-12}).$$

If $\Delta_3(t_r) \gg V > T^{18/67+\varepsilon}$ for $r \leqslant R$, then

$$(13.53) \qquad R \ll T^\varepsilon (T^2 V^{-4} + T^{57/13}V^{-132/13}).$$

Another proof of $\alpha_2 \leqslant \frac{1}{3}, \alpha_3 \leqslant \frac{1}{2}$ follows at once from Theorem 13.8 if we take $R = 1$, $t_R = T = x$, though the above estimates are naturally of greater interest if R is assumed to be large in some sense. Also we could have formulated Theorem 13.8 with the spacing condition $|t_r - t_s| \geqslant 1$ for $r \neq s \leqslant R$, and the only change would be that the exponent of V in (13.52) and (13.53) would be increased by unity. However, the spacing condition $|t_r - t_s| \geqslant V$ ($r \neq s$) imposed in the theorem seems more appropriate, since by an argument analogous to (13.35) and (13.36) we have $|\Delta_k(t')| \geqslant V/2$ if $|\Delta_k(t)| \geqslant V$ and $|t' - t| \leqslant Vt^{-\varepsilon}$.

Now we suppose that A is a fixed positive number (not necessarily an integer), and we formulate our power moment estimates for $\Delta_k(x)$ in the next two theorems.

THEOREM 13.9.

$$(13.54) \qquad \int_1^T |\Delta_2(t)|^A \, dt \ll T^{(A+4+\varepsilon)/4} \qquad \left(0 \leqslant A \leqslant \tfrac{35}{4}\right),$$

$$(13.55) \qquad \int_1^T |\Delta_2(t)|^A \, dt \ll T^{(35A+38+\varepsilon)/108} \qquad \left(A \geqslant \tfrac{35}{4}\right).$$

THEOREM 13.10.

$$(13.56) \qquad \int_1^T |\Delta_3(t)|^A \, dt \ll T^{(106A+253+\varepsilon)/279} \qquad \left(2 \leqslant A \leqslant \tfrac{2237}{607} = 3.685 \cdots \right),$$

$$(13.57) \qquad \int_1^T |\Delta_3(t)|^A \, dt \ll T^{(43A+63+\varepsilon)/96} \qquad \left(A \geqslant \tfrac{2237}{607}\right).$$

Theorem 13.9 shows that in a mean sense $\Delta_2(t)$ is of the conjectured order $t^{1/4+\varepsilon}$ for much higher powers than only the second, which followed from Theorem 13.5. The ranges for A in (13.54) and (13.56) both depend on the best-known values (13.13) and (13.14) for α_2 and α_3, respectively, and any improvement of these bounds for α_2 and α_3 would result in a wider range for A. The limit that (13.52) can theoretically give is

$$\text{(13.58)} \qquad \int_1^T |\Delta_2(t)|^{11}\, dt \ll T^{15/4+\varepsilon},$$

and this would in turn imply the (yet hypothetical) estimate $\alpha_2 \leqslant \frac{5}{16}$, which improves on (13.13) and differs from the best possible value $\alpha_2 = \frac{1}{4}$ by $\frac{1}{16}$. To see how $\alpha_2 \leqslant \frac{5}{16}$ follows from (13.58), suppose that for some $\varepsilon > 0$ we have

$$|\Delta(T)| = |\Delta_2(T)| > T^{5/16+\varepsilon} = G.$$

If $|t - T| \leqslant GT^{-\delta}$ ($\delta > 0$), then as in (13.36) we have $|\Delta(t)| \geqslant G/2$; hence by Hölder's inequality and (13.58) we obtain

$$G^2 T^{-\delta} \leqslant \int_{T-GT^{-\delta}}^{T+GT^{-\delta}} |\Delta(t)|\, dt$$

$$\leqslant \left(\int_{T-GT^{-\delta}}^{T+GT^{-\delta}} |\Delta(t)|^{11}\, dt \right)^{1/11} (2GT^{-\delta})^{10/11}$$

$$\ll T^{(15/4+\varepsilon)/11} G^{10/11} T^{-10\delta/11},$$

and this is a contradiction if δ is sufficiently small, in particular if $0 < \delta < 11\varepsilon$. This implies $\alpha_2 \leqslant \frac{5}{16}$, and the best unconditional estimate of α_2 which follows from (13.54) by this method of proof is only slightly weaker than (13.13).

Proof of Theorem 13.8. We start from (13.51) and use the Halász–Montgomery inequality to remove $d_k(n)$ from the sum in (13.51) and investigate the occurrence of large values of $\Delta_k(x)$. We use (A.40) and take $\xi = \{\xi_n\}_{n=1}^{\infty}$ with $\xi_n = d_k(n)n^{-(k+1)/2k}$ for $M < n \leqslant 2M$ and zero otherwise, and we let $\varphi_r = \{\varphi_{r,n}\}_{n=1}^{\infty}$ with $\varphi_{r,n} = e(k(nt_r)^{1/k})$ for $M < n \leqslant 2M$ and zero otherwise, where M is fixed and its range will be specified in a moment. We may restrict ourselves to the estimation of the number of points t_r lying in $[T/2, T]$, and we suppose that this interval is divided into subintervals of length not exceeding T_0 ($\gg V$). Denoting then by R_0 the number of t_r's lying in an interval of length not exceeding T_0 we have

$$\text{(13.59)} \qquad R \ll R_0(1 + T/T_0).$$

The idea behind this procedure (used already in Chapters 8 and 11) is that a suitable choice of T_0 will lead to (13.52) and (13.54), since $|t_r - t_s| \leqslant T_0$ for each of R_0^2 pairs of points (t_r, t_s). Choosing in (13.51) $N = T^{k-1+\varepsilon}V^{-k}$ we obtain by (A.40)

(13.60)

$$R_0 V^2 \ll T^{(k-1)/k} \log T \max_{M \leqslant N/2} \sum_{r \leqslant R_0} \left| \sum_{M < n \leqslant 2M} d_k(n) n^{-(k+1)/2k} e\left(k(nt_r)^{1/k} \right) \right|^2$$

$$\ll (TN)^{(k-1+\varepsilon)/k} + T^{(k-1+\varepsilon)/k} \max_{M \leqslant N/2} \max_{r \leqslant R_0} M^{-1/k}$$

$$\times \sum_{s \leqslant R_0, s \neq r} \left| \sum_{M < n \leqslant 2M} e\left(kn^{1/k}(t_r^{1/k} - t_s^{1/k}) \right) \right|,$$

since the contribution of the terms with $r = s$ is clearly $\ll N^{(k-1)/k}$. The last sum in (13.60) is an exponential sum of the form

$$(13.61) \quad S = \sum_{M < n \leqslant 2M} e(f(n)), \qquad f(x) = kx^{1/k}(t_r^{1/k} - t_s^{1/k}),$$

$$r \neq s, \ M \leqslant x \leqslant 2M.$$

This is similar to the sum (8.38) in Lemma 8.1, only (13.61) is somewhat simpler. Since $f'(x)$ is monotonic for $M \leqslant x \leqslant 2M$, we may suppose that $|f'(x)| < 1$ or $|f'(x)| \geqslant 1$ by splitting S into two subsums if necessary. If $|f'(x)| < 1$ holds, we estimate S by Lemmas 1.2 and 2.1 as $S \ll \max_{M \leqslant x \leqslant 2M} |f'(x)|^{-1}$ to obtain

(13.62)

$$M^{-1/k} \max_{r \leqslant R_0} \sum_{s \leqslant R_0, s \neq r} |S| \ll M^{1-2/k} \max_{r \leqslant R_0} \sum_{s \leqslant R_0, s \neq r} \left| t_r^{1/k} - t_s^{1/k} \right|^{-1}$$

$$\ll N^{(k-2)/k} T^{(k-1)/k} \max_{r \leqslant R_0} \sum_{s \leqslant R_0, s \neq r} \left| t_r - t_s \right|^{-1}$$

$$\ll N^{(k-2)/k} T^{(k-1)/k} V^{-1} \log T,$$

since $T/2 \leqslant t_r \leqslant T$ and $|t_r - t_s| \geqslant V$ for $r \neq s \leqslant R_0$.

If $f'(x) \gg 1$ holds for $M \leqslant x \leqslant 2M$, then observing that

$$f^{(m)}(x) \asymp |t_r - t_s| T^{(1-k)/k} M^{(1-mk)/k} \qquad (m = 1, 2, \ldots),$$

it follows that we may use the theory of exponent pairs (Section 2.3 of Chapter 2). Thus if $F = \max_{M \leqslant x \leqslant 2M} |f'(x)|$ and $(\varkappa, \lambda)$ is an exponent pair, then we have

$$S \ll F^\varkappa M^\lambda \ll T_0^\varkappa T^{(\varkappa - \varkappa k)/k} M^{(\lambda k + \varkappa - \varkappa k)/k}.$$

Therefore the contribution of these S is

(13.63)
$$\max_{M \le N/2} M^{-1/k} \max_{r \le R_0} \sum_{s \le R_0, s \ne r} |S|$$

$$\ll R_0 \max_{M \le N/2} M^{-1/k} M^{(\lambda k + \varkappa - \varkappa k)/k} T_0^\varkappa T^{(\varkappa - \varkappa k)/k}$$

$$\ll R_0 T_0^\varkappa T^{(\varkappa - \varkappa k)/k} N^{(\lambda k + \varkappa - 1 - \varkappa k)/k},$$

provided that

(13.64)
$$\lambda k \ge 1 + (k - 1)\varkappa.$$

If this condition is satisfied, then from (13.60), (13.62), and (13.63) we obtain

(13.65)
$$R_0 V^2 \ll (TN)^{(k-1+\varepsilon)/k} + T^{(2k-2+\varepsilon)/k} N^{(k-2)/k} V^{-1}$$

$$+ R_0 T_0^\varkappa T^{(k-1+\varepsilon)(1-\varkappa)/k} N^{(\lambda k + \varkappa - 1 - \varkappa k)/k}.$$

Now we consider first the case $k = 2$ where we choose $N = T^{k-1+\varepsilon} V^{-k} = T^{1+\varepsilon} V^{-2}$, and then the first two terms on the right-hand side of (13.65) are equal. We take $(\varkappa, \lambda) = (\frac{4}{18}, \frac{11}{18})$ and note that with this exponent pair equality holds in (13.64) for $k = 2$; hence

(13.66)
$$R_0 \ll T^{1+\varepsilon} V^{-3} + R_0 T_0^{2/9} T^{7/18+\varepsilon} V^{-2}.$$

Choosing $T_0 = V^9 T^{-7/4-\varepsilon}$ we have $T_0 \gg V$ for $V > T^{7/32+\varepsilon}$, and (13.66) reduces to

(13.67)
$$R_0 \ll T^{1+\varepsilon} V^{-3};$$

hence (13.52) follows from (13.59) with $T_0 = V^9 T^{-7/4-\varepsilon}$.

If $k = 3$ we choose $N = T^{k-1+\varepsilon} V^{-k} = T^{2+\varepsilon} V^{-3}$ and in (13.65) we take $(\varkappa, \lambda) = (\frac{13}{40}, \frac{22}{40})$. With this exponent pair equality holds in (13.64) for $k = 3$, and (13.65) gives

(13.68)
$$R_0 \ll T^{2+\varepsilon} V^{-4} + R_0 T_0^{13/40} T^{18/40+\varepsilon} V^{-2}.$$

Choosing $T_0 = V^{80/13} T^{-18/13-\varepsilon}$ we have $T_0 \gg V$ for $V > T^{18/67+\varepsilon}$; hence (13.68) gives $R_0 \ll T^{2+\varepsilon} V^{-4}$ and (13.53) follows again from (13.59). With a little more care we could replace T^ε in (13.52) and (13.53) by a suitable log power, and small improvements in the second terms on the right-hand sides of (13.52) and (13.53) could be obtained by a more elaborate choice of the exponent pair $(\varkappa, \lambda)$ in (13.65). From (13.65) one obtains for general k the estimate $R \ll T^{k-1+\varepsilon} V^{-k-1}$ for $|\Delta_k(t_r)| \ge V = V(T, k)$, $r \le R$, but in view of $\alpha_4 \le \frac{1}{2}$ (Theorem 13.2) this is weak already for $k = 4$.

Proof of Theorem 13.9 and Theorem 13.10. It is sufficient to prove our estimates for integrals over $[T/2, T]$ and then to sum over intervals of the form $[2^{-j}T, 2^{1-j}T]$, $j \geqslant 1$. We denote by τ_r the point for which

$$|\Delta_2(\tau_r)| = \sup_{t \in [T/2 + r - 1, T/2 + r]} |\Delta_2(t)| \qquad (r = 1, 2, \ldots),$$

and we consider first those τ_r for which

$$T^{1/4} \leqslant 2^m = V \leqslant |\Delta_2(\tau_r)| < 2V.$$

There are $O(\log T)$ choices for V [$\ll T^{35/108 + \varepsilon}$ by (13.13)], and by picking the maximal $|\Delta_2(\tau_r)|$ in τ-intervals of length V and by considering separately points with even and odd indices we may construct a system of points which we shall label $t_1, t_2, \ldots, t_R$, $R = R(V)$, and which satisfy

(13.69)

$$T^{1/4} \leqslant 2^m = V \leqslant |\Delta_2(t_r)| < 2V, \qquad |t_r - t_s| \geqslant V \quad \text{for} \quad r \neq s \leqslant R = R(V),$$

so that we may write

$$(13.70) \qquad \int_{T/2}^T |\Delta_2(t)|^A \, dt \leqslant T^{(A + 4 + \varepsilon)/4} + \sum_V V \sum_{r \leqslant R(V)} |\Delta_2(t_r)|^A.$$

Now we consider the range $2 \leqslant A \leqslant 11$ and we use (13.52) to bound $R = R(V)$, keeping in mind that (13.69) holds. Using $V \ll T^{35/108 + \varepsilon}$ we obtain

$$(13.71) \qquad V \sum_{r \leqslant R(V)} |\Delta_2(t_r)|^A \ll RV^{A+1} \ll T^\varepsilon (TV^{A-2} + T^{15/4} V^{A-11})$$

$$\ll T^{1 + 35(A - 2)/108 + \varepsilon} + T^{(A + 4 + \varepsilon)/4}.$$

Here the first term is larger than the second for $\frac{35}{4} \leqslant A \leqslant 11$, while the second is larger for $2 \leqslant A \leqslant \frac{35}{4}$, which in view of (13.70) proves (13.54) for $2 \leqslant A \leqslant \frac{35}{4}$, while the estimate for $0 \leqslant A < 2$ follows easily by Hölder's inequality for integrals and the estimate for $A = 2$. To obtain (13.55) for $A > \frac{35}{4}$ we proceed analogously, only now we have

$$V \sum_{r \leqslant R(V)} |\Delta_2(t_r)|^A \ll T^\varepsilon (TV^{A-2} + T^{15/4} V^{A-11})$$

$$\ll T^{1 + 35(A-2)/108 + \varepsilon} + T^{15/4 + 35(A - 11)/108 + \varepsilon} + T^{(A + 4 + \varepsilon)/4}$$

$$\ll T^{(35A + 38 + \varepsilon)/108}.$$

The proof of Theorem 13.10 is similar to the proof of Theorem 13.9 and uses (13.53) and (13.14); but while the proof of Theorem 13.9 is independent of (13.26), the proof of Theorem 13.10 will require a weak form of (13.43), namely, $\int_1^T \Delta_3^2(t)\, dt \ll T^{5/3+\varepsilon}$, which is equivalent to $\beta_3 = \frac{1}{3}$. Instead of (13.69) we impose for the proof of Theorem 13.10 a similar condition, namely,

$$U \leqslant 2^m = V \leqslant |\Delta_3(t_r)| < 2V, \qquad |t_r - t_s| \geqslant V \quad \text{for} \quad r \neq s \leqslant R = R(V).$$

Here the points t_r are constructed analogously as in the previous case and the optimal choice for U is $U = T^{106/279}$. For $2 \leqslant A \leqslant \frac{2237}{607} = 3.6853\ldots$ we then have

$$(13.72) \quad \int_{T/2}^T |\Delta_3(t)|^A\, dt \ll T^{5/3+\varepsilon}U^{A-2} + \sum_{V \geqslant U} V \sum_{r \leqslant R(V)} |\Delta_3(t_r)|^A,$$

and

$$V \sum_{r \leqslant R(V)} |\Delta_3(t_r)|^A \ll RV^{A+1}$$

$$\ll T^\varepsilon(T^2V^{A-3} + T^2U^{A-3} + T^{57/13}U^{A-119/13})$$

$$\ll T^{(43A+63+\varepsilon)/96} + T^{(106A+240+\varepsilon)/279} + T^{(106A+253+\varepsilon)/279}.$$

Here the third term is the largest one for $2 \leqslant A \leqslant \frac{2237}{607}$ and (13.72) gives

$$\int_{T/2}^T |\Delta_3(t)|^A\, dt \ll T^{5/3+106(A-2)/279+\varepsilon} + T^{(106A+253+\varepsilon)/279}$$

$$\ll T^{(106A+253+\varepsilon)/279}.$$

This proves (13.56). For $A > \frac{2237}{607}$ the analysis is analogous and gives (13.57).

13.8 THE CIRCLE PROBLEM

This chapter is concluded with a discussion of the classical circle problem, which was mentioned in Section 3.4 of Chapter 3. The problem of the estimation of

$$P(x) = R(x) - \pi x = \sum_{n \leqslant x} r(n) - \pi x$$

is known as the circle problem and bears many resemblances to the estimation of $\Delta(x) = \Delta_2(x)$ in the Dirichlet divisor problem. We recall G. H. Hardy's classical formula [(3.36) with $q = 1$]

$$(13.73) \qquad \sum_{n \leqslant x} {}' r(n) = \pi x - 1 + x^{1/2} \sum_{n=1}^{\infty} r(n) n^{-1/2} J_1(2\pi\sqrt{nx}),$$

and if we use the approximation

$$J_1(y) = -(2/\pi y)^{1/2} \cos(y + \pi/4) + O(y^{-3/2}),$$

we may write

$$(13.74) \quad P(x) = -\pi^{-1} x^{1/4} \sum_{n=1}^{\infty} r(n) n^{-3/4} \cos(2\pi\sqrt{nx} + \pi/4) + O(x^{\varepsilon}),$$

since in (13.73) we have to count $r(x)/2$ if $n = x$ is an integer, and obviously $r(n) \ll n^{\varepsilon}$. A trivial bound for $P(x)$ is $P(x) \ll x^{1/2}$, since $P(x)$ is clearly majorized by the circumference of a circle with radius $x^{1/2}$. One would expect that (13.74) would provide the analogue of the truncated Voronoi formula (3.16) for $\Delta(x)$, and this would be

$$(13.75) \qquad P(x) = -\pi^{-1} x^{1/4} \sum_{n \leqslant N} r(n) n^{-3/4} \cos(2\pi\sqrt{nx} + \pi/4)$$

$$+ O(x^{\varepsilon}) + O(x^{1/2+\varepsilon} N^{-1/2}).$$

A direct proof of (13.75) via (13.74) does not seem easy [as is also the case in the analogous problem of the truncated formula for $\Delta(x)$], but one may use the method of proof of (3.16) by considering $\sum_{n=1}^{\infty} r(n) n^{-s}$ for $\mathrm{Re}\, s > 1$ and using the inversion formula (A.10) to estimate $\sum_{n \leqslant x} r(n)$. A similar approach has been adopted by H.-E. Richert (1957), where general estimates for sums of the type $\sum_{n \leqslant x} f(n)(x - n)^{\varkappa}$ are considered for certain classes of arithmetical functions f. The sums in question are not developed into infinite series containing (generalized) Bessel functions, but into explicit exponential sums of length N plus error terms, and this is exactly what is needed for (13.75). Therefore instead of trying to obtain a direct proof of (13.75) we shall now briefly present Richert's discussion of the circle problem, and then obtain a result (Lemma 13.4) which is analogous to (13.75) and may be effectively used to obtain estimates of power moments with $P(x)$. Richert (1957) transforms the circle problem into a divisor problem by noting that

$$(13.76) \qquad r(n) = \sum_{a^2 + b^2 = n} 1 = 4 \sum_{d \mid n,\, d \equiv 1 (\mathrm{mod}\, 2)} (-1)^{(d-1)/2},$$

and writing

(13.77) $\quad D(x; k_1, l_1, k_2, l_2) = \sum\limits_{\substack{n_1 n_2 \leqslant x, \, n_j \equiv l_j \pmod{k_j}}} 1$

$$= x(k_1 k_2)^{-1} \log(x/k_1 k_2)$$

$$- \left(\frac{\Gamma'}{\Gamma}(l_1/k_1) + \frac{\Gamma'}{\Gamma}(l_2/k_2) \right) (k_1 k_2)^{-1} x + \Delta(x; k_1, l_1, k_2, l_2)$$

one has

(13.78) $\quad R(x) = 4D(x; 4, 1, 1, 1) - 4D(x; 4, 3, 1, 1)$

$$= \pi x + 4\Delta(x; 4, 1, 1, 1) - 4\Delta(x; 4, 3, 1, 1),$$

when one recalls that

$$\frac{\Gamma'}{\Gamma}\left(\frac{3}{4}\right) - \frac{\Gamma'}{\Gamma}\left(\frac{1}{4}\right) = \pi.$$

Thus (13.78) shows that $P(x)$ may be considered as four times the difference of two divisor problem error terms, and Richert obtained

(13.79) $\quad \Delta(x; k_1, l_1, k_2, l_2)$

$$= (\sqrt{2}\,\pi)^{-1} (x/k_1 k_2)^{1/4} \mathrm{Re}\left(e(-\tfrac{1}{8}) \sum\limits_{1 \leqslant n_1 n_2 \leqslant N} (n_1 n_2)^{-3/4} e(F) \right)$$

$$+ O(x^{1/5 + \varepsilon} N^{1/5}) + O(x^{1/2 + \varepsilon} N^{-1/2}),$$

where

(13.80) $\quad F = (4x n_1 n_2)^{1/2}/(k_1 k_2)^{1/2} - (l_1 n_1)/k_1 - (l_2 n_2)/k_2.$

Here for $N \leqslant x^{3/7}$ we have $x^{1/5} N^{1/5} \leqslant x^{1/2} N^{-1/2}$, so that in the range $x^{1/3} \leqslant N \leqslant x^{3/7}$ the first error term in (13.79) may be discarded. Except for the linear part $-(l_1 n_1)/k_1 - (l_2 n_2)/k_2$ the exponential term $e(F)$ in (13.79) is (up to a constant) the same as in the formula (3.17) for $\Delta(x)$. It is readily seen that the linear part poses no problem in the application of Kolesnik's method (Lemma 7.3), since the linear terms are small when compared with $(x n_1 n_2)^{1/2}$, and moreover the linear terms vanish already in the second partial derivatives of $F = F(n_1, n_2)$ in (13.80). As in the proof of Theorem 13.1 it follows then that we have

THEOREM 13.11.

$$(13.81) \qquad R(x) = \sum_{n \leqslant x} r(n) = \pi x + O(x^{35/108 + \varepsilon}).$$

Next to obtain (13.75) note that using (13.79) in (13.78) we have

$$e\big(-(l_1 n_1)/k_1 - (l_2 n_2)/k_2\big) = \exp(-2\pi i l_1 n_1/k_1)$$

$$= \begin{cases} e^{-\pi i n_1/2} & (k_1 = 4, k_2 = l_1 = 1) \\ e^{\pi i n_1/2} & (k_1 = 4, k_2 = 1, l_1 = 3) \end{cases}$$

This shows that for $x^{1/3} \leqslant N \leqslant x^{3/7}$ we have

$$(13.82) \quad P(x) = 4(\sqrt{2}\,\pi)^{-1}(x/4)^{1/4}\mathrm{Re}\!\left(e\!\left(-\frac{1}{8}\right) \sum_{n_1 n_2 \leqslant N} (n_1 n_2)^{-3/4}\right.$$

$$\times (-2i\sin(\pi n_1/2)) e\!\left(\sqrt{xn_1 n_2}\right)\Bigg)$$

$$+ O\big((x/N)^{1/2 + \varepsilon}\big)$$

$$= -4\pi^{-1}x^{1/4}\mathrm{Re}\!\left(e\!\left(\frac{1}{8}\right) \sum_{n_1 n_2 \leqslant N,\, n_1 \equiv 1 (\mathrm{mod}\, 2)} (n_1 n_2)^{-3/4}\right.$$

$$\times (-1)^{(n_1 - 1)/2} e\!\left(\sqrt{xn_1 n_2}\right)\Bigg)$$

$$+ O\big((x/N)^{1/2 + \varepsilon}\big)$$

$$= -\pi^{-1}x^{1/4} \sum_{n \leqslant N} r(n)n^{-3/4}\cos(2\pi\sqrt{xn} + \pi/4)$$

$$+ O\big((x/N)^{1/2 + \varepsilon}\big),$$

where we used (13.76). In view of $x^{1/3} \leqslant N \leqslant x^{3/7}$, the best that this approach could give is $P(x) \ll x^{2/7 + \varepsilon}$, which though better than Theorem 13.11 would still be poorer than the conjectured estimate $P(x) \ll x^{1/4 + \varepsilon}$ (the analogue of the conjecture $\alpha_2 = \frac{1}{4}$). We would like to use (13.75) to obtain a result analogous to Theorem 13.9 for power moments of $P(x)$, but as we have (13.75) for $x^{1/3} \leqslant N \leqslant x^{3/7}$ we would not obtain a result of the same strength as Theorem 13.9, since we need (13.75) in the range $x^{1/3} \leqslant N \leqslant x^{1/2}$ for that purpose. It would be possible to go through Richert's proof of (13.79) and improve the error term $O(x^{1/5+\varepsilon}N^{1/5})$; however, we find it more expedient to turn back to Hardy's formula (13.73) and use the technique of averaging with the exponential integral (A.38). This will enable us to obtain a result similar to (13.75), but without the restriction $N \leqslant x^{3/7}$.

LEMMA 13.4. For $T \leqslant x \leqslant 2T$, $T^{1/4} \leqslant G \leqslant T^{1/3}$ we have uniformly in x

$$(13.83) \quad P(x) \ll T^\varepsilon G$$

$$+ T^{1/4} \left| \sum_{n \leqslant TG^{-2} \log^2 T} r(n) n^{-3/4} \exp\left(2\pi i \sqrt{nx} - \tfrac{1}{4}\pi^2 n G^2 x^{-1}\right) \right|.$$

Proof of Lemma 13.4. Let $\|x\|$ denote the distance of x to the nearest integer and let the hypotheses of the lemma hold. The first step in the proof will be to show that

$$(13.84) \quad x^{1/4} \sum_{n > T^{3/2}} r(n) n^{-3/4} e(\sqrt{nx}) \ll T^\varepsilon G \qquad (\|x\| \gg G T^{-3/4}).$$

To see this write $U = [T^{3/2}] + 1$. Then

$$x^{1/4} \sum_{n > T^{3/2}} r(n) n^{-3/4} e(\sqrt{nx}) = x^{1/4} \int_{U-0}^\infty t^{-3/4} e(\sqrt{xt}) \, dR(t)$$

$$= \pi x^{1/4} \int_U^\infty t^{-3/4} e(\sqrt{xt}) \, dt$$

$$+ x^{1/4} \int_{U-0}^\infty t^{-3/4} e(\sqrt{xt}) \, dP(t)$$

$$= o(1) + x^{1/4} \int_{U-0}^\infty t^{-3/4} e(\sqrt{xt}) \, dP(t).$$

Integrating by parts and using $P(t) \ll t^{1/2}$ it is seen that the last expression above is

$$O(1) - x^{1/4} \int_U^\infty P(t) \left(-\tfrac{3}{4} t^{-7/4} + \pi i x^{1/2} t^{-5/4} \right) e(\sqrt{xt}) \, dt = O(1) - \pi i x^{3/4} I,$$

where

$$(13.85) \qquad\qquad I = \int_U^\infty P(t) t^{-5/4} e(\sqrt{xt}) \, dt.$$

Using (13.73) and the asymptotic formula for $J_1(y)$ we obtain

$$I = O(T^{-3/4}) - \pi^{-1} \sum_{n=1}^{\infty} r(n)n^{-3/4} \int_U^{\infty} t^{-1} \cos(2\pi\sqrt{nt} + \pi/4) e(\sqrt{xt}) \, dt.$$

The last integral may be estimated by (2.3). We obtain

$$I \ll T^{-3/4} + T^{-3/4} \sum_{n=1}^{\infty} r(n)n^{-3/4} |n^{1/2} - x^{1/2}|^{-1}$$

$$\ll T^{-3/4} + T^{-3/4} \left(\sum_{n \leqslant x/2} + \sum_{x/2 < n \leqslant 2x} + \sum_{n > 2x} \right)$$

$$\ll T^{-3/4} + T^{\varepsilon - 3/4} \sum_{x/2 < n \leqslant 2x} n^{-3/4} T^{1/2} |n - x|^{-1}$$

$$\ll T^{-3/4} + T^{\varepsilon - 1/4} G^{-1},$$

since $\|x\| \gg GT^{-3/4}$. Therefore in view of $G \geqslant T^{1/4}$ we obtain

$$x^{1/4} \sum_{n > T^{3/2}} r(n)n^{-3/4} e(\sqrt{nx}) \ll T^{\varepsilon}(1 + T^{1/2}G^{-1}) \ll T^{\varepsilon}G.$$

The next step in the proof of (13.83) is to derive a suitable averaged expression for $P(x)$. To do this we shall use the elementary integral

$$\int_{-\infty}^{\infty} \exp(-x^2) \, dx = \pi^{1/2}.$$

Abbreviating $L = \log T$ we have

(13.86) $\quad \pi^{1/2}P(x) - G^{-1} \int_{x-GL}^{x+GL} P(t) e^{-(x-t)^2 G^{-2}} \, dt$

$$= o(1) + G^{-1} \int_{x-GL}^{x+GL} (P(x) - P(t)) e^{-(x-t)^2 G^{-2}} \, dt$$

$$\ll 1 + L \max_{|x-t| \leqslant GL} |P(x) - P(t)|$$

$$\ll GT^{\varepsilon},$$

since

$$|P(x) - P(t)| \leqslant \pi|x - t| + \left| \sum_{t \leqslant n \leqslant x} r(n) \right| + O(T^{\varepsilon}) \ll (1 + |x - t|)T^{\varepsilon},$$

because $r(n) \ll n^\varepsilon$. To use (13.84) write

$$(13.87) \quad \int_{x-GL}^{x+GL} P(t)e^{-(x-t)^2 G^{-2}} \, dt$$

$$= \int_{x-GL}^{x+GL} \left(P(t) + \pi^{-1} t^{1/4} \sum_{n \leqslant T^{3/2}} r(n) n^{-3/4} \cos(2\pi\sqrt{nt} + \pi/4) \right) e^{-(x-t)^2 G^{-2}} \, dt$$

$$- \pi^{-1} \int_{x-GL}^{x+GL} t^{1/4} \sum_{n \leqslant T^{3/2}} r(n) n^{-3/4} \cos(2\pi\sqrt{nt} + \pi/4) e^{-(x-t)^2 G^{-2}} \, dt$$

$$= I_1 - \pi^{-1} I_2,$$

say. The main contribution in (13.87) [and hence the main contribution to $P(x)$ in (13.86)] comes from I_2, and to show this we shall prove

$$(13.88) \qquad\qquad\qquad I_1 \ll G^2 T^\varepsilon.$$

To achieve this let

$$(13.89) \quad \mathscr{A}_1 = \bigcup_{n=1}^\infty \{ [x - GL, x + GL] \cap [n - GT^{-3/4}, n + GT^{-3/4}] \},$$

$$\mathscr{A}_2 = [x - GL, x + GL] \setminus \mathscr{A}_1,$$

and split I_1 into integrals over $\mathscr{A}_1$ and $\mathscr{A}_2$, respectively. For $t \in \mathscr{A}_1$ we shall use the trivial estimate

$$P(t) + \pi^{-1} t^{1/4} \sum_{n \leqslant T^{3/2}} r(n) n^{-3/4} \cos(2\pi\sqrt{nt} + \pi/4) \ll T^{5/8 + \varepsilon}$$

to obtain

$$\int_{\mathscr{A}_1} \ll T^{5/8 + \varepsilon} G^2 T^{-3/4} \ll G^2,$$

since obviously $|\mathscr{A}_1| \ll G^2 T^{-3/4 + \varepsilon}$. For the integral over $\mathscr{A}_2$ we use (13.74) and (13.84) to obtain at once

$$\int_{\mathscr{A}_2} \ll G^2 T^\varepsilon;$$

hence (13.88) follows.

Therefore we are left with the evaluation of

(13.90)

$$I_2 = \int_{-GL}^{GL} (x + t)^{1/4} \sum_{n \leqslant T^{3/2}} r(n) n^{-3/4} \cos\left(2\pi\sqrt{n(x + t)} + \pi/4\right) \exp\left(-t^2 G^{-2}\right) dt,$$

and we can replace $(x + t)^{1/4}$ by $x^{1/4}$ with an error which is $\ll G^2 T^\varepsilon$. Combining previous estimates we obtain

(13.91)

$$P(x) \ll GT^\varepsilon$$

$$+ G^{-1} T^{1/4} \left| \sum_{n \leqslant T^{3/2}} r(n) n^{-3/4} \int_{-GL}^{GL} \exp\left(2\pi i \sqrt{n(x + t)} - t^2 G^{-2}\right) dt \right|.$$

Using (A.38) and Taylor's formula we obtain

$$\int_{-GL}^{GL} \exp\left(2\pi i \sqrt{n(x + t)} - t^2 G^{-2}\right) dt$$

$$= e(\sqrt{nx}) \int_{-\infty}^{\infty} \exp\left(\pi i t x^{-1/2} n^{1/2} - \tfrac{1}{4} \pi i t^2 x^{-3/2} n^{1/2} - t^2 G^{-2}\right) dt$$

$$+ O\left(G^4 L^4 T^{-5/2} n^{1/2}\right)$$

$$= (\pi/Y)^{1/2} \exp\left(2\pi i \sqrt{nx} - \tfrac{1}{4} \pi^2 n Y^{-1} x^{-1}\right) + O\left(G^4 L^4 T^{-5/2} n^{1/2}\right).$$

Here we have set

$$Y = G^{-2} + \tfrac{1}{4} \pi i x^{-3/2} n^{1/2} = (1 + o(1)) G^{-2} \qquad (T \to \infty),$$

since $n \leqslant T^{3/2}$. Therefore from (13.91) it follows that

(13.92)

$$P(x) \ll GT^\varepsilon$$

$$+ G^{-1} T^{1/4} |Y|^{-1/2} \left| \sum_{n \leqslant T^{3/2}} r(n) n^{-3/4} \exp\left(2\pi i \sqrt{nx} - \tfrac{1}{4} \pi^2 n Y^{-1} x^{-1}\right) \right|.$$

Now for $n \geqslant TG^{-2} L^2$ and any fixed $c > 0$

$$\left| \exp\left(-\tfrac{1}{4} \pi^2 n Y^{-1} x^{-1}\right) \right| \leqslant \exp\left(-\tfrac{1}{8} \pi^2 n x^{-1} G^2\right) \leqslant \exp\left(-\tfrac{1}{16} \pi^2 L^2\right) \ll T^{-c};$$

hence the contribution of the terms with $n > TG^{-2} L^2$ in (13.92) is negligible.

For $n \leqslant TG^{-2}L^2$ one can replace Y by G^{-2} with a total error $\ll GT^\varepsilon$, and (13.83) follows then from (13.92).

Lemma 13.4 is now completely analogous to the truncated Voronoi formula (3.17) [with $G = (TN^{-1}\log^2 T)^{1/2}$], since the exponential factor $\exp(-\frac{1}{4}\pi^2 nG^2 x^{-1})$, which appears in (13.83) does not exceed one and does not affect order results obtainable from (13.83). Combining Theorem 13.11 and Lemma 13.4 one can obtain easily the analogue of Theorem 13.9 virtually by repeating the same proof with $d(n)$ replaced by $r(n)$. The result will be

THEOREM 13.12.

$$\int_1^T |P(x)|^A \, dx \ll T^{(A+4+\varepsilon)/4} \qquad \left(0 \leqslant A \leqslant \tfrac{35}{4}\right),$$

$$\int_1^T |P(x)|^A \, dx \ll T^{(35A+38+\varepsilon)/108} \qquad \left(A \geqslant \tfrac{35}{4}\right).$$

This is of interest for $A > 2$, since K.-C. Tong (1956) proved an analogue of (13.33), namely,

$$(13.93) \quad \int_1^x P^2(t) \, dt = (3\pi^2)^{-1} \sum_{n=1}^\infty r^2(n) n^{-3/2} x^{3/2} + O(x \log^3 x).$$

As a further analogy between the Dirichlet divisor problem for $k = 2$ and the circle problem, note that (13.93) implies that $P(t) \ll t^\alpha$ cannot hold for $\alpha < \frac{1}{4}$ (this is the analogue of $\alpha_2 \geqslant \frac{1}{4}$). Also denoting the error term in (13.93) by $R^*(x)$ and following the proof of Theorem 13.6 we obtain that

$$R^*(x) \ll x^{3/4-\delta}$$

cannot hold for any $\delta > 0$.

Notes

The estimation of $\Delta_k(x)$ goes also under the name of the general divisor problem or simply the divisor problem, while the estimation of $\Delta_3(x)$ is often called the Piltz divisor problem. Some authors call the estimation of $\Delta(x) = \Delta_2(x)$ the Dirichlet divisor problem, but we found it preferable to treat here $\Delta_k(x)$ as the Dirichlet divisor problem in distinction with other divisor problems investigated in Chapter 14.

   P. G. L. Dirichlet proved

$$\sum_{n \leqslant x} d(n) = x(\log x + 2\gamma - 1) + O(x^{1/2})$$

by using (4.69). The method of using

$$(13.94) \qquad \sum_{mn \leqslant X} f(m,n) = \sum_{m \leqslant y} \sum_{n \leqslant X/m} f(m,n)$$

$$+ \sum_{n \leqslant z} \sum_{m \leqslant X/n} f(m,n) - \sum_{m \leqslant y} \sum_{n \leqslant z} f(m,n),$$

where $yz = X$ ($y, z > 1$) is known as the "hyperbola method" or the "convolution method" (see Sections 14.1 and 14.2 of Chapter 14 for more applications). The elementary proof of $\Delta_k(x) \ll x^{(k-1)/k} \log^{k-2} x$ which uses (13.94) and induction may be found in Chapter 12 of Titchmarsh (1951). This chapter of Titchmarsh's book is the analogue of the present chapter, but our treatment is more extensive and the results in most cases sharper.

Strictly speaking (13.3) holds for x not an integer, since in $\sum_{n \leqslant x} d_k(n)$ the last term is to be counted as $\frac{1}{2} d_k(x)$ if x is an integer; but as already remarked in the Notes of Chapter 3 concerning $\Delta(x) = \Delta_2(x)$, for most purposes this distinction is irrelevant. The same remark holds for $P(x)$, the error term in the circle problem, which was defined by (3.35).

General formulas for coefficients of the polynomial P_{k-1} in (13.2) may be found in the work of A. F. Lavrik (1979), and some integrals involving $\Delta_k(x)$ are evaluated by Lavrik et al. (1980).

The history of the estimation of α_k (and in particular of α_2) is at least as long and as rich as the history of estimates for $|\zeta(\frac{1}{2} + it)|$, which was given in the Notes of Chapter 7. Concerning α_2, we have already mentioned that Dirichlet proved $\alpha_2 \leqslant \frac{1}{2}$. Subsequent estimates of α_2 are as follows:

$$\alpha_2 \leqslant \tfrac{1}{3} = 0.33333\ldots \qquad \text{G. F. Voronoi (1904a, 1904b)}$$
$$\alpha_2 \leqslant \tfrac{33}{100} = 0.33 \qquad \text{J. G. van der Corput (1922)}$$
$$\alpha_2 \leqslant \tfrac{27}{82} = 0.329268\ldots \qquad \text{J. G. van der Corput (1928)}$$
$$\alpha_2 \leqslant \tfrac{15}{46} = 0.326086\ldots \qquad \text{H.-E. Richert (1953) and}$$
$$\text{Chih Tsung-tao (1950)}$$
$$\alpha_2 \leqslant \tfrac{12}{37} = 0.324324\ldots \qquad \text{G. Kolesnik (1969)}$$
$$\alpha_2 \leqslant \tfrac{346}{1067} = 0.324273\ldots \qquad \text{G. Kolesnik (1973)}$$
$$\alpha_2 \leqslant \tfrac{35}{108} = 0.324074\ldots \qquad \text{G. Kolesnik (1982)}$$

Kolesnik (1982) obtains actually $\Delta_2(x) \ll x^{35/108+\varepsilon}$ [so that (13.13) is slightly sharper], but his argument clearly gives also (13.13). Anyway the log factors are not so important since Kolesnik's method is not exhausted by the value $\alpha_2 \leqslant \tfrac{35}{108}$, and the method of his paper (in press) indicates the bound $\alpha_2 \leqslant \tfrac{139}{429} = 0.324009\ldots$.

The history of estimates of α_3 is as follows:

$$\alpha_3 \leqslant \tfrac{1}{2} = 0.5 \qquad \text{G. H. Hardy and J. E. Littlewood (1922)}$$
$$\alpha_3 \leqslant \tfrac{43}{87} = 0.494252\ldots \qquad \text{A. Walfisz (1925)}$$
$$\alpha_3 \leqslant \tfrac{37}{75} = 0.493333\ldots \qquad \text{F. V. Atkinson (1941a)}$$
$$\alpha_3 \leqslant \tfrac{14}{29} = 0.482758\ldots \qquad \text{Yüh Ming-i (1958)}$$
$$\alpha_3 \leqslant \tfrac{8}{17} = 0.470588\ldots \qquad \text{Yüh Ming-i and Wu Fang (1962)}$$
$$\alpha_3 \leqslant \tfrac{5}{11} = 0.454545\ldots \qquad \text{Chen Jing-run (1965b)}$$
$$\alpha_3 \leqslant \tfrac{43}{96} = 0.447916\ldots \qquad \text{G. Kolesnik (1981a)}$$

For several general estimates of α_k, all of which are poorer than those given by Theorem 13.2 when $k \geqslant 5$, the reader is referred to Chapter 12 of E. C. Titchmarsh (1951).

The estimate $\alpha_k \leqslant (3k - 4)/4k$, $4 \leqslant k \leqslant 8$ of Theorem 13.2 has been given first by D. R. Heath-Brown (1981b), who proved also $\alpha_k \leqslant (k - 3)/k$ for $k > 8$. This is superseded now by corresponding estimates of Theorem 13.2, which is due to the author (1980b).

The method of proof of Theorem 13.2 is based on the use of Theorem 8.4 and shows that one can obtain $\alpha_k \leqslant 1 - A/k$ for any fixed $A > 0$ and $k \geqslant k(A)$, but this is superseded by Theorem 13.3 for sufficiently large k. Besides choosing more carefully the exponent pairs in the proof of (8.80), there are other possibilities of improving Theorem 13.2 for large k, namely, the bound $\alpha_k \leqslant 1 - 34/7k$ for $k \geqslant 58$. Instead of $c(\theta) = (1 - \theta)/5$ ($\frac{5}{6} \leqslant \theta \leqslant 1$) one may use sharper bounds for $c(\theta)$ in (8.73) and Lemma 8.2 for appropriate ranges of θ. Thus from (7.59) with $l = 6$, we obtain $c(\theta) = (1 - \theta)/6$ for $\frac{28}{31} \leqslant \theta \leqslant 1$ and this will lead to $m(\sigma) \geqslant 5/(1 - \sigma)$ for $0.91 \leqslant \sigma \leqslant 1 - \varepsilon$, and consequently to $\alpha_k \leqslant 1 - 5/k$ for $k \geqslant 58$. Still a better result may be obtained if in bounding $\mu(\sigma)$ convexity is used for two consecutive values of l in (7.59). This was the idea used by A. Fujii (1976), who obtained a bound for α_k which does not depend explicitly on k, but on a parameter b, so that additional calculations are necessary to evaluate α_k, and a general formula for α_k is difficult to obtain. A calculation shows that Fujii's estimates lead to better values than $\alpha_k \leqslant 1 - 34/7k$ of Theorem 13.2 for $k \geqslant 109$, but his results may be further improved if instead of Theorem 7.10 of E. C. Titchmarsh (1951) one uses sharper bounds for $m(\sigma)$ (when σ is close to 1) obtained by the method of Theorem 8.4, as described above. Also a slight sharpening is obtained if instead of (7.59) one uses the sharper bound

$$\mu(\sigma) \leqslant \frac{1}{4Q - 2} \frac{240Qq - 16Q + 128}{240Qq - 15Q + 128}$$

$$\left(Q = 2^{q-1}, \; q = 3, 4, \ldots, \sigma = 1 - \frac{q + 1}{4Q - 2}\right),$$

which was proved long ago by E. Phillips (1933).

Theorem 13.3 is essentially due to H.-E. Richert (1960), who proved $\alpha_k \leqslant 1 - Ck^{-2/3}$ with an unspecified $C > 0$. His result was rediscovered by A. A. Karacuba (1971), who proved that

$$\Delta_k(x) \ll x^{1 - Ck^{-2/3}} (D \log x)^k$$

holds uniformly in k for some absolute constants $C, D > 0$. The reason that our proof of Theorem 13.3 does not yield a result uniform in k is that (13.16) contains the term $x^{1+\varepsilon}T^{-1}$, and it is the "ε" which vitiates uniformity. To obtain uniformity in k one can use the integrated variant of the inversion formula (A.8), namely,

$$(13.95) \quad \int_1^x \left(\sum_{n \leqslant t} a_n\right) dt = (2\pi i)^{-1} \int_{c-i\infty}^{c+i\infty} A(s) \frac{x^{s+1}}{s(s+1)} ds \quad \left(A(s) = \sum_{n=1}^{\infty} a_n n^{-s}\right),$$

and then replace the line of integration by the segment $[b - iT, b + iT]$ [as in (A.10)] with a suitable error term. The details of this method are to be found in Karacuba (1972).

A. Fujii (1976) showed that $\alpha_k \leqslant 1 - Ck^{-2/3}$ holds with

$$C = 2^{-1/2} (2^{3/2} - 1)^{-1/3} D^{-2/3},$$

where D is the constant which appears in (13.19), while our proof of Theorem 13.2 gives

$$C = \tfrac{1}{2} D^{-2/3},$$

which is sharper than Fujii's value (Fujii uses the value $D = 39$, which P. Turán (1971) attributes to L. Schoenfeld, but I wasn't able to find a published proof of this result).

Lemmas 13.1 and 13.2 are proved in Chapter 12 of E. C. Titchmarsh (1951). Concerning Theorem 13.4 it may be mentioned that $\beta_2 = \frac{1}{4}$, $\beta_3 = \frac{1}{3}$ are classical results which may be found in Titchmarsh (1951), while $\beta_4 = \frac{3}{8}$ has been proved by D. R. Heath-Brown (1981b) and the remaining bounds of Theorem 13.4 are due to the author (1984a). They improve on $\beta_5 \leqslant \frac{1}{2}$, $\beta_6 \leqslant \frac{35}{62}$, $\beta_7 \leqslant \frac{11}{18}$, $\beta_8 \leqslant \frac{149}{230}$ of K.-C. Tong (1953); indeed his bound for β_8 is poorer than our bound for β_7.

The form of Theorem 13.5 is due to H. Cramér (1922), and curiously enough no result of this type is to be found in Titchmarsh (1951). The results of Theorems 13.6 and 13.7 are proved in Chapter 10 of the author's work (1983b), while the theorems of Section 13.7 are proved by the author in (1983a).

Theorem 13.5 and its analogue for the circle problem provide weak omega results for $\Delta(x)$ and $P(x)$, namely, $\Delta(x) = \Omega(x^{1/4})$ and $P(x) = \Omega(x^{1/4})$. Some better results are known, and G. H. Hardy (1915, 1916a) proved

$$\Delta(x) = \begin{cases} \Omega_+\left((x \log x)^{1/4} \log \log x \right) \\ \Omega_-(x^{1/4}) \end{cases}$$

$$P(x) = \begin{cases} \Omega_-\left((x \log x)^{1/4} \right) \\ \Omega_+(x^{1/4}). \end{cases}$$

Hardy's Ω_- estimate for $\Delta(x)$ and Ω_+ estimate for $P(x)$ have been improved a little by K. S. Gangadharan (1961), and the best results in this direction seem to be due to K. Corrádi and I. Kátai (1967), who proved with some absolute $C_1, C_2 > 0$ that

$$\Delta(x) = \Omega_-\left\{ x^{1/4} \exp\left(C_1 (\log \log x)^{1/4} (\log \log \log x)^{-3/4} \right) \right\},$$

$$P(x) = \Omega_+\left\{ x^{1/4} \exp\left(C_2 (\log \log x)^{1/4} (\log \log \log x)^{-3/4} \right) \right\}.$$

Hardy's Ω_+ estimate for $\Delta(x)$ and Ω_- estimate for $P(x)$ withstood improvement for a very long time. Only recently J. L. Hafner (1981) succeeded in proving with some absolute constants $C_3, C_4 > 0$ that

$$\Delta(x) = \Omega_+\left\{ (x \log x)^{1/4} (\log \log x)^{(3 + \log 2)/4} \exp\left(-C_3 (\log \log \log x)^{1/2} \right) \right\},$$

$$P(x) = \Omega_-\left\{ (x \log x)^{1/4} (\log \log x)^{\log 2/4} \exp\left(-C_4 (\log \log \log x)^{1/2} \right) \right\}.$$

Hafner (1982) proves

$$\Delta_k(x) = \Omega^*\left\{ (x \log x)^{(k-1)/2k} (\log \log x)^c \exp\left(-C_5 (\log \log \log x)^{1/2} \right) \right\},$$

where $c = (k - 1)(k \log k - k + 1)/2k + k - 1$ and $\Omega^* = \Omega_+$ if $k = 3$ and $\Omega^* = \Omega_\pm$ if $k \geqslant 4$, which seems to be the best result known.

Although the circle problem, discussed in Section 13.8, is a digression from our main topic which is the zeta-function, I have nevertheless felt it appropriate to include this material [from Chapter 10 of my work (1983b)] for two reasons. First, the results are not without interest, and second they stress the intrinsic connection between the divisor problem for $\Delta(x) = \Delta_2(x)$ and the circle problem. Most earlier authors have investigated the circle problem, the divisor problem, and the problem of the order of $\zeta(\frac{1}{2} + it)$ separately and by different methods. The approach presented here shows a unified view of the circle and divisor problem, and the idea to use exponential averaging in the proof of Lemma 13.4 has been kindly suggested by M. Jutila, whose works (1983a, 1984b) (parts of which will be discussed in Chapter 15) show the intrinsic connection between $\Delta(x)$, $\zeta(\frac{1}{2} + it)$, and $E(T)$ mainly in the light of Atkinson's formula for $E(T)$. The problem of the estimation of $\zeta(\frac{1}{2} + it)$ was already discussed in Chapter 7 (Theorem 7.3), and the proof of Theorem 13.1 is very similar to the proof of Theorem 7.3. It turns out at present that all the best published exponents in the divisor problem, circle problem, and the problem of the order of $E(T)$ are the same one, namely, $\frac{35}{108} + \varepsilon$ by Kolesnik's method. Whether the real order of the functions in question (for which one naturally conjectures the exponent $\frac{1}{4} + \varepsilon$) is the same (up to ε's and log factors) is not possible to predict yet, though one expects the answer to be affirmative.

/ CHAPTER FOURTEEN /
VARIOUS OTHER DIVISOR PROBLEMS

14.1 SUMMATORY FUNCTIONS OF ARITHMETICAL CONVOLUTIONS

In Chapter 13 we considered the Dirichlet divisor problem, that is, the estimation of $\sum_{n \leqslant x} d_k(n)$, where $d_k(n)$ ($k \geqslant 2$ a fixed integer) is generated by $\zeta^k(s)$. There exist many other interesting problems involving summatory functions of arithmetical functions, where the generating series of the arithmetical function in question factors into a product, some of whose factors are of the form $\zeta^a(ms)$, where $m \neq 0$ is an integer and $a \neq 0$ is arbitrary. In this type of problem, which may be thought of as a general divisor problem, results from zeta-function theory may be often successfully employed and our aim in this chapter is to present several problems of this type. Naturally, we had to be selective, since the number of applications of zeta-function theory to summatory functions of multiplicative and additive functions seems almost endless.

We begin by considering summatory functions of arithmetical convolutions. Recall that if $f(n)$ and $g(n)$ are two arithmetical functions, then the convolution $h(n)$ of $f(n)$ and $g(n)$ was defined by (1.83) to be

$$(14.1) \qquad h(n) = \sum_{d|n} f(n/d)g(d) = \sum_{d|n} g(n/d)f(d).$$

A common procedure in dealing with the asymptotic formula for the sum $C(x) = \sum_{n \leqslant x} h(n)$ is to express suitably $h(n)$ in the form (14.1) and then to estimate $C(x)$ using asymptotic formulas for the summatory functions $A(x) = \sum_{n \leqslant x} f(n)$ and $B(x) = \sum_{n \leqslant x} g(n)$. If

$$(14.2) \qquad F(s) = \sum_{n=1}^{\infty} f(n)n^{-s}, \qquad G(s) = \sum_{n=1}^{\infty} g(n)n^{-s}$$

both converge absolutely in the half-plane $\sigma > \sigma_1$, then in this half-plane (14.2) is equivalent to

$$(14.3) \qquad H(s) = \sum_{n=1}^{\infty} h(n)n^{-s} = F(s)G(s).$$

If it is somehow known that the main contribution to $C(x)$ comes from $A(x)$, and $F(s)$ in (14.2) may be written as

$$(14.4) \qquad F(s) = \zeta^{m_1}(k_1 s)\zeta^{m_2}(k_2 s)\zeta^{m_3}(k_3 s)\ldots \qquad (\sigma > \sigma_1),$$

where $1 \leqslant k_1 < k_2 < \cdots$ are integers, then the estimation of $C(x)$ may be considered as a general divisor problem. In general $m_1, m_2,\ldots$ may be complex numbers (see Section 14.6 for an example), but in many applications they turn out to be integers. The product in (14.4) may be finite, in which case $F(s)$ possesses a finite zeta-product, or it may be infinite, when we say that $F(s)$ possesses an infinite zeta-product. The former case is illustrated by the Dirichlet divisor problem, where $F(s) = \zeta^k(s)$, $G(s) = 1$, and because of its importance and direct connections with the zeta-function this problem was treated separately in Chapter 13. As an example of an infinite zeta-product one may mention $\zeta(s)\zeta(2s)\zeta(3s)\ldots$, the generating function of $a(n)$, the number of nonisomorphic abelian groups with n elements. This function was introduced in Section 1.8 of Chapter 1, and some problems involving $a(n)$ will be dealt with in Section 14.4. Our aim in this section is to prove some reasonably general convolution results, which may be used to obtain asymptotics of various specific functions $C(x) = \sum_{n \leqslant x} h(n)$.

In the study of asymptotic formulas for arithmetical functions we invariably come upon such functions as $x^\alpha \log^\beta x$, $\operatorname{li} x$, $x^\alpha(\log x)^\beta(\log \log x)^\gamma$, $x^\alpha \exp(\beta \log^{\gamma_0} x)$. If α is complex and β and γ are real numbers ($\gamma_0 < 1$), then these functions are of the form $x^\alpha L(x)$, where $L(x)$ is called a slowly oscillating (or slowly varying) function, that is, $L(x)$ is a continuous, positive-valued function on (x_0, ∞) for some $x_0 \geqslant 1$ such that

$$(14.5) \qquad \lim_{x \to \infty} \frac{L(Cx)}{L(x)} = 1$$

for each $C > 0$. Although we have encountered slowly oscillating functions from the very beginning of this text, there was no need so far for general results involving summatory functions, and thus it seemed appropriate to defer introduction of slowly oscillating functions until this point. If $\delta(x)$ and $\rho(x)$ are continuous for $x \geqslant x_0$,

$$\lim_{x \to \infty} \delta(x) = 0, \qquad \lim_{x \to \infty} \rho(x) = \rho > 0,$$

then it is not difficult to see that the function

$$(14.6) \qquad L(x) = \rho(x)\exp\left(\int_{x_0}^{x} t^{-1}\delta(t)\, dt\right)$$

is slowly oscillating. Conversely, J. Karamata, who founded the theory of slowly oscillating functions, proved that every slowly oscillating function may be written in the form (14.6). From (14.6) it follows that $x^{-\varepsilon} < L(x) < x^{\varepsilon}$ for $x \geqslant x_1$, and that the limit (14.5) is uniform in C on each bounded interval $0 < k_1 \leqslant C \leqslant k_2 < \infty$.

In many applications of convolution results it seems more convenient to use

$$(14.7) \qquad h(n) = \sum_{d^k|n} f(n/d^k)g(d),$$

where $k \geqslant 1$ is a fixed integer, instead of employing directly (14.1). However, (14.7) may be at once reduced to the form (14.1) by defining

$$(14.8) \qquad G(n) = \begin{cases} g(m) & \text{if } n = m^k, \\ 0 & \text{if } n \neq m^k. \end{cases}$$

With this in mind we may formulate now

THEOREM 14.1. Let $f(n)$ be an arithmetical function for which

$$(14.9) \quad \sum_{n \leqslant x} f(n) = \sum_{j=1}^{l} c_j x^{a_j} L_j(x) + O(x^a), \qquad \sum_{n \leqslant x} |f(n)| = O(x^{a_1} P(x)),$$

where $a_1 \geqslant a_2 \geqslant \cdots \geqslant a_l > 1/k > a \geqslant 0$; $c_1, \ldots, c_l$ are constants; $k \geqslant 1$ is a fixed integer; $L_1(x), \ldots, L_l(x)$ are slowly oscillating functions; and $P(x)$ is a nondecreasing slowly oscillating function. Further let $g(n)$ be an arithmetical function for which

$$(14.10) \qquad \sum_{n \leqslant x} g(n) = O(x^b N(x)), \qquad \sum_{n \leqslant x} |g(n)| = O(x),$$

where $0 \leqslant b \leqslant 1$, $N(x)$ is a slowly oscillating function of the form

$$N(x) = \exp(C\omega(x)), \qquad \omega(x) = \int_{x_0}^{x} \eta(t)t^{-1}\, dt,$$

$\eta(x)$ is continuous and positive for $x \geqslant x_0$, $\lim_{x \to \infty} \eta(x) = 0$, and for every $A < 0$ $\lim_{x \to \infty} P(x)\exp(A\omega(x)) = 0$. Here $C < 0$ if $b = 1$ and $C > 0$ if $0 \leqslant b < 1$.

If $h(n)$ is defined by (14.7), then there exist functions $Q_1(x),\ldots,Q_l(x)$ such that $Q_j(x) \ll x^\varepsilon$ for $j = 1,\ldots,l$ and

$$(14.11) \qquad \sum_{n \leqslant x} h(n) = \sum_{j=1}^{l} c_j x^{a_j} Q_j(x) + \delta(x),$$

where for $b = 1$

$$(14.12) \qquad \delta(x) \ll x^{1/k}\exp(D\omega_1(x)), \qquad \omega_1(x) = \int_{x_1}^{x} \eta(t^u)t^{-1}\,dt,$$

and $x_1 = x_0^{1/u}$, $D < 0$ for every $u < 1/k$. If $0 \leqslant b < 1$, then for some $D > 0$

$$(14.13) \qquad \delta(x) \ll x^c \exp(D\omega(x)), \qquad c = \frac{a_1 - ab}{a_1 k - ak + 1 - b}.$$

Proof of Theorem 14.1. Let $y, z > 1$ and $yz = x$. Then we use (14.8) and a splitting-up argument which originated with Dirichlet to write

$$\sum_{n \leqslant x} h(n) = \sum_{n \leqslant x}\sum_{d|n} G(d)f(n/d) = \sum_{mn \leqslant x} G(m)f(n)$$

$$= \sum_{m \leqslant y} G(m) \sum_{n \leqslant x/m} f(n) + \sum_{n \leqslant z} f(n) \sum_{m \leqslant x/n} G(m)$$

$$- \sum_{m \leqslant y} G(m) \sum_{n \leqslant z} f(n) = S_1 + S_2 - S_3,$$

say. We have

$$\sum_{n \leqslant x} G(n) = \sum_{n \leqslant x^{1/k}} g(n) \ll x^{b/k}N(x^{1/k}), \qquad \sum_{n \leqslant x} |G(n)| \ll x^{1/k},$$

so that we obtain

$$S_1 = \sum_{m \leqslant y} G(m) \sum_{n \leqslant x/m} f(n) = \sum_{j=1}^{l} c_j x^{a_j} \sum_{m \leqslant y} G(m)m^{-a_j}L_j(x/m)$$

$$+ O\left(x^a \sum_{m \leqslant y} |G(m)|m^{-a}\right)$$

$$= \sum_{j=1}^{l} c_j x^{a_j} Q_j(x) + O(x^a y^{1/k-a}).$$

Here we used partial summation to obtain the O-term, and for $j = 1, \ldots, l$ we have set

$$(14.14) \quad Q_j(x) = \sum_{m \leqslant y} G(m) m^{-a_j} L_j(x/m) = \sum_{m \leqslant y^{1/k}} g(m) m^{-ka_j} L_j(x/m^k),$$

where $y = y(x)$ will be suitably chosen later. Using $ka_j > 1$, (14.10), and $L(x) \ll x^\varepsilon$ we obtain from (14.14) $Q_j(x) \ll x^\varepsilon$, as claimed.

It may be further shown that

$$\lim_{x \to \infty} Q_j(x)/L_j(x) = \sum_{n=1}^{\infty} g(n) n^{-ka_j},$$

which means that $Q_j(x)$ is slowly oscillating if it is continuous and the above limit is positive, for it is then asymptotic to a slowly oscillating function. A more detailed discussion of properties of $Q_j(x)$ is omitted, since in many applications to divisor problems the functions $Q_j(x)$ turn out to be polynomials in $\log x$, as will be the case in Theorem 14.2.

S_1 will therefore contribute the main terms in (14.11), and to estimate the error terms we distinguish the cases $b = 1$ and $0 \leqslant b < 1$ in (14.10).

The Case $b = 1$

If $b = 1$, then $N(x)$ is decreasing and therefore for $n \leqslant z$ we have $N((x/n)^{1/k}) \leqslant N(y^{1/k})$, which gives

$$S_2 = \sum_{n \leqslant z} f(n) \sum_{m \leqslant x/n} G(m) \ll x^{1/k} N(y^{1/k}) \sum_{n \leqslant z} |f(n)| n^{-1/k}$$

$$\ll x^{1/k} z^{a_1 - 1/k} P(z) N(y^{1/k}) = x^{a_1} y^{1/k - a_1} P(x/y) N(y^{1/k}),$$

$$S_3 = \sum_{m \leqslant y} G(m) \sum_{n \leqslant z} f(n) \ll z^{a_1} P(z) y^{1/k} N(y^{1/k})$$

$$= x^{a_1} y^{1/k - a_1} P(x/y) N(y^{1/k}).$$

Therefore we obtain

(14.15)

$$\sum_{n \leqslant x} h(n) = \sum_{j=1}^{l} c_j x^{a_j} Q_j(x) + O\left(x^a y^{1/k - a}\right) + O\left(x^{a_1} y^{1/k - a_1} P(x/y) N(y^{1/k})\right).$$

Let now $0 < u < 1/k$ and choose

$$y = x\left(N(x^u)\right)^{1/(a_1 - a)},$$

so that $y < x$ if $x \geqslant x_0$. We have $t^{-\varepsilon} < N(t) < t^\varepsilon$, since $N(t)$ is slowly oscillating, so that $x^u \leqslant x^{1/k - \varepsilon} \leqslant y^{1/k}$ for $0 < \varepsilon < 1/k - u$, and since $N(t)$ is decreasing this implies $N(y^{1/k}) \leqslant N(x^u)$. This means that the error terms in (14.15) are

$$\ll x^{1/k}\{N(x^u)\}^{(1/k - a)/(a_1 - a)}\left(1 + \frac{N(y^{1/k})}{N(x^u)}P(x/y)\right)$$

$$\ll x^{1/k}\{N(x^u)\}^{(1/k - a)/(a_1 - a)}P(x^u),$$

since $x/y < x^u$ if x is large enough. If $C_1 = (C/k - Ca)/(a_1 - a)$, then for every A satisfying $C_1 < A < 0$ we have

$$\{N(x^u)\}^{(1/k - a)/(a_1 - a)}P(x^u) = P(x^u)\exp(A\omega(x^u))\exp((C_1 - A)\omega(x^u))$$

$$\ll \exp(D\omega_1(x)),$$

where

$$D = (C_1 - A)u, \qquad \omega_1(x) = \int_{x_1}^{x}\eta(t^u)t^{-1}\,dt, \qquad x_1 = x_0^{1/u},$$

since by hypothesis

$$\lim_{x \to \infty} P(x^u)\exp(A\omega(x^u)) = 0.$$

The Case $0 \leqslant b < 1$

If $0 \leqslant b < 1$, then $N(x)$ is increasing and therefore $N(x^{1/k}n^{-1/k}) \leqslant N(x)$, so that

$$S_2 \ll x^{b/k}N(x)\sum_{n \leqslant z}|f(n)|n^{-b/k} \ll x^{b/k}N(x)P(x)z^{a_1 - b/k}$$

$$= x^{a_1}y^{b/k - a_1}N(x)P(x),$$

and the same estimate is found to hold for S_3 too. Therefore in this case

$$\sum_{n \leqslant x}h(n) = \sum_{j=1}^{l}c_j x^{a_j}Q_j(x) + O(x^a y^{1/k - a}) + O(x^{a_1}y^{b/k - a_1}N(x)P(x)).$$

If $D > C$, then

$$N(x)P(x) = \exp(D\omega(x))\exp((C - D)\omega(x))P(x) \ll \exp(D\omega(x)),$$

since for $A = C - D < 0$ we have

$$\lim_{x \to \infty} P(x)\exp(A\omega(x)) = 0.$$

Taking now

$$y = x^q, \qquad q = \frac{k(a_1 - a)}{1 - b + k(a_1 - a)},$$

we obtain finally

$$\sum_{n \leqslant x} h(n) = \sum_{j=1}^{l} c_j x^{a_j} Q_j(x) + O(x^c \exp(D\omega(x))),$$

$$c = \frac{a_1 - ab}{1 - b + k(a_1 - a)}$$

as asserted.

Theorem 14.1 is of a general nature, and now we specialize it to obtain a result which is better adapted for applications. This is

THEOREM 14.2. Let $f(n)$ be an arithmetical function for which

(14.16)

$$\sum_{n \leqslant x} f(n) = \sum_{j=1}^{l} x^{a_j} P_j(\log x) + O(x^a), \qquad \sum_{n \leqslant x} |f(n)| = O(x^{a_1}\log^r x),$$

where $a_1 \geqslant a_2 \geqslant \cdots \geqslant a_l > 1/k > a \geqslant 0$, $r \geqslant 0$, $P_1(t),\ldots, P_l(t)$ are polynomials in t of degrees not exceeding r, and $k \geqslant 1$ is a fixed integer. If

$$h(n) = \sum_{d^k | n} \mu(d)f(n/d^k),$$

then

(14.17)

$$\sum_{n \leqslant x} h(n) = \sum_{j=1}^{l} x^{a_j} R_j(\log x) + \delta(x),$$

where $R_1(t), \ldots, R_l(t)$ are polynomials in t of degrees not exceeding r, and for some $D > 0$

(14.18) $$\delta(x) \ll x^{1/k} \exp\left(-D(\log x)^{3/5}(\log\log x)^{-1/5}\right).$$

Furthermore, if the Riemann hypothesis is true, then for some $D > 0$

(14.19) $$\delta(x) \ll x^c \exp\left(\frac{D\log x}{\log\log x}\right), \qquad c = \frac{2a_1 - a}{2ka_1 - 2ka + 1}.$$

Proof of Theorem 14.2. The proof follows on applying Theorem 14.1 with $c_j L_j(x) = P_j(\log x)$, $P(x) = \log^r x$, $g(n) = \mu(n)$. For $M(x) = \sum_{n \leqslant x} \mu(n)$ we use Theorem 12.7, which gives with $\varepsilon(x) = (\log x)^{3/5}(\log\log x)^{-1/5}$

(14.20) $$M(x) \ll x \exp(-C\varepsilon(x)) \qquad (C > 0).$$

This corresponds to the case $b = 1$ of Theorem 14.1, while if the Riemann hypothesis holds we have by (1.140)

(14.21) $$M(x) \ll x^{1/2} \exp(C\omega(x)) \qquad [\omega(x) = \log x/\log\log x, \ C > 0],$$

which corresponds to the case $b = \frac{1}{2}$ of Theorem 14.1. If we could prove $M(x) \ll x^{b+\varepsilon}$ for some $\frac{1}{2} < b < 1$ [this is equivalent to $\zeta(s) \neq 0$ for $\sigma > b$], then Theorem 14.1 would give (14.17) with

$$\delta(x) \ll x^{c+\varepsilon}, \qquad c = \frac{a_1 - ab}{a_1 k - ak + 1 - b}.$$

It should be noted that for $j = 1, \ldots, l$

$$c_j \sum_{m \leqslant y} G(m) m^{-a_j} L_j(x/m) = c_j \sum_{m=1}^{\infty} \mu(m) m^{-ka_j} L_j(x/m^k)$$

$$-c_j \sum_{m^k > y} \mu(m) m^{-ka_j} L_j(x/m^k).$$

Now $L_j(x/m^k)$ may be written as a polynomial in $\log x$ of degree not exceeding r, so that

$$c_j \sum_{m=1}^{\infty} \mu(m) m^{-ka_j} L_j(x/m^k) = R_j(\log x),$$

where $R_j(t)$ is a polynomial in t of degree not exceeding r, and it remains to show that sums of the type

(14.22) $$\sum_{m > y^{1/k}} \mu(m) m^{-ka_j} \log^A m \qquad (A \geqslant 0)$$

contribute to the error term in (14.17). To this end, let $y^{1/k} = v$, $ka_j = c > 1$ and observe that

$$\sum_{m>v} \mu(m)m^{-c}\log^A m = \int_v^\infty t^{-c}\log^A t\, dM(t)$$

$$= v^{-c}M(v)\log^A v + O\left(\int_v^\infty |M(t)|t^{-c-1}\log^A t\, dt\right).$$

For $M(x)$ we use first (14.20), noting that $\exp(-C\varepsilon(x))$ is decreasing for $x \geqslant x_1$ and thus the sum in (14.22) is

$$\ll v^{1-c}\exp(-C\varepsilon(v))\log^A v + \exp(-C\varepsilon(v))\int_v^\infty t^{-c}\log^A t\, dt$$

$$\ll v^{1-c}\exp\left(-\frac{C}{2}\varepsilon(v)\right) \ll y^{1/k-a_j}\exp(-D\varepsilon(x)) \qquad (D > 0)$$

which multiplied by $c_j x^{a_j} Q_j(x)$ may be incorporated in the second O-term in (14.15).

If we use (14.21) for $M(x)$, then $x^{-1/2}\exp(C\omega(x))$ is decreasing for $x \geqslant x_2$; hence the sum in (14.22) is

$$\ll v^{1/2-c}\exp(C\omega(v))\log^A v + v^{-1/2}\exp(C\omega(v))\int_v^\infty t^{-c}\log^A t\, dt$$

$$\ll v^{1/2-c}\exp(C\omega(v))\log^A v \ll v^{1/2-c}\exp(2C\omega(v)),$$

and this contributes again to the error term. The remaining details of proof are the same as in Theorem 14.1.

14.2 SOME APPLICATIONS OF THE CONVOLUTION METHOD

In this section we shall apply Theorem 14.2 to obtain asymptotic formulas for several summatory functions of arithmetical functions, while some more involved problems, such as powerful numbers and finite abelian groups, will be discussed in Sections 14.4 and 14.5, respectively.

First, we consider the estimation of $Q_k(x)$, the number of k-free integers ($k \geqslant 2$ fixed) not exceeding x. From the product representation (1.88) we see that $f_k(n)$, the characteristic function of the set of k-free numbers, may be expressed as

$$(14.23) \qquad f_k(n) = \sum_{d^k|n} \mu(d).$$

Therefore Theorem 14.2 may be applied with $f(n) = 1$, $l = a_1 = 1$, $r = a$ $= 0$, and (14.17) gives

$$(14.24) \qquad Q_k(x) = \sum_{n \leqslant x} f_k(n) = \frac{x}{\zeta(k)}$$

$$+ O\left(x^{1/k}\exp\left(-C(\log x)^{3/5}(\log\log x)^{-1/5}\right)\right),$$

where the main term is most quickly obtained by residues. From (14.19) we infer that the Riemann hypothesis implies that the error term in (14.24) is

$$\ll x^{2/(2k+1)}\exp\left(\frac{C\log x}{\log\log x}\right) \qquad (C > 0).$$

Next we note that from the product representation (1.107), (1.105), and (1.106), we have respectively

$$(14.25) \qquad 2^{\omega(n)} = \sum_{\delta|n}\mu^2(\delta) = \sum_{\delta^2|n}\mu(\delta)d(n/\delta^2),$$

$$(14.26) \qquad d(n^2) = \sum_{\delta^2|n}\mu(\delta)d_3(n/\delta^2),$$

$$(14.27) \qquad d^2(n) = \sum_{\delta^2|n}\mu(\delta)d_4(n/\delta^2).$$

Therefore we may apply Theorem 14.2 with $k = 2$; $l = a_1 = 1$; $f(n) = d(n)$, $d_3(n)$, and $d_4(n)$, respectively. Using estimates for $D_k(x) = \sum_{n \leqslant x}d_k(n)$ of Chapter 13 we then have $a = \alpha_k + \varepsilon$, and since $\alpha_2 < \frac{1}{2}$, $\alpha_3 < \frac{1}{2}$, and $\alpha_4 = \frac{1}{2} + \varepsilon$ we obtain

$$(14.28) \quad \sum_{n \leqslant x} 2^{\omega(n)} = A_1 x \log x + A_2 x + O\left(x^{1/2}\exp(-C\varepsilon(x))\right),$$

$$(14.29) \quad \sum_{n \leqslant x} d(n^2) = B_1 x \log^2 x + B_2 x \log x + B_3 x + O\left(x^{1/2}\exp(-C\varepsilon(x))\right),$$

$$(14.30) \quad \sum_{n \leqslant x} d^2(n) = C_1 x \log^3 x + C_2 x \log^2 x + C_3 x \log x + C_4 x + O(x^{1/2+\varepsilon}),$$

where $\varepsilon(x) = (\log x)^{3/5}(\log\log x)^{-1/5}$ and $A_1,\ldots,C_4$ are constants which may be explicitly evaluated. More generally, we may consider the function

$$F_k(n) = \sum_{\delta^k|n}\mu(\delta)d_k(n/\delta^k)$$

and obtain, using Theorem 14.2,

$$(14.31) \qquad \sum_{n \leqslant x} F_k(n) = xQ_{k-1}(\log x) + O(x^{1/2}\exp(-C\varepsilon(x))),$$

where $Q_{k-1}(t)$ is a polynomial of degree $k - 1$ in t and we suppose that $\alpha_k < \frac{1}{2}$ holds.

For our next example we recall from elementary number theory that d is a unitary divisor of n if $d|n$ and $(d, n/d) = 1$. Let $(a, b)^{**}$ denote the greatest unitary divisor of both a and b. Then we may define d as a bi-unitary divisor of n if $d|n$ and $(d, n/d)^{**} = 1$. Let $\tau^{**}(n)$ denote the number of bi-unitary divisors of n. The function $\tau^{**}(n)$ is clearly multiplicative and satisfies

$$\tau^{**}(p_1^{\alpha_1} \cdots p_r^{\alpha_r}) = \prod_{\alpha_j even} \alpha_j \prod_{\alpha_j odd} (\alpha_j + 1),$$

where $n = p_1^{\alpha_1} \cdots p_r^{\alpha_r}$ is the canonical decomposition of n. Then for $\sigma > 1$ we have

$$(14.32) \qquad \sum_{n=1}^{\infty} \tau^{**}(n)n^{-s} = \prod_{p}(1 + 2p^{-s} + 2p^{-2s} + 4p^{-3s} + 4p^{-4s} + \cdots)$$

$$= \zeta^2(s)U(s)/\zeta(2s) = \sum_{n=1}^{\infty} 2^{\omega(n)}n^{-s}U(s),$$

and

$$(14.33) \qquad U(s) = \sum_{n=1}^{\infty} u(n)n^{-s}$$

is absolutely convergent for $\sigma > \frac{1}{3}$. Our aim is to prove that

$$(14.34) \qquad \sum_{n \leqslant x} \tau^{**}(n) = ax\left(\log x + 2\gamma - 1 + 2\sum_{p} \frac{(p^2 - p - 1)\log p}{p^4 + 2p^3 + 1}\right)$$

$$+ O(x^{1/2}\exp(-C\varepsilon(x))),$$

where $C > 0$, $\varepsilon(x) = (\log x)^{3/5}(\log\log x)^{-1/5}$ and

$$(14.35) \qquad a = \prod_{p}\left(1 - \frac{p-1}{p^3 + p^2}\right).$$

This may be achieved either by applying Theorem 14.2 with $k = 2$ to the function $f(n)$ generated by $\zeta^2(s)U(s)$, or by proceeding directly. From (14.32) we have

$$\sum_{n \leqslant x} \tau^{**}(n) = \sum_{n \leqslant x} u(n) \sum_{m \leqslant x/n} 2^{\omega(m)}.$$

Now using (14.28) with $R(x)$ for the error term and writing

$$\sum_{n \leqslant x} u(n) R(x/n) = \sum_{n \leqslant x^{1/2}} u(n) R(x/n) + \sum_{x^{1/2} < n \leqslant x} u(n) R(x/n)$$

$$\ll x^{1/2} \sum_{n \leqslant x^{1/2}} |u(n)| n^{-1/2} \exp\left(-C\varepsilon(x^{1/2})\right)$$

$$+ x^{1/2} \sum_{n > x^{1/2}} |u(n)| n^{-1/2}$$

$$\ll x^{1/2} \exp\left(-C_1 \varepsilon(x)\right) + x^{1/2 - \delta}$$

$$\ll x^{1/2} \exp\left(-C_1 \varepsilon(x)\right) \qquad \left(0 < \delta < \tfrac{1}{12}\right),$$

we obtain

(14.36) $$\sum_{n \leqslant x} \tau^{**}(n) = Ax \log x + Bx + O(x \exp(-C\varepsilon(x))),$$

where the main terms $Ax \log x + Bx$ are most quickly obtained by residues, so it remains only to evaluate A and B. A simple way to accomplish this is to write for $\sigma > \tfrac{1}{2}$

$$V(s) = \sum_{n=1}^{\infty} v(n) n^{-s} = U(s)/\zeta(2s).$$

Then we obtain

$$\sum_{n \leqslant x} \tau^{**}(n) = \sum_{n \leqslant x} v(n) \sum_{m \leqslant x/n} d(n)$$

$$= \sum_{n \leqslant x} v(n) \left(\frac{x}{n} \log \frac{x}{n} + (2\gamma - 1)\frac{x}{n} + O(x^{1/3} n^{-1/3})\right),$$

where $d(n)$ is the ordinary number of divisors function. Collecting terms and comparing with (14.36) we obtain

$$A = \sum_{n=1}^{\infty} v(n) n^{-1} = V(1),$$

$$B = (2\gamma - 1)V(1) - \sum_{n=1}^{\infty} v(n) n^{-1} \log n = (2\gamma - 1)V(1) + V'(1).$$

But for $\sigma > \frac{1}{2}$ we have the product representation

$$V(s) = \prod_p (1 - p^{-s})^2 (1 + 2p^{-s} + 2p^{-2s} + 4p^{-3s} + 4p^{-4s} + \cdots)$$

$$= \prod_p \left((1 - p^{-s})^2 + 2(p^s + 1)^{-1} \right).$$

Hence

$$V(1) = \prod_p \left(1 - 2p^{-1} + p^{-2} + \frac{2}{p+1} \right) = a$$

as given by (14.35), while

$$V'(s)/V(s) = (\log V(s))' = 2 \sum_p \log p \frac{(1 - p^{-s}) p^{-s} - p^s (p^s + 1)^{-2}}{(1 - p^{-s})^2 + 2(p^s + 1)^{-1}}.$$

Therefore

$$V'(1) = 2V(1) \sum_p \frac{p^2 - p - 1}{p^4 + 2p^3 + 1} \log p,$$

which shows that (14.34) indeed holds.

14.3   THREE-DIMENSIONAL DIVISOR PROBLEMS

Throughout this section we assume that $1 \leqslant a \leqslant b \leqslant c$ are fixed integers, and we denote by $d(a, b, c; k)$ the number of representations of k as $k = n_1^a n_2^b n_3^c$, where n_1, n_2, n_3 are natural numbers, that is,

$$(14.37) \qquad\qquad d(a, b, c; k) = \sum_{k = n_1^a n_2^b c_3^c} 1.$$

The general three-dimensional divisor problem consists of estimating the function $\Delta(a, b, c; x)$, which may be considered as the error term in the asymptotic formula

$$(14.38) \quad D(a, b, c; x) = \sum_{1 \leqslant k \leqslant x} d(a, b, c; k)$$

$$= \zeta(b/a)\zeta(c/a)x^{1/a} + \zeta(a/b)\zeta(c/b)x^{1/b}$$

$$+ \zeta(a/c)\zeta(b/c)x^{1/c} + \Delta(a, b, c; x).$$

This asymptotic formula plays an important role in many problems, and the main terms in (14.38) can be most conveniently obtained by residues, since

$$(14.39) \qquad \sum_{n=1}^{\infty} d(a,b,c;n)n^{-s} = \zeta(as)\zeta(bs)\zeta(cs) \qquad (\sigma > 1/a).$$

The product representation (14.39) reveals the close connection between $d(a,b,c;n)$ and zeta-function theory, and we see that the classical divisor problem for $d_3(n)$ is a special case of (14.38) when $a = b = c = 1$, that is, $\Delta(1,1,1;x) = \Delta_3(x)$ in customary notation. In (14.38) we have in fact to assume that $a < b < c$, and in case some of these numbers are equal we have to take the appropriate limit for (14.38) to hold in this case too. Our aim will be to provide a general estimate for $\Delta(a,b,c;x)$, and our main tools will be the formulas

$$(14.40) \qquad \sum_{n \leqslant z} n^{-s} = \zeta(s) + \frac{z^{1-s}}{1-s} - \psi(z)z^{-s} - sB(z)z^{-1-s}$$

$$+ O(z^{-2-s}),$$

if a, b, c are distinct and

$$(14.41) \qquad \sum_{n \leqslant z} n^{-1} = \log z + \gamma - \psi(z)z^{-1} + O(z^{-2})$$

if some of the numbers a, b, c are equal. Here $0 < s \neq 1$,

$$(14.42) \qquad \psi(z) = z - [z] - \tfrac{1}{2}, \qquad B(z) = \tfrac{1}{2}\psi^2(z) - \tfrac{1}{24},$$

and both (14.40) and (14.41) are easy consequences of the Euler–Maclaurin summation formula (A.24). For simplicity we shall henceforth assume that $1 \leqslant a < b < c$, but it is clear that only minor changes in the argument have to be made to deal with the case $a = b$, $b = c$ also. In accordance with (14.37) and (14.38) we define also

$$(14.43) \qquad d(a,b;k) = \sum_{n_1^a n_2^b = k} 1, \, D(a,b;x) = \sum_{1 \leqslant k \leqslant x} d(a,b;k),$$

aiming to reduce the estimation of $\Delta(a,b,c;x)$ in (14.38) to an expression involving the simpler function $\Delta(a,b;x)$, given by

(14.44)

$$D(a,b;x) = \sum_{1 \leqslant k \leqslant x} d(a,b;k) = \zeta(b/a)x^{1/a} + \zeta(a/b)x^{1/b} + \Delta(a,b;x).$$

For $1 \leqslant a < b < c, 1 \leqslant y \leqslant x$, we may write

(14.45)

$$D(a, b, c; x) = \sum_{nm^c \leqslant x} d(a, b; n)$$

$$= \sum_{m^c \leqslant y} \sum_{n \leqslant xm^{-c}} d(a, b; n) + \sum_{nm^c \leqslant x, \, y < m^c \leqslant x} d(a, b; n)$$

$$= \sum_{m^c \leqslant y} D(a, b; xm^{-c}) + \sum_{n \leqslant x/y} d(a, b; n) \sum_{y < m^c \leqslant x/n} 1$$

$$= \sum_{m^c \leqslant y} D(a, b; xm^{-c}) + \sum_{n \leqslant x/y} d(a, b; n) \{(x/n)^{1/c} - \psi((x/n)^{1/c})\}$$

$$- (y^{1/c} - \psi(y^{1/c})) D(a, b; x/y).$$

For further transformations of $D(a, b, c; x)$ we need the following

LEMMA 14.1. Let $1 \leqslant a < b < c$. Then if $\Delta(a, b; x)$ is defined by (14.44), we have

(14.46) $\quad \Delta(a, b; x) = - \sum_{n^{a+b} \leqslant x} \{\psi((x/n^b)^{1/a}) + \psi((x/n^a)^{1/b})\} + O(1),$

(14.47)

$$\sum_{n \leqslant x} (x/n)^{1/c} d(a, b; n) = \frac{c}{c-a} \varsigma\left(\frac{b}{a}\right) x^{1/a} + \frac{c}{c-b} \varsigma\left(\frac{a}{b}\right) x^{1/b}$$

$$+ \varsigma\left(\frac{a}{c}\right) \varsigma\left(\frac{b}{c}\right) x^{1/c} + \Delta(a, b; x) + O(1).$$

Proof of Lemma 14.1. We may write

$$D(a, b; x) = \sum_{n^{a+b} \leqslant x} \sum_{m^b \leqslant xn^{-a}} 1 + \sum_{m^{a+b} \leqslant x} \sum_{x^{1/(a+b)} < n \leqslant (xm^{-b})^{1/a}} 1$$

$$= \sum_{n^{a+b} \leqslant x} \{[(x/n^b)^{1/a}] + [(x/n^a)^{1/b}]\} - [x^{1/(a+b)}]^2$$

$$= \sum_{n^{a+b} \leqslant x} \{(x/n^b)^{1/a} + (x/n^a)^{1/b}\} - x^{2/(a+b)}$$

$$+ 2x^{1/(a+b)} \psi(x^{1/(a+b)})$$

$$- \sum_{n^{a+b} \leqslant x} \{\psi((x/n^b)^{1/a}) + \psi((x/n^a)^{1/b})\} + O(1).$$

Now if we use (14.40) with $sB(z)z^{-1-s} \ll z^{-1-s}$ and $z = x^{1/(a+b)}$, $s = b/a$ and $s = a/b$ and simplify, we obtain (14.46).

For (14.47) we use again Dirichlet's splitting-up argument and write

$$\sum_{n \leqslant x} (x/n)^{1/c} d(a, b; n) = \sum_{n^a m^b \leqslant x} (xn^{-a}m^{-b})^{1/c} = S_1 + S_2 - S_3,$$

say, where

$$S_1 = x^{1/c} \sum_{m^{a+b} \leqslant x} m^{-b/c} \sum_{n^a \leqslant xm^{-b}} n^{-a/c},$$

$$S_2 = x^{1/c} \sum_{n^{a+b} \leqslant x} n^{-a/c} \sum_{m^b \leqslant xn^{-a}} m^{-b/c},$$

$$S_3 = x^{1/c} \sum_{n^{a+b} \leqslant x} n^{-a/c} \sum_{m^{a+b} \leqslant x} m^{-b/c}.$$

Using (14.40) we obtain

$$S_1 = x^{1/c} \sum_{m^{a+b} \leqslant x} m^{-b/c} \left\{ \varsigma\left(\frac{a}{c}\right) + \frac{c}{c-a}\left(\frac{x}{m^b}\right)^{1/a-1/c} \right.$$

$$\left. -\left(\frac{x}{m^b}\right)^{-1/c} \psi\left(\left(\frac{x}{m^b}\right)^{1/a}\right) + O\left(\left(\frac{x}{m^b}\right)^{-1/c-1/a}\right) \right\}$$

$$= \frac{c}{c-a} \varsigma\left(\frac{b}{a}\right) x^{1/a} + \frac{c}{c-b} \varsigma\left(\frac{a}{c}\right) x^{(a+c)/c(a+b)}$$

$$- \frac{ac}{(c-a)(b-a)} x^{2/(a+b)} + \varsigma\left(\frac{a}{c}\right)\varsigma\left(\frac{b}{c}\right) x^{1/c}$$

$$- \varsigma\left(\frac{a}{c}\right) x^{a/c(a+b)} \psi\left(x^{1/(a+b)}\right) - \frac{c}{c-a} x^{1/(a+b)} \psi\left(x^{1/(a+b)}\right)$$

$$- \sum_{m^{a+b} \leqslant x} \psi\left(\left(\frac{x}{m^b}\right)^{1/a}\right) + O(1),$$

and S_2 is the same as S_1, only a and b are interchanged. Evaluating in a similar fashion S_3 and calculating $S_1 + S_2 - S_3$ we arrive at (14.47).

From (14.46) trivially $\Delta(a, b; x) \ll x^{1/(a+b)}$; hence using (14.46) and choosing

(14.48) $y = x^{c/(a+b+c)} = x^{c/d}, \qquad d = a + b + c,$

we find that

$$(14.49) \quad -\left(y^{1/c} - \psi(y^{1/c})\right)D(a,b;x/y)$$

$$= -\zeta(b/a)x^{1/a}y^{1/c-1/a} - \zeta(a/b)x^{1/b}y^{1/c-1/b} - y^{1/c}\Delta(a,b;x/y)$$

$$+\zeta(b/a)\psi(y^{1/c})x^{1/a}y^{-1/a} + \zeta(a/b)\psi(y^{1/c})x^{1/b}y^{-1/b} + O(y^{1/c}).$$

If we suppose additionally

$$(14.50) \qquad\qquad b \leqslant 2a,$$

then in (14.45) we may evaluate $\sum_{m^c \leqslant y} D(a,b;xm^{-c})$ and

$$\sum_{n \leqslant x/y} d(a,b;n)\left\{(x/n)^{1/c} - \psi\left((x/n)^{1/c}\right)\right\}$$

by using Lemma 14.1 with an error term $O(x^{1/d})$. The condition $b \leqslant 2a$ ensures that we have

$$x^{1/a}y^{-2/c-1/a} = x^{(b-a)/(a^2+ab+ac)} \leqslant x^{1/(a+b+c)} = x^{1/d},$$

and this term arises from $\sum_{m^c \leqslant y} D(a,b;xm^{-c})$ when (14.40) is applied with $z = y^{1/c}$, $s = c/a$. Collecting all the estimates in (14.45) and simplifying we obtain then

LEMMA 14.2. If $1 \leqslant a < b < c$, $b \leqslant 2a$, $y = x^{c/(a+b+c)}$, then

$$(14.51) \qquad \Delta(a,b,c;x) = -\frac{c}{a}\zeta\left(\frac{b}{a}\right)B\left(y^{1/c}\right)x^{1/a}y^{(-1/c-1/a)}$$

$$+ \sum_{m^c \leqslant y} \Delta\left(a,b;\frac{x}{m^c}\right) - S + O(y^{1/c}),$$

where

$$(14.52) \qquad\qquad S = \sum_{m^a n^b \leqslant x/y} \psi\left(\left(\frac{x}{m^a n^b}\right)^{1/c}\right).$$

We proceed now to obtain a more symmetric expression for the error term $\Delta(a,b,c;x)$ in the three-dimensional divisor problem. For this we shall need Lemma 14.2 and

LEMMA 14.3. For $y \geqslant 1$, $\rho > 0$ fixed, $1 \leqslant \mu \leqslant y^{1/(\rho+1)} \leqslant \nu$, $\mu^\rho \nu = y$ we have

$$(14.53) \qquad \sum_{y^{1/(\rho+1)} < n \leqslant \nu} \psi\left((y/n)^{1/\rho}\right) = \sum_{\mu < n \leqslant y^{1/(\rho+1)}} \psi(yn^{-\rho})$$

$$- \rho B(\mu) y\mu^{-1-\rho} + O(y\mu^{-2-\rho}) + O(1).$$

Proof of Lemma 14.3. Proceeding similarly as in the proof of Lemma 14.1 and using (14.40) in full we obtain

$$\sum_{m^\rho n \leqslant y} 1 = \zeta(\rho) y + \zeta(1/\rho) y^{1/\rho} - \sum_{n \leqslant \mu} \psi(yn^{-\rho}) - \sum_{n \leqslant \nu} \psi\left((y/n)^{1/\rho}\right)$$

$$- \rho B(\mu) y\mu^{-1-\rho} + O(y\mu^{-2-\rho}) + O(1).$$

Writing this equation once more with $\mu = \nu = y^{1/(\rho+1)}$ and subtracting we obtain (14.53).

To obtain a symmetric expression for $\Delta(a, b, c; x)$ we define, if (α, β, γ) is any permutation of (a, b, c),

$$(14.54) \qquad S_{\alpha,\beta,\gamma} = \sum_{\substack{m^{\alpha+\beta} n^\gamma \leqslant x \\ m > n}} \psi\left(\left(\frac{x}{m^\beta n^\gamma}\right)^{1/\alpha}\right)$$

$$= \sum_{n \leqslant x^{1/d}} \sum_{n < m \leqslant (xn^{-\gamma})^{1/(\alpha+\beta)}} \psi\left(\left(\frac{x}{m^\beta n^\gamma}\right)^{1/\alpha}\right),$$

$$(14.55) \qquad S^*_{\alpha,\beta,\gamma} = \sum_{m^{\alpha+\beta} n^\gamma \leqslant x} \psi\left(\left(\frac{x}{m^\beta n^\gamma}\right)^{1/\alpha}\right),$$

where summation is for $n \leqslant x^{1/d}$ if $\beta \leqslant \gamma$, and for $m > x^{1/d}$ if $\beta > \gamma$, and $d = a + b + c$ as before. The desired form of the expression for $\Delta(a, b, c; x)$ is given by

LEMMA 14.4. Let $1 \leqslant a < b < c$, $b \leqslant 2a$, $d = a + b + c$ and let (α, β, γ) be any permutation of (a, b, c). Then

$$(14.56) \qquad \Delta(a, b, c; x) = - \sum_{(\alpha,\beta,\gamma)} S_{\alpha,\beta,\gamma} + O(x^{1/d}),$$

where $S_{\alpha,\beta,\gamma}$ is given by (14.54).

Proof of Lemma 14.4. By considering the ranges of summation in the (m, n)-plane we see first that

$$(14.57) \qquad S^*_{\alpha,\beta,\gamma} + S^*_{\alpha,\gamma,\beta} = S_{\alpha,\beta,\gamma} + S_{\alpha,\gamma,\beta} + O(x^{1/d}),$$

so that (14.46) gives

(14.58) $$\sum_{m^c \leqslant y} \Delta(a, b; x/m^c) = -S^*_{b,a,c} - S^*_{a,b,c} + O(x^{1/d}).$$

For the sum S defined by (14.52) we have

(14.59) $$S = \sum_{n \leqslant (x/y)^{1/b}} \sum_{m \leqslant (x/y)^{1/a}n^{-b/a}} \psi\left(\left(\frac{x}{m^a n^b}\right)^{1/c}\right)$$

$$= \sum_{n \leqslant x^{1/d}} \sum_{m \leqslant (x/y)^{1/a}n^{-b/a}} \psi\left(\left(\frac{x}{m^a n^b}\right)^{1/c}\right)$$

$$+ S' = S^*_{c,a,b} + S' + S'',$$

where

$$S' = \sum_{x^{1/d} < n \leqslant (x/y)^{1/b}} \sum_{m \leqslant (x/y)^{1/a}n^{-b/a}} \psi\left(\left(\frac{x}{m^a n^b}\right)^{1/c}\right),$$

$$S'' = \sum_{n \leqslant x^{1/d}} \sum_{(xn^{-b})^{1/(a+c)} < m \leqslant (x/y)^{1/a}n^{-b/a}} \psi\left(\left(\frac{x}{m^a n^b}\right)^{1/c}\right).$$

Change of summation and an application of Lemma 14.3 with

$$\nu = x^{(a+b)/(a^2+ab+ac)}n^{-b/a}, \qquad \rho = c/a, \qquad \mu = x^{1/d}, \qquad y = (xn^{-b})^{1/a}$$

give

(14.60) $$S' = S^*_{c,b,a} + S^*_{b,c,a} + O(x^{1/d}),$$

(14.61) $$S'' = S^*_{a,c,b} - \frac{c}{a}\varsigma\left(\frac{b}{a}\right)B(x^{1/d})x^{1/a+(-1-c/a)/d} + O(x^{1/d}).$$

Combining the estimates (14.56)–(14.61) and using Lemma 14.2 we obtain (14.56).

Therefore we have obtained an expression for $\Delta(a, b, c; x)$ in terms of double sums $S_{\alpha,\beta,\gamma}$ involving the ψ-function. These sums can be further reduced to the estimation of ordinary exponential sums by means of the following

LEMMA 14.5. Let $q > 0$ and $g(n)$ [or $g(m, n)$] be any real-valued arith-metical function. Then for $1 \leqslant a < b \leqslant 2a$

$$(14.62) \quad \sum_{a < n \leqslant b} \psi(g(n)) \ll aq^{-1} + \sum_{\nu=1}^{\infty} \left| \sum_{a < n \leqslant b} e(\nu g(n)) \right| \min(\nu^{-1}, q\nu^{-2}),$$

while if $|\mathscr{D}|$ denotes the area of the region $\mathscr{D}$, then we have

$$(14.63) \quad \sum_{(m, n) \in \mathscr{D}} \psi(g(m, n)) \ll |\mathscr{D}| q^{-1}$$

$$+ \sum_{\nu=1}^{\infty} \left| \sum_{(m, n) \in \mathscr{D}} e(\nu g(m, n)) \right| \min(\nu^{-1}, q\nu^{-2}).$$

Proof of Lemma 14.5. The estimate (14.63) is a generalization of (14.62), and since both proofs are quite similar we shall prove (14.62) only. Observe that for $t > 0$

$$\psi(g(n) + t) - t \leqslant \psi(g(n)) \leqslant \psi(g(n) - t) + t,$$

which implies for $q > 0$

(14.64)

$$q \int_0^{1/q} \psi(g(n) + t) \, dt - \tfrac{1}{2} q^{-1} \leqslant \psi(g(n)) \leqslant q \int_0^{1/q} \psi(g(n) - t) \, dt + \tfrac{1}{2} q^{-1}.$$

Now if y is not an integer the Fourier series of $\psi(y)$ is

$$\psi(y) = -\pi^{-1} \sum_{\nu=1}^{\infty} \nu^{-1} \sin(2\pi\nu y),$$

and this can be integrated within finite limits to give

$$(14.65) \quad q \int_0^{1/q} \psi(y \pm t) \, dt$$

$$= -q(2\pi i)^{-1} \sum_{n=1}^{\infty} n^{-1} \int_0^{1/q} (e(ny \pm nt) - e(-ny \mp nt)) \, dt$$

$$= \sum_{\nu=-\infty}^{\infty} w_\nu e(\nu y),$$

where $w_0 = 0$, and if ν is an integer

$$w_\nu = -q(2\pi i\nu)^{-1} \int_0^{1/q} e(\pm \nu t) \, dt \qquad (\nu \neq 0)$$

so that

(14.66) $|w_\nu| \leqslant \min\left(|\nu|^{-1}, q\nu^{-2}\right)$ $(\nu \neq 0)$.

If in (14.65) we take $y = g(n)$ and then sum for $a < n \leqslant b$, we obtain (14.62) in view of (14.66).

We are now ready to prove the main result of this section, which is

THEOREM 14.3. For $1 \leqslant a < b < c$, $b \leqslant 2a$, $c \leqslant 2a$ or $c = 3$ we have

(14.67) $\Delta(a, b, c; x) \ll x^{577/348(a+b+c)}$.

Proof of Theorem 14.3. The proof will actually give

(14.68) $\Delta(a, b, c; x) \ll x^{(1+2\varkappa+\lambda)/(\varkappa+1)(a+b+c)}$,

where $(\varkappa, \lambda)$ is an exponent pair, and the value $\frac{577}{348} = 1.65804\ldots$ comes from choosing the exponent pair $(\varkappa, \lambda) = (\frac{97}{251}, \frac{132}{251})$. This value is not the optimal one obtainable by the method of (one-dimensional) exponent pairs, but it is nevertheless quite close to it. However, (14.56) and (14.62) show that $\Delta(a, b, c; x)$ is majorized by certain two-dimensional exponential sums, so that two-dimensional methods readily lead to improvements of (14.68). We start by proving

(14.69) $\sum_{a<n\leqslant b} \psi(yn^{-\sigma}) \ll y^{\varkappa/(1+\varkappa)} a^{(\lambda-\sigma\varkappa)/(1+\varkappa)} + y^{-1/2} a^{(2+\sigma)/2}$,

which is valid for $1 \leqslant a < b \leqslant 2a$, $y \geqslant 1$, $\sigma > 0$ fixed, and $(\varkappa, \lambda)$ an exponent pair. By (14.62) it is seen that we have to estimate the exponential sum

$$S = S(a, b, y, \nu) = \sum_{a<n\leqslant b} e(\nu y n^{-\sigma}).$$

Let now, for ν, y, σ fixed, $f(x) = \nu y x^{-\sigma}$. If $f'(x) = -\sigma\nu y x^{-\sigma-1} \gg 1$, we use the theory of exponent pairs (see Section 2.3) to obtain

$$S \ll \left(\nu y a^{-\sigma-1}\right)^{\varkappa} a^{\lambda} = \nu^{\varkappa} y^{\varkappa} a^{\lambda-\varkappa-\sigma\varkappa}.$$

However, if $|f'(x)| < 1$, we use Lemmas 1.2 and 2.2 to obtain in any case

$$S \ll \nu^{\varkappa} y^{\varkappa} a^{\lambda-\varkappa-\sigma\varkappa} + \nu^{-1/2} y^{-1/2} a^{(2+\sigma)/2}.$$

By splitting the series in Lemma 14.5 at $\nu = q$ we then obtain

$$\sum_{a<n\leqslant b} \psi(yn^{-\sigma}) \ll aq^{-1}$$

$$+ \sum_{\nu\leqslant q} \nu^{-1}\left(\nu^{\varkappa}y^{\varkappa}a^{\lambda-\varkappa-\sigma\varkappa} + \nu^{-1/2}y^{-1/2}a^{(2+\sigma)/2}\right)$$

$$+ \sum_{\nu>q} q\nu^{-2}\left(\nu^{\varkappa}y^{\varkappa}a^{\lambda-\varkappa-\sigma\varkappa} + \nu^{-1/2}y^{-1/2}a^{(2+\sigma)/2}\right)$$

$$\ll aq^{-1} + q^{\varkappa}y^{\varkappa}a^{\lambda-\varkappa-\sigma\varkappa} + y^{-1/2}a^{(2+\sigma)/2}.$$

Choosing

$$q = y^{-\varkappa/(\varkappa+1)}a^{(1-\lambda+\varkappa+\sigma\varkappa)/(\varkappa+1)}$$

to equalize the first two terms in the last estimate, we obtain (14.69). We recall that

$$S_{\alpha,\beta,\gamma} = \sum_{n\leqslant x^{1/d}} \sum_{n<m\leqslant(xn^{-\gamma})^{1/(\alpha+\beta)}} \psi\left(\left(\frac{x}{m^{\beta}n^{\gamma}}\right)^{1/\alpha}\right),$$

and we use (14.69) with $\sigma = \beta/\alpha$, $1 \leqslant a \leqslant (xn^{-\gamma})^{1/(\alpha+\beta)}$, $y = (xn^{-\gamma})^{1/\alpha}$ to estimate the inner sum above. Then

(14.70)
$$\sum_{a<m\leqslant 2a} \psi\left(\left(\frac{x}{m^{\beta}n^{\gamma}}\right)^{1/\alpha}\right) \ll (xn^{-\gamma})^{\varkappa/\alpha(\varkappa+1)}a^{(\alpha\lambda-\beta\varkappa)/\alpha(\varkappa+1)}$$

$$+ (xn^{-\gamma})^{1/(2\alpha+2\beta)},$$

since

$$y^{-1/2}a^{(2+\sigma)/2} \ll (xn^{-\gamma})^{1/(2\alpha+2\beta)}.$$

Summing over $O(\log x)$ values $a = 2^N$ we obtain, in view of the conditions $1 \leqslant a < b < c$, $b \leqslant 2a$, $c = 3$ or $c \leqslant 2a$,

$$S_{\alpha,\beta,\gamma} \ll \sum_{n\leqslant x^{1/d}} \left\{(xn^{-\gamma})^{1/(2\alpha+2\beta)}\log x + (xn^{-\gamma})^{\varkappa/(\alpha\varkappa+\alpha)}n^{(\alpha\lambda-\beta\varkappa)/(\alpha\varkappa+\alpha)}\right.$$

$$\left. + (xn^{-\gamma})^{\varkappa/(\alpha\varkappa+\alpha)+(\alpha\lambda-\beta\varkappa)/\alpha(\alpha+\beta)(\varkappa+1)}\right\}$$

$$\ll x^{1/(2a+2b)}\log^2 x + x^{3/(2a+2b+2c)}\log x$$

$$+ x^{\varkappa/\alpha(\varkappa+1)+(1/d)\{1+[\alpha\lambda-(\beta+\gamma)\varkappa]/\alpha(\varkappa+1)\}}$$

$$+ x^{(\varkappa+\lambda)/[(\varkappa+1)(\alpha+\beta)]}x^{(1/d)\{1-\gamma(\varkappa+\lambda)/[(\varkappa+1)(\alpha+\beta)]\}}$$

$$\ll x^{(1+2\varkappa+\lambda)/[(\varkappa+1)(a+b+c)]},$$

since $\frac{3}{2} \leqslant (1 + 2\varkappa + \lambda)/(\varkappa + 1)$ always. Note that the final bound does not depend on the permutation (α, β, γ), hence (14.68) follows from Lemma 14.4.

14.4 POWERFUL NUMBERS

Powerful numbers were defined in Section 1.7 of Chapter 1 as those natural numbers n whose canonical decomposition is

$$n = p_1^{\alpha_1} p_2^{\alpha_2} \cdots p_r^{\alpha_r} \qquad (\alpha_1 \geqslant k, \alpha_2 \geqslant k, \ldots, \alpha_r \geqslant k)$$

for a fixed integer $k \geqslant 2$. For $k = 2$ we have the set of square full numbers, and the generating function of their characteristic function is

$$(14.71) \qquad\qquad F_2(s) = \frac{\zeta(2s)\zeta(3s)}{\zeta(6s)},$$

as given by (1.94), while for general k the generating function $F_k(s)$ is defined by (1.91) and (1.96)–(1.98). Our aim is to derive an asymptotic formula for the summatory function

$$(14.72) \qquad\qquad A_k(x) = \sum_{n \leqslant x, n \in G(k)} 1 = \sum_{n \leqslant x} f_k(n),$$

where $G(k)$ denotes the set of powerful numbers. Therefore $A_k(x)$ represents the number of powerful integers not exceeding x. As k increases, the product representation for $F_k(s)$ becomes more and more complicated, so that it is natural to expect that sharp asymptotic formulas for $A_k(x)$ will be obtained for small values of k only. In particular the simple form (14.71) of the generating function $F_2(s)$ makes it possible to prove

THEOREM 14.4. There is a constant $C > 0$ such that

$$(14.73) \qquad A_2(x) = \frac{\zeta(\frac{3}{2})}{\zeta(3)} x^{1/2} + \frac{\zeta(\frac{2}{3})}{\zeta(2)} x^{1/3}$$

$$+ O\left\{ x^{1/6} \exp\left(-C \log^{3/5} x (\log \log x)^{-1/5} \right) \right\}.$$

Proof of Theorem 14.4. With $d(a, b; k)$ defined by (14.43) we see from (14.71) that

$$(14.74) \qquad\qquad f_2(n) = \sum_{\delta^6 | n} \mu(\delta) d(2, 3; n/\delta^6),$$

and to obtain (14.73) we shall apply Theorem 14.2 with $k = 6$, $l = 2$, $a_1 = \frac{1}{2}$, $a_2 = \frac{1}{3}$, $r = 0$, $f(n) = d(2, 3; n)$. The summatory function of $f(n)$ is then

$$\sum_{n \leqslant x} f(n) = \sum_{n \leqslant x} d(2, 3; n) = \zeta(\tfrac{3}{2})x^{1/2} + \zeta(\tfrac{2}{3})x^{1/3} + \Delta(2, 3; x),$$

where

$$\Delta(2, 3; x) = - \sum_{n^5 \leqslant x} \left\{ \psi\left(\left(\frac{x}{n^3}\right)^{1/2}\right) + \psi\left(\left(\frac{x}{n^2}\right)^{1/3}\right) \right\} + O(1),$$

and we used (14.44) and (14.46). For the application of Theorem 14.2 we need only

(14.75) $\Delta(2, 3; x) \ll x^\alpha, \qquad \alpha < \frac{1}{6}.$

This bound may be obtained from (14.69) by taking $(\varkappa, \lambda) = (\frac{2}{7}, \frac{4}{7})$, which gives then

(14.76) $\sum_{a < n \leqslant b} \psi(yn^{-\sigma}) \ll y^{2/9}a^{(4-2\sigma)/9} + y^{-1/2}a^{(2+\sigma)/2},$

where $1 \leqslant a < b \leqslant 2a$, $y \geqslant 1$, $\sigma > 0$. Using (14.76) with $y = x^{1/2}$, $\sigma = \frac{3}{2}$ and $y = x^{1/3}$, $\sigma = \frac{2}{3}$ we obtain

$$\Delta(2, 3; x) \ll 1 + \sum_{a = 2^k \leqslant x^{1/5}} (x^{1/9}a^{1/9} + x^{2/27}a^{8/27} + x^{-1/4}a^{7/4} + x^{-1/6}a^{4/3})$$

$$\ll x^{2/15} + x^{7/20}x^{-1/4} + x^{4/15}x^{-1/6} \ll x^{2/15},$$

hence (14.75) holds with $\alpha = \frac{2}{15} < \frac{1}{6}$. Analogously to (14.76) we would obtain in the general case $(1 \leqslant a < b)$

(14.77) $\Delta(a, b; x) \ll \begin{cases} x^{2/(3a+3b)} & \text{if } b < 2a, \\ x^{2/(9a)}\log x & \text{if } b = 2a, \\ x^{2/(5a+2b)} & \text{if } b > 2a. \end{cases}$

Thus all the conditions needed for the application of Theorem 14.2 are satisfied, and (14.73) follows from (14.17) since we may easily compute

$$\sum_{j=1}^{l} x^{a_j} R_j(\log x) = \frac{\zeta(\frac{3}{2})}{\zeta(3)}x^{1/2} + \frac{\zeta(\frac{2}{3})}{\zeta(2)}x^{1/3},$$

or even more conveniently we may observe that the main terms in (14.73) are the sum of residues of $F_2(s)x^s s^{-1}$ at its poles $s = \frac{1}{2}$ and $s = \frac{1}{3}$.

We proceed now to the discussion of the asymptotic formula for $A_k(x)$ when $k > 2$, and we make use of the product representations (1.96)–(1.98), that is,

$$(14.78) \quad F_k(s) = G_k(s) H_k(s),$$

$$(14.79) \quad H_k(s) = \sum_{n=1}^{\infty} h_k(n) n^{-s} = \zeta(ks) \zeta((k+1)s) \cdots \zeta((2k-1)s)$$

$$(\sigma > 1/k),$$

$$(14.80) \quad G_k(s) = \sum_{n=1}^{\infty} g_k(n) n^{-s} = \Phi_k(s)/\zeta((2k+2)s)$$

$$[\sigma > 1/(2k+2)].$$

We observe that $H_k(s)$ has poles of the first order for $s = 1/k, \ldots, 1/(2k-1)$, so it is reasonable to expect that

(14.81)

$$S_k(x) = \sum_{n \leqslant x} h_k(n) = \sum_{r=k}^{2k-1} C_{r,k} x^{1/r} + \Delta_k^*(x), \qquad C_{r,k} = \prod_{j=k, j \neq r}^{2k-1} \zeta(j/r),$$

if we sum residues of $H_k(s) x^s s^{-1}$ at its poles $s = 1/k, \ldots, 1/(2k-1)$. Here $\Delta_k^*(x)$ is to be considered as an error term, meaning that we expect

$$(14.82) \qquad \Delta_k^*(x) = o(x^{1/(2k-1)}) \qquad (x \to \infty).$$

From (14.78) it follows that $f_k(n)$ is the convolution of $g_k(n)$ and $h_k(n)$, and moreover we have

$$(14.83) \qquad \sum_{n \leqslant x} g_k(n) \ll x^{1/(2k+2)},$$

since $\Phi_k(s)$ converges absolutely for $\sigma > 1/(2k+3)$. Thus we may write

$$(14.84) \quad A_k(x) = \gamma_{0,k} x^{1/k} + \gamma_{1,k} x^{1/(k+1)} + \cdots + \gamma_{k-1,k} x^{1/(2k-1)} + \Delta_k(x),$$

where for $j = 0, 1, \ldots, k-1$

(14.85)

$$\gamma_{j,k} = \operatorname*{Res}_{s=1/(k+j)} F_k(s) s^{-1} = C_{k+j,k} \Phi_k(1/(k+j))/\zeta((2k+2)/(k+j)),$$

and in analogy with (14.81) $\Delta_k(x)$ is to be thought of as an error term [no confusion will arise with the notation $\Delta_k(x)$ used in Chapter 13 for the error term in the Dirichlet divisor problem]. Essentially, $\Delta_k(x)$ is of the same order of magnitude as $\Delta_k^*(x)$, so that the problem of the evaluation of the asymptotic formula for $A_k(x)$ reduces to the estimation of $\Delta_k^*(x)$. This fact is established by

LEMMA 14.6. If for $1/(2k + 2) \leqslant \eta_k < 1/(2k - 1)$ and some $\lambda_k \geqslant 0$

$$(14.86) \qquad \Delta_k^*(x) \ll x^{\eta_k}\log^{\lambda_k}x,$$

where $\Delta_k^*(x)$ is defined by (14.81), then for $\Delta_k(x)$ defined by (14.84) we have

$$(14.87) \qquad \Delta_k(x) \ll x^{\eta_k}\log^{\lambda'_k}x,$$

where $\lambda'_k = \lambda_k$ for $1/(2k + 2) < \eta_k < 1/(2k - 1)$ and $\lambda'_k = \lambda_k + 1$ for $\eta_k = 1/(2k + 2)$.

Proof of Lemma 14.6. Using (14.78) and (14.81) we have

$$(14.88)$$

$$A_k(x) = \sum_{n \leqslant x} f_k(n) = \sum_{n \leqslant x}\sum_{d|n} g_k(d)h_k(n/d) = \sum_{m \leqslant x} g_k(m) \sum_{n \leqslant x/m} h_k(n)$$

$$= \sum_{r=k}^{2k-1} C_{r,k}x^{1/r} \sum_{m \leqslant x} g_k(m)m^{-1/r} + \sum_{m \leqslant x} g_k(m)\Delta_k^*(x/m).$$

Using partial summation and (14.83) we find that, for $r = k, k + 1, \ldots, 2k - 1$,

$$(14.89) \qquad \sum_{m \leqslant x} g_k(m)m^{-1/r} = G_k(1/r) + \sum_{m > x} g_k(m)m^{-1/r}$$

$$= G_k(1/r) + O\left(x^{1/(2k+2)-1/r}\right).$$

Hence substituting (14.89) in (14.88) we obtain

$$A_k(x) = \sum_{r=k}^{2k-1} C_{r,k}G_k(1/r)x^{1/r} + O\left(x^{1/(2k+2)}\right) + \sum_{m \leqslant x} g_k(m)\Delta_k^*(x/m)$$

$$= \gamma_{0,k}x^{1/k} + \gamma_{1,k}x^{1/(k+1)} + \cdots + \gamma_{k-1,k}x^{1/(2k-1)}$$

$$+ O\left(x^{\eta_k}\log^{\lambda_k}x\right),$$

with

$$\gamma_{j,k} = C_{k+j,k}G_k(1/(k+j)) = C_{k+j,k}\Phi_k(1/(k+j))/\zeta((2k+2)/(k+j)),$$

since

$$\sum_{m \leqslant x} g_k(m) \Delta_k^*(x/m) \ll x^{\eta_k} \log^{\lambda_k} x \sum_{m \leqslant x} |g_k(m)| m^{-\eta_k} \ll x^{\eta_k} \log^{\lambda_k} x,$$

because the second term in the last estimate is $O(\log x)$ if $\eta_k = 1/(2k + 2)$ and it is bounded if $\eta_k > 1/(2k + 2)$.

This proves Lemma 14.6 and there is no need to consider the case $\eta_k < 1/(2k + 2)$, because it is not known to hold for $k > 3$ and because for $\eta_k < 1/(2k + 2)$ we could apply Theorem 14.2 and obtain an analogue of (14.73).

If we define ρ_k as the infimum of all $\rho > 0$ satisfying

$$(14.90) \qquad\qquad\qquad \Delta_k(x) \ll x^\rho$$

for a fixed $k \geqslant 3$, then the problem of the evaluation of the asymptotic formula for $A_k(x)$ reduces to proving upper bounds for ρ_k. We shall prove here

THEOREM 14.5. $\rho_3 \leqslant \frac{577}{4176} = 0.13817\ldots$, $\rho_k \leqslant 1/2k$ for $4 \leqslant k \leqslant 10$.

Proof of Theorem 14.5. From Theorem 14.3 we have

$$\Delta(3, 4, 5; x) \ll x^{577/4176},$$

hence

$$\Delta_3^*(x) = \Delta(3, 4, 5; x) \ll x^{\eta_3}, \qquad \eta_3 \leqslant \tfrac{577}{4176},$$

and $\rho_3 \leqslant \frac{577}{4176}$ follows from Lemma 14.6.

From Lemma 14.6 we also see that it will be sufficient to prove

$$\Delta_k^*(x) \ll x^{1/(2k)+\varepsilon} \qquad (4 \leqslant k \leqslant 10),$$

and with some more elaboration we could replace x^ε here by $\log^c x$ with some explicit $c = c(k) \geqslant 0$. Perron's inversion formula is applied to the function $H_k(s)$, which is seen to be regular for all values of s except for $s = 1/k$, $1/(k + 1), \ldots, 1/(2k - 1)$, where it has simple poles. Analogously to (A.10) we obtain for $b = 1/k + \varepsilon$, $T \ll x^C$,

$$\sum_{n \leqslant x} h_k(n) = (2\pi i)^{-1} \int_{b-iT}^{b+iT} H_k(s) x^s s^{-1} \, ds + O(x^{1/k+\varepsilon} T^{-1}) + O(x^\varepsilon),$$

since clearly $h_k(n) \ll n^\varepsilon$. If we move the line of integration to $\sigma = 1/(2k)$, then by the residue theorem we have

$$(2\pi i)^{-1} \int_{b-iT}^{b+iT} H_k(s) x^s s^{-1} ds$$

$$= \sum_{r=k}^{2k-1} \operatorname*{Res}_{s=1/r} H_k(s) x^s s^{-1} + (2\pi i)^{-1} (I_1 + I_2 + I_3)$$

say, where

$$I_1 = \int_{1/(2k)-iT}^{1/(2k)+iT} H_k(s) x^s s^{-1} ds$$

$$I_2 = \int_{b-iT}^{1/(2k)-iT} H_k(s) x^s s^{-1} ds, \qquad I_3 = \int_{1/(2k)+iT}^{b+iT} H_k(s) x^s s^{-1} ds.$$

If we use the bound

$$\zeta(\sigma + it) \ll t^{(1-\sigma)/3} \log t \qquad (\tfrac{1}{2} \leqslant \sigma \leqslant 1, t \geqslant t_0)$$

which comes from $\mu(\tfrac{1}{2}) \leqslant \tfrac{1}{6}, \mu(1) = 0$, and convexity, then we have for $k \leqslant 10$

$$I_2 + I_3 \ll \int_{1/2k}^{1/k+\varepsilon} x^\sigma T^{-1} \prod_{r=k}^{2k-1} |\zeta(r\sigma + irT)| d\sigma$$

$$\ll x^{1/k+\varepsilon} T^Q = x^{1/k+\varepsilon} T^{(k-11)/12+\varepsilon} \ll 1,$$

where

$$Q = \frac{1}{3} \sum_{r=k}^{2k-1} \left(1 - \frac{r}{2k}\right) + \varepsilon - 1,$$

provided that $T = x^C$ for some sufficiently large $C > 0$. Next, we have

$$I_1 \ll x^{1/(2k)} \int_1^T \prod_{r=k}^{2k-1} |\zeta(r/2k + irt)| t^{-1} dt + x^{1/(2k)},$$

and integration by parts shows that the proof will be finished if we can show that

(14.91) $$I_4 = \int_1^T \prod_{r=k}^{2k-1} |\zeta(r/2k + irt)| \, dt \ll T^{1+\varepsilon}.$$

Thus we have reduced the problem to the estimation of an integral involving a product of zeta-functions, which by Hölder's inequality for integrals (see

Section A.4) further reduces to estimates for higher power moments of the zeta-function. Namely, if $p_k, \ldots, p_{2k-1}$ are real numbers such that

$$\sum_{r=k}^{2k-1} 1/p_r = 1, \qquad p_k > 0, \ldots, p_{2k-1} > 0,$$

then by Hölder's inequality with $f_r(t) = \zeta(r/2k + irt)$ we obtain

$$I_4 = \int_1^T \prod_{r=k}^{2k-1} |f_r(t)| \, dt \leqslant \prod_{r=k}^{2k-1} \left(\int_1^T |f_r(t)|^{p_r} \, dt \right)^{1/p_r},$$

and if we choose $p_r = m(r/2k) \, (r = k, \ldots, 2k - 1)$, where $m(\sigma)$ is defined by (8.2), then (14.91) follows if we can show that

(14.92)
$$\sum_{r=k}^{2k-1} \frac{1}{m(r/2k)} \leqslant 1.$$

To check that (14.92) holds, we use the estimates given by Theorem 8.4. The most difficult case is $k = 10$, when we find that we may take [as bounds for $m(\sigma)$] the values

$$m\left(\tfrac{10}{20}\right) = 4, \, m\left(\tfrac{11}{20}\right) = 5, \, m\left(\tfrac{12}{20}\right) = \tfrac{20}{3}, \, m\left(\tfrac{13}{20}\right) = \tfrac{190}{21},$$

$$m\left(\tfrac{14}{20}\right) = 10, \, m\left(\tfrac{15}{20}\right) = \tfrac{528}{37}, \, m\left(\tfrac{16}{20}\right) = 19.8, \, m\left(\tfrac{17}{20}\right) = 36.8,$$

$$m\left(\tfrac{18}{20}\right) = 44.54, \, m\left(\tfrac{19}{20}\right) = 98.57;$$

hence adding we find that (14.92) holds for $k = 10$, since

$$\sum_{r=10}^{19} \frac{1}{m(r/20)} = 0.99087 \ldots .$$

This proves Theorem 14.5, but it may be remarked that if the Lindelöf hypothesis is true then our method of proof would give in fact $\rho_k \leqslant 1/(2k)$ for all k, and not only for $k \leqslant 10$.

14.5 NONISOMORPHIC ABELIAN GROUPS OF A GIVEN ORDER

As in Section 1.8 of Chapter 1 we define $a(n)$ to be the number of nonisomorphic abelian groups with n elements, and we employ the product representation

(14.93)
$$\sum_{n=1}^{\infty} a(n) n^{-s} = \zeta(s)\zeta(2s)\zeta(3s)\zeta(4s) \ldots \qquad (\sigma > 1)$$

to investigate the asymptotic behavior of the sum

(14.94) $$A(x) = \sum_{n \leqslant x} a(n).$$

Here we shall prove

THEOREM 14.6. If

(14.95) $$A(x) = \sum_{n \leqslant x} a(n) = A_1 x + A_2 x^{1/2} + A_3 x^{1/3} + R(x),$$

(14.96) $$A_j = \prod_{k=1, k \neq j}^{\infty} \zeta(k/j) \qquad (j = 1, 2, 3),$$

then

(14.97) $$R(x) \ll x^{577/2088}.$$

Proof of Theorem 14.6. In (14.95) $R(x)$ is to be considered as an error term, and the exponent $\frac{577}{2088} = 0.27634\ldots$ can be reduced to $\frac{97}{381} = 0.25459\ldots$ by using more refined techniques for the estimation of $\Delta(1, 2, 3; x)$. In fact the estimation of $R(x)$ is essentially the estimation of $\Delta(1, 2, 3; x)$, and our result follows from the estimate

$$\Delta(1, 2, 3; x) \ll x^{577/2088},$$

which is a consequence of Theorem 14.3. To see this, note that from (14.93) we have

(14.98) $$a(n) = \sum_{\delta \mid n} d(1, 2, 3; \delta) v(n/\delta),$$

where for $\sigma > \frac{1}{4}$

(14.99) $$V(s) = \sum_{n=1}^{\infty} v(n) n^{-s} = \zeta(4s) \zeta(5s) \zeta(6s) \ldots,$$

so that $V(s)$ converges absolutely for $\sigma > \frac{1}{4}$ and thus

(14.100) $$\sum_{n \leqslant x} |v(n)| \ll x^{1/4 + \varepsilon}.$$

If we suppose that

(14.101) $$\Delta(1, 2, 3; x) \ll x^{\theta}, \qquad \frac{1}{4} < \theta < \frac{1}{3},$$

then we have from (14.98)–(14.100) and (14.38)

$$\sum_{n\leqslant x} a(n) = \sum_{n\leqslant x}\sum_{\delta|n} d(1,2,3;\delta)v(n/\delta) = \sum_{n\leqslant x} v(n) \sum_{m\leqslant x/n} d(1,2,3;m)$$

$$= \sum_{n\leqslant x} v(n)D(1,2,3;x/n)$$

$$= \sum_{n\leqslant x} v(n)\{\zeta(2)\zeta(3)(x/n) + \zeta(1/2)\zeta(3/2)(x/n)^{1/2}$$

$$+\zeta(1/3)\zeta(2/3)(x/n)^{1/3} + \Delta(1,2,3;x/n)\}$$

$$= \zeta(2)\zeta(3)V(1)x + \zeta(1/2)\zeta(3/2)V(1/2)x^{1/2}$$

$$+\zeta(1/3)\zeta(2/3)V(1/3)x^{1/3}$$

$$+O(x^{1/4+\epsilon}) + O\left(\sum_{n\leqslant x} |v(n)|x^{\theta}n^{-\theta}\right)$$

$$= A_1 x + A_2 x^{1/2} + A_3 x^{1/3} + O(x^{\theta}).$$

Therefore we have shown that (14.101) implies

(14.102) $$R(x) \ll x^{\theta},$$

consequently (14.97) follows from Theorem 14.3, as asserted.

On the other hand it may be asked how well $\sum_j A_j x^{1/j}$ approximates $A(x)$, that is, what can be said about omega results for $A(x)$. A result of this type is given by

THEOREM 14.7. Let

(14.103) $$R_0(x) = \sum_{n\leqslant x} a(n) - \sum_{j=1}^{6} A_j x^{1/j}, \qquad A_j = \prod_{k=1,\,k\neq j}^{\infty} \zeta(k/j).$$

Then for any fixed $\delta > 0$

(14.104) $$R_0(x) \ll x^{1/6-\delta}$$

cannot hold.

Proof of Theorem 14.7. The proof is by contradiction. If (14.104) holds, then for some $\gamma < \frac{1}{6}$ (but close to $\frac{1}{6}$)

$$R_0(x) = (2\pi i)^{-1} \int_{\gamma-i\infty}^{\gamma+i\infty} F(s)x^s s^{-1}\,ds, \qquad F(s) = \prod_{j=1}^{\infty} \zeta(js),$$

and similarly as in the Dirichlet divisor problem (Section 13.5 of Chapter 13) the functions $g(x) = R_0(1/x)$ and $G(s) = s^{-1}F(s)$ are Mellin transforms. Hence by Parseval's identity

(14.105)

$$(2\pi)^{-1} \int_{-\infty}^{\infty} \frac{|F(\gamma + it)|^2}{|\gamma + it|^2} \, dt = \int_0^{\infty} R_0^2(1/x) x^{2\gamma-1} \, dx = \int_0^{\infty} R_0^2(x) x^{-2\gamma-1} \, dx,$$

where $\gamma > \gamma_0$, and γ_0 is the infimum of all positive numbers γ for which the first integral in (14.105) converges. But if (14.104) holds we may take $\gamma = \frac{1}{6} - \delta/2$ to obtain

(14.106) $\quad \displaystyle\int_T^{2T} |F(\gamma + it)|^2 \, dt \ll T^2 \int_{-\infty}^{\infty} \frac{|F(\gamma + it)|^2}{|\gamma + it|^2} \, dt$

$$\ll T^2 \left(1 + \int_1^{\infty} x^{1/3 - 2\delta - 2\gamma - 1} \, dx\right) \ll T^2.$$

However (14.106) is impossible, since we shall show that

(14.107) $$\int_T^{2T} |F(\tfrac{1}{6} + it)|^2 \, dt \gg T^2,$$

and in fact our method actually gives for $\gamma = \frac{1}{6} - \delta/2$

(14.108) $\quad \displaystyle\int_T^{2T} |F(\gamma + it)|^2 \, dt > T^{2+C_1(\delta)} \qquad [C_1(\delta) > 0],$

thus disproving (14.106) and proving Theorem 14.7. To obtain (14.107) let

$$U(s) = \zeta(\tfrac{2}{3} + s)\zeta(\tfrac{1}{3} + 2s) \prod_{j=3}^{18} \zeta(js).$$

Since for $\sigma \geq \frac{1}{6}$ we have

$$\prod_{j=19}^{\infty} \zeta(js) \gg 1,$$

we see that using the functional equation $\zeta(s) = \chi(s)\zeta(1-s)$ and $\chi(s) \asymp t^{1/2-\sigma}$ (14.107) will follow from

(14.109) $$\int_T^{2T} |U(\tfrac{1}{6} + it)|^2 \, dt \gg T,$$

which follows by the method of proof of Theorem 9.6 with $\zeta^k(s)$ replaced by $U^2(s)$. By more elaborate arguments one can obtain even the sharper lower bound $T \log T$ in (14.109).

We turn now to the investigation of the distribution of values of $a(n)$. We may define the nonnegative constant d_k as

$$(14.110) \qquad d_k = \lim_{x \to \infty} x^{-1} \sum_{n \leqslant x,\, a(n)=k} 1$$

for every fixed integer $k \geqslant 1$. One may call d_k the *local density* of $a(n)$, and we shall show the existence of d_k by proving

THEOREM 14.8. Let, for t real, $g_t(n) = \sum_{d|n} \mu(d) e^{ita(n/d)}$, and for $k \geqslant 1$ a fixed integer

$$(14.111) \qquad d_k = \sum_{n=1}^{\infty} (2\pi n)^{-1} \int_{-\pi}^{\pi} e^{-ikt} g_t(n)\, dt.$$

Then we have uniformly in k

$$(14.112) \qquad \sum_{n \leqslant x,\, a(n)=k} 1 = d_k x + O(x^{1/2} \log x).$$

Proof of Theorem 14.8. First we prove

$$(14.113) \qquad \sum_{n \leqslant x} e^{ita(n)} = \left(\sum_{n=1}^{\infty} g_t(n) n^{-1} \right) x + O(x^{1/2} \log x),$$

where the O-constant is uniform in t. Let

$$F_t(s) = \zeta(s) G_t(s), \quad G_t(s) = \sum_{n=1}^{\infty} g_t(n) n^{-s},$$

where $g_t(n)$ is as in the formulation of the theorem. For every n we have the trivial estimate

$$|g_t(n)| \leqslant \sum_{d|n} \mu^2(d) = 2^{\omega(n)}.$$

Suppose now that $n = pm$, where p is a prime not dividing m. Since $a(n)$ is multiplicative and $a(p) = 1$ for every prime p, we obtain

$$g_t(pm) = \sum_{d|pm} \mu(d) e^{ita(pm/d)} = \sum_{d|m} \mu(d) e^{ita(pm/d)}$$

$$+ \sum_{d|m} \mu(pd) e^{ita(pm/pd)}$$

$$= \sum_{d|m} \mu(d) e^{ita(p)a(m/d)} - \sum_{d|m} \mu(d) e^{ita(m/d)} = 0.$$

Therefore uniformly in t we have $|g_t(n)| \leqslant f(n)$, where

$$f(n) = \begin{cases} 0 & \text{if there is a } p \text{ such that } p\|n, \\ 2^{\omega(n)} & \text{otherwise.} \end{cases}$$

Now $f(n)$ is clearly multiplicative and for $\sigma > \frac{1}{2}$

$$F(s) = \sum_{n=1}^{\infty} f(n)n^{-s} = \prod_p (1 + 2p^{-2s} + 2p^{-3s} + \cdots) = \zeta^2(2s)H(s),$$

where

$$H(s) = \sum_{n=1}^{\infty} h(n)n^{-s}$$

is absolutely convergent for $\sigma > \frac{1}{3}$. Therefore

$$\sum_{n \leqslant x} e^{ita(n)} = \sum_{n \leqslant x} \sum_{d|n} g_t(d) = \sum_{n \leqslant x} g_t(n)[x/n]$$

$$= \left(\sum_{n=1}^{\infty} g_t(n)n^{-1} \right)x + O\left(x \sum_{n > x} |g_t(n)|n^{-1} \right)$$

$$+ O\left(\sum_{n \leqslant x} |g_t(n)| \right)$$

$$= \left(\sum_{n=1}^{\infty} g_t(n)n^{-1} \right)x + O\left(x \sum_{n > x} f(n)n^{-1} \right)$$

$$+ O\left(\sum_{n \leqslant x} f(n) \right)$$

$$= \left(\sum_{n=1}^{\infty} g_t(n)n^{-1} \right)x + O(x^{1/2}\log x),$$

where we have used

$$\sum_{n \leqslant x} f(n) = \sum_{n \leqslant x} \sum_{\delta^2|n} d(\delta)h(n/\delta) = \sum_{n \leqslant x} h(n) \sum_{m^2 \leqslant x/n} d(m)$$

$$\ll x^{1/2}\log x \sum_{n \leqslant x} |h(n)|n^{-1/2} \ll x^{1/2}\log x.$$

To obtain (14.112) from (14.113) recall that $a(n)$ takes only integer values and that, for m an integer,

$$\int_{-\pi}^{\pi} e^{imt} dt = \begin{cases} 0 & \text{if } m \neq 0, \\ 2\pi & \text{if } m = 0. \end{cases}$$

Therefore

$$(14.114) \quad \int_{-\pi}^{\pi} \sum_{n \leqslant x} e^{it(a(n)-k)} \, dt = \sum_{n \leqslant x} \int_{-\pi}^{\pi} e^{it(a(n)k)} \, dt = 2\pi \sum_{n \leqslant x, a(n)=k} 1.$$

On the other hand it follows from (14.113) that

(14.115)

$$\int_{-\pi}^{\pi} e^{-ikt} \sum_{n \leqslant x} e^{ita(n)} \, dt = x \int_{-\pi}^{\pi} e^{-ikt} \left(\sum_{n=1}^{\infty} g_t(n) n^{-1} \right) dt$$

$$+ O\left(x^{1/2} \log x\right)$$

$$= x \sum_{n=1}^{\infty} n^{-1} \left(\int_{-\pi}^{\pi} e^{-ikt} g_t(n) \, dt \right) + O\left(x^{1/2} \log x\right),$$

since the change of summation and integration is justified by absolute and uniform convergence. On comparing (14.114) and (14.115) we obtain (14.112).

One may derive a more explicit arithmetical expression for the local density d_k than (14.111). Observe that $a(qs) = a(s)$ if q is square free and s is square full [i.e., $s \in G(2)$] and $(q, s) = 1$. Since every n may be written uniquely as $n = qs$, $(q, s) = 1$ we have

$$(14.116) \quad \sum_{n \leqslant x, a(n)=k} 1 = \sum_{s \leqslant x, a(s)=k} \sum_{q \leqslant x/s, (q,s)=1} 1,$$

where q denotes square free and s denotes square full numbers. But for $s \geqslant 1$ fixed

$$\sum_{n \leqslant x, (n,s)=1} 1 = \sum_{n \leqslant x \, d|n, d|s} \mu(d) = \sum_{d|s} \mu(d)[x/d]$$

$$= x\varphi(s)s^{-1} + O(2^{\omega(s)}),$$

and moreover

$$\sum_{q \leqslant y, (q,s)=1} 1 = \sum_{n \leqslant y, (n,s)=1} \mu^2(n) = \sum_{n \leqslant y, (n,s)=1} \sum_{d^2|n} \mu(d)$$

$$= \sum_{d \leqslant y^{1/2}, (d,s)=1} \mu(d) \sum_{n \leqslant y/d^2, (n,s)=1} 1 = y\zeta^{-1}(2) \prod_{p|s}(1 + p^{-1})^{-1}$$

$$+ O\left(y^{1/2} 2^{\omega(s)}\right),$$

since

$$\varphi(s)s^{-1} = \prod_{p|s}(1 - p^{-1}), \qquad \sum_{d=1,(d,s)=1}^{\infty} \mu(d)d^{-2} = \zeta^{-1}(2)\prod_{p|s}(1 - p^{-2})^{-1}.$$

Using these estimates in (14.116) and simplifying we obtain

$$(14.117) \qquad d_k = \zeta^{-1}(2) \sum_{s=1, a(s)=k}^{\infty} s^{-1}\prod_{p|s}(1 + p^{-1})^{-1}$$

as the expression for the local density (14.110), where s denotes square full numbers. This method of approach would give (14.112) with the slightly weaker error term $O(x^{1/2}\log^2 x)$. From (14.117) and the order result (1.112) for $a(n)$ it follows that there exist absolute constants $C_1, C_2, C_3 > 0$ such that for $k \geqslant 3$

$$(14.118) \qquad d_k \leqslant C_1\exp(-C_2\log k \log\log k),$$

while for infinitely many k

$$(14.119) \qquad d_k \geqslant \exp(-C_3 \log k \log\log k).$$

14.6   THE GENERAL DIVISOR FUNCTION $d_z(n)$

So far we have studied the divisor function $d_k(n)$ ($k \geqslant 2$ a fixed integer), which may be defined by

$$(14.120) \qquad \sum_{n=1}^{\infty} d_k(n)n^{-s} = \zeta^k(s) = \prod_{p}(1 - p^{-s})^{-k} \qquad (\sigma > 1).$$

Now we define for an arbitrary complex z the general divisor function $d_z(n)$ by

$$(14.121) \qquad \sum_{n=1}^{\infty} d_z(n)n^{-s} = \zeta^z(s) = \prod_{p}(1 - p^{-s})^{-z} \qquad (\sigma > 1),$$

where a branch of $\zeta^z(s)$ is defined by

$$\zeta^z(s) = \exp\{z \log \zeta(s)\} = \exp\left(-z\sum_{p}\sum_{j=1}^{\infty} j^{-1}p^{-js}\right) \qquad (\sigma > 1).$$

This definition shows that $d_z(n)$ is a multiplicative function of n which generalizes $d_k(n)$. If p^α is an arbitrary prime power, then from (14.121) we have

$$d_z(p^\alpha) = (-1)^\alpha \binom{-z}{\alpha} = \frac{z(z+1)\cdots(z+\alpha-1)}{\alpha!},$$

so that by multiplicativity

(14.122)
$$d_z(n) = \prod_{p^\alpha \| n} \frac{z(z+1)\cdots(z+\alpha-1)}{\alpha!}.$$

Our main concern here will be the asymptotic formula for the summatory function

(14.123)
$$D_z(x) = \sum_{n \leqslant x} d_z(n).$$

To derive such a formula we shall proceed on well-established lines, using a variant of Perron's inversion formula (A.8). However, it should be remarked that the generating function $\zeta^z(s)$ of $d_z(n)$ is not known to be regular for $\frac{1}{2} < \sigma < 1$, since it has singularities at zeros of $\zeta(s)$. In establishing the asymptotic formulas for $D_k(x) = \sum_{n \leqslant x} d_k(n)$ in Chapter 13 this difficulty did not arise, and we could move the line of integration to $\sigma = \sigma_0$ for some suitable $\frac{1}{2} < \sigma_0 < 1$, and the main term $x P_{k-1}(\log x)$ in the asymptotic formula for $D_k(x)$ is the residue of $x^s \zeta^k(s) s^{-1}$ at the pole $s = 1$. In the general case of $D_z(x) = \sum_{n \leqslant x} d_z(n)$ the contour of integration may not encircle the point $s = 1$, which is now an algebraic singularity of the integrand, and the contour will be chosen to lie in the zero-free region

(14.124) $\quad \zeta(s) \neq 0$ for $\sigma \geqslant 1 - \dfrac{a}{\log|t|} \qquad (a > 0, |t| \geqslant t_0)$

and $\sigma \geqslant 1 - a/\log|t_0|$ ($|t| \leqslant t_0$). This is implied by (1.55), and the stronger zero-free region (6.1) would not give essentially anything more than the error term in (14.125). We shall prove

THEOREM 14.9. Let $A > 0$ be arbitrary but fixed, and let $N \geqslant 1$ be an arbitrary but fixed integer. If $|z| \leqslant A$, then uniformly in z

(14.125)

$$D_z(x) = \sum_{n \leqslant x} d_z(n)$$

$$= c_1(z) x \log^{z-1} x + c_2(z) x \log^{z-2} x + \cdots + c_N(z) x \log^{z-N} x$$

$$+ O(x \log^{\operatorname{Re} z - N - 1} x),$$

where $c_j(z) = B_{j-1}(z)/\Gamma(z - j + 1)$ $(j = 1, \ldots, N)$ and each $B_j(z)$ is regular for $|z| \leq A$.

Proof of Theorem 14.9. By the inversion formula (A.8) we have

$$(14.126) \qquad \int_0^x D_z(t)\, dt = \lim_{T \to \infty} (2\pi i)^{-1} \int_{2-iT}^{2+iT} \frac{x^{s+1}}{s(s+1)} \zeta^z(s)\, ds.$$

For convenience of notation we now introduce the functions

$$H(s, z) = \{\zeta(s)(s-1)\}^z, \qquad F(s, z, x) = \frac{x^{s+1}}{s(s+1)} \zeta^z(s),$$

and let $t_1 \geq t_0$ be a fixed number such that if $\eta = a/\log t_1$, then $0 < \eta < \frac{1}{2}$, where a and t_0 refer to (14.124), Furthermore, let r and ε be two arbitrary real numbers such that $0 < r < \eta$ and $0 < \varepsilon < \arctan(t_1/\eta)$. If $T > t_1$, then by Cauchy's theorem the integral on the right-hand side of (14.126) may be replaced by integrals $I_1, I_2, \ldots, I_9$ over the paths $L_1, L_2, \ldots, L_9$, which are defined as follows:

L_1 is the segment $\left[2 - iT, 1 - \dfrac{a}{\log T} - iT\right]$,

L_2 is the curve described by $1 - \dfrac{a}{\log|t|} + it$ for $-T \leq t \leq -t_1$,

L_3 is the segment $[1 - \eta - it_1, 1 - \eta - i\eta \tan \varepsilon]$,

L_4 is the segment $[1 - \eta - i\eta \tan \varepsilon, 1 - re^{i\varepsilon}]$,

L_5 is the arc of the circle $1 + re^{i\theta}$ for $-\pi + \varepsilon \leq \theta \leq \pi - \varepsilon$,

L_6 is the segment $[1 - re^{-i\varepsilon}, 1 - \eta + i\eta \tan \varepsilon]$,

L_7 is the segment $[1 - \eta + i\eta \tan \varepsilon, 1 - \eta + it_1]$,

L_8 is the curve described by $1 - \dfrac{a}{\log t} + it$ for $t_1 \leq t \leq T$,

L_9 is the segment $\left[1 - \dfrac{a}{\log T} + iT, 2 + iT\right]$.

Here L_1, L_2, L_8, and L_9 depend only on T and not on r or ε. For T and r fixed and for $\varepsilon \to 0$, L_3 and L_7 become the segments $[1 - \eta - it_1, 1 - \eta)$ and $(1 - \eta, 1 - \eta + it_1]$. These segments will be denoted by l_3 and l_7, and the corresponding integrals by J_3 and J_7, respectively. For I_4 we have

$$\lim_{\varepsilon \to 0} I_4 = \lim_{\varepsilon \to 0} \int_{L_4} H(s, z)(s-1)^{-z} \frac{x^{s+1}}{s(s+1)}\, ds$$

$$= \int_{1-\eta}^{1-r} H(\sigma, z)(1 - \sigma)^{-z} (e^{-i\pi})^{-z} \frac{x^{\sigma+1}}{\sigma(\sigma+1)}\, d\sigma,$$

and on L_6 the argument of s is increased by 2π, hence

$$\lim_{\varepsilon \to 0} I_6 = \lim_{\varepsilon \to 0} \int_{L_6} H(s, z)(s-1)^{-z} \frac{x^{s+1}}{s(s+1)} \, ds$$

$$= \int_{1-r}^{1-\eta} H(\sigma, z)(1-\sigma)^{-z}(e^{i\pi})^{-z} \frac{x^{\sigma+1}}{\sigma(\sigma+1)} \, d\sigma.$$

If γ_r denotes the circle $|s-1| = r$ without the point $s = 1 - r$, then adding the last two expressions and making the substitution $\sigma = 1 - u$ we obtain

(14.127)

$$\lim_{\varepsilon \to 0} (I_4 + I_5 + I_6) = \int_{\gamma_r} F(s, z, x) \, ds + 2i \sin \pi z \int_r^\eta \frac{H(1-u, z)u^{-z}}{(1-u)(2-u)} x^{2-u} \, du,$$

which does not depend on the choice of r.

If $T \to \infty$, then both I_1 and I_9 tend to zero, so that from (14.127) we have

(14.128)
$$\int_0^x D_z(t) \, dt = \Phi_z(x) + w(x, z),$$

where, for $0 < r < \eta$,

(14.129)

$$\Phi_z(x) = \frac{\sin \pi z}{\pi} \int_r^\eta \frac{H(1-u, z)u^{-z}}{(1-u)(2-u)} x^{2-u} \, du + (2\pi i)^{-1} \int_{\gamma_r} F(s, z, x) \, ds,$$

(14.130) $w(x, z) = J_2 + J_3 + J_7 + J_8,$

$$J_2 = (2\pi i)^{-1} \int_{-\infty}^{-t_1} F\left(1 - \frac{a}{\log|t|} + it, z, x\right)\left(i + \frac{a}{t \log^2|t|}\right) dt,$$

$$J_8 = (2\pi i)^{-1} \int_{t_1}^{\infty} F\left(1 - \frac{a}{\log t} + it, z, x\right)\left(i + \frac{a}{t \log^2 t}\right) dt,$$

$$J_3 = (2\pi i)^{-1} \int_{l_3} F(s, z, x) \, ds, \quad J_7 = (2\pi i)^{-1} \int_{l_7} F(s, z, x) \, ds.$$

If s lies on L_2 or L_8 then $\zeta(s) \ll \log|t|$ in view of (14.124), (6.50), and (6.51) [in fact any estimate $\zeta(s) \ll \log^C|t|$ would suffice here], so that

$$F(s, z, x) \ll x^{2-a/\log|t|} t^{-2} \log|t| \qquad (s \in L_2, L_8).$$

Now we fix $0 < \varepsilon < 1$ and observe that

$$-\varepsilon \log t - a\frac{\log x}{\log t} \leqslant -2(a\varepsilon \log x)^{1/2}$$

follows from

$$\left((\varepsilon \log t)^{1/2} - \left(a\frac{\log x}{\log t}\right)^{1/2}\right)^2 \geqslant 0.$$

Therefore setting $B = (a\varepsilon)^{1/2}$ we have

$$J_2 + J_8 \ll \int_{t_1}^{\infty} x^{2-a/\log t}\, t^{-2}\log t\, dt$$

$$\ll x^2 \int_{t_1}^{\infty} t^{-2+\varepsilon}\exp\left(-\varepsilon \log t - a\frac{\log x}{\log t}\right)\log t\, dt$$

$$\ll x^2\exp\left(-2B(\log x)^{1/2}\right).$$

On l_3 and l_7 we have $F(s, z, x) \ll x^{2-\eta}$, and since t_1 is fixed we obtain

$$J_3 + J_7 \ll t_1 x^{2-\eta} \ll x^2\exp\left(-2B(\log x)^{1/2}\right).$$

Thus we have proved that if $w(x, z)$ is defined by (14.130), then

(14.131) $$w(x, z) \ll x^2\exp\left(-2B(\log x)^{1/2}\right),$$

so that $w(x, z)$ represents the error term in (14.128).
    For $|z| \leqslant A$ the function $\Phi_z(x)$ is regular and

(14.132) $$\Phi_z'(x) = \frac{\sin \pi z}{\pi}\int_r^{\eta}\frac{H(1-u, z)}{1-u}u^{-z}x^{1-u}\, du$$

$$+ (2\pi i)^{-1}\int_{\gamma_r} H(s, z)s^{-1}(s-1)^{-z}x^s\, ds,$$

(14.133) $$\Phi_z''(x) = \frac{\sin \pi z}{\pi}\int_r^{\eta} H(1-u, z)u^{-z}x^{-u}\, du$$

$$+ (2\pi i)^{-1}\int_{\gamma_r} H(s, z)(s-1)^{-z}x^{s-1}\, ds.$$

Now $H(s, z)$ is bounded for $|s - 1| \leqslant \eta$, so that for $|z| \leqslant A$ and $0 < r < \eta < \frac{1}{2}$ we have

$$\int_r^\eta H(1 - u, z)u^{-z}x^{-u}\, du \ll \int_r^\eta u^{-A}x^{-u}\, du \ll \log^{A-1}x \int_{r\log x}^{\eta\log x} v^{-1}e^{-v}\, dv,$$

if $u = v/\log x$. For $|s - 1| = r$ we have $(s - 1)^{-z}x^{s-1} \ll x^r r^{-A}$, which gives

$$\int_{\gamma_r} H(s, z)(s - 1)^{-z}x^{s-1}\, ds \ll x^r r^{1-A}.$$

Now we choose $r = 1/\log x$, and then for $x \geqslant x_0$

(14.134) $$\Phi_z''(x) \ll \log^{A-1}x.$$

To obtain an asymptotic formula for $D_z(x)$ from (14.128) we shall use a differencing argument. From (14.122) we see that $|d_z(n)| \leqslant d_k(n)$ if $k = [A] + 1$ and $|z| \leqslant A$. If we use the weak asymptotic formula (see Chapter 13 for stronger results)

$$\sum_{n \leqslant x} d_k(n) = xP_{k-1}(\log x) + O(x^{k/(k+1)}),$$

then for $1 < \xi < x/2$ we have

$$\left| \xi^{-1} \int_x^{x+\xi} D_z(t)\, dt - D_z(x) \right| = \left| \xi^{-1} \int_x^{x+\xi} \{D_z(t) - D_z(x)\}\, dt \right|$$

$$\leqslant \xi^{-1} \int_x^{x+\xi} \sum_{x < n \leqslant t} d_k(n)\, dt \ll \xi \log^{k-1}x + x^{k/(k+1)}.$$

Next, we approximate to $D_z(x)$ by $\Phi_z'(x)$, where $\Phi_z'(x)$ is given by (14.132). We may write

$$|D_z(x) - \Phi_z'(x)| \leqslant \left| \xi^{-1} \int_x^{x+\xi} D_z(t)\, dt - \Phi_z'(x) \right|$$

$$+ \left| D_z(x) - \xi^{-1} \int_x^{x+\xi} D_z(t)\, dt \right|$$

$$\ll \left| \xi^{-1} \int_x^{x+\xi} D_z(t)\, dt - \Phi_z'(x) \right|$$

$$+ \xi \log^{k-1}x + x^{k/(k+1)}.$$

From (14.128) it follows on using (14.131) and (14.134) that

$$\int_x^{x+\xi} D_z(t)\, dt = \int_0^{x+\xi} D_z(t)\, dt - \int_0^x D_z(t)\, dt$$

$$= \Phi_z(x+\xi) - \Phi_z(x) + w(x+\xi, z) - w(x, z)$$

$$= \xi\Phi_z'(x) + \xi^2\int_0^1 (1-u)\Phi_z''(x+u\xi)\, du$$

$$+ O\left(x^2\exp\left(-2B(\log x)^{1/2}\right)\right)$$

$$= \xi\Phi_z'(x) + O(\xi^2\log^{A-1}x) + O\left(x^2\exp\left(-2B(\log x)^{1/2}\right)\right),$$

which implies

$$D_z(x) - \Phi_z'(x) \ll \xi\log^{A-1}x + x^2\xi^{-1}\exp\left(-2B(\log x)^{1/2}\right).$$

Finally, letting $\xi = x\exp(-B(\log x)^{1/2})$ it follows that

$$(14.135) \quad D_z(x) = \sum_{n\leqslant x} d_z(n) = \Phi_z'(x) + O\left(x\exp\left(-\frac{B}{2}(\log x)^{1/2}\right)\right),$$

and it remains to evaluate $\Phi_z'(x)$. For $r = 1/\log x$, $|z| \leqslant A$ and s as in (14.132) we have

$$(14.136) \quad H(s, z)s^{-1} = \sum_{j=0}^q B_j(z)(s-1)^j + R_q(s, z)(s-1)^{q+1}$$

for every fixed integer $q \geqslant 0$, where the $B_j(z)$'s are regular for $|z| \leqslant A$. Moreover, $R_q(s, z) = O(1)$ by Cauchy's classical inequality for coefficients of a power series; hence substituting (14.136) in (14.132) and noting that $(-1)^j\sin \pi z = \sin \pi(z-j)$ we obtain

(14.137)

$$\Phi_z'(x) = \sum_{j=0}^q xB_j(z)\left(\frac{\sin \pi(z-j)}{\pi}\int_r^\infty u^{j-z}x^{-u}\, du + (2\pi i)^{-1}\int_{\gamma_r}(s-1)^{j-z}x^{s-1}\, ds\right)$$

$$+ W(x, z),$$

where

$$(14.138) \quad W(x, z) = -\sum_{j=0}^{q} xB_j(z) \frac{\sin \pi(z-j)}{\pi} \int_{\eta}^{\infty} u^{j-z} x^{-u} du$$

$$+ x \frac{\sin \pi(z-q-1)}{\pi} \int_{r}^{\eta} R_q(1-u, z) u^{q+1-z} x^{-u} du$$

$$+ x(2\pi i)^{-1} \int_{\gamma_r} R_q(s, z)(s-1)^{q+1-z} x^{s-1} ds.$$

The first expression in (14.137) may be simplified if one uses the identity
(14.139)

$$\frac{\sin \pi(z-j)}{\pi} \int_{r}^{\infty} u^{j-z} x^{-u} du + (2\pi i)^{-1} \int_{\gamma_r} x^{s-1}(s-1)^{j-z} ds = \frac{(\log x)^{z-j-1}}{\Gamma(z-j)},$$

where $j = 0, 1, \ldots, q$. To derive this identity let $u = v/\log x$ in the first integral and $s = 1 + re^{i\theta} = 1 + e^{i\theta}/\log x$ in the second integral in (14.139) to obtain

$$(\log x)^{z-j-1} \left(\frac{\sin \pi(z-j)}{\pi} \int_{1}^{\infty} v^{j-z} e^{-v} dv + \frac{1}{2\pi i} \int_{-\pi}^{\pi} e^{i\theta(j-z)} \exp(e^{i\theta}) i e^{i\theta} d\theta \right)$$

$$= (\log x)^{z-j-1} \left(\frac{\sin \pi(z-j)}{\pi} \int_{1}^{\infty} v^{j-z} e^{-v} dv + \frac{1}{2\pi i} \int_{\mathscr{C}} w^{j-z} e^{w} dw \right),$$

where $\mathscr{C}$ is the circle $|w| = 1$ without the point $w = -1$.
However, if $\Delta = (-\infty, -1) \cup \mathscr{C} \cup [-1, -\infty)$, then

$$(14.140) \qquad\qquad \frac{1}{\Gamma(s)} = \frac{1}{2\pi i} \int_{\Delta} w^{-s} e^{w} dw,$$

and on the other hand

$$\int_{\Delta} w^{-s} e^{w} dw = -\int_{\infty}^{1} \exp\{-\log(xe^{-i\pi})s\} e^{-x} dx + \int_{\mathscr{C}} w^{-s} e^{w} dw$$

$$- \int_{1}^{\infty} \exp\{-\log(xe^{i\pi})s\} e^{-x} dx$$

$$= (e^{i\pi s} - e^{-i\pi s}) \int_{1}^{\infty} e^{-x} x^{-s} dx + \int_{\mathscr{C}} w^{-s} e^{w} dw$$

$$= 2i \sin \pi s \int_{1}^{\infty} e^{-x} x^{-s} dx + \int_{\mathscr{C}} w^{-s} e^{w} dw.$$

Hence taking $s = z - j$ we obtain (14.139). Therefore we may write

$$(14.141) \qquad D_z(x) = x \sum_{j=0}^{q} \frac{B_j(z)\log^{z-j-1}x}{\Gamma(z-j)} + O(|W(x,z)|)$$

$$+ O\left(x \exp\left(-\frac{B}{2}(\log x)^{1/2} \right) \right),$$

and so it remains to estimate $W(x, z)$, as given by (14.138). Letting $u = v/\log x$ we obtain, for $0 \leqslant j \leqslant q$ and uniformly in $|z| \leqslant A$,

$$\int_{\eta}^{\infty} u^{j-z}x^{-u}\,du \ll x^{-\eta} + \int_{1}^{\infty} u^{q+A}x^{-u}\,du$$

$$\ll x^{-\eta} + \log^{-A-q-1}x \int_{\log x}^{\infty} v^{q+A}e^{-v}\,dv$$

$$= x^{-\eta} + \log^{-A-q-1}x \int_{\log x}^{\infty} e^{-v/2}e^{-v/2}v^{q+A}\,dv$$

$$\ll x^{-\eta} + x^{-1/2}\log^{-A-q-1}x \ll x^{-\eta},$$

since $0 < \eta < \frac{1}{2}$. This means that the sum in (14.138) is $\ll x \log^{\mathrm{Re}\,z-q-2}x$ for $|z| \leqslant A$, and similarly we find that

$$\int_{r}^{\eta} R_q(1-u,z)u^{q+1-z}x^{-u}\,du \ll (\log x)^{\mathrm{Re}\,z-q-2},$$

$$\int_{\gamma_r} R_q(s,z)(s-1)^{q+1-z}x^{s-1}\,ds \ll (\log x)^{\mathrm{Re}\,z-q-2}.$$

Therefore setting $c_j(z) = B_{j-1}(z)/\Gamma(z-j+1)$ $(j = 1, \ldots, N)$, $N = q + 1$ we obtain (14.125), and it is easy to see that $c_1(z) = 1/\Gamma(z)$.

The method of proof of Theorem 14.9 may be used in other problems as well. For example, it may be proved by this method that

$$(14.142) \qquad B(x) = \sum_{j=1}^{N} b_j x(\log x)^{1/2-j} + O\left(x(\log x)^{-N-1/2} \right)$$

for any fixed integer $N \geqslant 1$, where $B(x)$ denotes the number of integers not exceeding x which are a sum of two integer squares. The constants $b_1, \ldots, b_N$ are computable, and in particular

$$(14.143) \qquad b_1 = 2^{-1/2} \prod_{p \equiv 3(\mathrm{mod}\,4)} \left(1 - p^{-2} \right)^{-1/2}.$$

Namely, if we set $b(n) = 1$ if $n = a^2 + b^2$ (a, b integers) and zero otherwise, then $b(n)$ is multiplicative, and for $\sigma > 1$

$$F(s) = \sum_{n=1}^{\infty} b(n)n^{-s} = (1 + 2^{-s} + 2^{-2s} + \cdots) \prod_q (1 + q^{-s} + q^{-2s} + \cdots)$$

$$\times \prod_r (1 + r^{-2s} + r^{-4s} + \cdots),$$

where q and r denote primes congruent to 1 and 3 (mod 4), respectively. Now if $L(s, \chi)$ is the L-function for the nonprincipal character mod 4

$$\chi(n) = \begin{cases} (-1)^{(n-1)/2} & \text{if } n \text{ is odd,} \\ 0 & \text{if } n \text{ is even,} \end{cases}$$

then for $\sigma > 1$ we have

$$\zeta(s)L(s, \chi) = (1 - 2^{-s})^{-1} \prod_q (1 - q^{-s})^{-2} \prod_r (1 - r^{-2s})^{-1},$$

which yields for $\sigma > 1$ the identity

(14.144) $\qquad F^2(s) = (1 - 2^{-s})^{-1} \prod_r (1 - r^{-2s})^{-1} \zeta(s)L(s, \chi).$

The function $L(s, \chi)$ is regular for $\sigma > 0$ and in particular

$$L(1, \chi) = 1 - \tfrac{1}{3} + \tfrac{1}{5} - \cdots = \pi/4,$$

while $\zeta(s)$ has a simple pole at $s = 1$. Moreover, (1.55) is a zero-free region not only for $\zeta(s)$ but also for $L(s, \chi)$, so that the method of proof of Theorem 14.9 (corresponding to the case $z = \tfrac{1}{2}$) yields (14.142), and (14.143) follows without difficulty then from the product representation (14.144).

14.7 SMALL ADDITIVE FUNCTIONS

By "small" additive functions we mean additive functions like $\omega(n) = \sum_{p|n} 1$ and $\Omega(n) = \sum_{p^\alpha \| n} \alpha$, whose average order is $C \log \log n$. More precisely, using the asymptotic formula (12.92) in the form

$$\sum_{p \leqslant x} 1/p = \log \log x + B + O(1/\log x)$$

and writing

$$\sum_{n \leqslant x} \omega(n) = \sum_{n \leqslant x} \sum_{p|n} 1 = \sum_{p \leqslant x} [x/p],$$

$$\sum_{n \leqslant x} \omega^2(n) = \sum_{pp' \leqslant x, p \neq p'} [x/pp'] + \sum_{p \leqslant x} [x/p]$$

we obtain

(14.145) $$\sum_{n \leqslant x} \omega(n) = x \log \log x + C_1 x + O(x/\log x)$$

and

(14.146) $$\sum_{n \leqslant x} \omega^2(n) = x(\log \log x)^2 + C_2 x \log \log x + O(x).$$

The asymptotic formula (14.145) shows that the average order of $\omega(n)$ is $\log \log n$ (or $\log \log x$), while (14.145) and (14.146) combined give for some $C_3 > 0$

(14.147) $$\sum_{n \leqslant x} (\omega(n) - \log \log x)^2 = C_3 x \log \log x + O(x),$$

and this also holds if $\omega(n)$ is replaced by $\Omega(n)$. The asymptotic formula (14.147) yields precise information about the closeness of $\omega(n)$ to $\log \log n$, since it implies

(14.148) $$\sum_{n \leqslant x, |\omega(n) - \log \log x| > (\log \log x)^\delta} 1 \ll x(\log \log x)^{1 - 2\delta},$$

and $x(\log \log x)^{1 - 2\delta} = o(x)$ as $x \to \infty$ for $\frac{1}{2} < \delta < 1$ fixed. Thus it is seen that for "almost all" n we have $|\omega(n) - \log \log n| \leqslant (\log \log n)^\delta$, $\frac{1}{2} < \delta < 1$, which is expressed by saying that the "normal" order of $\omega(n)$ is $\log \log n$.

It would be too great a digression here to go into the details of the theory of additive functions, and our purpose is different. Observe that for every additive function $f(n)$ and arbitrary complex z the function $z^{f(n)}$ is multiplicative, hence under suitable conditions on f it possesses an Euler product. For "small" additive functions this product is often dominated by $\zeta^z(s)$, which shows the importance of $\zeta(s)$ in the theory of additive functions, and therefore using the convolution method and Theorem 14.9 we may find a suitable asymptotic expression for the sum

(14.149) $$A(x, z) = \sum_{n \leqslant x} z^{f(n)} \qquad (|z| \leqslant R, R > 1).$$

If it turns out that $A(x, z)$, as a function of z, is regular for $|z| \leqslant R$, then this function may be differentiated, integrated, and so on, and then setting $z = 1$ one may obtain sharp asymptotic formulas for various sums involving $f(n)$, such as those appearing in (14.145) and (14.146). To avoid technical details and keep the exposition clear, we shall now evaluate $A(x, z)$ in (14.149) for $f(n) = \omega(n)$ only, but it is obvious that the method is very general. Our result will be

THEOREM 14.10. For every fixed $R > 0$ and every fixed integer $N \geqslant 0$ there exist functions $A_0(z), A_1(z), \ldots, A_N(z)$ regular in $|z| \leqslant R$ such that $A_0(0) = A_1(0) = \cdots = A_N(0) = 0$ and

$$(14.150) \quad \sum_{n \leqslant x} z^{\omega(n)} = x(\log x)^{z-1}\left(\sum_{j=0}^{N} A_j(z)\log^{-j}x + O(\log^{-N-1}x)\right),$$

where the O-constant is uniform for $|z| \leqslant R$.

Proof of Theorem 14.10. For $\sigma > 1$ we have

(14.151)

$$\sum_{n=1}^{\infty} z^{\omega(n)}n^{-s} = \prod_{p}(1 + zp^{-s} + zp^{-2s} + zp^{-3s} + \cdots) = \zeta^z(s)G(s, z),$$

where

(14.152)

$$G(s, z) = \sum_{n=1}^{\infty} g(n, z)n^{-s} = \prod_{p}(1 + zp^{-s} + zp^{-2s} + \cdots)(1 - p^{-s})^z$$

$$= \prod_{p}(1 + h(p, s, z)),$$

and

$$h(p, s, z) = \left(z - z^2 + \binom{z}{2}\right)p^{-2s} + \left(z - z^2 + z\binom{z}{2} - \binom{z}{3}\right)p^{-3s} + \cdots .$$

Therefore $G(s, z)$ converges absolutely and uniformly for $\sigma \geqslant \frac{1}{2} + \varepsilon, |z| \leqslant R$ and represents a regular function of both s and z. In particular

$$(14.153) \quad (-1)^M \frac{d^M}{ds^M}G(s, z)\bigg|_{s=1} = \sum_{n=1}^{\infty} g(n, z)n^{-1}\log^M n,$$

and each $g(n, z)$ is regular for $|z| \leqslant R$, and uniformly in z

$$(14.154) \quad \sum_{n \leqslant x} |g(n, z)| \ll x^{1/2+\varepsilon}.$$

We now use (14.151) and Theorem 14.9 to obtain

(14.155)

$$\sum_{n \leq x} z^{\omega(n)} = \sum_{n \leq x} \sum_{\delta | n} d_z(\delta) g(n/\delta, z) = \sum_{n \leq x} \sum_{m \leq x/n} d_z(m) g(n, z)$$

$$= x \sum_{n \leq x} g(n, z) n^{-1} \left(\sum_{j=1}^{N} c_j(z) \log^{z-j} x/n + O(\log^{\operatorname{Re} z - N - 1} x/n) \right).$$

By partial summation and (14.154) we have, for every $A \geq 0$ and $\varepsilon > 0$,

(14.156) $\quad \sum_{n \leq x} g(n, z) n^{-1} \log^A n = \sum_{n=1}^{\infty} g(n, z) n^{-1} \log^A n + O(x^{\varepsilon - 1/2} \log^A x),$

while

(14.157)

$$\sum_{n \leq x} g(n, z) n^{-1} \log^{\operatorname{Re} z - N - 1} x/n$$

$$= \sum_{n \leq x^{1/2}} + \sum_{x^{1/2} < n \leq x}$$

$$\ll \log^{\operatorname{Re} z - N - 1} x \sum_{n=1}^{\infty} |g(n, z)| n^{-1}$$

$$+ \sum_{x^{1/2} < n \leq x} |g(n, z)| n^{-1}$$

$$\ll \log^{\operatorname{Re} z - N - 1} x.$$

If we insert (14.156) and (14.157) in (14.155), use

$$\sum_{n \leq x} g(n, z) n^{-1} \log^{z-j} x/n = \log^{z-j} x \sum_{n \leq x} g(n, z) \left(1 - \frac{\log n}{\log x} \right)^{z-j},$$

expand the binomial and simplify, we arrive at (14.150). Each function $A_j(z)$ is seen to be an expression of the form $\sum_i D_i(z) c_i(z)$, where the D_i's and c_i's are regular functions, and from Theorem 14.9 we have $c_i(0) = 0$ for $i = 1, 2, \ldots$. Thus every $A_j(z)$ is regular in $|z| \leq R$ and satisfies $A_j(0) = 0$, which completes the proof of Theorem 14.10. A result analogous to (14.150) holds if $\omega(n)$ is replaced by $\Omega(n)$, but in this case one has to take $|z| \leq R < 2$.

The same method of approach may be used to evaluate asymptotically the sum $\sum_{n \leq x} z^{\Omega(n) - \omega(n)}$. The function $f(n) = \Omega(n) - \omega(n)$ is additive and moreover for every prime p and every integer $\alpha \geq 2$ we have $f(p) = 0$ and

$f(p^{\alpha}) = \alpha - 1$. Therefore for $\sigma > 1$

$$(14.158) \quad \sum_{n=1}^{\infty} z^{\Omega(n) - \omega(n)} n^{-s} = \prod_{p} (1 + p^{-s} + zp^{-2s} + z^2 p^{-3s} + \cdots)$$

$$= \zeta(s) \prod_{p} (1 + (z-1)p^{-2s} + (z^2 - z)p^{-3s}$$

$$+ (z^3 - z^2)p^{-4s} + \cdots)$$

$$= \zeta(s)\zeta^{z-1}(2s)V(s,z),$$

where

$$V(s,z) = \sum_{n=1}^{\infty} v(n,z)n^{-s}$$

is regular for $\operatorname{Re} s \geqslant \frac{1}{3} + \varepsilon, |z| \leqslant 1$. Now writing

$$H(s,z) = \zeta^{z-1}(2s)V(s,z) = \sum_{n=1}^{\infty} h(n,z)n^{-s}$$

we see that the estimation of $\sum_{n \leqslant x} h(n,z)$ bears resemblance to the estimation of $\sum_{n \leqslant x} z^{\omega(n)}$ in Theorem 14.10, only now x is replaced by $x^{1/2}$ (because s is replaced by $2s$ in the zeta factor) and z is replaced by $z - 1$ in the zeta-factor. Therefore we find that, for $|z| \leqslant 1$,

$$(14.159) \quad \sum_{n \leqslant x} h(n,z) = x^{1/2}\log^{z-2}x \left(\sum_{j=0}^{N} E_j(z)\log^{-j}x + O(\log^{-N-1}x) \right),$$

where the E_j's are regular for $|z| \leqslant 1$, $E_j(0) = 0$ and the O-constant is uniform in z. But then from (14.158) we have

$$(14.160) \quad \sum_{n \leqslant x} z^{\Omega(n) - \omega(n)} = \sum_{n \leqslant x} \sum_{d|n} h(d,z) = \sum_{n \leqslant x} h(n,z)[x/n]$$

$$= x \sum_{n=1}^{\infty} h(n,z)n^{-1} + O(x^{1/2}\log^{-1}x)$$

$$= xF(z) + O(x^{1/2}\log^{-1}x),$$

where from the product representation (14.158) we find that

(14.161)

$$F(z) = \prod_{p} \left(1 - \frac{1}{p}\right)\left(1 + \frac{1}{p-z}\right) = \frac{6}{\pi^2} \prod_{p} \left(1 - \frac{z}{p+1}\right) \bigg/ \left(1 - \frac{z}{p}\right),$$

while by more delicate analysis one can obtain

(14.162)

$$\sum_{n\leqslant x} z^{\Omega(n)-\omega(n)} = xF(z) + x^{1/2}\log^{z-2}x\left(\sum_{j=0}^{N} B_j(z)\log^{-j}x + O(\log^{-N-1}x)\right).$$

Here $|z| \leqslant 1$, $N \geqslant 0$ is an arbitrary but fixed integer, the B_j's are regular functions for $|z| \leqslant 1$ which satisfy $B_j(0) = 0$ for $j = 0, 1, \ldots, N$, and the O-constant is uniform in z.

We proceed now to the deduction of arithmetical corollaries from (14.150), (14.160), and (14.162), indicating only the main arguments. Observe first that the sum $\sum_{n\leqslant x} z^{\omega(n)}$ is a polynomial in z with the coefficient of z^q equal to

$$\sum_{n\leqslant x,\,\omega(n)=q} 1,$$

which represents the number of positive integers n not exceeding x for which $\omega(n) = q$. On the other hand, using

$$(\log x)^z = \sum_{n=0}^{\infty} (\log\log x)^n z^n/n!$$

and Cauchy's classical inequality for coefficients of a power series, we deduce, after equating coefficients of z^q in (14.150), the following

COROLLARY 14.1. For every fixed integer $q \geqslant 1$

$$(14.163) \quad \omega_q(x) = \sum_{n\leqslant x,\,\omega(n)=q} 1 = \sum_{j=0}^{N} xP_j(\log\log x)\log^{-j-1}x$$

$$+ O\left(x\log^{-N-2}x(\log\log x)^{q-1}\right),$$

where each $P_j(t)$ is a polynomial in t of degree not exceeding $q - 1$, and $N \geqslant 0$ is an arbitrary but fixed integer.

This formula is the sharpest known version of a classical result of E. Landau, and an analogous result holds if $\omega(n)$ is replaced by $\Omega(n)$. Similar analysis may be made regarding (14.160) or (14.162), which is the sharpest known result of its kind. The latter leads to

COROLLARY 14.2. For every fixed integer $q \geqslant 0$

$$(14.164) \qquad \sum_{n \leqslant x,\, \Omega(n) - \omega(n) = q} 1 = d_q x + x^{1/2} \sum_{j=1}^{N} Q_j(\log \log x) \log^{-j-1} x$$

$$+ O\left(x^{1/2} \log^{-N-2} x (\log \log x)^{q-1}\right),$$

where each $Q_j(t)$ is a polynomial in t of degree not exceeding $q - 1$, and $N \geqslant 1$ is an arbitrary but fixed integer. Here d_q is given by

$$(14.165) \qquad \sum_{q=0}^{\infty} d_q z^q = 6 \pi^{-2} \prod_p \frac{1 - z/(p+1)}{1 - z/p}.$$

Corollary 14.2 is due to H. Delange and represents the sharpest known asymptotic expansion for the sum on the left-hand side of (14.164); (14.160) would lead to the O-term $O(x^{1/2} \log^{-1} x)$ only. Note that d_q represents the local density of the function $f(n) = \Omega(n) - \omega(n)$, similar to the local density of $a(n)$, which was defined by (14.110). In particular, d_0 in (14.164) represents the local density of square-free integers (since $\Omega(n) - \omega(n) = 0$ is equivalent to n being square free), and (14.165) gives the well-known value $d_0 = 6\pi^{-2}$.

Taking $R = \frac{3}{2}$ in Theorem 14.10, forming the expressions

$$\frac{d}{dz}\left(\sum_{n \leqslant x} z^{\omega(n)} \right), \qquad \frac{d}{dz}\left[z \frac{d}{dz}\left(\sum_{n \leqslant x} z^{\omega(n)} \right) \right],$$

and taking $z = 1$ we obtain refinements of (14.145) and (14.146). In particular we have

COROLLARY 14.3. If $N \geqslant 1$ is an arbitrary but fixed integer, then

$$(14.166)$$

$$\sum_{n \leqslant x} \omega(n) = x \log \log x + bx + \sum_{j=1}^{N} \frac{(-1)^{j-1}}{j} G^{(j)}(1) \frac{x}{\log^j x} + O\left(\frac{x}{\log^{N+1} x}\right),$$

where

$$b = \gamma - \sum_p \left(\log \frac{p}{p-1} - \frac{1}{p} \right), \qquad G(s) = s^{-1}(s-1)\zeta(s).$$

Another possibility is to make the substitution $z = e^w$ in (14.150) and to compare the coefficients of w^q in the resulting expression. This leads to

COROLLARY 14.4. If $q \geqslant 1$ is a fixed integer and $N \geqslant 0$ is an arbitrary but fixed integer, then

(14.167)

$$\sum_{n \leqslant x} \omega^q(n) = \sum_{j=0}^{N} x R_j(\log \log x) \log^{-j} x + O\left(x \log^{-N-1} x (\log \log x)^{q-1}\right),$$

where $R_0(t)$ is a polynomial in t of degree q, and $R_j(t)$ for $j \geqslant 1$ is a polynomial of degree $\leqslant q - 1$ in t.

Finally, we mention that sums involving reciprocals of $\omega(n)$ may be often found by suitable integration of (14.150). If $\varepsilon(x)$ is a suitable function which tends to zero as $x \to \infty$, then

$$\int_{\varepsilon(x)}^{1} \sum_{2 \leqslant n \leqslant x} z^{\omega(n)-1} dz = \sum_{2 \leqslant n \leqslant x} \int_{\varepsilon(x)}^{1} z^{\omega(n)-1} dz = \sum_{2 \leqslant n \leqslant x} \frac{1}{\omega(n)}$$

$$- \sum_{2 \leqslant n \leqslant x} \frac{(\varepsilon(x))^{\omega(n)}}{\omega(n)}.$$

Now dividing (14.150) by z, integrating, and choosing $\varepsilon(x) = (\log x)^{-1/(2N+2)}$, we obtain

COROLLARY 14.5. If $N \geqslant 1$ is an arbitrary but fixed integer, then

(14.168) $$\sum_{2 \leqslant n \leqslant x} \frac{1}{\omega(n)} = x \sum_{j=1}^{N} \frac{A_j}{(\log \log x)^j} + O\left(\frac{x}{(\log \log x)^{N+1}}\right),$$

where $A_1 = 1$ and the other constants A_j are computable.

Notes

The Dirichlet convolution (14.1) is a fundamental operation in the theory of arithmetical functions, and it can be easily generalized to complex-valued functions over an arbitrary arithmetical semigroup G. For an extensive account of Dirichlet algebras over arithmetical semigroups and related topics see J. Knopfmacher (1975).

If $A(x) = \sum_{n \leqslant x} f(n)$, $B(x) = \sum_{n \leqslant x} g(n)$, $C(x) = \sum_{n \leqslant x} h(n)$ and h is the convolution of f and g, as given by (14.1), then

$$C(x) = \sum_{mn \leqslant x} f(m)g(n) = \sum_{m \leqslant x} f(m) \sum_{n \leqslant x/m} g(n) = \sum_{n \leqslant x} g(n) \sum_{m \leqslant x/n} f(m)$$

$$= \int_{1-0}^{x} A(x/t) \, dB(t) = \int_{1-0}^{x} B(x/t) \, dA(t),$$

and the integrals are called the Stieltjes resultant of A and B. Thus the summatory function of the convolution function may be regarded in the more general setting of Stieltjes resultants, and this viewpoint was adopted by J. P. Tull (1958, 1959) who proved some general convolution results which generalize earlier results of E. Landau.

Slowly oscillating functions were introduced by J. Karamata (1930), who made many contributions to the theory, and in particular he proved the canonical representation (14.6). A comprehensive survey, including all the facts about slowly oscillating functions mentioned in this text, is to be found in E. Seneta (1976).

Theorems 14.1 and 14.2 are proved in the author's paper (1978b), which also contains the results of Section 14.2 plus some additional applications of the convolution method. Applications of this method to the estimation of some other summatory functions connected with powers of the zeta-functions are to be found in the author's paper (1977b).

The estimate (14.24) for the number of k-free integers not exceeding x is the sharpest one known, and a proof may be found in A. Walfisz (1963). Walfisz in fact considers the problem when k does not have to be fixed, and proves then that the error term is

$$O\left\{ x^{1/k}\exp\left(-Ck^{-8/5}\log^{3/5}x\,(\log\log x)^{-1/5}\right)\right\}$$

for some absolute $C > 0$. Theorem 14.2 implied that, if the Riemann hypothesis is true, the error term is

$$O\left(x^{2/(2k+1)}\exp\left(\frac{C\log x}{\log\log x}\right)\right) \qquad (C > 0),$$

while H. L. Montgomery and R. C. Vaughan (1981) improved this to $O(x^{1/(k+1)+\varepsilon})$ if the Riemann hypothesis holds. For $k = 2$ they showed that

$$Q_2(x) = \sum_{n\leqslant x} \mu^2(n) = \frac{x}{\zeta(2)} - \sum_{n\leqslant N} \mu(n)\psi\left(x/n^2\right) + O\left(x^{1/2+\varepsilon}N^{-1/2}\right) + O\left(N^{1/2+\varepsilon}\right).$$

Here $N(\leqslant x)$ is a parameter, and choosing $N = x^{1/3}$ and estimating the sum over $n \leqslant N$ trivially, one obtains the error term $O(x^{1/3+\varepsilon})$. By using a nontrivial estimate, Montgomery and Vaughan showed that the error term for $Q_2(x)$ is $O(x^{9/28+\varepsilon})$. The exponent $\frac{9}{28}$ was reduced to $\frac{8}{25}$ by S. W. Graham (1981) and to $\frac{7}{22}$ by R. C. Baker and J. Pintz (in press), all assuming the Riemann hypothesis. On the other hand, it is not difficult to show unconditionally that the error term for $Q_k(x)$ in (14.24) is $\Omega(x^{1/(2k)})$, since the function $\zeta(s)/\zeta(ks)$ has poles on the line $\sigma = 1/2k$.

The asymptotic formulas (14.28) and (14.29) [also (14.30) with $k = 4$ if $\alpha_k < \frac{1}{2}$ holds] were obtained in a direct, more involved way by R. S. Rao and D. Suryanarayana (1970). Also our proof of the asymptotic formula (14.34) for $\tau^{**}(n)$ is much shorter than the one given by D. Suryanarayana and R. S. Rao (1975), and Theorem 14.2 can be used to provide simple proofs of several other results of the same authors.

Three-dimensional divisor problems were considered by several authors, including E. Krätzel (1969, 1982 in press), A. Ivić (1978a, 1981), A. Ivić and P. Shiu (1982), and M. Vogts (1981). E. Krätzel (1969) notes that E. Landau's classical methods give

$$\Delta(a,b,c;x) \ll x^{1/(2a)}, \qquad \Delta(a,b,c;x) = \Omega\left(x^{1/(a+b+c)}\right) \qquad (1 \leqslant a \leqslant b \leqslant c),$$

and he also proves $\Delta(a,b,c;x) = \Omega(x^{1/(2a+2b)})$, providing that $c > 2a + 2b$.

Lemma 14.1 is to be found in E. Krätzel (1969), Lemmas 14.2 and 14.4 are due to the author (1981) and independently to M. Vogts (1981), and Lemma 14.3 was proved by P. G. Schmidt (1968a). Schmidt in fact proved Lemma 14.4 for $\Delta(1,2,3;x)$, which is needed for the problem of nonisomorphic abelian groups [since (14.101) implies (14.102)], and our proof of Lemmas 14.2 and 14.4 is based on Schmidt's ideas. Lemma 14.5 is to be found in H.-E. Richert (1952).

Theorem 14.3 was proved by the author (1981), but it is certainly not the best possible result. Improvements were obtained by M. Vogts (1981) and A. Ivić and P. Shiu (1982). The latter work contains the estimate

$$\Delta(a,b,c;x) \ll x^{263/171(a+b+c)}\log^2 x$$

if $1 \leqslant a < b \leqslant c$, $c \leqslant a + b$, $92b \leqslant 171a$ or $(a,b,c) = (1,2,2)$. The proof of this bound depends on the theory of two-dimensional exponent pairs (see Section 2.4 of Chapter 2), and finer two-dimensional techniques may lead to further improvements. Thus the forthcoming work of E. Krätzel (in press) contains several new estimates for $\Delta(a,b,c;x)$, including

$$\Delta(3,4,5;x) \ll x^{22/177}\log^3 x;$$

hence using Theorem 14.2 Krätzel obtains for the number of cube-full integers not exceeding x the analogue of Theorem 14.4, namely,

$$A_3(x) = \gamma_{0,3}x^{1/3} + \gamma_{1,3}x^{1/4} + \gamma_{2,3}x^{1/5} + O\left\{ x^{1/8}\exp\left(-C\log^{3/5}x(\log\log x)^{-1/5}\right)\right\},$$

since $\frac{22}{177} < \frac{1}{8}$.

The investigation of powerful numbers has a long and rich history. Both powerful numbers and nonisomorphic abelian groups of finite order were first investigated in 1935 by P. Erdős and G. Szekeres (1935). They proved by an elementary argument

$$A_k(x) = \gamma_{0,k}x^{1/k} + O(x^{1/(k+1)}) \qquad (k \geqslant 2),$$

that is, $\rho_k \leqslant 1/(k+1)$ in the notation of (14.90). With this result the matter rested until 1958, when P. T. Bateman and E. Grosswald (1958) proved essentially Theorem 14.4, and besides that they obtained the estimates $\rho_3 \leqslant \frac{7}{46}$, $\rho_k \leqslant 1/(k+2)$ for $k \geqslant 2$ and

$$\rho_k \leqslant \max\left(\frac{r}{k(r+2)}, \frac{1}{k+r+1}\right), \qquad r = [\sqrt{2k}] \qquad (k \geqslant 4).$$

Later researches include the author's papers (1973a, 1978a) and the paper of E. Krätzel (1982), where he proved

$$\rho_k \leqslant \frac{1}{k+H(k)}, \qquad \sqrt{\frac{8k}{3}} < H(k) < \left(1 + \sqrt{7/3}\right)\sqrt{\frac{8k}{3}}$$

if k is sufficiently large, which seems to be the sharpest estimate for k large. A Ivić and P. Shiu (1982) proved

$$\rho_3 = \frac{263}{2052} = 0.128167\ldots, \qquad \rho_4 \leqslant \frac{3091}{25981} = 0.118971\ldots,$$

$$\rho_5 \leqslant \frac{1}{10}, \qquad \rho_6 \leqslant \frac{1}{12}, \qquad \rho_7 \leqslant \frac{1}{14},$$

and it may be noted that the bounds given above for ρ_3 and ρ_4 are sharper than the ones given by Theorem 14.5. The result that $\rho_k \leqslant 1/(2k)$ for $8 \leqslant k \leqslant 10$ is proved by the author in (1984a). For the number of square full integers between consecutive squares and in short intervals, see the papers of P. Shiu (1980, 1984).

For some additive problems involving powerful numbers, see A. Ivić and P. Shiu (1982). R. W. K. Odoni (1981) proved recently by complicated arguments from algebraic number theory that the number of integers $\leqslant x$ which are a sum of two square full numbers is

$$\gg x(\log x)^{-1/2}\exp\left(\frac{C\log\log x}{\log\log\log x}\right) \qquad (C > 0,\ x \geqslant x_0).$$

By (14.142) there are $\sim b_1 x(\log x)^{-1/2}$ integers $\leqslant x$ which are a sum of two integer squares, so that Odoni's result shows that there are many more numbers which are a sum of two square full numbers.

The problem of the estimation of the error term $R(x)$ in the asymptotic formula for $\sum_{n \leqslant x} a(n)$ was considered for the first time by P. Erdős and G. Szekeres (1935), in the same paper which initiated the study of powerful numbers too. Various estimates of the form $R(x) = O(x^a \log^b x)$ were proved by the following authors:

$a = \frac{1}{2},\ b = 0$	P. Erdős and G. Szekeres (1935)
$a = \frac{1}{3},\ b = 2$	D. G. Kendall and R. A. Rankin (1947)
$a = \frac{3}{10},\ b = \frac{9}{10}$	H.-E. Richert (1952)
$a = \frac{20}{69} = 0.28958\ldots,\ b = \frac{21}{23}$	W. Schwarz (1966)
$a = \frac{34}{123} = 0.27642\ldots,\ b = 0$	P. G. Schmidt (1968a)
$a = \frac{7}{27} = 0.2592592\ldots,\ b = 2$	P. G. Schmidt (1968b)
$a = \frac{105}{407} = 0.25798\ldots,\ b = 2$	B. R. Srinivasan (1973)
$a = \frac{97}{381} = 0.25459\ldots,\ b = 35$	G. Kolesnik (1981b)

H.-E. Richert (1952) was the fist to estimate $R(x)$ [or equivalently $\Delta(1,2,3; x)$] by sums involving the function $\psi(x) = x - [x] - \frac{1}{2}$, and he actually proved and used (14.69). Richert's method was later refined by W. Schwarz (1966), and even more important progress was achieved by P. G. Schmidt (1968a) who obtained the symmetric expression (14.56) for $\Delta(1,2,3; x)$. His paper (1968b) and the works of B. R. Srinivasan (1973) and G. Kolesnik (1981b) utilize two-dimensional exponential sum methods to estimate $R(x)$ in (14.95). Note that Kolesnik's exponent $a = \frac{97}{381}$ is close to $\frac{1}{4}$, so that one may expect that in the foreseeable future

$$\sum_{n \leqslant x} a(n) = \sum_{j=1}^{4} A_j x^{1/j} + o(x^{1/4}) \qquad \left(A_j = \prod_{k=1, k \neq j}^{\infty} \zeta(k/j)\right)$$

will be proved, which was conjectured long ago by H.-E. Richert (1952). In the other direction, R. Balasubramanian and K. Ramachandra (1981) have recently shown that

$$\sum_{n \leqslant x} a(n) = \sum_{j=1}^{5} A_j x^{1/j} + \Omega\left(x^{1/6}\log^{1/2}x\right),$$

which is somewhat stronger than Theorem 14.7, but our proof is much simpler. Theorem 14.7 was also proved earlier by W. Schwarz (1967) by a different method than ours, but his proof was conditional, since he has to assume the Riemann hypothesis for (14.104) to be false.

The existence of local densities d_k in (14.110) was proved by D. G. Kendall and R. A. Rankin (1947). The proof of Theorem 14.8 is due to the author (1978c). The error term in (14.112) can be slightly reduced further, and this was done by E. Krätzel (1982) and the author (1983c). The latter paper treats several problems connected with the distribution of values of $a(n)$ and related multiplicative functions, including the proofs of (14.118) and (14.119).

From Theorem 14.8 we immediately have, uniformly in k,

$$\sum_{x < n \leqslant x+h,\, a(n)=k} 1 = (d_k + o(1))h \qquad [h = o(x), x \to \infty],$$

if $h = x^\theta$, $\theta > \frac{1}{2}$. It was proved by the author (1981) that one can take here $\theta \geqslant \frac{581}{1744} = 0.33314\ldots$, and by an analysis based on the arithmetical structure of k, E. Krätzel (1980) succeeded in reducing further the value of θ.

There exist several proofs of Theorem 14.9, and the one given here is based on H. Delange (1973).

To see that (14.140) holds consider the contour $\mathscr{L} = (\infty, \varepsilon] \cup \lambda_\varepsilon \cup (\varepsilon, \infty)$, where λ_ε is the circle $|z| = \varepsilon$ without the point $z = \varepsilon$. Then the integral

$$(14.169) \qquad \int_{\mathscr{L}} e^{-z} z^{s-1} \, dz = \int_{\infty}^{\varepsilon} e^{-t} t^{s-1} \, dt + e^{(s-1)2\pi i} \int_{\varepsilon}^{\infty} e^{-t} t^{s-1} \, dt + \int_{\lambda_\varepsilon} e^{-z} z^{s-1} \, dz$$

does not depend on ε, and letting $\varepsilon \to 0$ we obtain

$$\left(e^{2\pi i s} - 1 \right) \Gamma(s) = \left(e^{2\pi i s} - 1 \right) \int_0^\infty e^{-z} z^{s-1} \, dz = \int_{\mathscr{L}} e^{-z} z^{s-1} \, dz.$$

If we replace s by $1 - s$, take $\varepsilon = 1$ in (14.169), and make the substitution $z = -w$, then we obtain

$$\int_\Delta w^{-s} e^w \, dw = \left(e^{\pi i s} - e^{-\pi i s} \right) \Gamma(1 - s) = 2i \sin \pi s \, \Gamma(1 - s).$$

Finally, recalling that $\Gamma(s)\Gamma(1 - s) = \pi/(\sin \pi s)$ we obtain (14.140).

The asymptotic formula (14.142) is a classical result of E. Landau. For facts about $L(s, \chi)$ mentioned in the text the reader is referred to K. Prachar (1957) or K. Chandrasekharan (1970).

The asymptotic formula (14.147) is due to P. Turán (1934), and similar types of results were later obtained for a wide class of additive functions by J. Kubilius and others. For this, and other aspects of the theory of additive arithmetical functions the reader is referred to the monograph of J. Kubilius (1964). In contrast with "small" additive functions such as $\omega(n)$ or $\Omega(n)$ one may consider also "large" additive functions, of which typical examples would be

$$\beta(n) = \sum_{p|n} p, \, B(n) = \sum_{p^\alpha \| n} \alpha p.$$

For a discussion of large additive functions see Chapter 6 of J.-M. De Koninck and A. Ivić (1980).

Theorem 14.10 and Corollaries 14.1–14.4 may be found in the works of H. Delange (1959, 1971, 1973), where various other generalizations are given, and also Delange (1973) contains a rigorous proof of the difficult formula (14.162). Delange's methods are partly based on A. Selberg (1954), who was the first to evaluate the summatory functions of $z^{\omega(n)}$ and $z^{\Omega(n)}$. The estimation of the sum

$$\sum_{n \leqslant x,\, \Omega(n) - \omega(n) = q} 1$$

is known as "Renyi's problem," after A. Rényi (1955) who obtained (14.165).

Corollary 14.5 is an example of a sum involving reciprocals of arithmetical functions. Such sums were systematically investigated by J.-M. De Koninck and A. Ivić (1980), and in Chapter 5 of their monograph one can find a refinement of the asymptotic formula (14.168).

/ *CHAPTER FIFTEEN* /
ATKINSON'S FORMULA
FOR THE MEAN SQUARE

15.1 INTRODUCTION

A classical problem in zeta-function theory is the investigation of the asymptotic behavior of the integral

$$I(T) = \int_0^T \left| \zeta\left(\tfrac{1}{2} + it\right) \right|^2 dt,$$

and the first nontrivial result has been obtained by G. H. Hardy and J. E. Littlewood in 1918, who showed that

(15.1) $I(T) = (1 + o(1))T \log T \qquad (T \to \infty).$

A substantial advance in this problem was made in 1922 by J. E. Littlewood, who proved that

(15.2) $E(T) \ll T^{3/4+\varepsilon},$

where

(15.3) $E(T) = I(T) - T\log(T/2\pi) - (2\gamma - 1)T.$

An explicit formula for $E(T)$ was discovered by F. V. Atkinson in 1949, and this formula is the main topic of this chapter. This important result of Atkinson was neglected for a long time, until first important applications were made by D. R. Heath-Brown in 1978, and it seems certain that the possibilities of Atkinson's formula are far from being exhausted. The depth and the scope of Atkinson's formula provide an adequate ending of this text, and the result will be formulated as

THEOREM 15.1. Let $0 < A < A'$ be any two fixed constants such that $AT < N < A'T$ and let $N' = N'(T) = T/2\pi + N/2 - (N^2/4 + NT/2\pi)^{1/2}$. Then

(15.4)

$$E(T) = 2^{-1/2} \sum_{n \leqslant N} (-1)^n d(n) n^{-1/2} \left\{ \text{ar sinh}\left((\pi n/2T)^{1/2}\right) \right\}^{-1}$$

$$\times (T/2\pi n + \tfrac{1}{4})^{-1/4} \cos(f(T, n)) - 2 \sum_{n \leqslant N'} d(n) n^{-1/2} (\log T/2\pi n)^{-1}$$

$$\times \cos\{T(\log T/2\pi n) - T + \pi/4\} + O(\log^2 T),$$

where

(15.5) $f(T, n) = 2T \,\text{ar sinh}\left((\pi n/2T)^{1/2}\right) + (2\pi nT + \pi^2 n^2)^{1/2} - \pi/4.$

We may rewrite (15.4) in the form

(15.6) $E(T) = \Sigma_1(T) + \Sigma_2(T) + O(\log^2 T),$

(15.7) $\Sigma_1(T) = 2^{1/2} (T/2\pi)^{1/4} \sum_{n \leqslant N} (-1)^n d(n) n^{-3/4} e(T, n) \cos(f(T, n)),$

where

(15.8)

$$e(T, n) = (1 + \pi n/2T)^{-1/4} \left\{ (2T/\pi n)^{1/2} \text{ar sinh}\left((\pi n/2T)^{1/2}\right) \right\}^{-1}$$

$$= 1 + O(nT^{-1}),$$

(15.9) $\Sigma_2(T) = -2 \sum_{n \leqslant N'} d(n) n^{-1/2} (\log T/2\pi n)^{-1} \cos(g(T, n)),$

where

(15.10) $\qquad\qquad g(T, n) = T \log(T/2\pi n) - T + \pi/4.$

Using the Taylor expansion

(15.11)

$$f(T, n) = -\pi/4 + 4\pi(nT/2\pi)^{1/2} + O(n^{3/2} T^{-1/2}), \qquad n = o(T),$$

it is seen that, apart from the oscillating factor $(-1)^n$, the first $o(T^{1/3})$ terms in $\Sigma_1(T)$ are asymptotically equal to the corresponding terms in the truncated Voronoi formula for $2\pi\Delta(T/2\pi)$, as given by (3.17). This deep analogy between the divisor problem and the mean square of the zeta-function on the critical line has been one of primary motivations of Atkinson's work concerning Theorem 15.1. This topic will be further pursued in Section 15.5.

There is another possibility of proving an explicit formula for $E(T)$. This has been found recently by R. Balasubramanian (1978), who used a complicated integration technique based on the Riemann–Siegel formula (4.3) to prove

$$(15.12) \quad E(T) = 2 \sum_{n \leqslant K} \sum_{m \neq n \leqslant K} \frac{\sin(T \log n/m)}{(mn)^{1/2} \log n/m}$$

$$+ 2 \sum_{n \leqslant K} \sum_{m \neq n \leqslant K} \frac{\sin(2\theta - T \log mn)}{(mn)^{1/2}(2\theta' - \log mn)} + O(\log^2 T),$$

where

$$(15.13) \quad \theta = \theta(T) = \frac{T}{2}\log(T/2\pi) - \frac{T}{2} - \pi/8, \qquad K = \left[(T/2\pi)^{1/2}\right].$$

Upper bounds for $E(T)$ may be obtained from (15.12), but it seems simpler to use Atkinson's formula and the averaging techniques similar to those of Chapter 7. In this way it will be seen that

$$(15.14) \qquad E(T) \ll T^{35/108 + \varepsilon},$$

which is completely analogous to corresponding estimates for $\Delta(x)$ and $P(x)$ furnished by Theorems 13.1 and 13.11, respectively, since the estimation will be reduced to very similar exponential sums. We reserve Section 15.2 of this chapter for the proof of the difficult Theorem 15.1, while some applications of Atkinson's formula will be presented in later sections.

15.2 PROOF OF ATKINSON'S FORMULA

We start from the obvious identity, valid for $\operatorname{Re} u > 1$, $\operatorname{Re} v > 1$,

$$(15.15) \quad \zeta(u)\zeta(v) = \sum_{m=1}^{\infty} \sum_{n=1}^{\infty} m^{-u} n^{-v} = \zeta(u+v) + f(u,v) + f(v,u),$$

where

$$(15.16) \qquad f(u,v) = \sum_{r=1}^{\infty} \sum_{s=1}^{\infty} r^{-u}(r+s)^{-v}.$$

We shall show first that $f(u, v)$ is a meromorphic function of u and v for $\mathrm{Re}(u + v) > 0$. Taking $\mathrm{Re}\, v > 1$ and writing $\psi(x) = x - [x] - \frac{1}{2}$, $\psi_1(x) = \int_1^x \psi(t)\,dt$, it follows on integrating by parts that

$$\sum_{s=1}^{\infty} (r + s)^{-v} = \int_{1-0}^{\infty} (r + t)^{-v} d[t] = v \int_r^{\infty} ([x] - r)x^{-v-1}\, dx$$

$$= r^{1-v}(v - 1)^{-1} - \tfrac{1}{2}r^{-v} - v\int_r^{\infty} \psi(x)x^{-v-1}\, dx$$

$$= r^{1-v}(v - 1)^{-1} - \tfrac{1}{2}r^{-v} - v(v + 1)\int_r^{\infty} \psi_1(x)x^{-v-2}\, dx$$

$$= r^{1-v}(v - 1)^{-1} - \tfrac{1}{2}r^{-v} + O(|v|^2 r^{-\mathrm{Re}\, v - 1}),$$

since $\psi_1(x) \ll 1$ uniformly in x. Hence

$$f(u, v) = (v - 1)^{-1}\sum_{r=1}^{\infty} r^{1-u-v} - \tfrac{1}{2}\sum_{r=1}^{\infty} r^{-u-v} + O\left(|v|^2 \sum_{r=1}^{\infty} r^{-\mathrm{Re}\, u - \mathrm{Re}\, v - 1}\right),$$

and therefore

$$f(u, v) - (v - 1)^{-1}\zeta(u + v - 1) + \tfrac{1}{2}\zeta(u + v)$$

is regular for $\mathrm{Re}(u + v) > 0$. Thus (15.15) holds by analytic continuation when u and v both lie in the critical strip, apart from the poles at $v = 1$, $u + v = 1$, and $u + v = 2$.

We consider next the case $\mathrm{Re}\, u < 0$, $\mathrm{Re}(u + v) > 2$. Using the Poisson summation formula (A.25) we obtain

$$(15.17) \quad \sum_{r=1}^{\infty} r^{-u}(r + s)^{-v} = \int_0^{\infty} x^{-u}(x + s)^{-v}\, dx$$

$$+ 2\sum_{m=1}^{\infty} \int_0^{\infty} x^{-u}(x + s)^{-v} \cos(2\pi m x)\, dx$$

$$= s^{1-u-v}\left(\int_0^{\infty} y^{-u}(1 + y)^{-v}\, dy \right.$$

$$\left. + 2\sum_{m=1}^{\infty} \int_0^{\infty} y^{-u}(1 + y)^{-v} \cos(2\pi m s y)\, dy \right),$$

after the change of variable $x = sy$. Summing over s and using (A.31) we have

$$(15.18) \quad g(u,v) = f(u,v) - \Gamma(u+v-1)\Gamma(1-u)\Gamma^{-1}(v)\zeta(u+v-1)$$

$$= 2\sum_{s=1}^{\infty} s^{1-u-v} \sum_{m=1}^{\infty} \int_0^{\infty} y^{-u}(1+y)^{-v}\cos(2\pi msy)\,dy.$$

To investigate the convergence of the last expression, we note that for $\mathrm{Re}\,u < 1$, $\mathrm{Re}(u+v) > 0$, $n \geqslant 1$,

$$(15.19) \quad 2\int_0^{\infty} y^{-u}(1+y)^{-v}\cos(2\pi ny)\,dy$$

$$= n^{u-1}\int_0^{\infty} y^{-u}(1+y/n)^{-v}(e(y)+e(-y))\,dy$$

$$= n^{u-1}\int_0^{i\infty} y^{-u}(1+y/n)^{-v}e(y)\,dy$$

$$+ n^{u-1}\int_0^{-i\infty} y^{-u}(1+y/n)^{-v}e(-y)\,dy \ll \frac{n^{\mathrm{Re}\,u-1}}{|u-1|}$$

uniformly for bounded u and v, which follows after integrating by parts. Thus the double series in (15.18) is absolutely convergent for $\mathrm{Re}\,u < 0$, $\mathrm{Re}\,v > 1$, $\mathrm{Re}(u+v) > 0$, by comparison with $\sum_{s=1}^{\infty}|s|^{-v}\sum_{m=1}^{\infty}|m^{u-1}|$, and represents an analytic function of both variables in this region. Hence (15.18) holds throughout this region and grouping terms with $ms = n$ together we have

$$(15.20) \quad g(u,v) = 2\sum_{n=1}^{\infty} \sigma_{1-u-v}(n)\int_0^{\infty} y^{-u}(1+y)^{-v}\cos(2\pi ny)\,dy,$$

where $\sigma_k(n) = \sum_{d|n}d^k$ is the sum of the kth powers of divisors of n, so that $\sigma_0(n) = d(n)$. Therefore, if $g(u,v)$ is the analytic continuation of the function given by (15.18), then for $0 < \mathrm{Re}\,u < 1$, $0 < \mathrm{Re}\,v < 1$, $u+v \neq 1$, we have

$$(15.21) \quad \zeta(u)\zeta(v) = \zeta(u+v) + \zeta(u+v-1)\Gamma(u+v-1)$$

$$\times \left(\frac{\Gamma(1-u)}{\Gamma(v)} + \frac{\Gamma(1-v)}{\Gamma(u)} \right) + g(u,v) + g(v,u).$$

It is, however, the exceptional case $u+v = 1$, in which we are interested. Here we may use the fact that $g(u,v)$ is continuous and write $u+v = 1+\delta$, $0 < |\delta| < \frac{1}{2}$, with the aim of letting $\delta \to 0$. Then the first terms on the right-hand side of (15.21) become

$$\zeta(1 + \delta) + \zeta(\delta)\Gamma(\delta)\left(\frac{\Gamma(1 - u)}{\Gamma(1 - u + \delta)} + \frac{\Gamma(u - \delta)}{\Gamma(u)} \right)$$

$$= \zeta(1 + \delta) + \zeta(1 - \delta)(2\pi)^{\delta}(2\cos\pi\delta/2)^{-1}\left(\frac{\Gamma(1 - u)}{\Gamma(1 - u + \delta)} + \frac{\Gamma(u - \delta)}{\Gamma(u)} \right)$$

$$= \delta^{-1} + \gamma + (\gamma - \delta^{-1})\left(\frac{1}{2} + \frac{\delta}{2}\log 2\pi \right)$$

$$\times\left(1 - \frac{\Gamma'(1 - u)}{\Gamma(1 - u)}\delta + 1 - \delta\frac{\Gamma'(u)}{\Gamma(u)} \right) + O(|\delta|)$$

$$= \frac{1}{2}\left(\frac{\Gamma'(1 - u)}{\Gamma(1 - u)} + \frac{\Gamma'(u)}{\Gamma(u)} \right) + 2\gamma - \log 2\pi + O(|\delta|),$$

where we used Taylor's formula for the gamma-function terms, the functional equation for the zeta-function, and

$$\zeta(s) = (s - 1)^{-1} + \gamma + O(|s - 1|).$$

Hence letting $\delta \to 0$ we have, for $0 < \operatorname{Re} u < 1$,

$$(15.22) \quad \zeta(u)\zeta(1 - u) = \frac{1}{2}\left(\frac{\Gamma'(1 - u)}{\Gamma(1 - u)} + \frac{\Gamma'(u)}{\Gamma(u)} \right) + 2\gamma - \log 2\pi$$

$$+ g(u, 1 - u) + g(1 - u, u),$$

with a view to the eventual application $u = \frac{1}{2} + it$ in mind. Reasoning as in (15.19) we have, for $\operatorname{Re} u < 0$,

$$(15.23) \quad g(u, 1 - u) = 2\sum_{n=1}^{\infty} d(n)\int_{0}^{\infty} y^{-u}(1 + y)^{u-1}\cos(2\pi ny)\, dy,$$

and so what we need is an analytic continuation of (15.23) valid when $\operatorname{Re} u = \frac{1}{2}$. At this point of the proof the Voronoi formula for $\Delta(x)$ comes into play, since it is a powerful tool which will provide the desired analytic continuation and enable us to integrate (15.22) over t when $u = \frac{1}{2} + it$ (t real), thus giving the expression $2i\int_{0}^{T}|\zeta(\frac{1}{2} + it)|^2\, dt$ on the left-hand side of (15.22). Using the Voronoi formula (3.1) and the asymptotic formulas (3.12) and (3.13) we have, when x is not an integer,

(15.24) $\quad \Delta(x) = (\pi\sqrt{2})^{-1} x^{1/4} \sum_{n=1}^{\infty} d(n) n^{-3/4}$

$$\times \left\{ \cos(4\pi\sqrt{nx} - \pi/4) - 3(32\pi\sqrt{nx})^{-1} \sin(4\pi\sqrt{xn} - \pi/4) \right\}$$

$$+ O(x^{-3/4}),$$

and the series is boundedly convergent in any finite x-interval.

Let now N be a positive integer, and let

(15.25) $\qquad h(u, x) = 2 \int_0^\infty y^{-u} (1 + y)^{u-1} \cos(2\pi xy) \, dy.$

Then we have with $D(x) = \sum_{n \leqslant x} d(n)$

$$\sum_{n > N} d(n) h(u, n) = \int_{N+1/2}^{\infty} h(u, x) \, dD(x)$$

$$= \int_{N+1/2}^{\infty} (\log x + 2\gamma) h(u, x) \, dx + \int_{N+1/2}^{\infty} h(u, x) \, d\Delta(x)$$

$$= -\Delta(N + \tfrac{1}{2}) h(u, N + \tfrac{1}{2})$$

$$+ \int_{N+1/2}^{\infty} (\log x + 2\gamma) h(u, x) \, dx$$

$$- \int_{N+1/2}^{\infty} \Delta(x) \frac{\partial h(u, x)}{\partial x} \, dx.$$

Hence (15.23) becomes

(15.26) $\quad g(u, 1 - u) = \sum_{n \leqslant N} h(u, n) d(n) - \Delta(N + \tfrac{1}{2}) h(u, N + \tfrac{1}{2})$

$$+ \int_{N+1/2}^{\infty} (\log x + 2\gamma) h(u, x) \, dx$$

$$- \int_{N+1/2}^{\infty} \Delta(x) \frac{\partial h(u, x)}{\partial x} \, dx$$

$$= g_1(u) - g_2(u) + g_3(u) - g_4(u),$$

say. Here $g_1(u)$ and $g_2(u)$ are analytic functions of u in the region $\operatorname{Re} u < 1$, since the right-hand side of (15.25) is analytic in this region. Consider next

$g_4(u)$. We have

$$h(u, x) = \int_0^{i\infty} y^{-u}(1 + y)^{u-1} e(xy) \, dy + \int_0^{-i\infty} y^{-u}(1 + y)^{u-1} e(-xy) \, dy,$$

$$\frac{\partial h(u, x)}{\partial x} = 2\pi i \int_0^{i\infty} y^{1-u}(1 + y)^{u-1} e(xy) \, dy$$

$$- 2\pi i \int_0^{-i\infty} y^{1-u}(1 + y)^{u-1} e(-xy) \, dy$$

$$= 2\pi i x^{u-2} \left(\int_0^{i\infty} y^{1-u}(1 + y/x)^{u-1} e(y) \, dy \right.$$

$$\left. - \int_0^{-i\infty} y^{1-u}(1 + y/x)^{u-1} e(-y) \, dy \right) \ll x^{\mathrm{Re}\, u - 2}$$

for $\mathrm{Re}\, u \leqslant 1$ and bounded u. Using only the estimate $\Delta(x) \ll x^{1/3+\varepsilon}$ it is seen that the integral defining $g_4(u)$ is an analytic function of u at any rate when $\mathrm{Re}\, u < \frac{2}{3}$.

It remains to consider $g_3(u)$. Let, for brevity, $X = N + \frac{1}{2}$. Then

(15.27) $$g_3(u) = \int_X^\infty (\log x + 2\gamma) \left(\int_0^{i\infty} y^{-u}(1 + y)^{u-1} e(xy) \, dy \right.$$

$$\left. + \int_0^{-i\infty} y^{-u}(1 + y)^{u-1} e(-xy) \, dy \right) dx.$$

For $\mathrm{Re}\, u < 0$ an integration by parts shows that the first two integrals in (15.27) are equal to

$$-(2\pi i)^{-1}(\log X + 2\gamma) \int_0^{i\infty} y^{-u-1}(1 + y)^{u-1} e(Xy) \, dy$$

$$-(2\pi i)^{-1} \int_X^\infty dx \int_0^{i\infty} y^{-u-1}(x + y)^{u-1} e(xy) \, dy$$

$$= -(2\pi i)^{-1}(\log X + 2\gamma) \int_0^\infty y^{-u-1}(1 + y)^{u-1} e(Xy) \, dy$$

$$+(2\pi i u)^{-1} \int_0^{i\infty} y^{-u-1}(X + y)^u e(y) \, dy.$$

In the last integral above the line of integration may be taken as $[0, \infty)$ and the variable y replaced by $y = Xz$. The other two integrals in (15.27) are treated similarly, and the results may be combined to produce

$$(15.28) \quad g_3(u) = -\pi^{-1}(\log X + 2\gamma) \int_0^\infty y^{-u-1}(1 + y)^{u-1} \sin(2\pi Xy) \, dy$$

$$+ (\pi u)^{-1} \int_0^\infty y^{-u-1}(1 + y)^u \sin(2\pi Xy) \, dy.$$

Noting that the integrals in (15.28) are uniformly convergent when $\mathrm{Re}\, u \leqslant 1 - \varepsilon$, it follows that (15.28) provides us with an analytic continuation which is valid when $\mathrm{Re}\, u = \frac{1}{2}$, and thus we may proceed to integrate (15.22). When $u = \frac{1}{2} + it$ we have $\zeta(u)\zeta(1 - u) = |\zeta(\frac{1}{2} + it)|^2$, so that the integration of (15.22) gives

$$2iI(T) = \int_{1/2-iT}^{1/2+iT} \zeta(u)\zeta(1 - u) \, du$$

$$= \tfrac{1}{2}(-d \log \Gamma(1 - u) + d \log \Gamma(u)) \Big|_{1/2-iT}^{1/2+iT} + 2iT(2\gamma - \log 2\pi)$$

$$+ \int_{1/2-iT}^{1/2+iT} (g(u, 1 - u) + g(1 - u, u)) \, du$$

$$= \log \frac{\Gamma(\frac{1}{2} + iT)}{\Gamma(\frac{1}{2} - iT)} + 2iT(2\gamma - \log 2\pi)$$

$$+ 2 \int_{1/2-iT}^{1/2+iT} g(u, 1 - u) \, du.$$

Using Stirling's formula in the form given by (A.34), this becomes

(15.29)

$$I(T) = T \log(T/2\pi) + (2\gamma - 1)T - i \int_{1/2-iT}^{1/2-iT} g(u, 1 - u) \, du + O(1)$$

$$= T \log(T/2\pi) + (2\gamma - 1)T + I_1 - I_2 + I_3 - I_4 + O(1),$$

where for $n = 1, 2, 3, 4$

$$(15.30) \qquad\qquad I_n = -i \int_{1/2-iT}^{1/2+iT} g_n(u) \, du,$$

so that using (15.26) and (15.28) we have

$$(15.31) \quad I_1 = 4 \sum_{n \leqslant N} d(n) \int_0^\infty \frac{\sin(T \log(1 + y)/y)\cos(2\pi n y)}{y^{1/2}(1 + y)^{1/2}\log(1 + y)/y} \, dy,$$

(15.32)

$$I_2 = 4\Delta(X) \int_0^\infty \frac{\sin(T \log(1 + y)/y)\cos(2\pi X y)}{y^{1/2}(1 + y)^{1/2}\log(1 + y)/y} \, dy,$$

$$(15.33) \quad I_3 = -\frac{2}{\pi}(\log X + 2\gamma) \int_0^\infty \frac{\sin(T \log(1 + y)/y)\sin(2\pi X y)}{y^{3/2}(1 + y)^{1/2}\log(1 + y)/y} \, dy$$

$$+ (\pi i)^{-1} \int_0^\infty y^{-1}\sin(2\pi X y) \, dy \int_{1/2-iT}^{1/2+iT} (1 + y^{-1})^u u^{-1} \, du,$$

and lastly

$$(15.34) \qquad I_4 = -i \int_X^\infty \Delta(x) \, dx \int_{1/2-iT}^{1/2+iT} \frac{\partial h(u, x)}{\partial x} \, du,$$

where N is a positive integer, $X = N + \frac{1}{2}$, and as in the formulation of the theorem we shall restrict N to the range $AT < N < A'T$. A more explicit formula for I_4 may be derived as follows. We have from (15.25)

$$\int_{1/2-iT}^{1/2+iT} \frac{\partial h(u, x)}{\partial x} \, du = 4i \frac{\partial}{\partial x} \left\{ \int_0^\infty \frac{\sin(T \log(1 + y)/y)\cos(2\pi x y)}{y^{1/2}(1 + y)^{1/2}\log(1 + y)/y} \, dy \right\}$$

$$= 4i \frac{\partial}{\partial x} \left\{ \int_0^\infty \frac{\sin(T \log(x + y)/y)\cos(2\pi y)}{y^{1/2}(x + y)^{1/2}\log(x + y)/y} \, dy \right\}$$

$$= 4i \int_0^\infty \frac{\cos 2\pi y}{y^{1/2}(x + y)^{3/2}\log(x + y)/y}$$

$$\times \left\{ T\cos(T \log(x + y)/y) - \sin(T \log(x + y)/y)\left(\frac{1}{2} + \log^{-1}\frac{x + y}{y}\right) \right\} dy.$$

Hence replacing y by xy we obtain

(15.35)

$$I_4 = 4 \int_X^\infty \frac{\Delta(x)}{x} \, dx \int_0^\infty \frac{\cos(2\pi xy)}{y^{1/2}(1+y)^{3/2}\log\{1+y\}/y}$$

$$\times \left\{ T\cos\left(T\log\frac{1+y}{y}\right) - \sin\left(T\log\frac{1+y}{y}\right)\left(\frac{1}{2} + \log^{-1}\frac{1+y}{y}\right) \right\} dy.$$

The main difficulty lies now in the evaluation of the integrals which represent I_n. We shall need two lemmas which will follow from Theorem 2.2. These are

LEMMA 15.1. Let $\alpha, \beta, \gamma, a, b, k, T$ be real numbers such that α, β, γ are positive and bounded, $\alpha \neq 1$, $0 < a < \frac{1}{2}$, $a < T/8\pi k$, $b \geqslant T$, $k \geqslant 1$, and $T \geqslant 1$. Then

(15.36) $\displaystyle \int_a^b y^{-\alpha}(1+y)^{-\beta}\left(\log\frac{1+y}{y}\right)^{-\gamma} \exp\left(iT\log\frac{1+y}{y} + 2\pi kiy\right) dy$

$$= (2k\pi^{1/2})^{-1} T^{1/2} V^{-\gamma} U^{-1/2} (U - 1/2)^{-\alpha}(U + 1/2)^{-\beta}$$

$$\times \exp\left(iTV + 2\pi ikU - \pi ik + \frac{\pi i}{4}\right) + O(a^{1-\alpha}T^{-1})$$

$$+ O(b^{\gamma-\alpha-\beta}k^{-1}) + R(T, k)$$

uniformly for $|\alpha - 1| > \varepsilon$, where

$$U = (T/2\pi k + 1/4)^{1/2}, \qquad V = 2 \operatorname{ar\,sinh}\left((\pi k/2T)^{1/2}\right),$$

$$R(T, k) \ll T^{(\gamma-\alpha-\beta)/2 - 1/4} k^{-(\gamma-\alpha-\beta)/2 - 5/4}, \qquad \text{for } 1 \leqslant k \leqslant T,$$

$$R(T, k) \ll T^{-1/2 - \alpha} k^{\alpha - 1}, \qquad \text{for } k \geqslant T.$$

A similar result holds for the corresponding integral with $-k$ in place of k, except that in that case the explicit term on the right-hand side of (15.36) is to be omitted.

LEMMA 15.2. For $AT^{1/2} < a < A'T^{1/2}, 0 < A < A', \alpha > 0$,

(15.37)

$$\int_a^\infty \frac{\exp i\{4\pi x\sqrt{n} - 2T\,\text{ar sinh}(x\sqrt{\pi/2T}) - (2\pi x^2 T + \pi^2 x^4)^{1/2} + \pi x^2\}}{x^\alpha \text{ar sinh}(x\sqrt{\pi/2T})((\frac{1}{2} + T/2\pi x^2 + \frac{1}{4})^{1/2})(\frac{1}{4} + T/2\pi x^2)^{1/4}}\,dx$$

$$= 4\pi T^{-1} n^{(\alpha-1)/2} (\log T/2\pi n)^{-1} (T/2\pi - n)^{3/2 - \alpha}$$

$$\times \exp i(T - T\log(T/2\pi n) - 2\pi n + \pi/4)$$

$$+ O\left(T^{-\alpha/2}\min\left(1, |2\sqrt{n} + a - (a^2 + 2T/n)^{1/2}|^{-1}\right)\right)$$

$$+ O\left(n^{(\alpha-1)/2}(T/2\pi - n)^{1-\alpha}T^{-3/2}\right),$$

provided that $n \geqslant 1$, $n < T/2\pi$, $(T/2\pi - n)^2 > na^2$. If the last two restrictions on n are not satisfied, or if $\sqrt{n}$ is replaced by $-\sqrt{n}$, then the main term and the last error term on the right-hand side of (15.37) are to be omitted.

Proof of Lemmas 15.1 and 15.2. To obtain Lemma 15.1 one may apply Theorem 2.2 with

$$\varphi(x) = x^{-\alpha}(1 + x)^{-\beta}\left(\log\frac{1+x}{x}\right)^{-\gamma}, \qquad f(x) = (T/2\pi)\log\frac{1+x}{x},$$

$$\Phi(x) = x^{-\alpha}(1 + x)^{\gamma-\beta}, \qquad F(x) = T/(1 + x), \qquad \mu(x) = x/2.$$

We have

$$f'(x) = -\frac{T}{2\pi x(1 + x)},$$

so that the saddle points of Theorem 2.2 are the roots of

$$x_0(x_0 + 1) = T/2\pi k,$$

hence $x_0 = U - \frac{1}{2}$ in the notation of Lemma 15.1. Thus

$$f_0'' = \frac{T(2x_0 + 1)}{2\pi x_0^2(x_0 + 1)^2} = 4\pi k^2 U T^{-1},$$

$$\log(1 + 1/x_0) = \log\frac{U + \frac{1}{2}}{U - \frac{1}{2}} = \log\frac{(2T/\pi k + 1)^{1/2} + 1}{(2T/\pi k + 1)^{1/2} - 1}$$

$$= \log\frac{(\pi k/2T + 1)^{1/2} + (\pi k/2T)^{1/2}}{(\pi k/2T + 1)^{1/2} - (\pi k/2T)^{1/2}}$$

$$= \log\left((\pi k/2T + 1)^{1/2} + (\pi k/2T)^{1/2}\right)^2$$

$$= 2\,\text{ar sinh}\left((\pi k/2T)^{1/2}\right).$$

Hence

$$f_0 + kx_0 = TV/2\pi + k(U - \tfrac{1}{2}),$$

and the main term furnished by Theorem 2.2 is

$$\varphi_0(f_0'')^{-1/2} e(f_0 + kx_0 + \tfrac{1}{8})$$

$$= (U - 1/2)^{-\alpha}(U + 1/2)^{-\beta}V^{-\gamma}(2k)^{-1}(T/\pi)^{1/2}$$

$$\times U^{-1/2}\exp i(TV + 2\pi kU - \pi k + \pi/4).$$

Consider now the error terms. If $1 \leqslant k \leqslant T$, we have then

$$A \leqslant A(T/k)^{1/2} < x_0 < A'(T/k)^{1/2}, \qquad \Phi_0 \ll x_0^{\gamma - \alpha - \beta},$$

$$\mu_0 \ll x_0, \qquad A(kT)^{1/2} < F_0 < A'(kT)^{1/2},$$

and thus for $1 \leqslant k \leqslant T$,

$$\Phi_0\mu_0 F^{-3/2} \ll T^{(\gamma - \alpha - \beta)/2 - 1/4}k^{-(\gamma - \alpha - \beta)/2 - 5/4},$$

while in case $k \geqslant T$ we obtain similarly

$$\Phi_0\mu_0 F^{-3/2} \ll T^{-\alpha - 1/2}k^{\alpha - 1}.$$

From $f'(x) = -T/(2\pi x(1 + x))$ we have, for $a < \max(\tfrac{1}{2}, T/8\pi k)$,

$$f'(x) + k < -ATa^{-1},$$

which gives

$$\Phi(a)(f'(a) + k)^{-1} \ll a^{1-\alpha}T^{-1}.$$

Likewise, if $b \geqslant T$,

$$\Phi(b)(f'(b) + k)^{-1} \ll b^{\gamma - \alpha - \beta}k^{-1}.$$

The error-term integral in Theorem 2.2 is

$$\ll \int_a^1 x^{-\alpha}e^{-Akx - AT}\, dx + \int_1^b x^{\gamma - \alpha - \beta}e^{-Akx - AT/x}\, dx,$$

and for $|\alpha - 1| > \varepsilon > 0$ the contribution of the above terms clearly does not exceed the order of the error terms given by Lemma 15.1. This establishes Lemma 15.1 for $k \geqslant 1$, while for $k \leqslant -1$ the argument differs only in that the terms in x_0 do not occur, since then there are no saddle points in Theorem 2.2.

For the proof of Lemma 15.2 we apply Theorem 2.2 with a, b as limits of integration, where $b > T$, and

$$\varphi(x) = x^{-\alpha}\left(\operatorname{ar\,sinh}\left(x\sqrt{\pi/2T}\right)\right)^{-1}\left(\left(T/2\pi x^2 + \tfrac{1}{4}\right)^{1/2} + \tfrac{1}{2}\right)^{-1}$$
$$\times\left(T/2\pi x^2 + \tfrac{1}{4}\right)^{-1/4},$$

$$f(x) = \tfrac{1}{2}x^2 - \left(Tx^2/2\pi + x^4/4\right)^{1/2} - \frac{T}{\pi}\operatorname{ar\,sinh}\left(x\sqrt{\pi/2T}\right).$$

We have then

$$f'(x) = x - \left(x^2 + 2T/\pi\right)^{1/2}, \qquad f''(x) = 1 - x\left(x^2 + 2T/\pi\right)^{-1/2},$$

so that we may take $\mu(x) = x/2$, $\Phi(x) = x^{-\alpha}$, $F(x) = T$. We dispose first of the error terms in a and b. We have

$$\Phi(a)\left(\left|f'_a + 2\sqrt{n}\right| + f''^{-1/2}_a\right)^{-1}$$

$$\ll T^{-\alpha/2}\min\left(1, \left|2\sqrt{n} + a - \left(a^2 + \frac{2}{\pi}T\right)^{1/2}\right|^{-1}\right),$$

and

$$\Phi(b)\left(f'_b + 2\sqrt{n}\right)^{-1} \ll b^{-\alpha}\left(\sqrt{n} + O(Tb^{-1})\right)^{-1}$$

which is $o(1)$ for $b \to \infty$. The error-term integral of Theorem 2.2 gives here

$$\ll \int_a^b x^{-\alpha}e^{-Ax\sqrt{n} - AT}\,dx \ll e^{-A\sqrt{nT} - AT},$$

while

$$\Phi_0 x_0 F_0^{-3/2} \ll x_0^{1-\alpha}T^{-3/2} \ll n^{(\alpha-1)/2}\left(T/2\pi - n\right)^{1-\alpha}T^{-3/2},$$

as x_0 is given by

$$f'(x_0) + 2\sqrt{n} = 0, \qquad x_0 = n^{-1/2}(T/2\pi - n).$$

Here if $\sqrt{n} \leqslant -1$, or $n > T/2\pi$ or $(T/2\pi - n) \leqslant na^2$ there will be no terms in x_0 and the lemma is proved. In other cases we find that

$$f''_0 = 2n(T/2\pi + n)^{-1}, \qquad \operatorname{ar\,sinh}\left(x_0\sqrt{\pi/2T}\right) = \tfrac{1}{2}\log(T/2\pi n),$$

$$\left(T/2\pi x_0^2 + \tfrac{1}{4}\right)^{1/2} - \tfrac{1}{2} = n(T/2\pi - n)^{-1},$$

$$\left(T/2\pi x_0^2 + \tfrac{1}{4}\right)^{1/2} + \tfrac{1}{2} = \frac{T}{2\pi}(T/2\pi - n)^{-1},$$

and so on, and calculating the main term, which is

$$\varphi_0(f_0'')^{-1/2}\exp(2\pi i f_0 + 4\pi i x_0\sqrt{n} + \tfrac{1}{4}\pi i),$$

we obtain Lemma 15.2.

Having now at disposal Lemmas 15.1 and 15.2 we proceed to evaluate I_n for $n \leqslant 4$, as given by (15.31)–(15.34). We consider first I_1, taking in Lemma 15.1 $0 < \alpha < 1$, $\alpha + \beta > \gamma$, so that we may let $a \to 0$, $b \to \infty$. Hence, if $\tfrac{1}{2} < \alpha < \tfrac{3}{4}$, $1 \leqslant k < AT$, we obtain

(15.38)
$$\int_0^\infty \frac{\sin(T\log(1+y)/y)\cos(2\pi ky)}{y^\alpha(1+y)^{1/2}\log(1+y)/y}\,dy$$

$$= (4k)^{-1}(T/\pi)^{1/2}\frac{\sin(TV + 2\pi kU - \pi k + \pi/4)}{VU^{1/2}(U - \tfrac{1}{2})^\alpha(U + \tfrac{1}{2})^{1/2}}$$

$$+ O(T^{-\alpha/2}k^{(\alpha-3)/2}),$$

and since this result holds uniformly in α we may put $\alpha = \tfrac{1}{2}$. Taking into account that $\sin(x - \pi k) = (-1)^k\sin x$ we obtain, after substituting (15.38) into (15.31),

(15.39)
$$I_1 = 2^{-1/2}\sum_{n \leqslant N}(-1)^n d(n)n^{-1/2}$$

$$\times\left(\frac{\sin(2T\operatorname{ar\,sinh}\sqrt{\pi n/2T} + \sqrt{2\pi nT + \pi^2 n^2} + \pi/4)}{(\operatorname{ar\,sinh}\sqrt{\pi n/2T})(T/2\pi n + \tfrac{1}{4})^{1/4}}\right) + O(T^{-1/4}),$$

taking $AT < N < A'T$. Similarly, from (15.32),

(15.40)
$$I_2 \ll |\Delta(X)|X^{-1/2} \ll T^{-1/6},$$

if we use $\Delta(X) \ll X^{1/3}$.

To deal with I_3 we write (15.33) in the form

(15.41)
$$I_3 = -\frac{2}{\pi}(\log X + 2\gamma)I_{31} + (\pi i)^{-1}I_{32}$$

and consider first I_{31}. We have

$$\int_0^\infty \frac{\sin(T\log(1+y)/y)\sin(2\pi Xy)}{y^{3/2}(1+y)^{1/2}\log(1+y)/y}\,dy = \int_0^{(2X)^{-1}} + \int_{(2X)^{-1}}^\infty \ll T^{-1/2},$$

if the first integral is estimated by the second mean value theorem for integrals as

$$2\pi X \int_0^\xi \frac{\sin(T\log(1+y)/y)}{y(1+y)} \left(\frac{y^{1/2}(1+y)^{1/2}}{\log(1+y)/y} \right) dy$$

$$= 2\pi X \xi^{1/2}(1+\xi)^{1/2}(\log(1+\xi)/\xi)^{-1} \int_\eta^\xi \frac{\sin(T\log(1+y)/y)}{y(1+y)} dy$$

$$= 2\pi X \xi^{1/2}(1+\xi)^{1/2}(\log(1+\xi)/\xi)^{-1} \left\{ T^{-1}\cos(T\log(1+y)/y) \right\} \Big|_\eta^\xi$$

$$\ll T^{-1/2},$$

where $0 \leqslant \eta \leqslant \xi \leqslant (2X)^{-1}$, and the integral $\int_{(2X)^{-1}}^\infty$ is estimated by Lemma 15.1 by treating the main terms on the right-hand side of (15.36) as an error term.

Take next I_{32} and write

$$I_{32} = \int_0^\infty y^{-1}\sin(2\pi Xy)\, dy \int_{1/2-iT}^{1/2+iT} \frac{1}{u}\left(\frac{1+y}{y} \right)^u du$$

$$= \int_0^1 \cdots\, dy + \int_1^\infty \cdots\, dy = I_{32}' + I_{32}'',$$

say. In I_{32}' we have $0 < y \leqslant 1$, hence by the residue theorem

$$\int_{1/2-iT}^{1/2+iT} \left(\frac{1+y}{y} \right)^u u^{-1}\, du = 2\pi i - \left(\int_{1/2+iT}^{-\infty+iT} + \int_{-\infty-iT}^{1/2-iT} \right)\left(\frac{1+y}{y} \right)^u u^{-1}\, du$$

$$= 2\pi i + O(T^{-1}y^{-1/2}),$$

since

$$\int_{1/2\pm iT}^{-\infty\pm iT} \left(\frac{1+y}{y} \right)^u u^{-1}\, du \ll T^{-1}\int_{-\infty}^{1/2}\left(\frac{1+y}{y} \right)^t dt \ll T^{-1}y^{-1/2}.$$

Hence

$$I_{32}' = 2\pi i \int_0^1 y^{-1}\sin(2\pi Xy)\, dy + O\left(T^{-1}\int_0^1 |\sin(2\pi Xy)|\, y^{-3/2}dy \right)$$

$$= 2\pi i\left(\frac{\pi}{2} \right) + O(X^{-1}) + O\left(T^{-1}\int_0^{X^{-1}} Xy^{-1/2}dy \right) + O\left(T^{-1}\int_{X^{-1}}^\infty y^{-3/2}dy \right)$$

$$= \pi^2 i + O(T^{-1/2}).$$

Next, an integration by parts gives

$$I_{32}'' = \int_1^\infty y^{-1}\sin(2\pi Xy)\, dy \int_{1/2-iT}^{1/2+iT}\left(\frac{1+y}{y}\right)^u u^{-1}\, du$$

$$= \left[-\frac{\cos(2\pi Xy)}{2\pi Xy}\int_{1/2-iT}^{1/2+iT}\left(\frac{1+y}{y}\right)^u u^{-1}\, du\right]_1^\infty$$

$$-\int_1^\infty \frac{\cos(2\pi Xy)}{2\pi Xy^2}\, dy \int_{1/2-iT}^{1/2+iT}\left(\frac{1+y}{y}\right)^u u^{-1}\, du$$

$$-\int_1^\infty \frac{\cos(2\pi Xy)}{2\pi Xy}\, dy \int_{1/2-iT}^{1/2+iT}\left(\frac{1+y}{y}\right)^{u-1} y^{-2}\, du \ll T^{-1}\log T,$$

since for $y \geqslant 1$

$$\int_{1/2-iT}^{1/2+iT}\left(\frac{1+y}{y}\right)^u u^{-1}\, du \ll \int_{1/2-iT}^{1/2+iT}|u^{-1}\, du| \ll \log T,$$

so that finally

$$I_3 = \pi + O(T^{-1/2}\log T).$$

It remains yet to evaluate I_4, as given by (15.35), which will produce the terms of $\Sigma_2(T)$ in (15.9) in the final result. We estimate first the inner integrals in (15.35), making $a \to 0$, $b \to \infty$ in Lemma 15.1. We have then in the notation of Lemma 15.1, for $k = x > AT$,

$$\int_0^\infty \frac{\cos(T\log(1+y)/y)\cos(2\pi xy)}{y^{1/2}(1+y)^{3/2}\log(1+y)/y}\, dy$$

$$= (4x)^{-1}(T/\pi)^{1/2}\frac{\cos(TV + 2\pi xU - \pi x + \pi/4)}{VU^{1/2}(U-1/2)^{1/2}(U+1/2)^{3/2}}$$

$$+ O(T^{-1}x^{-1/2}),$$

and similarly for $r = 1, 2$,

$$\int_0^\infty \frac{\sin(T\log(1+y)/y)\cos(2\pi xy)}{y^{1/2}(1+y)^{3/2}(\log(1+y)/y)^r}\, dy = O\left(T^{1/2}(U-1/2)^{-1/2}x^{-1}\right)$$

$$+ O(T^{-1}x^{-1/2}) = O(x^{-1/2}).$$

Thus we have

$$I_4 = \int_X^\infty x^{-1}\Delta(x)$$

$$\times \left(\frac{T\cos\left(2T \operatorname{ar\,sinh}\sqrt{\pi x/2T} + (2\pi xT + \pi^2 x^2)^{1/2} - \pi x + \pi/4\right)}{\left(\sqrt{2x}\operatorname{ar\,sinh}\sqrt{\pi x/2T}\right)\left((T/2\pi x + \tfrac{1}{4})^{1/2} + \tfrac{1}{2}\right)(T/2\pi x + \tfrac{1}{4})^{1/4}} \right.$$

$$\left. + O(x^{-1/2}) \right) dx.$$

Using $\Delta(x) \ll x^{1/3}$ and changing the variable x to $x^{1/2}$ in the above integral, we obtain with the aid of (15.24),

(15.42)

$$I_4 = \frac{T}{\pi} \sum_{n=1}^{\infty} d(n)n^{-3/4}$$

$$\times \int_{\sqrt{X}}^\infty \frac{\cos\left\{2T\operatorname{ar\,sinh}(x\sqrt{\pi/2T}) + (2\pi x^2 T + \pi^2 x^4)^{1/2} - \pi x^2 + \pi/4\right\}}{x^{3/2}\operatorname{ar\,sinh}(x\sqrt{\pi/2T})\left\{(T/2\pi x^2 + \tfrac{1}{4})^{1/2} + \tfrac{1}{2}\right\}(T/2\pi x^2 + \tfrac{1}{4})^{1/4}}$$

$$\times \left\{\cos(4\pi x\sqrt{n} - \pi/4) - 3(32\pi x\sqrt{n})^{-1}\sin(4\pi x\sqrt{n} - \pi/4)\right\} dx$$

$$+ O(T^{-1/6})$$

$$= \frac{T}{\pi} \sum_{n=1}^{\infty} d(n)n^{-3/4}J_n + O(T^{-1/6}),$$

say.

Now it is transparent why a result like Lemma 15.2 was formulated and proved; it is needed to estimate the integral J_n in (15.42). Indeed if $(T/2\pi - n)^2 > nX$, $n < T/2\pi$, that is to say if

(15.43) $n < (T/2\pi + X/2) - (X^2/4 + XT/2\pi)^{1/2} = Z,$

then an application of Lemma 15.2 gives, with $\alpha = \tfrac{3}{2}$, $\alpha = \tfrac{5}{2}$,

$$I_4 = 2 \sum_{n<Z} d(n)n^{-1/2}(\log T/2\pi n)^{-1}\cos(T(\log T/2\pi n) - T + \pi/4)$$

$$+ O\left(\sum_{n<Z} d(n)n^{-1/2}(T - 2\pi n)^{-1} \right)$$

$$+ O\left(T^{-1/2} \sum_{n<Z} d(n)n^{-1/2}(T - 2\pi n)^{-1/2} \right)$$

$$+ O\left(T^{1/4} \sum_{n=1}^{\infty} d(n)n^{-3/4}\min\left(1, |2\sqrt{n} + \sqrt{X} - \sqrt{X + 2T/\pi}|^{-1}\right) \right) + O(T^{-1/6})$$

$$= I_{41} + I_{42} + I_{43} + I_{44} + O(T^{-1/6}),$$

say. Now I_{41} contributes the main term in $-\Sigma_2(T)$ in (15.9), while the contribution of the other terms [I_{42} comes from applying Lemma 15.2 to estimate the sine terms in (15.42) with $\alpha = \frac{5}{2}$] is $\ll \log^2 T$. To see this, observe that in view of $AT < X < A'T$ we have

$$Z \ll T, \qquad T/2\pi - Z \gg T.$$

Hence

$$I_{42} \ll T^{-1} \sum_{n \leqslant Z} d(n) n^{-1/2} \ll T^{-1/2} \log T,$$

$$I_{43} \ll T^{-1/2} T^{-1/2} \sum_{n \leqslant Z} d(n) n^{-1/2} \ll T^{-1/2} \log T,$$

and it remains yet to deal with I_{44}. Since

$$\left(\tfrac{1}{2}\sqrt{X + 2T/\pi} - \tfrac{1}{2}\sqrt{X}\right)^2 = X/2 + T/2\pi - \sqrt{X^2/4 + XT/2\pi} = Z,$$

we have

$$I_{44} \ll T^{1/4} \sum_{n=1}^{\infty} d(n) n^{-3/4} \min\left(1, |n^{1/2} - Z^{1/2}|^{-1}\right)$$

$$= T^{1/4}\left(\sum_{n \leqslant Z/2} + \sum_{Z/2 < n \leqslant Z - Z^{1/2}} + \sum_{Z - Z^{1/2} < n \leqslant Z + Z^{1/2}} \right.$$

$$\left. + \sum_{Z + Z^{1/2} < n \leqslant 2Z} + \sum_{n > 2Z} \right)$$

$$= T^{1/4}(S_1 + S_2 + S_3 + S_4 + S_5),$$

say. Using partial summation and the crude estimate $\sum_{n \leqslant x} d(n) \sim x \log x$, we obtain

$$S_1 = \sum_{n \leqslant Z/2} d(n) n^{-3/4} \left(Z^{1/2} - n^{1/2}\right)^{-1}$$

$$\ll Z^{-1/2} \sum_{n \leqslant Z/2} d(n) n^{-3/4} \ll T^{-1/4} \log T,$$

$$S_2 = \sum_{Z/2 < n \leqslant Z - Z^{1/2}} d(n) n^{-3/4} \left(Z^{1/2} - n^{1/2}\right)^{-1}$$

$$\ll Z^{-1/4} \sum_{Z/2 < n \leqslant Z - Z^{1/2}} d(n)(Z - n)^{-1}$$

$$\ll T^{-1/4} \sum_{Z^{1/2} \leqslant k \leqslant Z/2} d([Z] - k) k^{-1}$$

$$\ll T^{-1/4}\left(Z(\log Z) Z^{-1} + \int_{Z^{1/2}}^{Z} t(\log t) t^{-2} dt \right) \ll T^{-1/4} \log^2 T,$$

$$S_3 = \sum_{Z - Z^{1/2} < n \leqslant Z + Z^{1/2}} d(n) n^{-3/4} \ll T^{-1/4} \log T,$$

while

$$S_4 \ll T^{-1/4}\log^2 T$$

follows analogously as the estimate for S_2. Finally

$$S_5 \ll \sum_{n>2Z} d(n)n^{-3/4}(n^{1/2} - Z^{1/2})^{-1} \ll \sum_{n>2Z} d(n)n^{-5/4} \ll T^{-1/4}\log T.$$

Therefore we obtain

$$I_4 = 2 \sum_{n\leqslant Z} d(n)n^{-1/2}(\log T/2\pi n)^{-1}\cos\{T(\log T/2\pi n) - T + \pi/4\}$$

$$+ O(\log^2 T),$$

and here the limit of summation Z may be replaced by

$$N' = N'(T) = T/2\pi + N/2 - (N^2/4 + NT/2\pi)^{1/2},$$

as in the formulation of Theorem 15.1, with a total error which is $O(\log^2 T)$. This proves Theorem 15.1 if N is an integer, and if N is not an integer then in (15.4) we replace N by $[N]$ again with an error $\ll \log^2 T$.

15.3 MODIFIED ATKINSON'S FORMULA

Atkinson's formula for $E(T)$, as given by (15.4), has the restriction that N should satisfy $AT < N < A'T$. So far this restriction has not proved to be important in applications, and D. R. Heath-Brown employed Atkinson's formula to derive Theorem 7.2, which enabled him to obtain the 12th power moment estimate $M(12) \leqslant 2$. Another application of Atkinson's formula, due also to Heath-Brown, concerns the asymptotic formula for $\int_2^T E^2(t)\, dt$ and will be presented in Section 15.4 of this chapter. For both of these applications the range $AT < N < A'T$ has proved to be quite sufficient, but it seems desirable to have a more flexible form of Atkinson's formula available. M. Jutila found recently a method for transforming Dirichlet polynomials containing the divisor function $d(n)$ by the use of Voronoi's summation formula. Jutila's method was used in our proof of Theorem 7.2, and it can also be successfully used in connection with Atkinson's formula to yield

THEOREM 15.2. Let $T^\delta \ll N \ll T^2$, and let N and $f(T, n)$ be as in Theorem 15.1. Then

(15.44)

$$E(T) = 2^{-1/2} \sum_{n \leqslant N} (-1)^n d(n) n^{-1/2} \left\{ \text{ar sinh}(\pi n/2T)^{1/2} \right\}^{-1}$$

$$\times (T/2\pi n + \tfrac{1}{4})^{-1/4} \cos(f(T, n))$$

$$-2 \sum_{n \leqslant N'} d(n) n^{-1/2} (\log T/2\pi n)^{-1} \cos\{T(\log T/2\pi n) - T + \pi/4\}$$

$$+ O\big((1 + T^{1/2}N^{-1} + T^{1/4}N^{-1/4})\log^2 T\big).$$

This formula differs from Atkinson's original formula (15.4) in the error term, which is now a function of N also, but this is compensated by the wide range $T^\delta \ll N \ll T^2$, where $\delta > 0$ is arbitrary. If $AT < N < A'T$, then the above error terms reduce to $O(\log^2 T)$, that is, one obtains exactly Atkinson's formula (15.4). A proof of (15.44) it given by M. Jutila (1984a), based on the method of his Theorem 1. To prove (15.44) it suffices to show that if $T^\delta \ll N_1 < N_2 \ll T^2$, $N_1 \asymp N_2$, then with $L = \log T$ we have

(15.45) $$2 \sum_{N'(T, N_2) \leqslant n \leqslant N'(T, N_1)} d(n) n^{-1/2} (\log T/2\pi n)^{-1}$$

$$\times \cos(T \log(T/2\pi n) - T + \pi/4)$$

$$= -2^{-1/2} \sum_{N_1 \leqslant n \leqslant N_2} (-1)^n d(n) n^{-1/2} \big(\text{ar sinh}\sqrt{\pi n/2T} \big)^{-1}$$

$$\times (T/2\pi n + \tfrac{1}{4})^{-1/4} \cos(f(T, n))$$

$$+ O\big(T^{1/2}N_1^{-1}L^2\big) + O\big(L^2 \min\big((T/N_1)^{1/2}, (T/N_1)^{1/4}\big)\big)$$

$$+ O\big(N_1^{1/2} T^{-1} (T^2/N_1)^\varepsilon\big).$$

Here $N'(T, N) = N' = T/2\pi + N/2 - (N^2/4 + NT/2\pi)^{1/2}$, and the idea is to start from (15.4) with $N \asymp T$ and use the Voronoi summation formula to shorten one sum in (15.4) and to lengthen the other. The details of the proof are similar to the proof of Theorem 7.2 and thus will be omitted, but some remarks, however, will be offered. The case $N_2 \leqslant N_0$ is considered first, where

N_0 is fixed and satisfies $T/4\pi \leqslant N'(T, N_0) \leqslant 3T/8\pi$. As in the proof of Theorem 7.2 the summands are multiplied by $e(n) = 1$, which will regulate the distribution of the saddle points coming from the application of Theorem 2.2. After this the sum is transformed by the Voronoi formula (3.2), and the integral $\int_{N'(T, N_2)}^{N'(T, N_1)}(\log x + 2\gamma)f(x)\,dx$ estimated by Lemma 2.2. Since $\exp(iT (\log T/2\pi n)) = n^{-iT}\exp(iT \log T/2\pi)$, the saddle points will be the same as those given by (7.48), except now in the sum on the left-hand side of (15.45) we shall have an extra factor $(\log T/2\pi n)^{-1}$, and as in the proof of Theorem 7.2 we see that

$$\log(T/2\pi x_0) = 2\log\big((\pi n/2T)^{1/2} + (1 + \pi n/2T)^{1/2}\big)$$

$$= 2\,\mathrm{ar}\,\sinh\big((\pi n/2T)^{1/2}\big).$$

Therefore calculating

$$\sum \varphi(x_0)f_n''(x_0)^{-1/2}e\big(f(x_0) + kx_0 + \tfrac{1}{8}\big)$$

by Theorem 2.2 we obtain the right-hand side of (15.45). The error terms in (15.45) are obtained by reasoning analogous to the one given in the proof of Theorem 7.2, when one observes that

$$T/2\pi - N'(T, N_1) \asymp (TN_1)^{1/2}, \qquad \big(\log(T/2\pi x)\big)^{-1} \ll (T/N_1)^{1/2}$$

for $N'(T, N_2) \leqslant x \leqslant N'(T, N_1)$. In the case when $T \ll N_1 \ll T^2$ it seems easier to transform the sum on the right-hand side of (15.45) by Voronoi's formula, using actually the averaged sum

(15.46)
$$U^{-1}\int_0^U \sum_{N_1 + u \leqslant n \leqslant N_2 - u} \cdots \, du.$$

This sum is similar to the one appearing in the proof of Theorem 7.2, only here we take $U = T^{1/2} + N_1T^{-1}$. The terms arising from saddle points of the sum in (15.46) will be exactly those on the left-hand side of (15.45), and the total contribution of the error terms is given by (15.44). This approach seems less difficult than attempts to adapt Atkinson's original proof of Theorem 15.1, where one encounters considerable difficulties when $N = o(T)$. Furthermore, the approach via Voronoi's summation formula may be used to yield an explicit formula for $|\zeta(\tfrac{1}{2} + it)|^2$ itself, which corresponds to a differentiated form of (15.4) in a certain sense. This result is also given by M. Jutila (1984a), and it will be stated here as

THEOREM 15.3. Let $t \geq t_0$, $t^\delta \ll N \leq t/4$, and let $N' = N'(t, N)$ and $f(t, n)$ be as in Theorem 15.1. Then

(15.47)

$$\left| \zeta(\tfrac{1}{2} + it) \right|^2 = -2^{1/2} \sum_{n \leq N} (-1)^n d(n) n^{-1/2} (\tfrac{1}{4} + t/2\pi n)^{-1/4} \sin(f(t, n))$$

$$+2 \sum_{n \leq N'} d(n) n^{-1/2} \cos(t \log(t/2\pi n) - t - \pi/4)$$

$$+ O(N^{1/4} t^{-1/4} \log^2 t) + O(\log t).$$

The equation (15.47) may be considered as an approximate functional equation for $|\zeta(\tfrac{1}{2} + it)|^2$, different from the one that follows from (4.10) with $s = \tfrac{1}{2} + it$. However, this difference is in some sense not essential, since (4.10) may be used to prove (15.47), as will be explained now. First, recall that

$$\left| \zeta(\tfrac{1}{2} + it) \right|^2 = \zeta^2(\tfrac{1}{2} + it) \chi^{-1}(\tfrac{1}{2} + it),$$

where by (1.25) for $t \geq t_0$

(15.48) $$\chi(\tfrac{1}{2} + it) = (2\pi/t)^{it} e^{it + i\pi/4} (1 + O(t^{-1})),$$

so that (4.10) will give

(15.49)

$$\left| \zeta(\tfrac{1}{2} + it) \right|^2 = \chi^{-1}(\tfrac{1}{2} + it) \sum_{n \leq N'} d(n) n^{-1/2 - it}$$

$$+ \chi(\tfrac{1}{2} + it) \sum_{n \leq N'} d(n) n^{-1/2 + it}$$

$$+ \chi(\tfrac{1}{2} + it) \sum_{N' < n \leq t^2/(4\pi^2 N')} d(n) n^{-1/2 + it} + O(\log t).$$

Now using (15.48) we have

$$\chi(\tfrac{1}{2} + it) \sum_{n \leq N'} d(n) n^{-1/2 + it} + \chi^{-1}(\tfrac{1}{2} + it) \sum_{n \leq N'} d(n) n^{-1/2 - it}$$

$$= 2 \operatorname{Re} \left\{ \exp(it \log(t/2\pi) - it - i\pi/4) \sum_{n \leq N'} d(n) n^{-1/2 - it} \right\}$$

$$+ O(\sqrt{N'} t^{-1} \log t)$$

$$= 2 \sum_{n \leq N'} d(n) n^{-1/2} \cos(t \log(t/2\pi n) - t - \pi/4) + O(\sqrt{N'} t^{-1} \log t).$$

Here the error term is trivially dominated by the error terms in (15.47), and so it is seen that (15.47) reduces to the proof of

$$(15.50) \qquad \sum_{N' \leqslant n \leqslant t^2/(4\pi^2 N')} d(n) n^{-1/2 + it}$$

$$= -2^{1/2} \exp(-it \log(2\pi/t) - it - i\pi/4)$$

$$\times \sum_{n \leqslant N} (-1)^n d(n) n^{-1/2} \left(\tfrac{1}{4} + t/2\pi n \right)^{-1/4} \sin(f(t, n))$$

$$+ O\left(N^{1/4} t^{-1/4} \log^2 t \right) + O(\log t).$$

This is again achieved via the Voronoi summation formula (3.2) and the use of the proof of Theorem 7.2. The terms of the sum on the left-hand side of (15.50) are again multiplied by $e(n) = 1$, and an averaged form of the sum, as in (15.46), is considered. The series which appears in Voronoi's formula is split into two parts at $N(1 + \varepsilon)$. The terms with $n > N(1 + \varepsilon)$ will have no saddle points in view of the range of summation, which is $N' < n \leqslant t^2/(4\pi^2 N')$, while the terms for $n \leqslant N$ will give rise to saddle points x_0 [given again by (7.48)], which will contribute the main terms on the right-hand side of (15.50). The error terms in (15.47) are small for $N \ll T$, and thus this formula can also be used for the derivation of a variant of Theorem 7.2, and then also for higher power moments of the zeta-function. The proof of (15.47) is notably simpler than the proof of Atkinson's formula (15.4), and (15.47) can be also examined from another viewpoint in light of Atkinson's formula. Namely, starting from (15.22) we have

$$(15.51) \quad \left| \zeta\left(\tfrac{1}{2} + it\right) \right|^2 = 2 \operatorname{Re}\{ g(u, 1 - u)\} + O(\log t), \quad u = \tfrac{1}{2} + it,$$

where $g(u, 1 - u)$ is defined by (15.23). Using Voronoi's formula we have

$$(15.52) \quad g(u, 1 - u) = 2 \sum_{n \leqslant N} d(n) n \int_0^\infty y^{-u} (1 + y)^{u-1} \cos(2\pi n y) \, dy$$

$$+ \int_N^\infty (\log x + 2\gamma) h(u, x) \, dx$$

$$+ \sum_{n=1}^\infty d(n) \int_N^\infty h(u, x) \alpha(nx) \, dx,$$

where $\alpha(nx)$ is given by (3.15) and $h(u, x)$ by (15.25). A direct application of

Theorem 2.2 gives, for $1 \ll N \ll t^2$,

(15.53) $\quad 4 \operatorname{Re} \sum_{n \leqslant N} d(n) \int_0^\infty y^{-u}(1+y)^{u-1} \cos(2\pi ny) \, dy$

$$= -2^{1/2} \sum_{n \leqslant N} (-1)^n d(n) n^{-1/2} (\tfrac{1}{4} + t/2\pi n)^{-1/4} \sin(f(t,n))$$

$$+ O(N^{1/4} t^{-3/4} \log t) + O(N^{1/2} t^{-1}) + O(\log t),$$

so that combining (15.52) and (15.53) we obtain the main term on the right-hand side of (15.47). However, difficulties arise with this approach when one tries to estimate the series on the right-hand side of (15.52), and therefore the first proof of (15.47) seems preferable.

15.4 THE MEAN SQUARE OF $E(t)$

Let, as before,

$$E(T) = \int_0^T \left| \zeta(\tfrac{1}{2} + it) \right|^2 dt - T \log(T/2\pi) - (2\gamma - 1)T.$$

Atkinson's formula (15.4) for $E(T)$ provides the means for obtaining a mean-square estimate for $E(t)$ which is analogous to Theorem 13.5. The method of proof, which is due to D. R. Heath-Brown, is similar in nature to Cramér's proof of (13.26) and the result is contained in

THEOREM 15.4.

(15.54)

$$\int_2^T E^2(t) \, dt = \left\{ \frac{2}{3} (2\pi)^{-1/2} \sum_{n=1}^\infty d^2(n) n^{-3/2} \right\} T^{3/2} + O(T^{5/4} \log^2 T).$$

Proof of Theorem 15.4. It will be sufficient to prove

(15.55) $\quad \int_T^{2T} E^2(t) \, dt = \frac{2}{3} (2\pi)^{-1/2} \sum_{n=1}^\infty d^2(n) n^{-3/2} \left((2T)^{3/2} - T^{3/2} \right)$

$$+ O(T^{5/4} \log^2 T),$$

and then to replace T by $T/2$, $T/2^2$, and so on, and to sum all the results. We use Atkinson's formula in the form

(15.56) $\quad\quad\quad E(T) = \Sigma_1(T) + \Sigma_2(T) + R(T),$

where $\Sigma_1(T)$ and $\Sigma_2(T)$ are given by (15.7) and (15.9), and $R(T) \ll \log^2 T$.

Then

$$(15.57) \quad \int_T^{2T} E^2(t)\, dt = \int_T^{2T} \Sigma_1^2(t)\, dt + 2\int_T^{2T} \Sigma_1(t)(\Sigma_2(t) + R(t))\, dt$$
$$+ \int_T^{2T} (\Sigma_2(t) + R(t))^2\, dt.$$

The main term on the right-hand side of (15.55) will come from the first integral on the right-hand side of (15.57). We choose $N = T$ in Atkinson's formula and proceed to show that

(15.58)

$$\int_T^{2T} \Sigma_1^2(t)\, dt = \frac{2}{3}(2\pi)^{-1/2} \sum_{n=1}^{\infty} d^2(n)n^{-3/2}\big((2T)^{3/2} - T^{3/2}\big) + O(T^{1+\varepsilon}).$$

To demonstrate this we merely square out $\Sigma_1(t)$ and integrate term by term, estimating the nondiagonal terms (i.e., those for which $m \neq n$) by the following

LEMMA 15.3. Let $g_j(t)$ $(1 \leqslant j \leqslant k)$ and $f(t)$ be continuous, monotonic real-valued functions on $[a, b]$ and let $f(t)$ have a continuous, monotonic derivative on $[a, b]$. If $|g_j(t)| \leqslant M_j$ $(1 \leqslant j \leqslant k)$, $|f'(t)| \geqslant M_0^{-1}$ on $[a, b]$, then

$$(15.59) \quad \left| \int_a^b \prod_{j=1}^k g_j(t) \exp(if(t))\, dt \right| \leqslant 2^{k+3} \prod_{j=0}^k M_j.$$

Proof of Lemma 15.3. The lemma is a straightforward generalization of Lemma 2.1. Recall that if $F(x)$, $G(x)$ are real-valued on $[a, b]$ and $F(x)$ is monotonic, then the second mean-value theorem for integrals states that

$$(15.60) \quad \int_a^b F(x)G(x)\, dx = F(a) \int_a^\xi G(x)\, dx + F(b) \int_\xi^b G(x)\, dx$$

for some $a \leqslant \xi \leqslant b$. Applying (15.60) k times to the real and imaginary part of the integral in (15.59) we obtain

$$\left| \int_a^b \prod_{j=1}^k g_j(t) \exp(if(t))\, dt \right|$$

$$\leqslant 2^k \prod_{j=1}^k M_j \left(\max_{a \leqslant a_0 < b_0 \leqslant b} \left| \int_{a_0}^{b_0} \cos(f(t))\, dt \right| \right.$$

$$\left. + \max_{a \leqslant a_1 < b_1 \leqslant b} \left| \int_{a_1}^{b_1} \sin(f(t))\, dt \right| \right)$$

$$\leqslant 2^{k+1} \prod_{j=0}^k M_j \left(\max_{a \leqslant \beta_0 < b} \left| \int_{\alpha_0}^{\beta_0} d\sin(f(t)) \right| + \max_{a \leqslant \alpha_1 < \beta_1 \leqslant b} \left| \int_{\alpha_1}^{\beta_1} d\cos(f(t)) \right| \right)$$

$$\leqslant 2^{k+3} \prod_{j=0}^k M_j.$$

Now we return to the proof of Theorem 15.4, noting that the terms of $\Sigma_1^2(t)$ are of the form

$$\tfrac{1}{4}(-1)^{m+n}d(n)d(m)(mn)^{-1/2}g(t)\cos(f(t)),$$

where with $f(T, n)$ given by (15.5) we have

$$f(t) = f(t, n) \mp f(t, m),$$

$$g(t) = g_1(t)g_2(t)g_3(t)g_4(t),$$

$$g_1(t) = \left(t/2\pi n + \tfrac{1}{4}\right)^{-1/4}, \qquad g_2(t) = \left(t/2\pi m + \tfrac{1}{4}\right)^{-1/4},$$

$$g_3(t) = \left(\operatorname{ar\,sinh}\sqrt{\pi n/2t}\,\right)^{-1}, \qquad g_4(t) = \left(\operatorname{ar\,sinh}\sqrt{\pi m/2t}\,\right)^{-1}.$$

The contribution of the terms with $m \neq n$ is estimated by Lemma 15.3, where we take $M_1 \ll (n/T)^{1/4}$, $M_2 \ll (m/T)^{1/4}$, $M_3 \ll (T/n)^{1/2}$, $M_4 \ll (T/m)^{1/2}$. Also, since

$$f'(t, n) = 2\operatorname{ar\,sinh}\sqrt{\pi n/2t},$$

we may take

$$M_0 \ll T^{1/2}\left|n^{1/2} \mp m^{1/2}\right|^{-1}.$$

Thus the contribution of the nondiagonal terms is

$$\ll T \sum_{m \neq n \leqslant T} d(m)d(n)(mn)^{-3/4}\left|n^{1/2} - m^{1/2}\right|^{-1} + T \sum_{n=1}^{\infty} d^2(n)n^{-2} \ll T^{1+\varepsilon}$$

by repeating the estimate of (13.31), where the second sum above comes from those terms for which $m = n$ but $f(t) \neq 0$.

The contribution of the diagonal terms $m = n$ to the left-hand side of (15.58) is

$$\frac{1}{4} \sum_{n \leqslant N} d^2(n)n^{-1} \int_T^{2T} g(t)\,dt,$$

where we recall that $N = T$. For $|x| < 1$ we have $(\operatorname{ar\,sinh} x)^{-2} = x^{-2} + O(1)$, and for $n \leqslant N = T$ we thus have

$$g(t) = 2^{3/2}t^{1/2}(\pi n)^{-1/2} + O(n^{1/2}T^{-1/2}),$$

which gives

$$\int_T^{2T} \Sigma_1^2(t)\, dt = \tfrac{1}{4}(2^{3/2}\pi^{-1/2}) \sum_{n \leqslant T} d^2(n) n^{-3/2} \int_T^{2T} t^{1/2}\, dt$$

$$+ O\left(\sum_{n \leqslant T} d^2(n) n^{-1/2} T^{1/2} \right) + O(T^{1+\varepsilon})$$

$$= (2\pi)^{-1/2} \sum_{n=1}^{\infty} d^2(n) n^{-3/2} \int_T^{2T} t^{1/2}\, dt$$

$$+ O\left(T^{3/2} \sum_{n > T} d^2(n) n^{-3/2} \right) + O(T^{1+\varepsilon})$$

$$= \tfrac{2}{3}(2\pi)^{-1/2} \sum_{n=1}^{\infty} d^2(n) n^{-3/2} \left((2T)^{3/2} - T^{3/2}\right) + O(T^{1+\varepsilon}).$$

This proves (15.58), and it remains to consider the mean value of $\Sigma_2(t)$. We shall prove

(15.61) $$\int_T^{2T} \Sigma_2^2(t)\, dt \ll T \log^4 T.$$

The method is similar to the one used in proving (15.58), except that in this case there is no main term in our estimate. In proving (15.58) we chose $N = T$ to be independent of t. Here

$$N' = N'(t) = t/(2\pi) + N/2 - (N^2/4 + Nt/2\pi)^{1/2}$$

will vary with t. However, for $T \leqslant t \leqslant 2T$ and $n \leqslant N'(t)$ we have $(\log(t/2\pi n))^{-1} \ll 1$, so that the logarithmic factor in $\Sigma_2(t)$ will cause no trouble. The terms of $\Sigma_2^2(t)$ are of the form

$$2d(m)d(n)(mn)^{-1/2} g(t)$$

$$\times \left\{ \cos\left(t \log(t^2/4\pi^2 mn) - 2t + \pi/2\right) + \cos(t \log m/n) \right\},$$

where

$$g(t) = (\log t/2\pi m)^{-1}(\log t/2\pi n)^{-1} \ll 1.$$

For each pair m, n we have to integrate over that subinterval of $[T, 2T]$ for which $N'(t) \geqslant \max(m, n)$. Since

$$\left\{ t\left(\log t^2/4\pi^2 mn\right) - 2t + \pi/2 \right\}' \gg |\log m/n|, \qquad (t \log m/n)' \gg |\log m/n|,$$

an application of Lemma 15.1 shows that the contribution of the terms $m \neq n$ is

$$\ll \sum_{n \neq m \leqslant T} d(m)d(n)(mn)^{-1/2}|\log m/n|^{-1}$$

$$\ll \sum_{m \neq n \leqslant T} (d^2(m)m^{-1} + d^2(n)n^{-1})|\log m/n|^{-1}$$

$$\ll \sum_{m \neq n \leqslant T} d^2(n)n^{-1}|\log m/n|^{-1}$$

$$\ll \sum_{n \leqslant T} d^2(n)n^{-1} \sum_{m \leqslant T, m \neq n} |\log m/n|^{-1}$$

$$\ll \sum_{n \leqslant T} d^2(n)n^{-1}(T + n \log T) \ll T \log^4 T,$$

since

$$\sum_{m \leqslant T, m \neq n} |\log m/n|^{-1} = \sum_{m \leqslant n-1} (\log n/m)^{-1} + \sum_{n < m \leqslant T} (\log m/n)^{-1}$$

$$= \sum_{r \leqslant n-1} \left(\log \frac{n}{n-r}\right)^{-1} + \sum_{r \leqslant T-n} \left(\log \frac{n+r}{n}\right)^{-1}$$

$$\ll \sum_{r \leqslant n-1} nr^{-1} + \sum_{r \leqslant T-n} (1 + nr^{-1}) \ll T + n \log T.$$

The terms $m = n$ trivially contribute

$$\ll T \sum_{n \leqslant T} d^2(n)n^{-1} \ll T \log^4 T,$$

and therefore (15.61) follows.

The proof of (15.54) is finally obtained by combining (15.57), (15.58), (15.61), and using the Cauchy–Schwarz inequality, since

$$\left\{ \int_T^{2T} \Sigma_1(t)(\Sigma_2(t) + R(t)) \, dt \right\}^2 \ll \int_T^{2T} \Sigma_1^2(t) \, dt \int_T^{2T} (\Sigma_2^2(t) + R^2(t)) \, dt$$

$$\ll T^{3/2}(T \log^4 T + T \log^4 T) \ll T^{5/2} \log^4 T,$$

and

$$\int_T^{2T} (\Sigma_2(t) + R(t))^2 \, dt \leqslant 2 \int_T^{2T} (\Sigma_2^2(t) + R^2(t)) \, dt \ll T \log^4 T.$$

This finishes the proof of Theorem 15.4, which gives immediately

COROLLARY 15.1.

$$E(T) = \Omega(T^{1/4}).$$

This is analogous to $\Delta(x) = \Omega(x^{1/4})$ which follows from Theorem 13.5, but the sharper Ω-results known to hold for $\Delta(x)$ are not known yet to hold for $E(T)$. This should not be surprising, as Atkinson's formula for $E(T)$ was derived with the aid of a formula for $\Delta(x)$, embodied in Voronoi's formula. Thus it is natural to expect that problems involving $E(T)$ will be at least as difficult as those involving $\Delta(x)$, and more about the connection between $E(T)$ and $\Delta(x)$ will be found in the next section. Going through the proof of Theorem 15.4 it may be observed that the proof enables one to estimate the integral of $E^2(t)$ over a short interval, and we obtain

COROLLARY 15.2. For $T^\varepsilon \ll G \leqslant T$ uniformly in G

$$\int_{T-G}^{T+G} E^2(t)\, dt \ll T^\varepsilon (GT^{1/2} + T).$$

This estimate is analogous to (13.48) for $\Delta(x)$, and the main interest in estimates of this sort is that they provide us with a way of estimating $\zeta(\tfrac{1}{2} + iT)$, and Corollary 15.2 leads to the classical estimate $\zeta(\tfrac{1}{2} + iT) \ll T^{1/6+\varepsilon}$. To see this, observe that with $L = \log T$ and the notation of (15.3) we have

(15.62)

$$\int_{T-G}^{T+G} \left|\zeta(\tfrac{1}{2} + it)\right|^2 dt \ll \int_{-GL}^{GL} \exp(-t^2 G^{-2})\, dI(T + t)$$

$$= \int_{-GL}^{GL} \exp(-t^2 G^{-2})\left(\log\frac{T+t}{2\pi} + 2\gamma\right) dt$$

$$+ O(1) + \int_{-GL}^{GL} E(T+t) t G^{-2} \exp(-t^2 G^{-2})\, dt$$

$$\ll GL + G^{-1} L\left(\int_{T-GL}^{T+GL} E^2(t)\, dt\right)^{1/2}\left(\int_{T-GL}^{T+GL} dt\right)^{1/2}$$

$$\ll T^\varepsilon (G + T^{1/4} + G^{-1/2}T^{1/2}),$$

if we use Corollary 15.2 and the Cauchy–Schwarz inequality. In view of Lemma 7.1 [(7.2) with $k = 2$] and $\int_{-L^2}^{L^2} \ll \int_{-G}^{G}$, the estimate $\zeta(\tfrac{1}{2} + iT) \ll T^{1/6+\varepsilon}$ follows from (15.62) with the choice $G = T^{1/3}$. Therefore if we define $F(T)$ by

$$(15.63) \quad \int_{2}^{T} E^2(t)\, dt = \tfrac{2}{3}(2\pi)^{-1/2} \sum_{n=1}^{\infty} d^2(n) n^{-3/2} T^{3/2} + F(T),$$

any order estimate $F(T) \ll T^{c+\varepsilon}$, $\frac{3}{4} \leq c \leq \frac{5}{4}$, would give $\zeta(\frac{1}{2} + iT) \ll T^{c/6+\varepsilon}$ by the above method. In analogy with Theorem 13.6 it may be conjectured that

$$(15.64) \qquad\qquad F(T) = \Omega(T^{3/4-\delta})$$

for any $\delta > 0$. By the method of proof of Theorem 13.6 this may be obtained if the truth of the Lindelöf hypothesis is assumed, in which case trivially

$$|E(T_1) - E(T_2)| \leq \left| \int_{T_1}^{T_2} |\zeta(\tfrac{1}{2} + it)|^2 \, dt \right| + \left| T_1(\log T_1/2\pi) + T_1(2\gamma - 1) \right.$$

$$\left. - T_2(\log T_2/2\pi) - T_2(2\gamma - 1) \right|$$

$$\ll T^\varepsilon |T_1 - T_2|$$

for $T \leq T_1, T_2 \leq 2T$. Thus it is seen that the method of proof of Theorem 13.6 may be applied and (15.64) follows, but it would be interesting to obtain an unconditional proof of (15.64).

15.5   THE CONNECTION BETWEEN E(T) AND Δ(x)

As mentioned in Section 15.1, a comparison between (15.4) and (3.17) shows a similarity between $\Sigma_1(T)$ and $2\pi\Delta(T/2\pi)$, since apart from the oscillating factor $(-1)^n$ the first $o(T^{1/3})$ terms are asymptotically equal to each other. The influence of $\Sigma_2(T)$ in Atkinson's formula (15.4) usually may be made small by some averaging process, so that there is in a certain sense also an analogy between $E(T)$ and $2\pi\Delta(T/2\pi)$, pointed out already by Atkinson (1949).

Furthermore, if α_2 and θ_2 are the infima of constants a_2 and c_2 such that $\Delta(x) \ll x^{a_2+\varepsilon}$, $E(T) \ll T^{c_2+\varepsilon}$ for every $\varepsilon > 0$, then one would expect $\alpha_2 = \theta_2 = \frac{1}{4}$ in view of Theorems 13.5 and 15.4. Moreover, these theorems show that the inequalities $\alpha_2 < \frac{1}{4}$ and $\theta_2 < \frac{1}{4}$ cannot hold. Albeit the equality $\alpha_2 = \theta_2$ is still not known to hold, the best upper bounds $\alpha_2 \leq \frac{35}{108}$ and $\theta_2 \leq \frac{35}{108}$ are indeed equal. The bound $\alpha_2 \leq \frac{35}{108}$ is Theorem 13.1, while it was shown by R. Balasubramanian (1978) from (15.12) that the estimation of $E(T)$ may be reduced to the estimation of exponential sums to which the methods of G. Kolesnik used for the proof of (13.13) equally apply. Therefore the bound $\theta_2 \leq \frac{35}{108}$ (given here as Corollary 15.4 by another approach) follows similarly as $\alpha_2 \leq \frac{35}{108}$. Following the method of M. Jutila (1983a) it will be shown that $E(T)$ may be majorized by an expression which is very similar to the one given by (3.17) for $2\pi\Delta(T/2\pi)$, only the expression for $E(T)$ will contain the additional factor $(-1)^n$. Thus it turns out that the three problems of estimating the order of $\Delta(x)$, $E(T)$, and $\zeta(\frac{1}{2} + iT)$ [and in view of Section 13.8 of Chapter 13 one might add $P(x)$ also] may be unified in more or less one problem, with similar exponential sums appearing in each case. Previously these problems have been primarily treated separately and by different methods. Although we emphasize that $\alpha_2 = \theta_2$ is not yet known to be true, it is hard to image a method for the estimation of exponential sums in question which would yield $\alpha_2 \neq \theta_2$.

Our first task will be technical and consists of introducing a new function $\Delta^*(x)$. This function will be similar to $\Delta(x)$, but will contain the oscillating factor $(-1)^n$, thus providing a more exact analogy between $E(T)$ and the Dirichlet divisor problem for $k = 2$. Let us for this purpose consider the function

(15.65) $$D^*(x) = -D(x) + 2D(2x) - \tfrac{1}{2}D(4x),$$

where

$$D(x) = \sum_{n \leqslant x} d(n) = x \log x + (2\gamma - 1)x + \Delta(x).$$

Then we may write

(15.66) $$D^*(x) = x \log x + (2\gamma - 1)x + \Delta^*(x),$$

(15.67) $$\Delta^*(x) = -\Delta(x) + 2\Delta(2x) - \tfrac{1}{2}\Delta(4x).$$

Now it will turn out that $2\pi\Delta^*(T/2\pi)$ is the "right" analogue of $\Sigma_1(T)$ in Atkinson's formula, since for $N \ll x$ we have

(15.68) $$\Delta^*(x) = (\pi\sqrt{2})^{-1}x^{1/4} \sum_{n \leqslant N} (-1)^n d(n) n^{-3/4} \cos(4\pi\sqrt{nx} - \pi/4)$$

$$+ O(x^{1/2+\varepsilon}N^{-1/2}).$$

To see that (15.68) holds use (3.17) with $N \ll x$, namely,

$$\Delta(x) = (\pi\sqrt{2})^{-1}x^{1/4} \sum_{n \leqslant N} d(n) n^{-3/4} \cos(4\pi\sqrt{nx} - \pi/4) + O(x^{1/2+\varepsilon}N^{-1/2})$$

with $x, 2x, 4x$ and $N, N/2, N/4$, respectively. From (15.67) we have then

(15.69) $$\pi\sqrt{2}\, x^{-1/4}\Delta^*(x) = -\sum_{n \leqslant N} d(n) n^{-3/4} \cos(4\pi\sqrt{nx} - \pi/4)$$

$$+ 2^2 \sum_{2k \leqslant N} d(k)(2k)^{-3/4} \cos(4\pi\sqrt{2kx} - \pi/4)$$

$$- 2 \sum_{4m \leqslant N} d(m)(4m)^{-3/4} \cos(4\pi\sqrt{4mx} - \pi/4)$$

$$+ O(x^{1/4+\varepsilon}N^{-1/2}).$$

The sums on the right-hand side of (15.69) will give one sum

$$\sum_{r \leqslant N} f(r) r^{-3/4} \cos(4\pi\sqrt{rx} - \pi/4)$$

over natural numbers r, and it remains to consider $f(r)$. If r is odd, then obviously $f(r) = -d(r) = (-1)^r d(r)$, since $2k$ and $4m$ are even. If $r = 2s$, but s is odd, then $d(2s) = 2d(s)$ and so $f(r)$ comes from the first two sums on the right-hand side of (15.69) and equals $f(r) = -d(r) + 2d(r) = (-1)^r d(r)$. Finally, if $r = 4q$, observe that from $d(2^a) = a + 1$ we always have $d(4q) = 2d(2q) - d(q)$, so that in this case

$$f(r) = -d(4q) + 2^2 d(2q) - 2d(q) = d(4q) = (-1)^r d(r),$$

and thus (15.68) follows from (15.69).

Next, we need an averaged expression for $E(T)$. This will be accomplished by integrating $E(T)$ over very short intervals, the precise meaning of "very short" being given below. Because of the square roots in the expression for $E(T)$ it will be technically more convenient to work with the function

$$(15.70) \qquad\qquad E_0(x) = E(x^2)$$

than with $E(x)$ directly, and with this in mind we define the averaged integral

$$(15.71) \qquad\qquad E_1(x) = G^{-1} \int_{-H}^{H} E_0(x + u) e^{-u^2 G^{-2}} \, du.$$

Here $H = GL = G \log T$, $T^{-a} \leqslant G \leqslant T^{-b}$ for some $\frac{1}{2} > a > b > 0$. The estimate that we need is contained in

LEMMA 15.4. For $\frac{1}{2}T^{1/2} \leqslant x \leqslant \frac{3}{2}T^{1/2}$, $M = G^{-2}L^2$ we have

$$(15.72)$$

$$E_1(x) = (2\pi x^2)^{1/4} \sum_{n \leqslant M} (-1)^n d(n) n^{-3/4} e(x^2, n) r(x, n) \cos(f(x^2, n))$$

$$+ O(T^\varepsilon).$$

Here

$$(15.73) \qquad r(x, n) = \exp\left\{-4G^2 \left(x \operatorname{ar} \sinh(\sqrt{\pi n/2}\, x^{-1})\right)^2\right\},$$

and the expressions for e and f are given by Atkinson's formula, that is,

(15.74)

$$e(x, n) = (1 + \pi n/2x)^{-1/4}\left(\sqrt{2x/\pi n} \text{ ar } \sinh(\sqrt{\pi n/2x})\right)^{-1}$$

$$= 1 - \frac{\pi n}{24x} + O(n^2 x^{-2}),$$

(15.75) $\quad f(x, n) = 2x \text{ ar } \sinh\sqrt{\pi n/2x} + (\pi^2 n^2 + 2\pi n x)^{1/2} - \pi/4$

$$= -\pi/4 + (8\pi n x)^{1/2} + O(n^{3/2} x^{-1/2}),$$

where $n \ll x$ in both (15.74) and (15.75).

Proof of Lemma 15.4. Take $N = T$ in Atkinson's formula. By (15.6) we have

(15.76) $\quad E_1(x) = \sum_{j=1}^{2} G^{-1} \int_{-H}^{H} \Sigma_j\left((x + u)^2\right) e^{-u^2 G^{-2}} du + O(L^2).$

Consider here first the term with $j = 1$. By (15.7) this is

(15.77) $\quad (2/\pi)^{1/4} G^{-1} \int_{-H}^{H} (x + u)^{1/2} \sum_{n \leqslant T} (-1)^n d(n) n^{-3/4}$

$$\times e\left((x + u)^2, n\right)\cos\left(f\left((x + u)^2, n\right)\right) e^{-u^2 G^{-2}} du.$$

As in many previous proofs we shall use here the exponential integral (A.38), namely,

$$\int_{-\infty}^{\infty} \exp(At - Bt^2) \, dt = (\pi/B)^{1/2}\exp(A^2/4B), \qquad \text{Re } B > 0.$$

The choice $H = GL$ in (15.71) makes it possible to replace the limits of integration in (15.77) by $(-\infty, \infty)$ with a negligible error. However, before doing this we use Taylor's formula to replace $(x + u)^{1/2}$ by $x^{1/2}$ and likewise $e((x + u)^2, n)$ by $e(x^2, n)$ with a total error which is $\ll 1$. Also by Taylor's formula using $f'(t, n) = 2 \text{ ar } \sinh\sqrt{\pi n/2t}$ we have

(15.78) $\quad f\left((x + u)^2, n\right) = f(x^2, n) + 4xu \text{ ar } \sinh\left(\sqrt{\pi n/2} \, x^{-1}\right)$

$$+ A(n, x)u^2 + O(T^{-1/2} G^3 L^3).$$

Here $A(n, x) \asymp (n/T)^{3/2}$, since in view of

$$\text{ar sinh } z = z - \frac{1}{2}\left(\frac{z^3}{3}\right) + \frac{1 \times 3}{2 \times 4}\left(\frac{z^5}{5}\right) - \frac{1 \times 3 \times 5}{2 \times 4 \times 6}\left(\frac{z^7}{7}\right) + \cdots,$$

$$|z| \leqslant 1,$$

we have with $F(x) = f(x^2, n)$ that $F''(x) \asymp (n/T)^{3/2}$, $F^{(3)}(x) \ll n^{3/2}T^{-2}$ holds. Now we substitute (15.78) into (15.77), using $\exp(iy) = 1 + O(|y|)$ for real y, so that the error term in (15.78) makes a total contribution $\ll G^3L^4 \ll 1$. Then we use (A.38), noting that with the abbreviation $B(n,x) = G^{-2} - A(n, x)i$ the expression in (15.77) becomes

$$(15.79) \quad (2/\pi)^{1/4}x^{1/2}G^{-1} \sum_{n \leqslant T} (-1)^n d(n)n^{-3/4}e(x^2, n)$$

$$\times \text{Re}\left\{ e^{if(x^2, n)}(\pi/B(n, x))^{1/2} \right.$$

$$\left. \times \exp\left(-\frac{4\left[x \text{ ar sinh}(\sqrt{\pi n/2}\, x^{-1})\right]^2}{B(n, x)} \right) \right\} + O(1).$$

Here the terms with $n > M = G^{-2}L^2$ make a negligible contribution because of the presence of the exponential factor containing $(x \text{ ar sinh} \cdots)^2$, and if we replace $B(n, x)$ by G^{-2} using Taylor's formula we make a total error which is $\ll T^{-5/4}G^{-3/2}L^5 \ll 1$.

In this fashion the main term in (15.72) is obtained, and to complete the proof of Lemma 15.4 it remains to show that the term with $j = 2$ in (15.76) is $\ll T^{\varepsilon}$. Since $N = T$ was fixed in the definition of $\Sigma_1(T)$ in Atkinson's formula, then N' in the definition of $\Sigma_2(T)$ in Atkinson's formula will depend on $(x + u)^2$. However, it is convenient to replace $N'((x + u)^2, T)$ by $N'(x^2, T)$. Recalling that

$$N' = N'(x, T) = T/2\pi + x/2 - (x^2/4 + xT/2\pi)^{1/2},$$

we have

$$N'(x^2, T) - N'((x + u)^2, T) \ll T^{1/2}GL.$$

For $n \leqslant N'((x + u)^2, T)$ we have $\log((x + u)^2/2\pi n) \gg 1$ and

$$\left(\log\frac{(x + u)^2}{2\pi n}\right)^{-1} = \left(\log\frac{x^2}{2\pi n}\right)^{-1} + O(T^{-1/2}GL).$$

Therefore by Atkinson's formula, for $|u| \leqslant H$,

$$(15.80) \quad \Sigma_2\big((x + u)^2\big) = -2 \sum_{n \leqslant N'(x^2, T)} d(n) n^{-1/2}\bigg(\log\frac{x^2}{2\pi n}\bigg)^{-1}$$

$$\times \cos\Big(g\big((x + u)^2, n\big)\Big) + O(T^\varepsilon G),$$

where from (15.10) we obtain

$$g\big((x + u)^2, n\big) = g(x^2, n) + 2x \log\big(x^2/2\pi n\big) u$$

$$+ \big(\log(x^2/2\pi n) + 2\big) u^2 + O(G^3 L^3 T^{-1/2}).$$

We substitute the expression for $g((x + u)^2, n)$ in (15.80) and argue as in the case $j = 1$, using the integral (A.38). The exponential factor, analogous to the one in (15.79) with $(x \operatorname{ar sinh} \cdots)^2$, will make each term in the sum $\ll T^{-c}$ for any fixed $c > 0$, while the error term in (15.80) will make the contribution $O(T^\varepsilon)$ in (15.72) so that Lemma 15.4 follows.

Having proved Lemma 15.4 we shall use it to obtain an expression for $E(T)$ analogous to the expression for $2\pi\Delta^*(T/2\pi)$ which follows from (15.68), except that $\cos(4\pi\sqrt{nx} - \pi/4)$ will be replaced by $\cos(f(T, n))$. We suppose that $T/2 \leqslant t_1 \leqslant T \leqslant t_2 \leqslant 2T$, and with $I(T) = \int_0^T |\zeta(\frac{1}{2} + it)|^2 dt$ we have trivially

$$I(t_1) \leqslant I(T) \leqslant I(t_2).$$

This gives easily

$$(15.81) \quad E(t_1) + O((T - t_1)\log T) \leqslant E(T) \leqslant E(t_2) + O((t_2 - T)\log T)$$

by (15.3), and the idea is to integrate (15.81) over a very short interval using Lemma 15.4. We shall consider the first inequality in (15.81) only, since the other one is treated in exactly the same way. Since the relevant range for the order of $E(T)$ is $T^{1/4} \ll E(T) \ll T^{1/3}$, we suppose that Y is a parameter which satisfies $T^{1/4}L^{-1} \leqslant Y \leqslant T^{1/3}L^{-1}$ and let $G = T^{-1/2}YL^{-2}$, so that G clearly satisfies the condition assumed in Lemma 15.4. Letting

$$t_1 = T - Y + 2(T - Y)^{1/2}u + u^2, \qquad |u| \leqslant GL,$$

it is seen that with our choice $G = T^{-1/2}YL^{-2}$ we have $t_1 \leqslant T$ as needed in (15.81). Therefore integrating (15.81) we obtain

(15.82)

$$G^{-1}\int_{-GL}^{GL} E\Big(T - Y + 2(T - Y)^{1/2}u + u^2\Big) e^{-u^2 G^{-2}} du + O(YL) \leqslant \sqrt{\pi}\, E(T).$$

But the integral in (15.82) is just $E_1((T - Y)^{1/2})$ by (15.70), and thus (15.81) gives in fact

$$E_1\big((T - Y)^{1/2}\big) + O(YL) \leqslant \sqrt{\pi}\, E(T) \leqslant E_1\big((T + Y)^{1/2}\big) + O(YL).$$

The integrals $E_1((T \pm Y)^{1/2})$ are evaluated by Lemma 15.4, and setting $X = YL$ it is seen that we obtain

THEOREM 15.5. Let $T \leqslant \tau \leqslant 2T$, $T^{1/4} \leqslant X \leqslant T^{1/3}$. Then uniformly in τ

(15.83)

$$E(\tau) \ll X + T^{1/4}$$

$$\times \sup_{|t-\tau| \leqslant X} \bigg| \sum_{n \leqslant TX^{-2}L^8} (-1)^n d(n) n^{-3/4} e(t, n) r(t^{1/2}, n) \cos(f(t, n)) \bigg|.$$

The value $TX^{-2}L^8$ appears because $M = G^{-2}L^2 = TY^{-2}L^6 = TX^{-2}L^8$, and the presence of the exponential factors in the proof of Lemma 15.4 which come from the application of (A.38) make it possible to obtain the result for $T \leqslant \tau \leqslant 2T$. Using partial summation we may remove the factors $e(t, n)$ and $r(t^{1/2}, n)$ to obtain the analogue of (15.68), which may be stated as

COROLLARY 15.3. Let $T^{1/4} \leqslant X \leqslant T^{1/3}$ and $M = TX^{-2}L^8$. Then

(15.84)

$$E(T) \ll X + T^{1/4} \sup_{|t-T| \leqslant X} \sup_{u \leqslant M} \bigg| \sum_{n \leqslant u} (-1)^n d(n) n^{-3/4} \cos(f(t, n)) \bigg|.$$

This is a restricted analogue of (15.68), with N corresponding to $M = TX^{-2}L^8$ here. Since for $n \leqslant t$ we have

$$f(t, n) = -\pi/4 + 4\pi(nt/2\pi)^{1/2} + O(n^{3/2}t^{-1/2}),$$

it is seen that (15.84) corresponds to $2\pi\Delta^*(T/2\pi)$, and so using Kolesnik's method we obtain easily from (15.84) the analogue of Theorem 13.1, namely,

COROLLARY 15.4.

(15.85) $$E(T) \ll T^{35/108+\varepsilon}.$$

The analogy between $E(t)$ and $2\pi\Delta^*(t/2\pi)$ can be pursued even further. From Theorem 15.4 we have

(15.86) $$\int_2^T E^2(t)\, dt = (C_1 + o(1))T^{3/2} \qquad (T \to \infty),$$

while squaring and integrating (15.68) in the way Theorem 13.5 was derived we obtain

$$(15.87) \qquad \int_2^T \Delta^{*2}(t)\, dt = (C_2 + o(1)) T^{3/2} \qquad (T \to \infty),$$

which shows that the average order of both $|E(t)|$ and $|\Delta^*(t)|$ is $\ll t^{1/4}$. However, if we define

$$(15.88) \qquad E^*(t) = E(t) - 2\pi\Delta^*(t/2\pi),$$

then it can be shown that the average order of $|E^*(t)|$ is $\ll t^{1/6}\log^{3/2}t$. This follows from

THEOREM 15.6.

$$(15.89) \qquad \int_2^T E^{*2}(t)\, dt \ll T^{4/3}\log^3 T.$$

Proof of Theorem 15.6. The general idea of the proof is the same one that was used in the proof of Theorem 13.5. It will be sufficient to prove (15.89) for the integral over $[T, 2T]$, and we apply Atkinson's formula with $N = T$ in $\Sigma_1(t)$. The quality of the final result in (15.89) is being regulated by the size of the error term in the expansion for $f(t, n)$ in (15.75), which is small for $n = o(t^{1/3})$. We write

$$(15.90) \qquad \Sigma_1(t) = \Sigma_{11}(t, X) + \Sigma_{12}(t, X),$$

where in Σ_{11} summation is over $n \leqslant X$, and in Σ_{12} over $X < n \leqslant T$. If we set

$$(15.91) \quad S(t, X) = 2^{1/2}(t/2\pi)^{1/4} \sum_{n \leqslant X} (-1)^n d(n) n^{-3/4} \cos(f(t, n)),$$

then from (15.7) and (15.8) we infer

$$\Sigma_{11}(t, X) - S(t, X) \ll T^{-3/4} \sum_{n \leqslant X} d(n) n^{1/4} \ll T^{-3/4} X^{5/4}\log X \ll \log T$$

with the choice

$$(15.92) \qquad X = T^{1/3}.$$

We use now (15.68) with $N = T$, $x = t/2\pi$ and decompose the sum similarly as the sum in (15.90):

$$(15.93) \qquad \Delta^*(t/2\pi) = \Delta_1^*(t/2\pi, X) + \Delta_2^*(t/2\pi, X) + O(T^\varepsilon).$$

Therefore we obtain

$$\int_T^{2T} E^{*2}(t)\,dt \ll \int_T^{2T}\left(S(t,X) - 2\pi\Delta_1^*(t/2\pi, X)\right)^2 dt + \int_T^{2T}\Sigma_{12}^2(t,X)\,dt$$

$$+ \int_T^{2T}\Sigma_2^2(t)\,dt + \int_T^{2T}\Delta_2^{*2}(t/2\pi, X)\,dt + T^{1+\varepsilon}$$

$$= \sum_{j=1}^4 I_4 + O(T^{1+\varepsilon}),$$

say. By (15.61) we have $I_3 \ll T\log^4 T$, and likewise the nondiagonal terms (those with $m \neq n$ when the sum is squared) of I_2 contribute $\ll T^{1+\varepsilon}$. The diagonal terms give trivially

$$\ll T^{3/2}\sum_{n>X} d^2(n)n^{-3/2} \ll T^{4/3}\log^3 T$$

with the choice $X = T^{1/3}$, and the same argument applies to I_4 as well. Hence

$$I_2 + I_3 + I_4 \ll T^{4/3}\log^3 T.$$

It remains to estimate I_1. Using $\cos a - \cos b = -2\sin((a+b)/2)\sin((a-b)/2)$ and defining

$$h_{\mp}(t,n) = \tfrac{1}{2}\left\{f(t,n) \mp \left(-\pi/4 + 2(2\pi nt)^{1/2}\right)\right\},$$

we have

$$S(t,X) - 2\pi\Delta_1^*(t/2\pi, X) = -2(2t/\pi)^{1/4}\sum_{n\leqslant X}(-1)^n d(n)n^{-3/4}$$

$$\times \sin(h_-(t,n))\sin(h_+(t,n)).$$

Hence

$$I_1 \ll T^{1/2}\sum_{m,n\leqslant X} d(m)d(n)(mn)^{-3/4}$$

$$\times\left|\int_T^{2T}\sin(h_-(t,m))\sin(h_-(t,n))\sin(h_+(t,m))\sin(h_+(t,n))\,dt\right|.$$

As in the proof of Theorems 13.5 and 15.4 we may estimate the nondiagonal terms $m \neq n$ above by Lemma 2.1 to obtain a total contribution which is $\ll T^{1+\varepsilon}$. As for the diagonal terms, observe that by (15.75)

$$h_-(t,n) \ll n^{3/2}t^{-1/2} \ll n^{3/2}T^{-1/2},$$

and thus using $|\sin x| \leqslant |x|$ for x real we get a contribution which is

$$\ll T^{1/2} \sum_{n \leqslant X} d^2(n) n^{-3/2} \int_T^{2T} \sin^2(h_-(t, n)) \, dt \ll T^{1/2} \sum_{n \leqslant X} d^2(n) n^{3/2}$$

$$\ll T^{1/2} X^{5/2} \log^3 X \ll T^{4/3} \log^3 T.$$

This completes the proof of Theorem 15.6.

15.6 LARGE VALUES AND POWER MOMENTS OF $E(T)$

Pursuing further the analogy between $E(T)$ and the divisor problem we now present estimates for power moments of $E(T)$. These estimates are the analogues of Theorems 13.9 and 13.12, and the result is contained in

THEOREM 15.7.

$$(15.94) \qquad \int_2^T |E(t)|^A \, dt \ll T^{(A+4+\varepsilon)/4} \qquad \left(0 \leqslant A \leqslant \tfrac{35}{4}\right),$$

$$(15.95) \qquad \int_2^T |E(t)|^A \, dt \ll T^{(35A+38+\varepsilon)/108} \qquad \left(A \geqslant \tfrac{35}{4}\right).$$

The proof of Theorem 15.7 is completely analogous to the proof of Theorem 13.9, using (15.85) instead of (13.13) and (15.83) as the analogue of the truncated Voronoi formula (3.17) with $TX^{-2}L^8$ corresponding to N. A large values estimate for $E(T)$, namely,

$$(15.96) \qquad R \ll T^\varepsilon (TV^{-3} + T^{15/4} V^{-12}) \qquad (T^{1/4} \ll V \ll T^{1/3})$$

is deduced for $E(T)$ in the same way as (13.52) was derived. Here the restriction $T^{1/4} \ll V \ll T^{1/3}$ is not essential, since $V \gg T^{1/3}$ cannot hold because of Corollary 15.4, and for $V \ll T^{1/4}$ one will trivially obtain (15.94) for the corresponding discrete sum. For (15.96) we suppose that $T/2 \leqslant t_1 < \cdots < t_R \leqslant T$ are points which satisfy $|t_r - t_s| \geqslant CV$ ($r \neq s \leqslant R$) for some suitable $C > 0$, $T^{1/4} \ll V \ll T^{1/3}$, and $E(t_r) \gg V$ for $r = 1, \ldots, R$. Choosing $X = CV$ we have then from (15.83)

(15.97)

$$R \ll T^{1/2+\varepsilon} V^{-2} \max_{M \leqslant CTV^{-2}L^8} \sum_{r \leqslant R}$$

$$\times \left| \sum_{M < n \leqslant 2M} (-1)^n d(n) n^{-3/4} e(t_r', n) r(t_r'^{1/2}, n) \exp(if(t_r', n)) \right|^2,$$

where t_r' is the point for which the supremum in (15.83) is attained. Considering separately $t'_{4m}, t'_{4m+1}, t'_{4m+2}, t'_{4m+3}$ we may suppose that $|t'_r - t'_s| \geqslant CV$ when $r \neq s$. From this point the proof of (15.96) is almost identical with the proof of (13.52), since after the application of the Halász–Montgomery inequality (A.40) the functions e and r which appear in (15.97) may be easily removed by partial summation keeping in mind that $r(x, n)$ is monotonic and $\leqslant 1$, and that (15.74) holds. Similarly, one has (15.75) for $f(t, n)$, and the theory of exponent pairs that was used in the proof of (13.52) may be equally well applied here, producing (15.96). Theorem 15.7 follows then from (15.96) in the same way Theorem 13.9 followed from (13.52).

In analogy with (13.58) it may be noted that the theoretical limit for power moments that (15.96) can give is

$$\int_2^T |E(t)|^{11} \, dt \ll T^{15/4+\varepsilon},$$

which would then give

(15.98) $$\zeta\left(\tfrac{1}{2} + iT\right) \ll T^{5/32+\varepsilon}.$$

Using Lemma 7.1, Hölder's inequality, and arguing as in (15.62) we have with $L = \log T$

$$\left|\zeta\left(\tfrac{1}{2} + iT\right)\right|^2 \ll GT^\varepsilon\left(1 + G^{-2}\int_{T-GL}^{T+GL} |E(t)| \, dt\right)$$

$$\ll GT^\varepsilon\left\{1 + G^{-2}\left(\int_2^{2T} |E(t)|^{11} \, dt\right)^{1/11} G^{10/11}\right\} \ll T^{5/16+\varepsilon},$$

for $G = T^{5/16}$.

Notes

G. H. Hardy and J. E. Littlewood (1918) proved (15.1), while J. E. Littlewood (1922) obtained (15.2) by means of results connected with the approximate functional equation for the zeta-function. E. C. Titchmarsh's book (1951) contains a proof of $E(T) \ll T^{1/2+\varepsilon}$, while a proof that $E(T) \ll T^{5/12}\log^2 T$ has been given by Titchmarsh (1934b).

For the proof of (15.12) R. Balasubramanian (1978) uses the Riemann–Siegel formula (4.3) and (4.4) with $N = 5$. The idea of his proof is to square the expression for $e^{i\theta}\zeta(\tfrac{1}{2} + it)$ [where θ is given by (15.13)] and to estimate carefully the resulting integrals, some of which are technically quite complicated. His paper also contains the result $\gamma_{n+1} - \gamma_n \ll \gamma_n^{1/6+\varepsilon}$ about the differences of consecutive zeros on the critical line, and this was already discussed in Chapter 10.

In Section 15.2 we have followed closely Atkinson's original proof (1949) of (15.4), where curiously in 1.3 on p. 375 he makes a mistake in sign, obtaining $+2\sum_{n \leqslant N'} d(n) \cdots$ in place of $-2\sum_{n \leqslant N'} d(n) \cdots$. The corrected form of (15.4) was stated by M. Jutila (1984b) without comment. For technical reasons [to avoid the last term in $\sum'_{n \leqslant x} d(n)$ in (3.1)] one takes $X = N + \tfrac{1}{2}$,

N an integer, from (15.27) onwards, and it is easily seen that this restriction does not affect the final result.

Heath-Brown's derivation (1978a) of Theorem 7.2 starts from (15.62) in the form

$$\int_{T-G}^{T+G}\left|\zeta\left(\tfrac{1}{2}+it\right)\right|^2 dt \ll GL + \int_{-GL}^{GL} E(T+t)\,tG^{-2}e^{-t^2G^{-2}}\,dt$$

and uses Atkinson's formula. The contribution of $\Sigma_2(T+t)$ to the above integral is small, and the main contribution comes from $\Sigma_1(T+t)$, producing a sum of length $\ll TG^{-2}\log^2 T$. In analyzing the difference in the proofs of Theorem 7.2, it should be noted that in Heath-Brown's proof one uses first the Voronoi summation formula, which is implicit in Atkinson's formula, and then the exponential integral (A.38), while in our proof one makes first an exponential averaging of the approximate functional equation (4.11) and then applies the Voronoi summation formula.

The results of Section 15.3 are to be found in M. Jutila's paper (1984a), where more general applications of Voronoi-type summation formulas are considered.

Theorem 15.4 is due to D. R. Heath-Brown (1978b). The result $E(T) = \Omega(T^{1/4})$, stated here as Corollary 15.1, was obtained a little before Heath-Brown by A. Good (1977) who used a complicated technique which was not based on Atkinson's formula. A plausible conjecture is that $F(T) \ll T^{3/4+\varepsilon}$, where $F(T)$ is defined by (15.63), and this would lead to the hypothetical $\zeta(\tfrac{1}{2}+iT) \ll T^{1/8+\varepsilon}$, a result out of reach at present. The same bound would of course follow also from the conjectural estimate $E(T) \ll T^{1/4+\varepsilon}$, which in view of (15.84) seems to be of the same degree of difficulty as the classical conjecture $\alpha_2 = \tfrac{1}{4}$ in the Dirichlet divisor problem. There seems to be no method available at present which would permit one to deduce from the (global) estimate $E(T) \ll T^{c+\varepsilon}$, $\tfrac{1}{4} \leqslant c \leqslant \tfrac{35}{108}$, anything better than the obvious (local) estimate $\zeta(\tfrac{1}{2}+iT) \ll T^{c/2+\varepsilon}$.

Using the techniques of K. S. Gangadharan (1961), K. Corrádi and I. Kátai (1967), and J. L. Hafner (1981a) which give Ω-results for $\Delta(x)$, one should be able to improve the result $E(T) = \Omega(T^{1/4})$, given by Corollary 15.1.

The analysis between local and global estimates of $\Delta(x)$ and $\zeta(\tfrac{1}{2}+iT)$ has been thoroughly discussed by M. Jutila (1983a, 1984b), where the results of Section 15.5 may be found. Jutila's proof (1983a) of (15.68) has been, however, replaced here by a short elementary argument which uses (3.17). Jutila (1984b) proves a more general estimate than Theorem 15.6, namely,

$$\int_{T-H}^{T+H} E^{*2}(t)\,dt \ll HT^{1/3}\log^3 T + T^{1+\varepsilon} \qquad (2 \leqslant H \leqslant T),$$

but the proof of this more general result is almost the same as the proof of (15.89).

Interesting conditional results are obtained by M. Jutila (1983a), and they seem to be the first ones of their kind. For example, if the conjecture that $\alpha_2 = \tfrac{1}{4}$ in the divisor problem is true, then Jutila proved

$$\zeta\left(\tfrac{1}{2}+iT\right) \ll T^{3/20+\varepsilon}, \qquad E(T) \ll T^{5/16+\varepsilon},$$

and the exponents in the above estimates are better than the best published unconditional values $\tfrac{35}{216}$ and $\tfrac{35}{108}$, respectively. These results may be compared with the conditional estimate (15.98), which is the limit of (15.96).

Theorem 15.7, which is the analogue of Theorems 13.9 and 13.12 for $E(t)$, has been given by the author (1983a).

APPENDIX

A.1 INTRODUCTION

This appendix contains miscellaneous results which are repeatedly used in the text. Most of the material consists of well-known analytic facts given here as a reference for the sake of completeness of the exposition. This seemed preferable to quoting these results from the literature each time the need for such a result arises at a specific place in the text. The material presented here is only very loosely connected, and its inclusion was solely motivated by the needs of previous chapters. Thus detailed proofs are not given, and sometimes only a reference to a standard text is offered, where proofs and a more detailed account may be found.

A.2 MELLIN TRANSFORMS

Let $f(x)x^{\sigma-1}$ belong to $L(0, \infty)$ and let $f(x)$ have bounded variation on every finite x-interval. Then

$$(A.1) \qquad F(s) = \int_0^\infty x^{s-1} f(x)\, dx, \qquad s = \sigma + it \qquad (\sigma, t \text{ real})$$

is defined as the Mellin transform of $f(x)$. From (A.1) we can recover $f(x)$ in terms of $F(s)$ by Mellin's inversion formula

$$(A.2) \qquad \tfrac{1}{2}\big(f(x+0) + f(x-0)\big) = (2\pi i)^{-1} \lim_{T \to \infty} \int_{\sigma - iT}^{\sigma + iT} F(s) x^{-s}\, ds.$$

In the case when $f(x)$ is continuous (A.2) can be obtained without difficulty directly from (A.1), while in the general case it seems more suitable to write (A.1) as a Fourier transform by a change of variable and then to appeal to results from the theory of Fourier transforms and integrals. A detailed account of (A.1) and (A.2) is to be found in E. C. Titchmarsh's book (1948) on Fourier integrals. The relations given by (A.1) and (A.2) are inverse to one another. Namely, if

(A.3) $$f(x) = (2\pi i)^{-1} \int_{\sigma-i\infty}^{\sigma+i\infty} F(s) x^{-s} ds,$$

where $F(\sigma + iu)$ belongs to $L(-\infty, \infty)$ and it of bounded variation in the neighborhood of the point $u = t$ and (A.3) holds, then

(A.4) $$\tfrac{1}{2}\{F(\sigma + i(t + 0)) + F(\sigma + i(t - 0))\} = \lim_{a \to \infty} \int_{1/a}^{a} f(x) x^{\sigma+it-1} dx,$$

and in most applications (A.4) will reduce to (A.1).

An analogue of the well-known Parseval's identity for Fourier integrals holds also for Mellin transforms, namely, if f and F are connected by (A.1), then

(A.5) $$(2\pi)^{-1} \int_{-\infty}^{\infty} |F(\sigma + it)|^2 dt = \int_{0}^{\infty} f^2(x) x^{2\sigma-1} dx.$$

As in the case of (A.2) this identity may be derived from Parseval's identity for Fourier transforms, or one may argue directly by writing

$$(2\pi i)^{-1} \int_{\sigma-i\infty}^{\sigma+i\infty} F(s)\overline{F(s)} \, ds = \int_{0}^{\infty} f(x)\left((2\pi i)^{-1} \int_{\sigma-i\infty}^{\sigma+i\infty} F(s) x^{\sigma-it-1} ds\right) dx$$

$$= \int_{0}^{\infty} f(x) x^{2\sigma-1}\left((2\pi i)^{-1} \int_{\sigma-i\infty}^{\sigma+i\infty} F(s) x^{-s} ds\right) dx$$

$$= \int_{0}^{\infty} f^2(x) x^{2\sigma-1} dx,$$

where (A.1) and (A.2) were used under the assumption that f is continuous. Setting $s = \sigma + it$ we obtain (A.5).

Formulas analogous to (A.5) holds also for two or more functions. As an example, suppose that $F(s)$ and $G(s)$ are Mellin transforms of two continuous functions $f(x)$ and $g(x)$, respectively. Then

(A.6) $$(2\pi i)^{-1} \int_{\sigma-i\infty}^{\sigma+i\infty} F(s)G(1 - s) \, ds$$

$$= (2\pi i)^{-1} \int_{\sigma-i\infty}^{\sigma+i\infty} G(1 - s)\left(\int_{0}^{\infty} f(x) x^{s-1} dx\right) ds$$

$$= (2\pi i)^{-1} \int_{0}^{\infty} f(x) \, dx \int_{\sigma-i\infty}^{\sigma+i\infty} G(1 - s) x^{s-1} ds$$

$$= \int_{0}^{\infty} f(x)g(x) \, dx.$$

Finally, it may be mentioned that the inversion of the gamma-integral (see Section A.7)

$$\Gamma(s) = \int_0^\infty e^{-x} x^{s-1} dx \qquad (\mathrm{Re}\, s > 0),$$

gives by (A.2) the useful relation

(A.7) $$e^{-x} = (2\pi i)^{-1} \int_{c-i\infty}^{c+i\infty} \Gamma(s) x^{-s} ds \qquad (c, x > 0).$$

A.3 INVERSION FORMULAS FOR DIRICHLET SERIES

We shall consider Dirichlet series of the form $A(s) = \sum_{n=1}^\infty a_n n^{-s}$ which have a finite abscissa of absolute convergence, and we shall set $f(x) = \sum'_{n \leqslant x} a_n$. General theory of Dirichlet series will not be discussed here, since our main interest lies in inversion formulas, which represent formulas expressing $f(x)$ (or some similar function involving the a_n's) by series or integrals containing $A(s)$. Sometimes all these formulas go under the name of "Perron's formula," although this terminology is most often used for one particular formula of this sort, namely,

(A.8) $$\sum_{n \leqslant x}{}' a_n = (2\pi i)^{-1} \int_{c-i\infty}^{c+i\infty} A(s) x^s s^{-1} ds,$$

where $c > 0$ is such a number that $A(s)$ is absolutely convergent for $\mathrm{Re}\, s = c$. Here $\sum'$ means that if x is an integer then $\frac{1}{2} a_x$ comes instead of a_x into the sum. One obtains (A.8) easily from

(A.9) $$(2\pi i)^{-1} \int_{c-i\infty}^{c+i\infty} Y^s s^{-1} ds = \begin{cases} 0, & 0 \leqslant Y < 1 \\ \frac{1}{2}, & Y = 1 \\ 1, & Y > 1 \end{cases} \qquad (c > 0)$$

since in view of absolute convergence of $A(s)$ one may integrate term by term the right-hand side of (A.8). Using (A.9) we obtain

$$(2\pi i)^{-1} \int_{c-i\infty}^{c+i\infty} A(s) x^s s^{-1} ds = \sum_{n=1}^\infty a_n (2\pi i)^{-1} \int_{c-i\infty}^{c+i\infty} (x/n)^s s^{-1} ds = \sum_{n \leqslant x}{}' a_n.$$

To see that (A.9) holds one may evaluate the integral directly by the residue theorem, or defining $f(x) = 0$ for $0 \leqslant x < 1$, $f(1) = \frac{1}{2}$, $f(x) = 1$ for $x > 1$ one has the Mellin transform

$$F(s) = \int_0^\infty f(x) x^{s-1} dx = x^s s^{-1}\big|_1^\infty = -s^{-1} \qquad (\mathrm{Re}\, s < 0)$$

and (A.9) follows from (A.2) on replacing x by Y and s by $-s$. Instead of (A.8) it is often desirable to have a truncated form of the inversion formula, namely, a formula where the integral is over a finite segment whose length may be suitably chosen. Such a formula may be obtained [see K. Prachar (1967) for all the details of proof] if instead of (A.9) we use Lemma 12.1. Following the proof of Theorem 12.1 we obtain then the following inversion formula:

Let $A(s) = \sum_{n=1}^{\infty} a_n n^{-s}$ converge absolutely for $\sigma = \operatorname{Re} s > 1$ and let $|a_n| < C\Phi(n)$, where $C > 0$ and for $x \geqslant x_0$ $\Phi(x)$ is monotonically increasing. Let further

$$\sum_{n=1}^{\infty} |a_n| n^{-\sigma} \ll (\sigma - 1)^{-\alpha}$$

as $\sigma \to 1 + 0$ for some $\alpha > 0$. If $w = u + iv$ (u, v real) is arbitrary, $b > 0$, $T > 0$, $u + b > 1$, then

(A.10)

$$\sum_{n \leqslant x} a_n n^{-w} = (2\pi i)^{-1} \int_{b-iT}^{b+iT} A(s + w) x^s s^{-1} \, ds + O\left(x^b T^{-1}(u + b - 1)^{-\alpha}\right)$$

$$+ O\left(T^{-1}\Phi(2x) x^{1-u} \log 2x\right) + O\left(\Phi(2x) x^{-u}\right),$$

and the estimate is uniform in x, T, b, and u provided that b and u are bounded.

Another inversion formula for Dirichlet series is

(A.11) $\quad \displaystyle\sum_{n \leqslant x} a_n n^{-w} \log^{k-1} x/n = (2\pi i)^{-1}(k-1)! \int_{c-i\infty}^{c+i\infty} A(s + w) x^s s^{-k} \, ds$

$$(c > 0),$$

where $k \geqslant 2$ is a fixed integer, w is an arbitrary complex number, and $c + \operatorname{Re} w$ exceeds the abscissa of absolute convergence of $A(s)$. This formula may be obtained if one integrates term by term the right-hand side of (A.11) with the use of

(A.12)

$$(2\pi i)^{-1} \int_{c-i\infty}^{c+i\infty} Y^s s^{-k} \, ds = \begin{cases} 0 & 0 \leqslant Y \leqslant 1 \\ \dfrac{1}{(k-1)!} \log^{k-1} Y & Y > 1 \end{cases} \quad (c > 0),$$

where $k \geqslant 2$ is a fixed integer. To see that (A.12) holds one may start from

$$\int_0^{\infty} e^{-sx} x^{k-1} \, dx = s^{-k}(k-1)! \qquad (\operatorname{Re} s > 0),$$

and make the change of variable $u = e^{-x}$ to obtain

$$\frac{1}{(k-1)!} \int_0^1 u^{s-1}(-1)^{k-1} \log^{k-1} u \, du = s^{-k}.$$

The inversion formula (A.2) gives then

$$(A.13) \quad (2\pi i)^{-1} \int_{c-i\infty}^{c+i\infty} u^{-s} s^{-k} \, ds = \begin{cases} \dfrac{(-1)^{k-1}}{(k-1)!} \log^{k-1} u & 0 < u < 1, \\ 0 & u \geqslant 1, \end{cases}$$

so that (A.12) follows with $Y = u^{-1}$.

Finally, we present an inversion formula for a weighted sum which differs from the one appearing in (A.11). We suppose that q is a fixed positive number and that $A(s)$ converges absolutely for $\operatorname{Re} s = c > 0$. Then

$$(A.14) \quad \frac{1}{\Gamma(q+1)} \sum_{n \leqslant x} a_n (x-n)^q = (2\pi i)^{-1} \int_{c-i\infty}^{c+i\infty} \frac{\Gamma(s) A(s)}{\Gamma(s+q+1)} x^{s+q} \, ds,$$

and we note that for $q = 0$ (A.14) reduces to (A.8) since $\Gamma(s+1) = s\Gamma(s)$. One may obtain (A.14) by termwise integration of the right-hand side with the aid of

$$(A.15) \quad (2\pi i)^{-1} \int_{c-i\infty}^{c+i\infty} \frac{u^{-s} \Gamma(s) \, ds}{\Gamma(s+q+1)} = f(u)$$

$$= \begin{cases} \dfrac{(1-u)^q}{\Gamma(q+1)} & 0 < u \leqslant 1, \\ 0 & u > 1, \end{cases} \quad (c > 0)$$

when one replaces u by n/x. To see that (A.15) holds we may use the well-known beta integral [see (A.31)], namely

$$\int_0^1 x^{a-1}(1-x)^{b-1} \, dx = B(a,b) = \frac{\Gamma(a)\Gamma(b)}{\Gamma(a+b+1)} \quad (\operatorname{Re} a > 0, \operatorname{Re} b > 0),$$

to obtain

(A.16)

$$F(s) = \int_0^\infty f(x) x^{s-1} \, dx = \int_0^1 \frac{(1-x)^q}{\Gamma(q+1)} x^{s-1} \, dx = \frac{\Gamma(q+1)\Gamma(s)}{\Gamma(q+1)\Gamma(q+s+1)}.$$

This shows that $\Gamma(s)/\Gamma(q+s+1)$ is the Mellin transform of $f(x)$, and consequently (A.15) follows from the inversion formula (A.2) for Mellin transforms.

A.4 HÖLDER'S INEQUALITY

This is a classical inequality, which may be stated in a general form as follows. Let all numbers which appear in this section be positive, and let $p_1 + p_2 + \cdots + p_k = 1$. Hölder's inequality then states that

$$(A.17) \qquad \sum_{j=1}^{N} x_{j,1} \cdots x_{j,k} \leqslant \left(\sum_{j=1}^{N} x_{j,1}^{1/p_1} \right)^{p_1} \cdots \left(\sum_{j=1}^{N} x_{j,k}^{1/p_k} \right)^{p_k}.$$

The special case $k = 2$, $p_1 = p_2 = \frac{1}{2}$ is known as the Cauchy–Schwarz inequality and may be written as

$$(A.18) \qquad \left(\sum_{j=1}^{N} x_j y_j \right)^2 \leqslant \sum_{j=1}^{N} x_j^2 \sum_{j=1}^{N} y_j^2.$$

The inequality (A.17) follows easily from the case $k = 2$ by mathematical induction. To obtain (A.17) for $k = 2$ note that for $\alpha + \beta = 1$ and $x \geqslant 1$ $x^\alpha \leqslant \alpha x + \beta$, since $f(x) = x^\alpha - \alpha x - \beta$ is easily seen to be decreasing by considering its derivative. Then for $a, b \in [0, 1]$ we obtain

$$a^\alpha b^\beta \leqslant \alpha a + \beta b,$$

and taking $\alpha = p_1$, $\beta = p_2$,

$$a = x_{j,1}^{1/\alpha} \left(\sum_{j=1}^{N} x_{j,1}^{1/\alpha} \right)^{-1}, \qquad b = x_{j,2}^{1/\beta} \left(\sum_{j=1}^{N} x_{j,2}^{1/\beta} \right)^{-1},$$

and summing over $j = 1, \ldots, N$ we obtain (A.17) for $k = 2$. A result similar to (A.17) holds if the sums are replaced by an integral.

A.5 PARTIAL SUMMATION FORMULAS

Partial summation is a standard elementary technique for transforming sums into more manageable sums or integrals, and some of these useful formulas are recorded here as a reference.

Let $\{a_n\}_{n=1}^{\infty}$ be a sequence of complex numbers and $\{b_n\}_{n=1}^{\infty}$ a sequence of real numbers. If $b_1 \geqslant b_2 \geqslant \cdots \geqslant 0$ and $M \geqslant 1$ is an integer, then

$$(A.19) \qquad \left| \sum_{M < n \leqslant N} a_n b_n \right| \leqslant b_M \max_{M < n \leqslant N} \left| \sum_{M < m \leqslant n} a_m \right|,$$

while if $0 \leqslant b_1 \leqslant b_2 \leqslant \cdots$, then

(A.20)
$$\left| \sum_{M < n \leqslant N} a_n b_n \right| \leqslant 2 b_N \max_{M < n \leqslant N} \left| \sum_{M < m \leqslant n} a_m \right|.$$

These simple inequalities show that monotonic sequences may be removed from sums, and they are both proved analogously. To obtain (A.19) we define

$$A_n = \sum_{M < m \leqslant n} a_m.$$

Then

$$\left| \sum_{M < n \leqslant N} a_n b_n \right| = \left| \sum_{M < n \leqslant N} (A_n - A_{n-1}) b_n \right|$$

$$\leqslant |A_N| b_N + \sum_{M < n \leqslant N-1} |A_n| (b_n - b_{n+1}) \leqslant b_M \max_{M < n \leqslant N} |A_n|.$$

To transform sums into integrals it is often convenient to write sums as Stieltjes integrals and then to integrate by parts. For example, if $\{a_n\}_{n=1}^{\infty}$ is a sequence of real numbers, $g(x) \in C^1[\lambda_1, x]$, $\lambda_1 \leqslant \lambda_2 \leqslant \cdots$ is a sequence of real numbers tending to infinity, then

(A.21)
$$\sum_{\lambda_1 \leqslant \lambda_n \leqslant x} a_n g(\lambda_n) = A(x) g(x) - \int_{\lambda_1}^{x} A(t) g'(t) \, dt,$$

where

(A.22)
$$A(t) = \sum_{\lambda_1 \leqslant \lambda_n \leqslant t} a_n.$$

Namely, we can write

$$\sum_{\lambda_1 \leqslant \lambda_n \leqslant x} a_n g(\lambda_n) = \int_{\lambda_1 - 0}^{x+0} g(t) \, dA(t),$$

since $A(t)$ has jumps of weight a_n for $t = \lambda_n$ and otherwise it is a constant function. An integration by parts yields immediately (A.21), since

$$g(t) A(t) \Big|_{\lambda_1 - 0}^{x+0} = g(x) A(x) - g(\lambda_1 - 0) A(\lambda_1 - 0) = g(x) A(x).$$

Similarly, we can obtain

(A.23)
$$\sum_{X < n \leqslant Y} f(n) = \int_{X}^{Y} f(t) \, dt - \psi(Y) f(Y) + \psi(X) f(X) + \int_{X}^{Y} \psi(t) f'(t) \, dt,$$

where $f(x) \in C^1[X, Y]$ and $\psi(t) = t - [t] - \frac{1}{2}$. This is a special case of the so-called Euler–Maclaurin summation formula, and essentially only a variant of (A.21). The general Euler–Maclaurin formula is (for simplicity we shall assume here that a and b are integers)

$$(A.24) \qquad \sum_{a \leqslant k \leqslant b} f(k) = \int_a^b f(t)\, dt + \tfrac{1}{2}(f(a) + f(b))$$

$$+ \sum_{m=1}^n \frac{B_{2m}}{(2m)!} \left(f^{(2m-1)}(b) - f^{(2m-1)}(a) \right)$$

$$+ \int_a^b P_{2n+1}(t) f^{(2n+1)}(t)\, dt.$$

Here $n \geqslant 0$ is a fixed integer, $f(x) \in C^{2n+1}[a, b]$, B_m is the mth Bernoulli number, and P_m is the mth periodic Bernoulli function defined by $P_m(x) = B_m(x - [x])$, where $B_m(x)$ is the mth Bernoulli polynomial defined by

$$\frac{ze^{xz}}{e^z - 1} = \sum_{m=0}^{\infty} B_m(x) z^m / m! \qquad (|z| < 2\pi),$$

so that $B_m = B_m(0)$, $B_1(x) = x - \frac{1}{2}$, $B_2(x) = x^2 - x + \frac{1}{6}$, and so on. A proof of (A.24) may be obtained as follows: by the Stieltjes-integral representation and integration by parts we have

$$\sum_{a \leqslant k \leqslant b} f(k) = \int_{a-0}^{b+0} f(t)\, d([t]) = \int_a^b f(t)\, dt + \int_{a-0}^{b+0} f(t)\, d([t] - t + \tfrac{1}{2})$$

$$= \int_a^b f(t)\, dt - \int_{a-0}^{b+0} f(t)\, dP_1(t)$$

$$= \int_a^b f(t)\, dt + \tfrac{1}{2}(f(a) + f(b)) + \int_a^b P_1(t) f'(t)\, dt,$$

which is (A.23) for $a = [X] + 1$, $b = [Y]$. From the defining property of Bernoulli polynomials [see T. M. Apostol (1976a), Chapter 12] $B'_{n+1}(x) = (n + 1)B_n(x)$, so that one may take

$$\int P_n(x)\, dx = (n + 1)^{-1} P_{n+1}(x),$$

and repeated integration by parts of $\int_a^b P_1(t) f'(t)\, dt$ leads to (A.24), since for any integer r we have $P_m(r) = B_m(0) = B_m$ and $B_{2m+1} = 0$ for $m \geqslant 1$.

A.6   THE POISSON SUMMATION FORMULA

There exist several variants of this useful formula. We shall state the following version: let a, b be integers and let $f(x)$ be a function of the real variable x

with bounded first derivative on $[a, b]$. Then

$$(A.25) \qquad \sideset{}{'}\sum_{a \leqslant n \leqslant b} f(n) = \int_a^b f(x)\, dx + 2 \sum_{n=1}^{\infty} \int_a^b f(x)\cos(2n\pi x)\, dx.$$

Here as usual Σ' means that $\tfrac{1}{2}f(a)$ and $\tfrac{1}{2}f(b)$ are to be taken instead of $f(a)$ and $f(b)$, respectively. To derive (A.25) we use (A.23) in the form

$$\sideset{}{'}\sum_{a \leqslant n \leqslant b} f(n) = \int_a^b f(x)\, dx + \int_a^b \psi(x) f'(x)\, dx,$$

and thus we have to show that

$$(A.26) \qquad \int_a^b \psi(x) f'(x)\, dx = 2 \sum_{n=1}^{\infty} \int_a^b f(x)\cos(2n\pi x)\, dx.$$

This is achieved by using the Fourier-series expansion

$$(A.27) \qquad \psi(x) = -\pi^{-1} \sum_{n=1}^{\infty} n^{-1}\sin(2n\pi x),$$

which is valid if x is not an integer. The series in (A.27) is equal to zero if x is an integer, and moreover by partial summation it is seen that its partial sums are uniformly bounded for any real x. Therefore using (A.27) in (A.26) and integrating by parts we obtain the right-hand side of (A.26), since $\sin(2n\pi a) = \sin(2n\pi b) = 0$ if a and b are integers.

A more detailed account of Poisson's summation formula may be found, for example, in Chapter 10 of M. N. Huxley's book (1972a), where a good bound for the tails of the series in (A.25) is given.

A.7 THE GAMMA-FUNCTION

Several standard properties of the gamma-function will be stated now, and the proofs are readily found in books on analysis.

For $\operatorname{Re} s > 0$ the gamma-function is defined as

$$(A.28) \qquad \Gamma(s) = \int_0^{\infty} e^{-x} x^{s-1}\, dx,$$

and for other values of s by analytic continuation. $\Gamma(s)$ is a regular function of s in the whole plane, except for points $s = 0, -1, -2, \ldots, -n, \ldots$, which are poles of the first order with residues $(-1)^n/n!$ $(n = 0, 1, 2, \ldots)$. The gamma-function satisfies the functional equation

$$(A.29) \qquad \Gamma(s + 1) = s\Gamma(s),$$

and the useful relations

(A.30) $\Gamma(s)\Gamma(1 - s) = \pi/\sin \pi s, \qquad \Gamma(s)\Gamma(s + \tfrac{1}{2}) = 2\sqrt{\pi}\, 2^{-2s}\Gamma(2s).$

Other common properties of the gamma-function are

(A.31)

$$B(a, b) = \int_0^1 x^{a-1}(1 - x)^{b-1} \, dx = \frac{\Gamma(a)\Gamma(b)}{\Gamma(a + b)} \qquad (\mathrm{Re}\, a > 0, \mathrm{Re}\, b > 0)$$

and

(A.32) $\Gamma'(1) = \int_0^\infty e^{-x}\log x \, dx = -\gamma = -0.5772157\ldots,$

by which the Euler constant γ may be defined.

From the theory of the asymptotic approximation of the gamma-function we shall need the so-called Stirling's formula in the form

(A.33) $\log \Gamma(s + b) = (s + b - \tfrac{1}{2})\log s - s + \tfrac{1}{2}\log 2\pi + O(|s|^{-1}),$

which is valid for b constant and $|\arg s| \leqslant \pi - \delta$ $(\delta > 0)$, if $s = 0$ and neighborhoods of poles of $\Gamma(s + b)$ are excluded. Another variant of Stirling's formula states that

(A.34) $|\Gamma(s)| = (2\pi)^{1/2}|t|^{\sigma-1/2}e^{-\pi|t|/2}\bigl(1 + O(|t|^{-1})\bigr) \qquad (|t| \geqslant t_0),$

and this is valid for $C_1 \leqslant \mathrm{Re}\, s \leqslant C_2$, and the O-constant depends on C_1 and C_2. Also for $\delta > 0$ fixed, $|\arg s| \leqslant \pi - \delta$, $|s| \geqslant \delta$, we have

(A.35) $\Gamma'(s)/\Gamma(s) = \log s - 1/2s + O(|s|^{-2}).$

A.8 THE PHRAGMÉN – LINDELÖF PRINCIPLE

The well-known maximum modulus principle for regular functions states that if $f(s)$ is regular and $|f(s)| \leqslant C$ on the boundary of a domain, then the same inequality holds for all points of the domain. The Phragmén–Lindelöf principle is a generalization of the maximum modulus principle for an unbounded domain. Most often it is used for a strip $\sigma_1 \leqslant \sigma \leqslant \sigma_2$ in the s-plane and in this case it may be formulated as follows:

Let $f(s)$ be regular in the region $\mathscr{D}$: $\sigma_1 \leqslant \sigma \leqslant \sigma_2$, $-\infty < t < \infty$, where

$$|f(s)| < A\exp(e^{C|t|}) \qquad \left(A > 0, 0 < C < \frac{\pi}{\sigma_2 - \sigma_1}\right).$$

If $|f(s)| \leqslant B$ holds for $\sigma = \sigma_1$ and $\sigma = \sigma_2$ with some absolute $B > 0$, then it holds for all points of $\mathcal{D}$.

A.9 HADAMARD'S THREE-CIRCLES THEOREM

Let $f(s)$ be regular and one-valued in the strip $\sigma_1 \leqslant \sigma \leqslant \sigma_2$, where it satisfies $|f(s)| \leqslant C$. If $L(\alpha)$ is the supremum of $|f(s)|$ on the line $\sigma = \alpha$ $(\sigma_1 \leqslant \alpha \leqslant \sigma_2)$, then

$$(A.36) \qquad L(\sigma) \leqslant L(\sigma_1)^{(\sigma_2 - \sigma)/(\sigma_2 - \sigma_1)} L(\sigma_2)^{(\sigma - \sigma_1)/(\sigma_2 - \sigma_1)}.$$

This is a convexity result which follows by considering $g(s) = e^{\theta s} f(s)$, where θ is a suitable real number, and where we may suppose that $f(s) \not\equiv 0$. In the strip $\sigma_1 \leqslant \sigma \leqslant \sigma_2$ $g(s)$ is regular and $|g(s)|$ is bounded by $e^{\theta \sigma_1} L(\sigma_1)$ and $e^{\theta \sigma_2} L(\sigma_2)$ on $\sigma = \sigma_1$ and $\sigma = \sigma_2$, respectively. Now by the Phragmén–Lindelöf principle

$$e^{\theta \sigma} L(\sigma) \leqslant \max\{ e^{\theta \sigma_1} L(\sigma_1), e^{\theta \sigma_2} L(\sigma_2)\} \qquad (\sigma_1 \leqslant \sigma \leqslant \sigma_2),$$

and taking

$$\theta = \frac{1}{\sigma_2 - \sigma_1} \log \frac{L(\sigma_1)}{L(\sigma_2)}$$

the last inequality yields (A.36). With the aid of the function $s = \log z$ the region $0 < r_1 \leqslant |z| \leqslant r_2$ is mapped on the strip $\log r_1 \leqslant \sigma \leqslant \log r_2$. Therefore the result stated above gives the following one, which is known as Hadamard's three-circles theorem:

Let $0 < r_1 < r_2$ and let $f(z)$ be regular and one-valued for $r_1 \leqslant |z| \leqslant r_2$. If

$$M(r) = \max_{|z| = r} |f(z)|,$$

then for $r_1 \leqslant r \leqslant r_2$

$$(A.37) \qquad M(r) \leqslant M(r_1)^{\log(r_2/r)/\log(r_2/r_1)} M(r_2)^{\log(r/r_1)/\log(r_2/r_1)}.$$

A.10 AN EXPONENTIAL INTEGRAL

Very often integrals are smoothed by introducing a certain exponential weight which simplifies subsequent estimations. One such integral is

$$(A.38) \qquad \int_{-\infty}^{\infty} \exp(At - Bt^2)\, dt = (\pi/B)^{1/2} \exp(A^2/4B) \qquad (\operatorname{Re} B > 0),$$

which in fact represents a regular function of A and B provided that $\operatorname{Re} B > 0$. By the principle of analytic continuation it is sufficient to prove (A.38) for B real and positive, when the change of variable

$$t = A/(2B) + xB^{-1/2}$$

gives

$$\int_{-\infty}^{\infty} \exp(At - Bt^2)\, dt = B^{-1/2}\exp(A^2/4B)\int_{-\infty}^{\infty} e^{-x^2}\, dx$$

$$= (\pi/B)^{1/2}\exp(A^2/4B).$$

A.11 THE HALÁSZ – MONTGOMERY INEQUALITIES

The inequalities in question are certain general inequalities for vectors in inner-product spaces which have found many applications recently in analytic number theory. Their connection with large sieve inequalities is very close, and the whole subject is extensively treated by H. L. Montgomery (1971, 1978, 1982), where detailed references are given. To formulate the inequalities, suppose that ξ, $\varphi_1, \ldots, \varphi_R$ are arbitrary vectors in an inner-product vector space over $\mathbb{C}$, where (a, b) will be the notation for the inner product and $\|a\|^2 = (a, a)$. Then

(A.39)
$$\sum_{r \leqslant R} |(\xi, \varphi_r)| \leqslant \|\xi\| \left(\sum_{r,\, s \leqslant R} |(\varphi_r, \varphi_s)| \right)^{1/2},$$

(A.40)
$$\sum_{r \leqslant R} |(\xi, \varphi_r)|^2 \leqslant \|\xi\|^2 \max_{r \leqslant R} \sum_{s \leqslant R} |(\varphi_r, \varphi_s)|.$$

Both of these inequalities are derived by a judicious use of the Cauchy–Schwarz inequality for vector spaces. To see this observe that from $(a, b) = \overline{(b, a)}$ one has

$$\sum_{r \leqslant R} c_r(\xi, \varphi_r) = \left(\xi, \sum_{r \leqslant R} \bar{c}_r \varphi_r \right)$$

for any scalars c_r. Thus

(A.41)
$$\left| \sum_{r \leqslant R} c_r(\xi, \varphi_r) \right|^2 \leqslant \|\xi\|^2 \left\| \sum_{r \leqslant R} \bar{c}_r \varphi_r \right\|^2 = \|\xi\|^2 \sum_{r,\, s \leqslant R} \bar{c}_r c_s(\varphi_r, \varphi_s).$$

If we take $c_r = \exp(-i \arg(\xi, \varphi_r))$, then $|c_r| = 1$ and

$$\sum_{r \leqslant R} c_r(\xi, \varphi_r) = \sum_{r \leqslant R} |(\xi, \varphi_r)|,$$

so that (A.39) follows at once from (A.41). For (A.40) we use the elementary inequality

$$|\bar{c}_r c_s| \leq \tfrac{1}{2}|c_r|^2 + \tfrac{1}{2}|c_s|^2$$

to obtain

(A.42)

$$\sum_{r,\,s\leq R} \bar{c}_r c_s(\varphi_r, \varphi_s) \leq \sum_{r\leq R}|c_r|^2 \sum_{s\leq R}|(\varphi_r, \varphi_s)| \leq \sum_{r\leq R}|c_r|^2 \max_{r\leq R} \sum_{s\leq R}|(\varphi_r, \varphi_s)|,$$

so that combining (A.41) and (A.42) we have (A.40) if we take $c_r = \overline{(\xi, \varphi_r)}$.

If $a = \{a_n\}_{n=1}^\infty$ and $b = \{b_n\}_{n=1}^\infty$ are two (vector) sequences of complex numbers, then the standard inner product of a and b is defined as

(A.43)
$$(a, b) = \sum_{n=1}^\infty a_n \bar{b}_n.$$

REFERENCES

R. J. Anderson. Simple zeros of the Riemann zeta-function, *J. Number Theory* **17**, 176–182 (1983).

R. J. Anderson and H. M. Stark. Oscillation theorems, in *Analytic Number Theory* (ed. M. I. Knopp), LNM 899, Springer-Verlag, Berlin-Heidelberg-New York, 1981, pp. 79–106.

R. Apéry. Interpolation de fractions continues et irrationalité de certaines constantes, *Math. CTHS Bull. Sec. Sci. III* (Bibl. Nat. Paris), 37–53 (1981).

T. M. Apostol. *Introduction to Analytic Number Theory*, Springer-Verlag, Berlin-Heidelberg-New York, 1976a.

T. M. Apostol. *Modular Functions and Dirichlet Series in Number Theory*, GTM 41, Springer-Verlag, Berlin-Heidelberg-New York, 1976b.

G. I. Arhipov and A. A. Karacuba. On I. M. Vinogradov's integral, *Sov. Math. Dokl.* **19**, 389–391 (1978).

F. V. Atkinson. A divisor problem, *Quart. J. Math. (Oxford)* **12**, 193–200 (1941a).

F. V. Atkinson. The mean value of the zeta-function on the critical line, *Proc. London Math. Soc.* **47** (2), 174–200 (1941b).

F. V. Atkinson. The mean value of the Riemann zeta-function, *Acta Math.* **81**, 353–376 (1949).

R. Backlund. Sur les zéros de la fonction $\zeta(s)$ de Riemann, *C. R. Acad. Sci. (Paris)* **158**, 1979–1982 (1914).

R. C. Baker and J. Pintz. The distribution of square-free numbers, *Acta Arith.* (in press).

R. Balasubramanian. An improvement of a theorem of Titchmarsh on the mean square of $|\zeta(1/2 + it)|$, *Proc. London Math. Soc.* **36**, 540–576 (1978).

R. Balasubramanian and K. Ramachandra. On the frequency of Titchmarsh's phenomenon for $\zeta(s)$. III, *Proc. Indian Acad. Sci.* **86A**, 341–351 (1977).

R. Balasubramanian and K. Ramachandra. Some problems of analytic number theory II, *Hardy-Ramanujan J.* **4**, 13–40 (1981).

P. T. Bateman and E. Grosswald. On a theorem of Erdős and Szekeres, *Illinois J. Math.* **2**, 88–98 (1958).

B. C. Berndt. Arithmetical identities and Hecke's functional equation, *Proc. Edinburgh Math. Soc.* **16**, 221–226 (1969a).

B. C. Berndt. Identities involving the coefficients of a class of Dirichlet series I, *Trans. Am. Math. Soc.* **137**, 345–359 (1969b); V, **160**, 139–156 (1971); and VII **201**, 247–261 (1975a).

B. C. Berndt. The number of zeros for $\zeta^{(k)}(s)$, *J. London Math. Soc.* **2** (2), 577–580 (1970).

B. C. Berndt. On the Hurwitz zeta-function, *Rocky Mountain J. Math.* **2**, 151–157 (1972a).

B. C. Berndt. Two new proofs of Lerch's functional equation, *Proc. Am. Math. Soc.* **32**, 403–408 (1972b).

B. C. Berndt. Elementary evaluation of $\zeta(2n)$, *Am. Math. Monthly* **48**, 148–154 (1975b).

B. C. Berndt. Modular transformations and generalizations of several formulas of Ramanujan, *Rocky Mountain J. Math.* **7**, 147–189 (1977).

B. C. Berndt. Ramanujan's notebooks, *Math. Mag.* **51**, 147–164 (1978).

B. C. Berndt. Chapter 14 of Ramanujan's second notebook, *L'Enseignment Math.* **26**, 1–65 (1980).

B. C. Berndt. Chapter 8 of Ramanujan's second notebook, *J. Reine Angew. Math.* **338**, 1–55 (1983).

B. C. Berndt and R. J. Evans. Chapter 7 of Ramanujan's second notebook, *Math. Proc. Indian Acad. Sci.* (1983) (in press).

B. C. Berndt, P. T. Joshi and B. M. Wilson. Chapter 2 of Ramanujan's second notebook, *Glasgow Math. J.* **22**, 199–216 (1981).

B. C. Berndt and B. M. Wilson. Chapter 4 of Ramanujan's second notebook, *Proc. Roy. Soc. Edinburgh* **89A**, 87–109 (1981a).

B. C. Berndt and B. M. Wilson. Chapter 5 of Ramanujan's second notebook, in *Analytic Number Theory* (ed. M. I. Knopp), LNM 899, Springer-Verlag, Berlin-Heidelberg-New York 1981b, pp. 49–78.

E. Bombieri. Sulle formule di A. Selberg generalizzate per classi di funzioni aritmetiche e le applicazioni al problema del resto nel "Primzahlsatz," *Riv. Mat. Univ. Parma* **3** (2), 393–440 (1962).

R. Brent. On the zeros of the Riemann zeta-function in the critical strip, *Math. Comp.* **33** (148), 1361–1372 (1979).

K. Chandrasekharan. *Arithmetical Functions*, Springer-Verlag, Berlin-Heidelberg-New York, 1970.

K. Chandrasekharan and R. Narasimhan. Hecke's functional equation and arithmetical identities, *Ann. Math.* **74** (2), 1–23 (1961).

K. Chandrasekharan and R. Narasimhan. Functional equations with multiple gamma factors and the average order of arithmetical functions, *Ann. Math.* **76** (2), 93–136 (1962).

Chen Jing-run. On the order of $\zeta(1/2 + it)$, *Chinese Math. Acta* **6**, 463–478 (1965a).

Chen Jing-run. On the divisor problem for $d_3(n)$, *Sci. Sinica* **14**, 19–29 (1965b).

Chih Tsung-tao. A divisor problem, *Acad. Sinica Sci. Record* **3**, 177–182 (1950).

B. Conrey. Zeros of derivatives of the Riemann zeta-function on the critical line, *J. Number Theory* **16**, 49–74 (1983).

R. J. Cook. On the occurrence of large gaps between prime numbers, *Glasgow Math. J.* **20**, 43–48 (1979).

R. J. Cook. An upper bound for the sum of large differences between prime numbers, *Proc. Am. Math. Soc.* **81**, 33–40 (1981).

J. G. van der Corput. Zahlentheoretische Abschätzungen, *Math. Ann.* **84**, 53–79 (1921).

J. G. van der Corput. Verschärfung der Abschätzung beim Teilerproblem, *Math. Ann.* **87**, 39–65 (1922).

J. G. van der Corput. Zum Teilerproblem, *Math. Ann.* **98**, 697–716 (1928).

K. Corrádi and I. Kátai. A comment on K. S. Gangadharan's paper entitled "Two classical lattice point problems," *Magyar Tud. Akad. Mat. Fiz. Oszt. Kozl.* **17**, 89–97 (1967).

H. Cramér. Über zwei Sätze von Herrn G. H. Hardy, *Math. Zeit.* **15**, 200–210 (1922).

H. Cramér. On the order of magnitude of the difference between consecutive primes, *Acta Arith.* **2**, 23–46 (1937).

H. Davenport. *Multiplicative Number Theory* (2nd edition), GTM 74, Springer-Verlag, Berlin-Heidelberg-New York, 1980.

J.-M. De Koninck and A. Ivić. *Topics in Arithmetical Functions*, Mathematics Studies 43, North-Holland, Amsterdam-New York-Oxford, 1980.

H. Delange. Sur des formules dues à Atle Selberg, *Bull. Sci. Math.* 2, **83**, 101–111 (1959).

H. Delange. Sur des formules de Atle Selberg, *Acta Arith.* **19**, 105–146 (1971).

H. Delange. Sur un théorème de Rényi III, *Acta Arith.* **23**, 153–182 (1973).

P. Deligne. La conjecture de Weil, *Inst. Hautes Études Sci. Publ. Math.* **53**, 273–307 (1975).

J.-M. Deshouillers and H. Iwaniec. Power mean values of the Riemann zeta-function, *Mathematika* **29**, 202–212 (1982) and II, *Acta Arith.* **48**, 305–312 (1984).

H. G. Diamond. Elementary methods in the study of the distribution of prime numbers, *Bull. Am. Math. Soc.* **7**, 553–589 (1982).

H. G. Diamond and J. Steinig. An elementary proof of the prime number theorem with a remainder term, *Invent. Math.* **11**, 199–258 (1970).

A. L. Dixon and W. L. Ferrar. Lattice point summation formulae, *Quart. J. Math.* (*Oxford*) **2**, 31–54 (1931).

A. L. Dixon and W. L. Ferrar. On divisor transforms, *Quart. J. Math.* (*Oxford*) **3**, 43–59 (1932).

H. M. Edwards. *Riemann's Zeta-Function*, Academic, New York-London, 1974.

P. Epstein. Zur Theorie allgemeiner Zetafunktionen II, *Math. Ann.* **63**, 205–216 (1907).

P. Erdős. On the difference of consecutive primes, *Quart. J. Math.* (*Oxford*) **6**, 124–128 (1935).

P. Erdős. The difference of consecutive primes, *Duke Math. J.* **6**, 438–441 (1940).

P. Erdős. On a new method in elementary number theory which leads to an elementary proof of the prime number theorem, *Proc. Natl. Acad. Sci. USA* **35**, 374–384 (1949).

P. Erdős. Problems and results on combinatorial number theory III, in *Number Theory Day* (ed. M. B. Nathanson), LNM 626, Springer-Verlag, Berlin-Heidelberg-New York, 1977, pp. 43–73.

P. Erdős. Many old and some new problems of mine in number theory, *Congressus Numerantium* **30**, 3–27 (1981).

P. Erdős and G. Szekeres. Über die Anzahl der Abelschen Gruppen gegebener Ordnung und über ein verwandtes zahlentheoretisches Problem, *Acta Sci. Math.* (*Szeged*) **7**, 95–102 (1935).

L. Euler. Remarques sur un beau rapport entre les séries des puissances tout directes que réciproques, *Mém. Acad. Roy. Sci. Belles Lettres* **17**, 83–106 (1768).

M. Forti and C. Viola. Density estimates for the zeros of *L*-functions, *Acta Arith.* **23**, 379–391 (1973).

A. Fujii. On the problem of divisors, *Acta Arith.* **31**, 355–360 (1976).

W. Gabcke. Neue Herleitung und explizite Restabschätzung der Riemann-Siegel-Formel, Mathematisch-Naturwissenschaftliche Fakultät der Georg-August Universität zu Göttingen, Dissertation, Göttingen, 1979.

R. M. Gabriel. Some results concerning the integrals of moduli of regular functions along certain curves, *J. London Math. Soc.* **2**, 112–117 (1927).

K. S. Gangadharan. Two classical lattice point problems, *Proc. Cambridge Phil. Soc.* **57**, 699–721 (1961).

D. A. Goldston. Large differences between consecutive prime numbers, Dissertation, University of California at Berkeley, 1981.

D. A. Goldston. On a result of Littlewood concerning prime numbers II, *Acta Arith.* **43**, 49–51 (1983).

A. Good. Ein Ω-Resultat für quadratische Mittel der Riemannschen Zetafunktion auf der kritische Linie, *Invent. Math.* **41**, 233–251 (1977).

S. W. Graham. The distribution of squarefree numbers, *J. London Math. Soc.* **24** (2), 54–64 (1981).

S. W. Graham. Large values of the Riemann zeta-function, Austin Number Theory Conference Proceedings 1982 (in press).

J. Gram. Sur les zéros de la fonction $\zeta(s)$ de Riemann, *Acta Math.* **27**, 289–304 (1903).

E. Grosswald. Sur l'ordre de grandeur des differences $\psi(x) - x$ et $\pi(x) - \mathrm{li}x$, *Compt. Rend. Acad. Sci.* (*Paris*) **260**, 3813–3816 (1965).

J. Hadamard. Etude sur les propriétés des fonctions entières et en particulier d'une fonction considérée par Riemann, *J. Math.* **9** (4), 171–215 (1893).

J. Hadamard. Sur la distribution des zéros de la fonction $\zeta(s)$ et ses conséquences arithmétiques, *Bull. Soc. Math. France* **24**, 199–220 (1896).

J. L. Hafner. New omega theorems for two classical lattice point problems, *Invent. Math.* **63**, 181–186 (1981a).

J. L. Hafner. On the representation of the summatory function of a class of arithmetical functions, in *Analytic Number Theory* (ed. M. I. Knopp), LNM 899, Springer-Verlag, Berlin-Heidelberg-New York, 1981b, pp. 145–165.

J. L. Hafner. On the average order of a class of arithmetical functions, *J. Number Theory* **15**, 36–76 (1982).

G. Halász. The number-theoretic work of P. Turán, *Acta Arith.* **37**, 9–19 (1980).

G. Halász and P. Turán. On the distribution of roots of Riemann zeta and allied functions I, *J. Number Theory* **1**, 121–137 (1969).

H. Hamburger. Über die Riemannsche Funktionalgleichung der ζ-Funktion I, II, III, *Math. Zeit.* **10**, 240–254 (1921); **11**, 224–245 (1922a); **13**, 283–311 (1922b).

W. Haneke. Verschärfung der Abschätzung von $\zeta(1/2 + it)$, *Acta Arith.* **8**, 357–430 (1962–1963).

G. H. Hardy. Sur les zéros de la fonction $\zeta(s)$ de Riemann, *Compt. Rend. Acad. Sci.* (*Paris*) **158**, 1012–1014 (1914).

G. H. Hardy. On the expression of a number as a sum of two squares, *Quart. J. Math.* (*Oxford*) **46**, 263–283 (1915).

G. H. Hardy. On Dirichlet's divisor problem, Proc. London Math. Soc. **15** (2), 1–25 (1916a).

G. H. Hardy. The average order of the functions $P(x)$ and $\Delta(x)$, *Proc. London Math. Soc.* **15** (2), 192–213 (1916b).

G. H. Hardy. *Ramanujan*, Chelsea, New York, 1962.

G. H. Hardy and J. E. Littlewood. Contributions to the theory of the Riemann zeta-function and the distribution of primes, *Acta Math.* **41**, 119–196 (1918).

G. H. Hardy and J. E. Littlewood. The zeros of Riemann's zeta-function on the critical line, *Math. Zeit.* **10**, 283–317 (1921).

G. H. Hardy and J. E. Littlewood. The approximate functional equation in the theory of the zeta-function, with applications to the divisor problems of Dirichlet and Piltz, *Proc. London Math. Soc.* **21** (2), 39–74 (1922).

G. H. Hardy and J. E. Littlewood. The approximate functional equation for $\zeta(s)$ and $\zeta^2(s)$, *Proc. London Math. Soc.* **29** (2), 81–97 (1929).

G. Harman. Almost-primes in short intervals, *Math. Ann.* **258**, 107–112 (1981).

G. Harman. Primes in short intervals, *Math. Zeit.* **180**, 335–348 (1982).

D. R. Heath-Brown. The twelfth power moment of the Riemann zeta-function, *Quart. J. Math. (Oxford)* **29**, 443–462 (1978a).

D. R. Heath-Brown. The mean value theorem for the Riemann zeta-function, *Mathematika* **25**, 177–184 (1978b).

D. R. Heath-Brown. The fourth power moment of the Riemann zeta-function, *Proc. London Math. Soc.* **38** (3), 385–422 (1979a).

D. R. Heath-Brown. Zero-density estimates for the Riemann zeta-function and Dirichlet *L*-functions, *J. London Math. Soc.* **19** (2), 221–232 (1979b).

D. R. Heath-Brown. The differences between consecutive primes II, III, *J. London Math. Soc.* **19** (2), 207–220 (1979c); **20** (2), 177–178 (1979d).

D. R. Heath-Brown. Simple zeros of the Riemann zeta-function on the critical line, *Bull. London Math. Soc.* **11**, 17–18 (1979e).

D. R. Heath-Brown. A large values estimate for Dirichlet polynomials, *J. London Math. Soc.* **20** (2), 8–18 (1979f).

D. R. Heath-Brown. Fractional moments of the Riemann zeta-function, *J. London Math. Soc.* **24** (2), 65–78 (1981a).

D. R. Heath-Brown. Mean values of the zeta-function and divisor problems, in *Recent Progress in Analytic Number Theory*, Symposium Durham 1979 (Vol. 1), Academic, London, 1981b, pp. 115–119.

D. R. Heath-Brown. Mean Value Theorems for the Riemann zeta-function, *Séminaire Delange-Pisot-Poitou*, Paris 1979/1980, Birkhaüser-Verlag, Basel-Stuttgart, 1981c, pp. 123–134.

D. R. Heath-Brown. Gaps between primes, and the pair correlation of zeros of the zeta-function, *Acta Arith.* **41**, 85–99 (1982).

D. R. Heath-Brown and H. Iwaniec. On the difference between consecutive primes, *Invent. Math.* **55**, 49–69 (1979).

H. Heilbronn. Über den Primzahlsatz von Herrn Hoheisel, *Math. Zeit.* **36**, 394–423 (1933).

D. A. Hejhal. A note on the Voronoi summation formula, *Monatsh. Math.* **87**, 1–14 (1979).

G. Hoheisel. Primzahlprobleme in der Analysis, *Sitz. Preuss. Akad. Wiss.* **33**, 580–588 (1930).

J. I. Hutchinson. On the roots of the Riemann zeta-function, *Trans. Am. Math. Soc.* **27**, 49–60 (1925).

M. N. Huxley. *The Distribution of Prime Numbers*, Oxford University Press, Oxford, 1972a.

M. N. Huxley. On the difference between consecutive primes. *Invent. Math.* **15**, 155–164 (1972b).

M. N. Huxley. Small differences between consecutive primes, *Mathematika* **20**, 229–232 (1973a) and II, **24**, 142–152 (1977).

M. N. Huxley. Large values of Dirichlet polynomials, *Acta Arith.* **24**, 329–346 (1973b); II, **27**, 159–169 (1975a); III, **26**, 435–444 (1975b).

M. N. Huxley. Large gaps between prime numbers, *Acta Arith.* **38**, 63–68 (1980).

A. E. Ingham. Mean value theorems in the theory of the Riemann zeta-function, *Proc. London Math. Soc.* **27** (2), 273–300 (1926).

A. E. Ingham. On the difference between consecutive primes, *Quart. J. Math. (Oxford)* **8**, 255–265 (1937).

A. E. Ingham. On the estimation of $N(\sigma, T)$, *Quart. J. Math. (Oxford)* **11**, 291–292 (1940).

M. I. Israilov. Coefficients of the Laurent expansion of the Riemann zeta-function (Russian), *Dokl. Akad. Nauk SSSR* **12**, 9–10 (1979).

M. I. Israilov. The Laurent expansion of the Riemann zeta-function (Russian), *Trudy Mat. Inst. Steklova* **158**, 98–104 (1981).

A. Ivić. An asymptotic formula for the elements of a semigroup of integers, *Mat. Vesnik* (*Belgrade*) **10** (25), 255–257 (1973a).

A. Ivić. An application of Dirichlet series to certain arithmetical functions, *Math. Balkanica* **3**, 158–165 (1973b).

A. Ivić. On certain functions that generalize von Mangoldt's function $\Lambda(n)$, *Mat. Vesnik* (*Belgrade*) **12** (27), 361–366 (1975).

A. Ivić. On the asymptotic formulas for a generalization on von Mangoldt's function, *Rendiconti Mat. Roma* **10** (1), Serie VI, 51–59 (1977a).

A. Ivić. On the asymptotic formulae for some functions connected with powers of the zeta-function, *Mat. Vesnik* (*Belgrade*) **14** (29), 79–90 (1977b).

A. Ivić. On the asymptotic formulas for powerful numbers. *Publ. Inst. Math.* (*Belgrade*) **23** (37), 85–94 (1978a).

A. Ivić. A convolution theorem with applications to some divisor problems, *Publ. Inst. Math.* (*Belgrade*) **24** (38), 67–78 (1978b).

A. Ivić. The distribution of values of the enumerating function of non-isomorphic abelian groups of finite order, *Arch. Math.* **30**, 374–379 (1978c).

A. Ivić. On sums of large differences between consecutive primes, *Math. Ann.* **241**, 1–9 (1979a).

A. Ivić. A note on the zero-density estimates for the zeta-function, *Arch. Math.* **33**, 155–164 (1979b).

A. Ivić. On certain sums involving von Mangoldt's function in short intervals, *Publ. Inst. Math.* (*Belgrade*) **27** (41), 91–98 (1980a).

A. Ivić. Exponent pairs and the zeta-function of Riemann, *Studia Sci. Math. Hung.* **15**, 157–181 (1980b).

A. Ivić. On the number of finite non-isomorphic abelian groups in short intervals, *Math. Nachr.* **101**, 257–271 (1981).

A. Ivić. Large values of the error term in the divisor problem, *Invent. Math.* **71**, 513–520 (1983a).

A. Ivić. *Topics in Recent Zeta-Function Theory*, Publ. Math. d'Orsay, Université de Paris-Sud, Orsay, 1983b.

A. Ivić. On the number of abelian groups of a given order and on certain related multiplicative functions, *J. Number Theory* **16**, 119–137 (1983c).

A. Ivić. Exponent pairs and power moments of the zeta-function, in *Proc. Coll. Soc. János Bolyai*, Vol. 34 (Budapest, 1981), North-Holland, Amsterdam, 1984a, pp. 749–768.

A. Ivić. A zero-density theorem for the Riemann zeta-function, in *Proc. I. M. Vinogradov Conference* (Moscow, 1981), *Trudy Mat. Inst. AN SSSR* **163**, 85–89 (1984b).

A. Ivić and P. Shiu. The distribution of powerful numbers, *Illinois J. Math.* **26**, 576–590 (1982).

H. Iwaniec. *Fourier Coefficients of Cusp Forms and the Riemann Zeta-Function*, Exposé No. 18, Séminaire de Théorie des Nombres, Université Bordeaux, 1979/80.

H. Iwaniec. On mean values for Dirichlet's polynomials and the Riemann zeta-function, *J. London Math. Soc.* **22** (2), 29–45 (1980a).

H. Iwaniec. A new form of the error term in the linear sieve, *Acta Arith.* **37**, 307–320 (1980b).

H. Iwaniec and M. Jutila. Primes in short intervals, *Arkiv Mat.* **17**, 167–176 (1979).

H. Iwaniec and M. Laborde. P_2 in short intervals, *Ann. Inst. Fourier* (*Grenoble*) **31**, 37–56 (1981).

H. Iwaniec and J. Pintz. Primes in short intervals, *Monatshefte Math.* **98**, 115–143 (1984).

W. Jurkat and A. Peyerimhoff. A constructive approach to Kronecker approximations and its applications to the Mertens conjecture, *J. Reine Angew. Math.* **286 / 287**, 322–340 (1976).

M. Jutila. On a density theorem of H. L. Montgomery for L-functions, *Ann. Sci. Fenn. Ser. A I*, 520 (1972).

M. Jutila. On large values of Dirichlet polynomials, in *Coll. Soc. János Bolyai*, Vol. 13, Topics in Number Theory, North-Holland, Amsterdam, 1976, pp. 129–140.

M. Jutila. Zero-density estimates for L-functions, *Acta. Arith.* **32**, 52–62 (1977).

M. Jutila. Zeros of the zeta-function near the critical line, in *Studies in Pure Mathematics, To the Memory of Paul Turán*, Birkhaüser Verlag, Basel-Stuttgart, 1982, pp. 385–394.

M. Jutila. Riemann's zeta-function and the divisor problem, *Arkiv Mat.* **21**, 75–96 (1983a).

M. Jutila. On the value distribution of the zeta-function on the critical line, *Bull. London Math. Soc.* **15**, 513–518 (1983b).

M. Jutila. Transformation formulae for Dirichlet polynomials, *J. Number Theory* **18**, 135–156 (1984a).

M. Jutila. On a formula of Atkinson, in *Proc. Coll. Soc. János Bolyai*, Vol. 34 (Budapest, 1981), North-Holland, Amsterdam, 1984b, pp. 807–823.

M. Jutila. On the approximate functional equation for $\zeta^2(s)$ (unpublished).

A. A. Karacuba. Estimates of trigonometric sums by Vinogradov's method, and some applications (Russian), *Proc. Steklov Inst. Math.* **119**, 241–255 (1971).

A. A. Karacuba. Uniform estimates of the error term in Dirichlet's divisor problem (Russian), *Izv. Akad. Nauk SSSR* **36**, 475–483 (1972).

A. A. Karacuba. *Elements of Analytic Number Theory* (Russian), Nauka, Moscow, 1975.

A. A. Karacuba. On the distance between consecutive zeros of the Riemann zeta-function on the critical line (Russian), *Trudy Mat. Inst. Steklova AN SSSR* **157**, 49–63 (1981).

A. A. Karacuba. On the zeros of $\zeta(s)$ lying in short intervals on the critical line (Russian), *Izv. Akad. Nauk SSSR* **48**, 569–584 (1984).

J. Karamata. Sur un mode de croissance régulière des fonctions, *Mathematica (Cluj)* **4**, 38–53 (1930).

D. G. Kendall and R. A. Rankin. On the number of Abelian groups of a given order, *Quart. J. Math. (Oxford)* **18**, 197–208 (1947).

J. Knopfmacher. *Abstract Analytic Number Theory*, North-Holland, Amsterdam, 1975.

J. Koekoek. A note on $\zeta'(0)$, *Nieuw Arch. voor Wiskunde* **4**, 75–77 (1983).

G. Kolesnik. The distribution of prime numbers in sequences of the form $[n^c]$ (Russian), *Mat. Zametki* **2**, 117–128 (1967).

G. Kolesnik. The improvement of the error term in the divisor problem (Russian), *Mat. Zametki* **6**, 545–554 (1969).

G. Kolesnik. On the estimation of certain trigonometric sums (Russian), *Acta Arit.* **25**, 7–30 (1973).

G. Kolesnik. On the order of Dirichlet L-functions, *Pacific J. Math.* **82**, 479–484 (1979).

G. Kolesnik. On the estimation of multiple exponential sums, in *Recent Progress in Analytic Number Theory*, Symposium Durham 1979 (Vol. 1), Academic, London, 1981a, pp. 231–246.

G. Kolesnik. On the number of abelian groups of a given order, *J. Reine Angew. Math.* **329**, 164–175 (1981b).

G. Kolesnik. On the order of $\zeta(1/2 + it)$ and $\Delta(R)$, *Pacific J. Math.* **98**, 107–122 (1982).

G. Kolesnik. On the method of exponent pairs, *Acta Arith.* (in press).

N. M. Korobov. Estimates of trigonometric sums and their applications (Russian), *Usp. Mat. Nauk* **13**, 185–192 (1958a).

N. M. Korobov. Estimates of Weyl sums and the distribution of prime numbers (Russian), *Dokl. Akad. Nauk SSSR* **123**, 28–31 (1958b).

N. M. Korobov. On zeros of $\zeta(s)$ (Russian), *Dokl. Akad. Nauk SSSR* **118**, 231–232 (1958c).

E. Krätzel. Teilerprobleme in drei Dimensionen, *Math. Nachr.* **42**, 275–288 (1969).

E. Krätzel. Die maximale Ordnung der Anzahl der wesentlich vershiedenen Abelschen Gruppen *n*-ter Ordnung, *Quart. J. Math. (Oxford)* **21** (2), 273–275 (1970).

E. Krätzel. Zahlen *k*-ter Art, *Am. J. Math.* **94**, 309–328 (1972).

E. Krätzel. Die Werteverteilung der Anzahl der nicht-isomorphen Abelschen Gruppen endlicher Ordnung in kurzen Intervallen, *Math. Nachr.* **98**, 135–144 (1980).

E. Krätzel. Die Werteverteilung der Anzahl der nicht-isomorphen Abelschen Gruppen endlicher Ordnung und ein verwandtes zahlentheoretisches Problem, *Publ. Inst. Math. (Belgrade)* **31** (45), 93–101 (1982).

E. Krätzel. Double exponential sums and three-dimensional divisor problems, *Proc. Banach Center Symposium in Analytic Number Theory*, Warszawa 1982, Vol. XVII (in press).

J. Kubilius. *Probabilistic Methods in the Theory of Numbers*, Transl. Math. Monogr., American Mathematical Society, Providence, RI, 1964.

E. Landau. Euler und Functionalgleichung der Riemannsche Zeta-Funktion, *Biblio. Math.* (3) Bd. 7, Leipzig, 1906, pp. 69–79.

E. Landau. *Einführung in die Elementare und Analytische Theorie der Algebraischen Zahlen und der Ideale*, Chelsea, New York, 1949.

A. F. Lavrik. On the principal term in the divisor problem and power series of the Riemann zeta-function in a neighborhood of its pole (Russian), *Trudy Mat. Inst. Steklova* **142**, 165–173 (1976) and reprinted in *Proc. Steklov Inst. Math. 1979* **3**, 175–183 (1979).

A. F. Lavrik, M. I. Israilov, and Ž. Edgorov. On integrals containing the error term in the divisor problem (Russian), *Acta Arith.* **37**, 381–389 (1980).

A. F. Lavrik and A. Š. Sobirov. On the remainder term in the elementary proof of the prime number theorem (Russian), *Dokl. Akad. Nauk SSSR* **211**, 534–536 (1973).

D. H. Lehmer. On the roots of the Riemann zeta-function, *Acta Math.* **95**, 291–298 (1956a).

D. H. Lehmer. Extended computation of the Riemann zeta-function, *Mathematika* **3**, 102–108 (1956b).

M. Lerch. Note sur la fonction $\Re(w, x, s) = \sum_{k=0}^{\infty} e^{2k\pi i x}/(w + x)^s$, *Acta Math.* **11**, 19–24 (1887).

N. Levinson. More than one third of the zeros of Riemann's zeta-function are on $\sigma = 1/2$, *Adv. Math.* **13**, 383–436 (1974).

N. Levinson. A simplification of the proof that $N_0(T) > \frac{1}{3}N(T)$ for Riemann's zeta-function, *Adv. Math.* **18**, 239–242 (1975a).

N. Levinson. Deduction of semi-optimal mollifier for obtaining lower bounds for $N_0(T)$ for Riemann's zeta-function, *Proc. Natl. Acad. Sci. USA* **72**, 294–297 (1975b).

N. Levinson and H. L. Montgomery. Zeros of the derivative of the Riemann zeta-function, *Acta Math.* **133**, 49–65 (1974).

J. E. Littlewood. Quelques conséquences de l'hypothèse que la fonction $\zeta(s)$ de Riemann n'a pas de zéros dans le démi-plan $R(s) > 1/2$, *Comp. Rend. Acad. Sci. (Paris)* **154**, 263–266 (1912).

J. E. Littlewood. Sur la distribution des nombres premiers, *Compt. Rend. Acad. Sci. (Paris)* **158**, 1869–1872 (1914).

J. E. Littlewood. Researches in the theory of the Riemann zeta-function, *Proc. London Math. Soc.* **20**, Records XXII–XXVIII (1922).

J. van de Lune. A note on a formula of van der Pol, *Acta Arith.* **35**, 361–366 (1979).

J. van de Lune and H. J. J. te Riele. *On the Zeros of the Riemann Zeta-Function III*, Report NW 146/1983, Mathematical Centre, Amsterdam, 1983.

J. van de Lune, H. J. J. te Riele, and D. T. Winter, *Rigorous High-Speed Separation of Zeros of Riemann's Zeta-Function*, Report NW 113/81, Mathematical Centre, Amsterdam, 1981.

H. von Mangoldt. Zu Riemann's Abhandlung "Über die Anzahl...," *Crelle's J.* **114**, 255–305 (1895).

H. von Mangoldt. Beweis der Gleichung $\sum_{k=1}^{\infty} \mu(k)/k = 0$, *Sitz. Preuss. Akad. Wiss.*, *Berlin*, 835–852 (1897).

N. A. Meller. Computations connected with the check of Riemann's hypothesis (Russian), *Dokl. Akad. Nauk SSSR* **123**, 246–248 (1958).

F. Mertens. Ueber eine zahlentheoretische Funktion, *Sitzber. Akad. Wien* **106**, 761–830 (1897).

S.-H. Min. On the order of $\zeta(1/2 + it)$, *Trans. Am. Math. Soc.* **65**, 448–472 (1949).

H. L. Montgomery. Zeros of L-functions, *Invent. Math.* **8**, 346–354 (1969).

H. L. Montgomery. *Topics in Multiplicative Number Theory*, LNM 227, Springer-Verlag, Berlin-Heidelberg-New York, 1971.

H. L. Montgomery. The pair correlation of zeros of the zeta-function, in *Proceedings of the Symposium on Pure Mathematics*, 24 AMS, Providence, RI, 1973, pp. 181–193.

H. L. Montgomery. Extreme values of the Riemann zeta-function, *Comment. Math. Helv.* **52**, 511–518 (1977).

H. L. Montgomery. The analytic principle of the large sieve, *Bull. Amer. Math. Soc.* **84**, 547–567 (1978).

H. L. Montgomery. Maximal variants of the large sieve, *J. Fac. Sci. Tokyo* **28**, 805–812 (1982).

H. L. Montgomery and R. C. Vaughan. Hilbert's inequality, *J. London Math. Soc.* **8** (2), 73–82 (1974).

H. L. Montgomery and R. C. Vaughan. The distribution of squarefree numbers, in *Recent Progress in Analytic Number Theory*, Symposium Durham 1979 (Vol. 1), Academic, London, 1981, pp. 247–256.

J. Moser. On a certain sum in the theory of the Riemann zeta-function (Russian), *Acta Arith.* **31**, 34–43 (1976a).

J. Moser. On a theorem of Hardy-Littlewood in the theory of the Riemann zeta-function (Russian), *Acta Arith.* **31**, 45–51 (1976b) and **35**, 403–404 (1979).

J. Moser. The proof of the Titchmarsh hypothesis in the theory of the Riemann zeta-function (Russian), *Acta Arith.* **36**, 147–156 (1980).

J. Moser. New consequences of the Riemann–Siegel formula (Russian), *Acta Arith.* **42**, 1–10 (1982).

J. Moser. An improvement on a density theorem of Hardy and Littlewood on zeros of $\zeta(1/2 + it)$ (Russian), *Acta Math. Univ. Com.* **42-43**, 41–50 (1983a).

J. Moser. On a biquadratic sum in the theory of the Riemann zeta-function (Russian), *Acta Math. Univ. Com.* **42-43**, 35–39 (1983b).

Y. Motohashi. On the sum of the Möbius function in a short segment, *Proc. Japan Acad. Ser. A Math. Sci.* **52**, 477–479 (1976).

Y. Motohashi. On Vinogradov's zero-free region for the Riemann zeta-function, *Proc. Japan Acad. Ser. A Math. Sci.* **54**, 300–302 (1978).

Y. Motohashi. A note on almost primes in short intervals, *Proc. Japan Acad. Ser. A Math. Sci.* **55**, 225–226 (1979).

Y. Motohashi. An elementary proof of Vinogradov's zero-free region for the Riemann zeta-function, in *Recent Progress in Analytic Number Theory*, Symposium Durham 1979 (Vol. 1), Academic, London, 1981 pp. 257–267.

Y. Motohashi. A note on the approximate functional equation for $\zeta^2(s)$, *Proc. Japan Acad.* **59A**. 392–396 (1983) and II, **59A**, 469–472 (1983).

J. Mueller. On the difference between consecutive primes, in *Recent Progress in Analytic Number Theory*, Symposium Durham 1979 (Vol. 1), Academic, London, 1981, pp. 269–273.

M. Nair. On Chebyshev-type inequalities for primes, *Am. Math. Monthly* **89**, 126–129 (1982).

R. W. K. Odoni. A problem of Erdős on sums of two squarefull numbers, *Acta Arith.* **39**, 145–162 (1981).

E. Phillips. The zeta-function of Riemann; further developments of van der Corput's method, *Quart. J. Math. (Oxford)* **4**, 209–225 (1933).

J. Pintz. On the remainder term of the prime number formula III, *Studia Sci. Math. Hung.* **12**, 345–369 (1977); IV, **13**, 29–42 (1978); V, **15**, 215–223 (1980a); VI, **15**, 225–230 (1980b).

J. Pintz. On the remainder term of the prime number formula I, *Acta Arith.* **36**, 341–365 (1980c); and II, **37**, 209–220 (1980d).

J. Pintz. Oscillatory properties of $M(x) = \sum_{n \leq x} \mu(n)$ II, *Studia Sci. Math. Hung.* **15**, 491–496 (1980e).

J. Pintz. On Legendre's prime number formula, *Am. Math. Monthly* **87**, 733–735 (1980f).

J. Pintz. On sign changes of $M(x) = \sum_{n \leq x} \mu(n)$, *Analysis* **1**, 191–195 (1981a).

J. Pintz. On primes in short intervals I, *Studia Sci. Math. Hung.* **16**, 395–414 (1981b); II (in press).

J. Pintz. Oscillatory properties of $M(x) = \sum_{n \leq x} \mu(n)$, *Acta Arith.* **42**, 49–55 (1982a).

J. Pintz. Oscillatory properties of the remainder term of the prime number formula, in *Studies in Pure Mathematics, To the Memory of Paul Turán*, Birkhaüser Verlag, Basel-Stuttgart, 1982b, pp. 551–560.

J. Pintz. On the mean-value of the remainder term of the prime number formula, *Proc. Banach Center Symposium in Analytic Number Theory*, Warszawa 1982c, Vol. XVII (in press).

J. Pintz. On the remainder term of the prime number formula and the zeros of Riemann's zeta-function, *Proc. Journées Arithmétiques* (Nordwijkerhout, Netherlands 1983) LNM 1068, Springer-Verlag, Berlin-Heidelberg-New York, 1984, pp. 186–197.

K. Prachar. Primzahlverteilung, Springer-Verlag, Berlin-Göttingen-Heidelberg, 1957 (Russian translation; Mir, Moscow, 1967).

K. Ramachandra. On the frequency of Titchmarsh's phenomenon for $\zeta(s)$, *J. London Math. Soc.* **8** (2), 683–690 (1974a).

K. Ramachandra. A simple proof of the mean fourth power estimate for $\zeta(1/2 + it)$ and $L(1/2 + it, \chi)$, *Ann. Scuola Norm. Sup. (Pisa)* **1**, 81–97 (1974b).

K. Ramachandra. Some new density estimates for the zeros of the Riemann zeta-function, *Ann. Acad. Sci. Fenn. Ser. A I Math.* **1**, 177–182 (1975a).

K. Ramachandra. Applications of a theorem of Montgomery and Vaughan to the zeta-function, *J. London Math. Soc.* **10** (2), 482–486 (1975b).

K. Ramachandra. On the frequency of Titchmarsh's phenomenon for $\zeta(s)$ II, *Acta Math. Sci. Hung.* **30**, 7–13 (1977).

K. Ramachandra. Some remarks on the mean value of the Riemann zeta-function and other Dirichlet series I, *Hardy-Ramanujan J.* **1**, 1–15 (1978) and II, **3**, 1–24 (1980a).

K. Ramachandra. Some remarks on a theorem of Montgomery and Vaughan, *J. Number Theory* **11**, 465–471 (1980b).

K. Ramachandra. Progress towards a conjecture on the mean value of Titchmarsh series, in *Recent Progress in Analytic Number Theory*, Symposium Durham 1979 (Vol. 1), Academic, London, 1981, pp. 303–318.

K. Ramachandra. Mean value of the Riemann zeta-function and other remarks I, in *Proc. Coll. Soc. János Bolyai*, Vol. 34 (Budapest 1981), North-Holland, Amsterdam, 1984a, pp. 1317–1347; II, in *Proc. I. M. Vinogradov Conference* (Moscow, 1981), *Trudy Mat. Inst.*

AN SSSR **163**, 200–204 (1984b); and III, *Hardy-Ramanujan J.* **6**, 1–21 (1983).

S. Ramanujan. *Notebooks of Srinivasa Ramanujan* (2 vols.), Tata Institute of Fundamental Research, Bombay, 1957.

S. Ramanujan. *Collected Papers*, Chelsea, New York, 1962.

M. Ram Murty. Some Ω-results for Ramanujan's τ-function, in *Number Theory*, LNM 938, Springer-Verlag, Berlin-Heidelberg-New York, 1982, pp. 123–137.

R. A. Rankin. Van der Corput's method and the theory of exponent pairs, *Quart. J. Math. (Oxford)* **6** (2), 147–153 (1955).

R. A. Rankin. The difference between consecutive prime numbers V, *Proc. Edinburgh Math. Soc.* **13** (2), 331–332 (1962/63).

R. S. Rao and D. Suryanarayana. The number of pairs of integers with l.c.m. $\leq x$, *Arch. Math.* **21**, 490–497 (1970).

A. Rényi. On the density of certain sequences of integers, *Publ. Inst. Math. (Belgrade)* **8**, 157–162 (1955).

H.-E. Richert. Über die Anzahl Abelscher Gruppen gegebener Ordnung I, *Math. Zeit.* **56**, 21–32 (1952).

H.-E. Richert. Verschärfung der Abschätzung beim Dirichletschen Teilerproblem, *Math. Zeit.* **58**, 204–218 (1953).

H.-E. Richert. Über Dirichletreihen mit Funktionalgleichungen, *Publ. Inst. Math. (Belgrade)* **11**, 73–124 (1957).

H.-E. Richert. Einführung in die Theorie der starken Rieszschen Summierbarkeit von Dirichletreihen, *Nachr. Akad. Wiss. Göttingen (Math.-Physik)*, 17–75 (1960).

H.-E. Richert. Zur Abschätzung der Riemannschen Zetafunktion in der Nähe der Vertikalen $\sigma = 1$, *Math. Ann.* **169**, 97–101 (1967).

H. te Riele. Mertens' conjecture disproved, *CWI Newsletter*, No. 1, November 1983, pp. 23–24.

B. Riemann. Über die Anzahl der Primzahlen unter eine gegebener Grösse, *Monatsber. Akad. Berlin*, 671–680 (1859).

J. B. Rosser, J. M. Yohe, and L. Schoenfeld, Rigorous computation and the zeros of the Riemann zeta-function, *Cong. Proc. Int. Federation Information Process 1968*, (Washington-New York, 1969), North-Holland, Amsterdam, 1969, pp. 70–76.

E. Schmidt. Über die Anzahl der Primzahlen unter gegebener Grenze, *Math. Ann.* **57**, 195–204 (1903).

P. G. Schmidt. Zur Anzahl Abelscher Gruppen gegebener Ordnung, *J. Reine Angew. Math.* **229**, 34–42 (1968a), and II, *Acta Arith.* **13**, 405–417 (1968b).

W. Schwarz. Über die Anzahl Abelscher Gruppen gegebener Ordnung I, *Math. Zeit.* **92**, 314–320 (1966), and II, *J. Reine Angew. Math.* **228**, 133–138 (1967).

W. Schwarz and E. Wirsing. The maximal order of non-isomorphic abelian groups of order n, *Arch. Math.* **24**, 59–62 (1973).

A. Selberg. On the zeros of Riemann's zeta-function, *Skr. Norske Vid. Akad. Oslo* **10**, 1–59 (1942).

A. Selberg. On the normal density of primes in short intervals and the difference between consecutive primes, *Arch. Math. Naturvid.* **47**, 87–105 (1943).

A. Selberg. Contributions to the theory of the Riemann zeta-function, *Arch. Math. Naturvid. B* **48**, 89–155 (1946).

A. Selberg. An elementary proof of the prime number theorem, *Ann. Math.* **50** (2), 305–313 (1949).

A. Selberg. Note on a paper by L. G. Sathe, *J. Indian Math. Soc.* **18**, 83–87 (1954).

E. Seneta. *Regularly Varying Functions*, LNM 508, Springer-Verlag, Berlin-Heidelberg-New York, 1976.

R. Sherman Lehman. On the difference $\pi(x) - \mathrm{li}\,x$, *Acta Arith*. **11**, 397–410 (1966a).

R. Sherman Lehman. Separation of zeros of the Riemann zeta-function, *Math. Comp.* **20**, 523–541 (1966b).

Shi-Tuo Lou. A lower bound for the number of zeros of Riemann's zeta-function on $\sigma = 1/2$, in *Recent Progress in Analytic Number Theory*, Symposium Durham 1979 (Vol. 1), Academic, London, 1981, pp. 319–325.

P. Shiu. On the number of square-full integers between consecutive squares, *Mathematika* **27**, 171–178 (1980).

P. Shiu. On square-full integers in a short interval, *Glasgow Math. J.* **25**, 127–134 (1984).

C. L. Siegel. Über Riemann's Nachlass zur analytischen Zahlentheorie, *Quell. Stud. Gesch. Mat. Astr. Physik* **2**, 45–80 (1932).

S. Skewes. On the difference $\pi(x) - \mathrm{li}\,x$, *Proc. London Math. Soc.* **5** (3), 48–70 (1955).

B. R. Srinivasan. Lattice point problems of many-dimensional hyperboloids II, *Acta Arith.* **8**, 173–204 (1963), and III, *Math. Ann.* **160**, 280–311 (1965).

B. R. Srinivasan. On the number of abelian groups of a given order, *Acta Arith.* **23**, 195–205 (1973).

T. J. Stieltjes. *Correspondance d'Hermite et de Stieltjes*, Tome 1, Gauthier-Villars, Paris, 1905.

M. N. Subbarao. On some arithmetic convolutions, in *The Theory of Arithmetic Functions*, LNM 251, Springer-Verlag, Berlin-Heidelberg-New York, 1972, pp. 247–271.

D. Suryanarayana and R. Sitaramachandra Rao. The distribution of square-full integers, *Arkiv Mat.* **11** (2), 195–201 (1973b).

D. Suryanarayana and R. Sitaramachandra Rao. The distribution of square-full integers, *Arkiv. Mat.* **11** (2), 195–201 (1973b).

D. Suryanarayana and R. Sitaramachandra Rao. The number of bi-unitary divisors of an integer II, *J. Indian Math. Soc.* **39**, 261–280 (1975).

E. C. Titchmarsh. On van der Corput's method and the zeta-function of Riemann I, *Quart. J. Math.* (*Oxford*) **2**, 161–173 (1931a); II, **2**, 313–320 (1931b); III, **3**, 133–141 (1932); IV, **5**, 98–105 (1934a); and V, **5**, 195–210 (1934b).

E. C. Titchmarsh. On Epstein's zeta-function, *Proc. London Math. Soc.* **36**, 485–500 (1934c).

E. C. Titchmarsh. The lattice points in a circle, *Proc. London Math. Soc.* **38**, 96–115 (1935a).

E. C. Titchmarsh. The zeros of the Riemann zeta-function, *Proc. Roy. Soc.* (*London*) **151**, 234–235 (1935b) and II, **157**, 261–263 (1936).

E. C. Titchmarsh. The approximate functional equation for $\zeta^2(s)$, *Quart. J. Math.* (*Oxford*) **9**, 109–114 (1938).

E. C. Titchmarsh. On the order of $\zeta(1/2 + it)$, *Quart. J. Math.* (*Oxford*) **13**, 11–17 (1942).

E. C. Titchmarsh. *Introduction to the Theory of Fourier Integrals*, Oxford University Press, Oxford, 1948.

E. C. Titchmarsh. *The Theory of the Riemann Zeta-Function*, Oxford University Press, Oxford, 1951.

K.-C. Tong. On divisor problems, *J. Chinese Math. Soc.* **2**, 258–266 (1953).

K.-C. Tong. On divisor problems III, *Acta Math. Sinica* **6**, 515–541 (1956).

J. P. Tull. Dirichlet multiplication in lattice point problems, *Duke Math. J.* **26**, 73–80 (1958); and II, *Pacific J. Math.* **9**, 609–615 (1959).

P. Turán. On a theorem of Hardy and Ramanujan, *J. London Math. Soc.* **9**, 274–276 (1934).

P. Turán. On the remainder term of the prime-number formula II, *Acta Math. Acad. Sci. Hung.* **1**, 155–166 (1950).

P. Turán. On some recent results in the analytical theory of numbers, *Proc. Symp. Pure Math.*

XX 1969, Institute of Number Theory, AMS, 1971, pp. 359–374.

P. Turán. *A New Method in the Analysis and Its Applications*, Wiley-Interscience, New York, 1984.

R. T. Turganaliev. The asymptotic formula for fractional mean value moments of the zeta-function of Riemann (Russian), *Trudy Mat. Inst. AN SSSR* **158**, 203–226 (1981).

C. J. de la Vallée-Poussin. Recherches analytiques sur la théorie des nombres; Première partie: la fonction $\zeta(s)$ de Riemann et les nombres premiers en général, *Ann. Soc. Sci. Brux.* **20**, 183–256 (1896).

C. J. de la Vallée-Poussin. Sur la fonction $\zeta(s)$ de Riemann et le nombre des nombres premiers inférieurs à une limite donnée, *Mém. Acad. Roy. Sci. Belg.* **59**, 74 (1899).

I. M. Vinogradov. A new estimate for $\zeta(1 + it)$ (Russian), *Izv. Akad. Nauk SSSR, Ser. Mat.* **22**, 161–164 (1958).

I. M. Vinogradov. *Special Variants of the Method of Trigonometric Sums* (Russian), Nauka, Moscow, 1976.

I. M. Vinogradov. *The Method of Trigonometric Sums in Number Theory* (Russian, 2nd edition), Nauka, Moscow, 1980.

M. Vogts. Teilerprobleme in drei Dimensionen, *Math. Nachr.* **101**, 243–256 (1981).

G. F. Voronoi. Sur une fonction transcendante et ses applications à la sommation de quelques séries, *Ann. École Normale* **21** (3), 207–268 (1904a); **21** (3), 459–534 (1904b).

G. F. Voronoi. *Sur le Développement, à l'Étude des Fonctions Cylindriques, des Sommes Doubles $\Sigma f(pm^2 + 2qmn + rn^2)$, où $pm^2 + 2qmn + rn^2$ est une Forme Positive à Coefficients Entiers* (Verh. III Math. Kong. Heidelberg), Teubner, Leipzig, 1905, pp. 241–245.

A. Walfisz. Zur Abschätzung von $\zeta(1/2 + it)$, *Nachr. Ges. Wiss. Göttingen Math.-Physik. Kl.*, 155–158 (1924).

A. Walfisz. Über zwei Gitterpunktprobleme, *Math. Ann.* **95**, 69–83 (1925).

A. Walfisz. *Weylsche Exponentialsummen in der Neueren Zahlentheorie*, VEB Deutscher Verlag, Berlin, 1963.

G. N. Watson, *A Treatise on the Theory of Bessel Functions* (2nd edition), Cambridge University Press, Cambridge, 1944.

R. Wiebelitz. Über approximative Funktionalgleichungen der Potenzen der Riemannschen Zeta-funktion, *Math. Nachr.* **6**, 263–270 (1951–52).

E. Wirsing. Elementare Beweise des Primzahlsatzes mit Restglied I, *J. Reine Angew. Math.* **211**, 205–214 (1962); and II, **214 / 215**, 1–18 (1964).

D. Wolke. Grosse Differenzen zwischen aufeinanderfolgenden Primzahlen, *Math. Ann.* **218**, 269–271 (1975).

D. Wolke. Fast-Primzahlen in kurzen Intervallen, *Math. Ann.* **244**, 233–242 (1979).

D. Wolke. On the explicit formula of Riemann–von Mangoldt, *Studia Sci. Math. Hung.* **15**, 231–239 (1980) and II, *J. London Math. Soc.* **28**(2), 406–416 (1983).

Yüh Ming-i. A divisor problem, *Acta Math. Sinica* **8**, 496–506 (1958).

Yüh Ming-i and Wu Fang. On the divisor problem for $d_3(n)$, *Sci. Sinica* **11**, 1055–1060 (1962).

AUTHOR INDEX

Anderson, R. J., 53, 265
Apéry, R., 7, 50
Apostol, T. M., 51, 490
Arhipov, G. I., 168
Atkinson, F. V., 80, 125, 129, 139, 171, 172, 178, 228, 381, 441, 443, 462, 471, 481

Backlund, R., 52
Baker, R. C., 437
Balasubramanian, R., 248, 249, 260, 261, 264, 265, 439, 443, 471, 481
Bateman, P. T., 438
Berndt, B. C., 49, 50, 51, 94, 265
Bombieri, E., 51, 347
Brent, R. P., 52

Chandrasekharan, K., 50, 51, 94, 96, 169, 264, 347, 440
Chen, Jing-run, 197, 381
Chih Tsung-tao, 381
Chowla, S., 52
Conrey, B., 264, 265
Cook, R. J., 350
Corput, J. G. van der, 55, 79, 190, 381
Corrádi, K., 383, 482
Cramér, H., 299, 300, 349, 383, 465

Davenport, H., 50, 347
De Koninck, J.-M., 440
Delange, H., 435, 440
Deligne, P., 44, 51
Deshouillers, J.-M., 195
Diamond, H. G., 51, 347
Dirichlet, P. G. L., 125, 232, 351, 380, 381, 388, 400
Dixon, A. L., 89, 92, 96

Edwards, H. M., 52, 264
Epstein, P., 51
Erdős, P., 347, 350, 438, 439
Euler, L., 1, 6, 49
Evans, R. J., 49, 51

Ferrar, W. L., 89, 92, 96
Forti, F., 295
Fujii, A., 382, 383

Gabcke, W., 124
Gabriel, R. M., 229
Gallagher, P. X., 195
Gangadharan, K. S., 383, 482
Gauss, C. F., 44, 346
Goldston, D. A., 51, 347, 349
Good, A., 482
Graham, S. W., 80, 228, 437
Gram, J., 52
Grosswald, E., 347, 438

Hadamard, J., 7, 12, 13, 21, 49, 346
Hafner, J. L., 94, 383, 482
Halász, G., 52, 294
Halberstam, H., 350
Hamburger, H., 50
Haneke, W., 52, 80, 197
Hardy, G. H., 51, 94, 97, 105, 110, 124, 126, 190, 197. 251, 254, 260, 264, 265, 266, 363, 372, 376, 381, 383, 441, 481
Harman, G., 349, 350
Heath-Brown, D. R., 51, 121, 129, 140, 171, 178, 194, 195, 206, 228, 229, 249, 265, 294, 295, 300, 324, 334, 349, 350, 382, 383, 441, 460, 465, 482
Heilbronn, H., 349
Hejhal, D., 96

Hoheisel, G., 349
Hutchinson, J. I., 52
Huxley, M. N., 128, 228, 273, 289, 293, 295, 349, 350, 491

Ingham, A. E., 129, 139, 140, 273, 274, 275, 293, 294, 349
Israilov, M. I., 49
Ivić, A., 349, 350, 437, 438, 440
Iwaniec, H., 113, 172, 194, 195, 197, 228, 299, 324, 332, 348, 349, 350

Jurkat, W., 53
Jutila, M., 80, 124, 125, 128, 196, 197, 281, 293, 294, 295, 299, 324, 348, 349, 384, 460, 461, 462, 471, 481, 482

Karacuba, A. A., 50, 168, 260, 261, 264, 265, 267, 382
Karamata, J., 387, 437
Kátai, I., 383, 482
Kendall, D. G., 439
Knopfmacher, J., 50, 51, 436
Koekoek, J., 50
Kolesnik, G., 78, 80, 81, 172, 191, 193, 196, 197, 353, 354, 374, 381, 384, 439, 471, 477
Korobov, N. M., 167, 168, 347
Krätzel, E., 51, 437, 439, 440
Kronecker, L., 232, 237
Kubilius, J., 440

Laborde, M., 332, 350
Landau, E., 49, 51, 298, 300, 347, 434, 437, 440
Lavrik, A. F., 347, 381
Legendre, A.-M., 346
Lehmer, D. H., 52
Lerch, M., 51
Levinson, N., 251, 252, 253, 254, 264, 265
Linnik, Yu. V., 168
Littlewood, J. E., 44, 51, 52, 97, 105, 110, 126, 190, 197, 253, 260, 265, 307, 381, 441, 481
Lune, J. van de, 50, 52, 348

Mangoldt, H. von, 17, 49, 51, 298, 346
Meller, N. A., 52
Mertens, F., 47, 53
Min, S.-H., 80, 126, 197
Montgomery, H. L., 48, 51, 140, 141, 195, 249, 293, 294, 349, 437, 494
Moser, J., 260, 261, 264, 265, 266, 267
Motohashi, Y., 125, 169, 349, 350
Mueller, J., 51, 349

Nair, M., 346
Narasimhan, R., 94

Odlyzko, A., 47, 53
Odoni., R. W. K., 438

Peyerimhoff, A., 53
Phillips, E., 79, 197, 382
Pintz, J., 53, 324, 346, 347, 348, 349, 437
Prachar, K., 50, 51, 168, 346, 347, 348, 440, 486

Rademacher, H., 51
Ramachandra, K., 128, 140, 248, 249, 295, 439
Ramanujan, S., 44, 49, 50, 51
Ram Murty, M., 51
Rankin, R. A., 77, 81, 350, 439
Rao, R. S., 437
Rényi, A., 440
Richert, H.-E., 80, 96, 167, 228, 294, 346, 350, 373, 374, 376, 381, 382, 437, 439
Riele, H. J. J. te, 47, 52, 53, 348
Riemann, B., 1, 8, 17, 49, 52, 97, 122, 298, 300, 346
Rosser, J. B., 52

Schmidt, E., 347
Schmidt, P. G., 437, 439
Schoenfeld, L., 383
Schwarz, W., 51, 439
Selberg, A., 251, 264, 265, 293, 300, 347, 349, 440
Seneta, E., 437
Sherman Lehman, R., 52
Shi-Tuo, Lou, 264
Shiu, P., 51, 437, 438
Siegel, C. L., 49, 98, 124
Skewes, S., 44, 52
Sobirov, A. Š., 347
Srinivasan, B. R., 55, 78, 81, 439
Stark, H. M., 53
Steinig, J., 347
Stieltjes, T. J., 49
Subbarao, M. V., 50
Suryanarayana, D., 437
Szekeres, G., 438, 439

Titchmarsh, E. T., 50, 51, 52, 79, 80, 95, 124, 125, 139, 168, 169, 196, 197, 227, 229, 248, 249, 264, 266, 267, 293, 348, 381, 382, 383, 481, 483
Tong, K.-C., 363, 364, 380, 383

Tull, J. P., 437
Turán, P., 52, 294, 306, 347, 348, 440
Turganaliev, R. T., 229

Vallée-Poussin, C. J. de la, 7, 21, 346
Vaughan, R. C., 140, 437
Vinogradov, I. M., 167, 168, 347
Viola, C., 295
Vogts, M., 437, 438
Voronoi, G. F., 79, 83, 93, 94, 95, 381

Walfisz, A., 167, 169, 197, 346, 348, 381, 437
Watson, G. N., 95
Weierstrass, C., 1
Weyl, H., 80
Wiebelitz, R., 110, 126, 127
Wilson, B. M., 51
Wirsing, E., 51, 347
Wolke, D., 347, 350
Wu, Fang, 381

Yüh Ming-i, 381

SUBJECT INDEX

Abscissa of (absolute) convergence, 30, 34
Almost prime numbers, 299, 300, 330
Analytic continuation, 3, 4, 9, 10, 11, 23, 39, 41, 42, 43, 91, 100, 106, 444, 445, 446, 449, 491, 494
Arithmetical functions:
  additive functions, 429–436
  characteristic function of k-free numbers, 32, 393
  characteristic function of powerful numbers, 33, 407, 438
  characters, 38, 39, 40
  divisor function, 26, 420
  Euler function, 36
  general divisor function, 420
  generalized von Mangoldt function, 34, 51, 299, 309
  Liouville function, 315
  Möbius function, 32, 269, 309
  number of distinct prime divisors of integer, 36
  number of finite nonisomorphic abelian groups, 37, 38, 386, 413, 438
  number of finite nonisomorphic semisimple rings, 38
  partition function, 37
  Ramanujan's function, 43, 51
  sum of divisors function, 37, 445
  von Mangoldt function, 3, 299
Atkinson's formula, 129, 139, 172, 178, 354, 384, 441, 443, 460, 461, 464, 465, 466, 470, 471, 472, 474, 475, 476, 478, 482

Bernoulli functions, 490
Bernoulli numbers, 6, 490
Bernoulli polynomials, 490

Bessel functions, 83, 84, 85, 89, 90, 95, 125, 373
Bi-unitary divisor, 395

Cauchy-Schwarz inequality, see Inequalities
Cauchy's theorem (formula), 59, 61, 62, 65, 66, 138, 301, 422
Chebyshev function, 297
Circle method, 350
Circle problem, 96, 352, 372, 373, 383, 384
Congruences, systems of, 147, 148
Convexity estimates, 202, 203, 228, 229, 360
Convolution of arithmetical functions, 31, 385

Dedekind zeta-function, see Zeta-functions
Density hypothesis, see Hypothesis
Diophantine approximation, 231, 232
Dirichlet algebra, 436
  box principle, 232
  convolution, 31, 50, 436
  divisor problem, 45, 200, 351, 359, 372, 380, 385, 386, 410, 416, 472, 482
  polynomials, 130, 134, 141, 176, 196, 214, 227, 267, 271, 273, 274, 275, 283, 325, 328, 341, 460
  series 25, 30, 31, 34, 37, 48, 96, 137, 141, 249, 300, 315, 485, 486
Double zeta sums, 270, 283, 290, 295, 343

Elementary proof of prime number theorem, 347
Elementary symmetric functions, 147
Epstein zeta-function, see Zeta-functions
Euler constant, 5, 352, 492
  identity, 3
  product, 2, 41, 44, 430

Euler-Maclaurin summation formula *see*
  Summation formulas
Exponent pairs (definition and properties), 73–79

Fourier coefficients, 43
  series expansion, 10, 15, 22, 67, 96, 404, 491
  transforms (integrals), 95, 96, 337, 483, 484

General divisor problem, 385, 386
Global problems of prime number theory, 298

Hadamard product formula, 12
  three circles theorem, 46, 493
Halász-Montgomery inequality, *see* Inequalities
Hardy-Ramanujan formula, 37, 51
Hecke congruence subgroups, 195
Hilbert's inequality, *see* Inequalities
Hölder's inequality, *see* Inequalities
Hurwitz zeta function, *see* Zeta-functions
Hyperbola method, 381
Hypothesis:
  density hypothesis, 47, 52, 269, 283, 289,
    294, 295, 321, 350
  Lindelöf hypothesis, 45, 47, 51, 52, 188, 199,
    265, 269, 294, 359, 413, 471
  Mertens hypothesis, 47, 51, 52
  Montgomery's pair correlation hypothesis, 48,
    51, 322
  Riemann hypothesis, 44–49, 51, 52, 144, 168,
    188, 229, 248, 249, 251, 252, 255, 269,
    299, 300, 305, 306, 321, 322, 333, 349,
    392, 394, 437, 439

Ideal theory, 51
Inequalities:
  Cauchy's inequality for power series, 426, 434
  Cauchy-Schwarz inequality, 27, 69, 80, 81,
    131, 133, 134, 135, 138, 139, 195, 206,
    211, 212, 218, 225, 227, 245, 256, 261,
    266, 286, 287, 291, 293, 295, 307, 325,
    328, 344, 358, 363, 364, 469, 470, 488,
    494
  Halász-Montgomery inequality, 211, 228, 273,
    274, 275, 276, 279, 280, 325, 329, 366,
    368, 481, 494, 495
  Hilbert's inequality, 130
  Hölder's inequality, 149, 151, 156, 204, 218,
    223, 228, 276, 278, 282, 284, 335, 368,
    371, 412, 413, 481, 488
  large sieve inequality, 140, 494
Integral functions, 13, 14, 15, 16, 50, 100

Jensen's formula, 12, 13, 14, 17, 253

Kloosterman sums, 140, 194, 195

Laplacian (non-Euclidean), 195
Large sieve inequality, *see* Inequalities
Laurent series expansion, 4, 41, 111, 352
Leibnitz's rule, 76
Lerch zeta-function, *see* Zeta-functions
L-functions, *see* Zeta-functions
Lindelöf hypothesis, *see* Hypothesis
Linear sieve, *see* Sieve
Local density (of abelian groups), 417, 419, 420,
  435, 439
Local problems of prime number theory, 298

Maximum modulus principle, 24, 164, 239, 492
Mellin integral, 122, 242
  inversion formula, 483
  transforms, 88, 95, 416, 483, 484, 485, 487
Mertens hypothesis, *see* Hypothesis
Methods for estimation of exponential sums:
  of Corput, J. G. van der, 144, 168
  of Hardy-Littlewood, 144, 168
  of Kolesnik, G., 80, 81, 172, 191, 197, 353,
    374, 384, 471, 477
  of Titchmarsh, E. C., 80
  of Vinogradov-Korobov, 144–152, 156, 162,
    168
  of Weyl, H., 144, 168
Möbius inversion formula, 35, 36
Modular functions, 51
Montgomery's pair correlation hypothesis, *see*
  Hypothesis

Parseval's identity, 357, 416, 484
Perron's formula, 51, 411, 421, 485
Phragmén-Lindelöf principle, 25, 236, 492, 493
Piltz divisor problem, 380
Poisson summation formula, *see* Summation
  formulas
Prime number theorem, 241, 298, 299, 304,
  305, 306, 314, 323, 329, 346, 347, 348

Ramanujan-Petersson conjecture, 51
Ramanujan zeta-function, *see* Zeta-functions
Reflection principle, 122, 123, 124, 128, 135,
  188, 195, 205, 281, 286, 292
Residue theorem, 87, 89, 100, 114, 128, 136,
  172, 173, 222, 245, 247, 302, 313, 325,
  333, 352, 354, 412, 456
Riemann hypothesis, *see* Hypothesis
Riemann-Siegel formula, 49, 98, 99, 104, 124,
  125, 253, 443, 481

Riemann-von Mangoldt formula, 14, 17, 20, 40, 52, 251, 265, 269, 271, 276, 303

Riesz means, 229

Saddle point (method), 55, 57, 64, 65, 109, 125, 462

Second mean value theorem for integrals, 22, 56, 456, 466

Selberg sieve, *see* Sieve

Siegel's theorem, 40

Siegel-Walfisz theorem for arithmetic progressions, 40

Sieve:
  linear, 299, 323, 332, 349
  Selberg, 169

Slowly oscillating functions, 386, 387, 389, 390, 437

Slowly varying functions, 386

Standard inner product, 495

Stieltjes constants, 49
  integral, 5, 490
  resultant, 436, 437

Stirling's formula, 9, 13, 14, 18, 24, 89, 122, 123, 136, 173, 206, 242, 245, 252, 449, 492

Summation formulas:
  Euler-Maclaurin summation formula, 4, 67, 398, 490

Poisson summation formula, 96, 444, 490, 491

Voronoi summation formula, 55, 80, 83–96, 105, 106, 110, 125, 172, 181, 186, 196, 361, 373, 380, 443, 446, 460, 461, 462, 464, 470, 480, 482

Tauberian argument (theorem), 139

Taylor's formula (expansion), 58, 60, 62, 64, 65, 90, 127, 154, 155, 180, 257, 379, 442, 446, 474, 475

Uniqueness property of Dirichlet series, 31

Unitary divisor, 395

Voronoi summation formula, *see* Summation formulas

Zero-detection method, 270, 279, 293

Zeta-functions other than $\zeta(s)$:
  Dedekind zeta-function, 42, 51
  Epstein zeta-function, 42, 51
  Hurwitz zeta-function, 40, 41
  Lerch zeta-function, 41, 51
  L-functions (Dirichlet L-series), 38, 51, 293, 429
  Ramanujan zeta-function, 43, 44, 51
  zeta-functions of cusp forms, 43, 51

A CATALOG OF SELECTED DOVER
BOOKS IN ALL FIELDS OF INTEREST

CONCERNING THE SPIRITUAL IN ART, Wassily Kandinsky. Pioneering work by father of abstract art. Thoughts on color theory, nature of art. Analysis of earlier masters. 12 illustrations. 80pp. of text. 5⅜ x 8½.   23411-8

ANIMALS: 1,419 Copyright-Free Illustrations of Mammals, Birds, Fish, Insects, etc., Jim Harter (ed.). Clear wood engravings present, in extremely lifelike poses, over 1,000 species of animals. One of the most extensive pictorial sourcebooks of its kind. Captions. Index. 284pp. 9 x 12.   23766-4

CELTIC ART: The Methods of Construction, George Bain. Simple geometric techniques for making Celtic interlacements, spirals, Kells-type initials, animals, humans, etc. Over 500 illustrations. 160pp. 9 x 12. (Available in U.S. only.)   22923-8

AN ATLAS OF ANATOMY FOR ARTISTS, Fritz Schider. Most thorough reference work on art anatomy in the world. Hundreds of illustrations, including selections from works by Vesalius, Leonardo, Goya, Ingres, Michelangelo, others. 593 illustrations. 192pp. 7⅛ x 10¼.   20241-0

CELTIC HAND STROKE-BY-STROKE (Irish Half-Uncial from "The Book of Kells"): An Arthur Baker Calligraphy Manual, Arthur Baker. Complete guide to creating each letter of the alphabet in distinctive Celtic manner. Covers hand position, strokes, pens, inks, paper, more. Illustrated. 48pp. 8¼ x 11.   24336-2

EASY ORIGAMI, John Montroll. Charming collection of 32 projects (hat, cup, pelican, piano, swan, many more) specially designed for the novice origami hobbyist. Clearly illustrated easy-to-follow instructions insure that even beginning papercrafters will achieve successful results. 48pp. 8¼ x 11.   27298-2

THE COMPLETE BOOK OF BIRDHOUSE CONSTRUCTION FOR WOODWORKERS, Scott D. Campbell. Detailed instructions, illustrations, tables. Also data on bird habitat and instinct patterns. Bibliography. 3 tables. 63 illustrations in 15 figures. 48pp. 5¼ x 8½.   24407-5

BLOOMINGDALE'S ILLUSTRATED 1886 CATALOG: Fashions, Dry Goods and Housewares, Bloomingdale Brothers. Famed merchants' extremely rare catalog depicting about 1,700 products: clothing, housewares, firearms, dry goods, jewelry, more. Invaluable for dating, identifying vintage items. Also, copyright-free graphics for artists, designers. Co-published with Henry Ford Museum & Greenfield Village. 160pp. 8¼ x 11.   25780-0

HISTORIC COSTUME IN PICTURES, Braun & Schneider. Over 1,450 costumed figures in clearly detailed engravings–from dawn of civilization to end of 19th century. Captions. Many folk costumes. 256pp. 8⅜ x 11¾.   23150-X

THE WIT AND HUMOR OF OSCAR WILDE, Alvin Redman (ed.). More than 1,000 ripostes, paradoxes, wisecracks: Work is the curse of the drinking classes; I can resist everything except temptation; etc. 258pp. 5⅜ x 8½. 20602-5

SHAKESPEARE LEXICON AND QUOTATION DICTIONARY, Alexander Schmidt. Full definitions, locations, shades of meaning in every word in plays and poems. More than 50,000 exact quotations. 1,485pp. 6½ x 9¼. 2-vol. set.
Vol. 1: 22726-X
Vol. 2: 22727-8

SELECTED POEMS, Emily Dickinson. Over 100 best-known, best-loved poems by one of America's foremost poets, reprinted from authoritative early editions. No comparable edition at this price. Index of first lines. 64pp. 5³⁄₁₆ x 8¼. 26466-1

THE INSIDIOUS DR. FU-MANCHU, Sax Rohmer. The first of the popular mystery series introduces a pair of English detectives to their archnemesis, the diabolical Dr. Fu-Manchu. Flavorful atmosphere, fast-paced action, and colorful characters enliven this classic of the genre. 208pp. 5³⁄₁₆ x 8¼. 29898-1

THE MALLEUS MALEFICARUM OF KRAMER AND SPRENGER, translated by Montague Summers. Full text of most important witchhunter's "bible," used by both Catholics and Protestants. 278pp. 6⅝ x 10. 22802-9

SPANISH STORIES/CUENTOS ESPAÑOLES: A Dual-Language Book, Angel Flores (ed.). Unique format offers 13 great stories in Spanish by Cervantes, Borges, others. Faithful English translations on facing pages. 352pp. 5⅜ x 8½. 25399-6

GARDEN CITY, LONG ISLAND, IN EARLY PHOTOGRAPHS, 1869–1919, Mildred H. Smith. Handsome treasury of 118 vintage pictures, accompanied by carefully researched captions, document the Garden City Hotel fire (1899), the Vanderbilt Cup Race (1908), the first airmail flight departing from the Nassau Boulevard Aerodrome (1911), and much more. 96pp. 8⅞ x 11¾. 40669-5

OLD QUEENS, N.Y., IN EARLY PHOTOGRAPHS, Vincent F. Seyfried and William Asadorian. Over 160 rare photographs of Maspeth, Jamaica, Jackson Heights, and other areas. Vintage views of DeWitt Clinton mansion, 1939 World's Fair and more. Captions. 192pp. 8⅞ x 11. 26358-4

CAPTURED BY THE INDIANS: 15 Firsthand Accounts, 1750-1870, Frederick Drimmer. Astounding true historical accounts of grisly torture, bloody conflicts, relentless pursuits, miraculous escapes and more, by people who lived to tell the tale. 384pp. 5⅜ x 8½. 24901-8

THE WORLD'S GREAT SPEECHES (Fourth Enlarged Edition), Lewis Copeland, Lawrence W. Lamm, and Stephen J. McKenna. Nearly 300 speeches provide public speakers with a wealth of updated quotes and inspiration–from Pericles' funeral oration and William Jennings Bryan's "Cross of Gold Speech" to Malcolm X's powerful words on the Black Revolution and Earl of Spenser's tribute to his sister, Diana, Princess of Wales. 944pp. 5⅜ x 8⅜. 40903-1

THE BOOK OF THE SWORD, Sir Richard F. Burton. Great Victorian scholar/adventurer's eloquent, erudite history of the "queen of weapons"–from prehistory to early Roman Empire. Evolution and development of early swords, variations (sabre, broadsword, cutlass, scimitar, etc.), much more. 336pp. 6⅛ x 9¼. 25434-8

CATALOG OF DOVER BOOKS

THE STORY OF THE TITANIC AS TOLD BY ITS SURVIVORS, Jack Winocour (ed.). What it was really like. Panic, despair, shocking inefficiency, and a little heroism. More thrilling than any fictional account. 26 illustrations. 320pp. 5⅜ x 8½.
20610-6

FAIRY AND FOLK TALES OF THE IRISH PEASANTRY, William Butler Yeats (ed.). Treasury of 64 tales from the twilight world of Celtic myth and legend: "The Soul Cages," "The Kildare Pooka," "King O'Toole and his Goose," many more. Introduction and Notes by W. B. Yeats. 352pp. 5⅜ x 8½.
26941-8

BUDDHIST MAHAYANA TEXTS, E. B. Cowell and others (eds.). Superb, accurate translations of basic documents in Mahayana Buddhism, highly important in history of religions. The Buddha-karita of Asvaghosha, Larger Sukhavativyuha, more. 448pp. 5⅜ x 8½.
25552-2

ONE TWO THREE . . . INFINITY: Facts and Speculations of Science, George Gamow. Great physicist's fascinating, readable overview of contemporary science: number theory, relativity, fourth dimension, entropy, genes, atomic structure, much more. 128 illustrations. Index. 352pp. 5⅜ x 8½.
25664-2

EXPERIMENTATION AND MEASUREMENT, W. J. Youden. Introductory manual explains laws of measurement in simple terms and offers tips for achieving accuracy and minimizing errors. Mathematics of measurement, use of instruments, experimenting with machines. 1994 edition. Foreword. Preface. Introduction. Epilogue. Selected Readings. Glossary. Index. Tables and figures. 128pp. 5⅜ x 8½. 40451-X

DALÍ ON MODERN ART: The Cuckolds of Antiquated Modern Art, Salvador Dalí. Influential painter skewers modern art and its practitioners. Outrageous evaluations of Picasso, Cézanne, Turner, more. 15 renderings of paintings discussed. 44 calligraphic decorations by Dalí. 96pp. 5⅜ x 8½. (Available in U.S. only.)
29220-7

ANTIQUE PLAYING CARDS: A Pictorial History, Henry René D'Allemagne. Over 900 elaborate, decorative images from rare playing cards (14th–20th centuries): Bacchus, death, dancing dogs, hunting scenes, royal coats of arms, players cheating, much more. 96pp. 9¼ x 12¼.
29265-7

MAKING FURNITURE MASTERPIECES: 30 Projects with Measured Drawings, Franklin H. Gottshall. Step-by-step instructions, illustrations for constructing handsome, useful pieces, among them a Sheraton desk, Chippendale chair, Spanish desk, Queen Anne table and a William and Mary dressing mirror. 224pp. 8⅛ x 11¼.
29338-6

THE FOSSIL BOOK: A Record of Prehistoric Life, Patricia V. Rich et al. Profusely illustrated definitive guide covers everything from single-celled organisms and dinosaurs to birds and mammals and the interplay between climate and man. Over 1,500 illustrations. 760pp. 7½ x 10⅛.
29371-8